ArcGIS Pro 3.x Cookbook

Create, manage, analyze, maintain, and visualize geospatial data using ArcGIS Pro

Tripp Corbin, GISP

ArcGIS Pro 3.x Cookbook

Copyright © 2024 Packt Publishing

All rights reserved. No part of this book may be reproduced, stored in a retrieval system, or transmitted in any form or by any means, without the prior written permission of the publisher, except in the case of brief quotations embedded in critical articles or reviews.

Every effort has been made in the preparation of this book to ensure the accuracy of the information presented. However, the information contained in this book is sold without warranty, either express or implied. Neither the author, nor Packt Publishing or its dealers and distributors, will be held liable for any damages caused or alleged to have been caused directly or indirectly by this book.

Packt Publishing has endeavored to provide trademark information about all of the companies and products mentioned in this book by the appropriate use of capitals. However, Packt Publishing cannot guarantee the accuracy of this information.

Group Product Manager: Kaustubh Manglurkar
Publishing Product Manager: Deepesh Patel
Book Project Manager: Kirti Pisat
Senior Editor: Shrishti Pandey
Technical Editor: Seemanjay Ameriya
Copy Editor: Safis Editing
Proofreader: Safis Editing
Indexer: Manju Arasan
Production Designer: Ponraj Dhandapani
DevRel Marketing Coordinator: Nivedita Singh

First published: February 2018

Second edition: April 2024

Production reference: 1050424

Published by Packt Publishing Ltd.
Grosvenor House
11 St Paul's Square
Birmingham
B3 1RB, UK.

ISBN 978-1-83763-170-4

www.packtpub.com

To my amazing and supportive family, especially my wife Polly, who has always believed I can accomplish anything I set my mind to and given me the strength to do so.

– Tripp Corbin, GISP

Foreword

Tripp Corbin is a true leader within the geospatial community. With over 25 years of experience, he is a recognized expert on multiple geospatial platforms including Esri, Autodesk, Trimble, and more. This claim is backed by the numerous certifications Tripp holds including **GIS Professional (GISP)**, Esri ArcGIS Pro Associate, Esri Enterprise Systems Design Associate, **CompTIA Technical Trainer (CTT+)**, and Microsoft Certified Professional, among others. This is in addition to being a past President of the **Urban and Regional Systems Association (URISA)**, a contributing member of multiple geospatial user groups and serving on the xyHt Editorial Board. It is an honor to know and work with him. This book will prove to be an amazing resource for anyone using ArcGIS Pro, whether you are a beginner or migrating from the older ArcMap application. You will learn how to perform not only some of the most common tasks but also explore more advanced capabilities such as 3D analysis and validating data to ensure accuracy and completeness.

The scenarios and recipes included in this book represent real-world situations encountered by ArcGIS Pro users every day. Completing this book will arm you with the skills and knowledge needed to successfully use ArcGIS Pro to address similar questions or issues you too will face. I can attest to that, having over 20 years of GIS experience myself supporting operations for local governments, utilities, and private companies.

Become an ArcGIS Pro power user creating clean accurate data, 2D and 3D visualizations, performing spatial analysis, and more.

Kirk Larson

Vice President of GIS Consulting Services, Surveying And Mapping, LLC (SAM).

Contributors

About the author

Tripp Corbin, GISP has over 25 years of surveying, mapping, and GIS-related experience and is recognized as a geospatial industry leader with expertise in a variety of geospatial software packages, including Esri, Autodesk, and Trimble products. His experience is backed by multiple industry certifications—he is a Certified **GIS Professional** (**GISP**), Esri Certified Enterprise System Design Associate, and ArcGIS Pro Associate, among others. He is a graduate of the URISA GIS Leadership Academy.

I want to thank my employer, Surveying And Mapping LLC (SAM), for their support and for allowing me the time to write this book. I also want to thank Tim Gaunt, GISP for helping me when I ran into issues and allowing me to bounce ideas off him.

About the reviewer

Chase Young is from Maryville, MO, and graduated from Northwest Missouri State University with a bachelor's degree in GIS. He has been in the GIS industry for 10 years and has worked for two geospatial companies in that time. He has had many roles over the years at these firms, ranging from GIS technician to GIS developer, which is his current role. Chase has spent the last 4 years as lead for his firm's custom Arc Online applications and Esri enterprise deployment. Since starting his career in the GIS field, he has worked with Esri mapping software.

Table of Contents

Preface — xv

1

ArcGIS Pro Capabilities and Terminology — 1

Determining whether your computer can run ArcGIS Pro — 2
Getting ready — 2
How to do it... — 3
How it works... — 5

Determining your ArcGIS Pro license level — 5
Getting ready — 5
How to do it... — 6
How it works... — 8
There's more... — 8

Opening an existing ArcGIS Pro project — 8
Getting ready — 9
How to do it... — 9
How it works... — 10

There's more... — 11

Opening and navigating a map — 12
Getting ready — 12
How to do it... — 12
How it works... — 17

Creating a project with a template — 17
Getting ready — 17
How to do it... — 17
How it works... — 24
There's more... — 24

Creating a project without a template — 25
Getting ready — 26
How to do it... — 26
How it works... — 28

2

Adding and Configuring Layers — 29

Adding a layer from a geodatabase — 29
Getting ready — 30
How to do it... — 30
How it works... — 40

Adding a layer from ArcGIS Online — 41
Getting ready — 41
How to do it... — 42
How it works... — 46

Plotting X, Y points from a table	47	Geocoding addresses	51
Getting ready	47	Getting ready	51
How to do it…	47	How to do it…	51
How it works…	50	How it works…	62

3

Linking Data Together — 63

Joining two tables	64	How it works…	94
Getting ready	64	There's more…	94
How to do it…	64	Joining features spatially	96
How it works…	74	Getting ready	97
Labeling features using a joined table	74	How to do it…	97
Getting ready	74	How it works…	101
How to do it…	74	Creating feature-linked annotation	101
How it works…	81	Getting ready	102
Querying data in a joined table	82	How to do it…	102
Getting ready	82	How it works…	112
How to do it…	83	Creating and using a relationship class using existing data	112
How it works…	87	Getting ready	113
Creating and using a relate	88	How to do it…	113
Getting ready	88	How it works….	119
How to do it…	88		

4

Editing Existing Spatial Features — 121

Configuring editing options	122	How it works…	136
Getting ready	122	Splitting a line feature	136
How to do it…	122	Getting ready	137
How it works…	130	How to do it…	137
Reshaping an existing feature	130	How it works…	146
Getting ready	130	Merging features	146
How to do it…	131	Getting ready	146

How to do it…	147
How it works…	152
Aligning features	**153**
Getting ready	153
How to do it…	153
How it works…	160

5
Creating New Spatial Data   161

Creating new point features	**161**
Getting ready	162
How to do it…	162
How it works…	168
Creating new line features	**168**
Getting ready	169
How to do it…	169
How it works…	184
Creating polygon features	**184**
Getting ready	184
How to do it…	185
How it works…	191
Creating a new polygon feature using the Autocomplete Polygon tool	**192**
Getting ready	192
How to do it…	192
How it works…	198

6
Editing Tabular Data   199

Editing individual attributes using the Attributes pane	**199**
Getting ready	200
How to do it…	200
How it works…	204
Editing multiple attributes with a single edit using the Attributes pane	**204**
Getting ready	204
How to do it…	205
How it works…	210
Editing individual attributes in the table view	**210**
Getting ready	211
How to do it…	211
How it works…	215
Using the Calculate Field tool to populate multiple features	**216**
Getting ready	216
How to do it…	216
How it works…	222
Using the Calculate Geometry tool to populate values for multiple features	**223**
Getting ready	223
How to do it…	223
How it works…	227

7

Projection and Coordinate System Basics — 229

Determining the coordinate system for an existing map	231
Getting ready	231
How to do it...	231
How it works…	235
Setting the coordinate system for a new map	235
Getting ready	235
How to do it	235
How it works…	240
Changing the coordinate system of a map	240
Getting ready	241
How to do it	241
How it works…	245
Defining a coordinate system for data	245
Getting ready	246
How to do it	246
How it works…	252
Projecting data to a different coordinate system	252
Getting ready	252
How to do it	253
How it works…	265

8

Creating a Geodatabase — 267

Creating a new geodatabase and feature classes	268
Getting ready	268
How to do it...	268
How it works…	276
Creating a feature dataset	277
Getting ready	277
How to do it...	277
How it works…	280
Creating a coded values domain	281
Getting ready	281
How to do it...	282
How it works…	286
Creating subtypes	286
Getting ready	287
How to do it...	287
How it works…	295

9

Enabling Advanced Functionality in a Geodatabase — 297

Enabling editor tracking	298
Getting ready	298
How to do it...	298
How it works…	303

Adding GPS metadata capture	304	How to do it…	311
Getting ready	304	How it works…	320
How to do it…	305	**Creating contingent values**	**321**
How it works…	310	Getting ready	322
Creating attribute rules	**310**	How to do it…	322
Getting ready	310	How it works…	332

10

Validating and Editing Data with Topologies — 333

Creating a new geodatabase topology	334	Correcting spatial features with topology tools	349
Getting ready	334	Getting ready	349
How to do it…	335	How to do it…	349
How it works…	342	How it works…	352
Validating spatial data using a geodatabase topology	342	Editing data with a map topology	353
Getting ready	342	Getting ready	353
How to do it...	342	How to do it…	353
How it works…	348	How it works…	362

11

Converting Data — 363

Converting shapefiles to a geodatabase feature class	364	There's more…	380
		Exporting tabular data to an Excel spreadsheet	385
Getting ready	364	Getting ready	385
How to do it…	364	How to do it…	385
How it works…	372	How it works…	393
There's more…	373	Importing an Excel spreadsheet into ArcGIS Pro	393
Merging multiple shapefiles into a single geodatabase feature class	375		
Getting ready	376	Getting ready	394
How to do it…	376	How to do it…	394
How it works…	380	How it works…	401

There's more…	402	Getting ready	406
Importing selected features into an existing layer	**406**	How to do it…	406
		How it works…	410

12

Proximity Analysis — 411

Selecting features within a specific distance	**411**	**Determining the nearest feature using the Near tool**	**424**
Getting ready	412	Getting ready	424
How to do it…	412	How to do it…	424
How it works…	417	There's more…	428
Creating buffers	**417**	How it works…	431
Getting ready	417	**Calculating how far apart features are using the Generate Near Table tool**	**431**
How to do it…	418		
Challenge	420	Getting ready	432
There's more…	420	How to do it…	432
		There's more…	438

13

Spatial Statistics and Hotspots — 441

Selecting features within a specific distance	**442**	**Identifying the central feature based on geographic distribution**	**458**
Getting ready	442	Getting ready	458
How to do it…	442	How to do it…	459
How it works…	451	How it works…	463
Finding the mean center of a geographic distribution	**451**	**Calculating the geographic dispersion of data**	**464**
Getting ready	452	Getting ready	465
How to do it…	452	How to do it…	465
There's more…	454	How it works…	473

14

3D Maps and Analysis — 475

Creating a 3D scene	477	Creating 3D features	507
Getting ready	478	Getting ready	507
How to do it…	478	How to do it…	507
How it works…	489	How it works…	520
Enabling your data to store Z coordinates (elevation)	490	Calculating lines of sight	520
Getting ready	490	Getting ready	520
How to do it…	491	How to do it…	521
How it works…	499	How it works…	524
Creating multipatch features from 2D	499	Calculating the volume of a polygon	525
Getting ready	500	Getting ready	525
How to do it…	500	How to do it…	525
How it works…	507	How it works…	530

Index — 531

Other Books You May Enjoy — 540

Preface

ArcGIS Pro is Esri's newest desktop **Geographic Information System** (**GIS**) application with powerful tools for visualizing, maintaining, and analyzing data. It is replacing the venerable ArcMap and ArcCatalog applications, forcing users to migrate to this new application, which makes use of a new modern ribbon interface and 64-bit processing to increase the speed and efficiency of using GIS. This cookbook will help existing ArcMap users transition to this new application while also teaching new users how to make use of the powerful GIS tools that it contains.

Beginning with a refresher on ArcGIS Pro and how to work with projects, this book will quickly take you through recipes to create a geodatabase and convert various data formats supported by the application. You will learn how to link tables from outside sources to existing GIS data to expand the amount of data used in ArcGIS. You will learn methods to edit spatial and tabular data using ArcGIS Pro and how to use various capabilities in ArcGIS Pro to ensure data integrity. Lastly, the book will show you how to enable advanced behavior in a geodatabase, such as editor tracking and GPS metadata.

Upon completing the recipes in this book, you will be able to effectively use ArcGIS Pro as your primary desktop GIS application. You will be able to maintain, analyze, and display data using common methods and tools in ArcGIS Pro.

Who this book is for

This book is for those who wish to learn how to use ArcGIS Pro and the powerful data editing, analysis, and creation tools it contains. This includes GIS professionals, architects, specialists, analysts, and technicians wishing to migrate to or expand their skills with ArcGIS Pro, as well as those just getting started.

Some basic GIS experience is helpful but not required. For beginners, this book will provide you with the skills needed to leverage ArcGIS Pro's powerful tools daily to manage, visualize, maintain, and analyze your GIS data.

What this book covers

Chapter 1, *Getting to Know ArcGIS Pro Terminology and the User Interface*, explores ArcGIS Pro, learning its associated terminology and navigating the user interface.

Chapter 2, *Adding Layers to a Map from Different Sources*, examines ArcGIS Pro's support for a wide range of data formats. You will learn how to add new layers to a map that reference different data formats and how to configure them.

Chapter 3, Linking Data Together, delves into how ArcGIS Pro allows you to link your GIS with other external and internal data. You will learn about the different methods that ArcGIS Pro supports to link data to your GIS layers and how you can then use it.

Chapter 4, Editing Existing Spatial Features, teaches you how to edit features in your GIS that already exist. This will include reshaping, splitting, combining, and aligning them.

Chapter 5, Creating New Spatial Data, discusses how to use tools in ArcGIS Pro to create new features, including points, lines, and polygons, using feature templates and various editing tools.

Chapter 6, Editing Tabular Data, explores how GIS data typically includes both spatial and tabular data. You will learn how to edit the tabular or attribute data associated with the features you create or edit. You will examine methods to edit those values one at a time or through mass updates.

Chapter 7, Projections and Coordinate System Basics, focuses on how GIS relies on coordinate systems to properly display data. You will learn how to use tools in ArcGIS Pro to determine the coordinate system assigned to objects and how to change them.

Chapter 8, Creating a Geodatabase, examines how to create a new geodatabase to store your GIS data and how to add new items to it, such as feature classes and feature datasets.

Chapter 9, Enabling Advanced Functionality in a Geodatabase, teaches you how to enable advanced functionality supported by the geodatabase format, including editor tracking, GPS metadata, attribute rules, and contingent values.

Chapter 10, Validating and Editing Data with Topologies, shows you how to create and use topologies to help ensure the quality of the spatial data included in your GIS. GIS is only as good as the accuracy of the data it contains.

Chapter 11, Converting Data, explores ArcGIS Pro's support for many data types. You will learn how to convert data formats using tools in ArcGIS Pro, including shapefiles, spreadsheets, and CAD files.

Chapter 12, Proximity Analysis, delves into how to use several tools that analyze how far features are from one another. This will include tools such as buffer, nearest feature, and selecting features within a set distance.

Chapter 13, Spatial Statistics and Hot Spots, focuses on the tools that can help you identify clusters within your data or identify centers of geographically dispersed data points.

Chapter 14, 3D Maps and 3D Analysis, teaches you how to create 3D maps called scenes, create 3D data, and use various tools included in the 3D Analyst extension to perform 3D analysis. Displaying and analyzing 3D data is relatively new to GIS.

To get the most out of this book

You should have some basic GIS understanding and experience. Experience using other Esri products is helpful but not required. Also, understanding how to access computer files and directories will make following the directions in this book easier.

Software/hardware covered in the book	Operating system requirements
ArcGIS Pro 3.x (x=latest version)	Windows 10 or 11 – 64-bit
3D Analyst extension for ArcGIS Pro	
ArcGIS Online	
Microsoft Excel or another spreadsheet program	

Chapter 1 of this book will provide you with the system requirements to run ArcGIS Pro and walk you through the process of checking your system to verify that it meets those requirements.

Download the recipe data files

You can download the recipe data files for this book from GitHub at `https://github.com/PacktPublishing/ArcGIS-Pro-3.x-Cookbook`. On opening the link, click on the **Code** button and select **Download Zip File**. If there's an update to the code, it will be updated in the GitHub repository.

Once you have the data downloaded, you need to extract the zip file to your local `C:` drive in the `Student` folder. This folder may need to be created if it does not exist. The steps are as follows:

1. Open **File Explorer** on your computer. It is typically located on your **Taskbar** at the bottom of your display. The icon looks like an old fashion file folder in a metal holder.
2. In the left panel of **File Explorer**, scroll down until you see **This PC** and expand it. Then locate Local Disk (`C:`) and select it.
3. Select the **Home** tab at the top of **File Explorer**. Then click on **New Folder**. This will create a new folder that you need to name.
4. In the box that appears with the text **New Folder**, type `Student` and press your *Enter* key. You have named your new folder **Student**.

Next, extract the zip file you downloaded to the **Student** folder you just created.

5. Locate the **ArcGIS-Pro-3.x-Cookbook-main.zip** file You downloaded that contains the data for the book. This is typically found in your **Downloads** folder in **File Explorer**.
6. Right-click on the zip file and select **Extract All** from the menu that appears.
7. Click the **Browse** button and navigate to `C:\Student`. Then click **Extract**.

8. In **File Explorer**, navigate to the `C:\Student` folder to verify the data extracted properly. You should see a folder named **ArcGISPro3Cookbook** or **ArcGIS-Pro-3.x-Cookbook-main**.

9. If the folder is named **ArcGIS-Pro-3.x-Cookbook-main**, you need to rename it to **ArcGISPro3Cookbook**. To do this right-click on the folder and select **Rename**. They type in **ArcGISPro3Cookbook** and press your *Enter* key when done.

We also have other code bundles from our rich catalog of books and videos available at `https://github.com/PacktPublishing/`. Check them out!

Conventions used

There are a number of text conventions used throughout this book.

`Code in text`: Indicates code words in text, database table names, folder names, filenames, file extensions, pathnames, dummy URLs, user input, and Twitter handles. Here is an example: "Select the `City_Limit` feature class."

Bold: Indicates a new term, an important word, or words that you see on screen. For instance, words in menus or dialog boxes appear in **bold**. Here is an example: "Click on the **Create** button located on the **Editing** tab in the ribbon. This will open the **Create Features** pane on the right-hand side of the interface."

> **Tips or important notes**
> Appear like this.

Get in touch

Feedback from our readers is always welcome.

General feedback: If you have questions about any aspect of this book, email us at `customercare@packtpub.com` and mention the book title in the subject of your message.

Errata: Although we have taken every care to ensure the accuracy of our content, mistakes do happen. If you have found a mistake in this book, we would be grateful if you would report this to us. Please visit `www.packtpub.com/support/errata` and fill in the form.

Piracy: If you come across any illegal copies of our works in any form on the internet, we would be grateful if you would provide us with the location address or website name. Please contact us at `copyright@packt.com` with a link to the material.

If you are interested in becoming an author: If there is a topic that you have expertise in and you are interested in either writing or contributing to a book, please visit `authors.packtpub.com`.

Share Your Thoughts

Once you've read *ArcGIS Pro 3.x Cookbook*, we'd love to hear your thoughts! Scan the QR code below to go straight to the Amazon review page for this book and share your feedback.

https://packt.link/r/1-837-63170-0

Your review is important to us and the tech community and will help us make sure we're delivering excellent quality content.

Download a free PDF copy of this book

Thanks for purchasing this book!

Do you like to read on the go but are unable to carry your print books everywhere?

Is your eBook purchase not compatible with the device of your choice?

Don't worry, now with every Packt book you get a DRM-free PDF version of that book at no cost.

Read anywhere, any place, on any device. Search, copy, and paste code from your favorite technical books directly into your application.

The perks don't stop there, you can get exclusive access to discounts, newsletters, and great free content in your inbox daily

Follow these simple steps to get the benefits:

1. Scan the QR code or visit the link below

`https://packt.link/free-ebook/9781837631704`

2. Submit your proof of purchase
3. That's it! We'll send your free PDF and other benefits to your email directly

1
ArcGIS Pro Capabilities and Terminology

ArcGIS Pro represents a huge step forward in desktop **geographic information systems** (**GIS**). This new 64-bit solution allows GIS professionals to take full advantage of modern computer hardware, which brings increased performance and capabilities that have not previously been available for the desktop. It also has a brand-new modern ribbon interface. This is completely different from the toolbar-based interface we have become accustomed to in ArcMap or ArcCatalog. While more intuitive for completely new users, it can be a bit challenging for existing ArcGIS Desktop users.

In this chapter, you will begin exploring ArcGIS Pro. You will first determine whether your computer has the capability to run this powerful software. Then, you will determine which license levels are available to you. This is important, as it will affect your ability to complete some of the recipes in this book.

You will then move on to working in ArcGIS Pro. You will start by learning how to open an existing ArcGIS Pro project. Then, you will open and navigate a map. From there, you will learn methods for adding new layers and configuring some of their properties. Finally, you will learn how to create a new project from the beginning. This will include adding new maps and importing ArcMap map documents.

In this chapter, we will cover the following recipes:

- Determining whether your computer can run ArcGIS Pro
- Determining your ArcGIS Pro license level
- Opening an existing ArcGIS Pro project
- Opening and navigating a map
- Creating a project with a template
- Creating a project without a template

Determining whether your computer can run ArcGIS Pro

Unlike the 32-bit ArcMap and ArcCatalog applications, ArcGIS Pro supports multithreading/hyperthreading (use of multiple core processors), **graphics processing units** (**GPUs**), and more than 4 GB of RAM. This also means that ArcGIS Pro requires more computer resources to run properly.

The minimum requirements for ArcGIS Pro 3.x are the following:

- Windows 64-bit OS: Windows 10 or 11 Home, Pro, or Enterprise
- Windows Server 2016, 2019, and 2022 Standard and Datacenter
- Multithreaded/hyperthreaded dual-core processor
- 8 GB RAM (*Note*: this is increased from past versions)
- 32 GB hard drive space
- 4 GB dedicated video memory (*Note*: this is increased from past versions)
- Microsoft .NET Framework 6.0.5
- Microsoft Edge
- DirectX 11
- OpenGL 4.3
- Shader Model 5.0

Keep in mind that these are the minimum requirements. The more hardware, the better where ArcGIS Pro is concerned. In general, I would recommend at least 12 GB of RAM, an i5 dual-core processor, and a separate video card with its own GPU and memory. For a complete set of recommended hardware specifications, you may want to look at my other book from Packt Publishing, *Learning ArcGIS Pro 2*.

In this recipe, you will learn how to use the system requirements tool to verify whether your computer can run ArcGIS Pro. This is a free tool provided by Esri.

Getting ready

To work through this recipe, you will need to make sure you have access to the internet and sufficient permissions to install software on your computer.

How to do it...

So now, it is time to verify whether your computer meets the required specifications to run ArcGIS Pro. You will use the Esri-provided tool to do this. Open your favorite web browser, such as Google Chrome, Microsoft Edge, or Firefox:

1. Go to http://pro.arcgis.com.
2. Click the **Get Started** tab, as shown in the following screenshot:

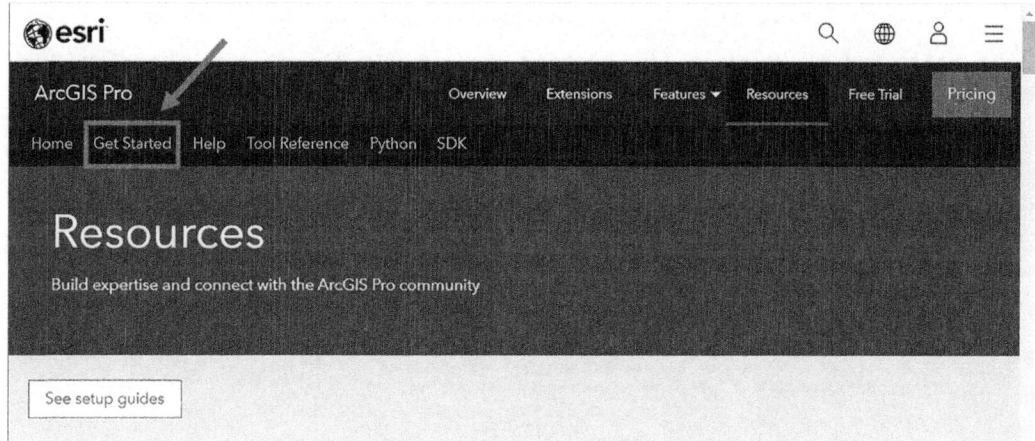

Figure 1.1 – ArcGIS Pro Resources page

3. Click on the small arrow located next to **Set up** in the left panel of the web page, as shown in the following screenshot:

Figure 1.2 – Introduction to ArcGIS Pro

4. Click on **ArcGIS Pro 3.x system requirements**, which appears below **Set up**. The actual version number will depend on the most current release of the application from Esri.

5. Click on the **Verify your computer's ability** hyperlink located below the first paragraph on the **ArcGIS Pro 3.x system requirements** page. (X = version number. The version is 3.2 at the time of publishing.)

6. A new tab will open in your browser that takes you to the **Can your computer run ArcGIS Pro 3.x** page powered by **System Requirements Lab**.

7. Click on the **Run Tech Check** button located on the lower-right side of the page.

8. If prompted with **This type of file can harm your computer. Do you want to keep Detection.exe anyway?** click on the **Keep** button.

9. Once you have downloaded the `Detection.exe` file, click it to run the file. This executable will check the specifications of your computer and generate a report indicating whether your computer is capable of running ArcGIS Pro.

10. When the hardware detection application is complete, return to your web browser to see the results. It hopefully will indicate that your system passed, similar to the following screenshot:

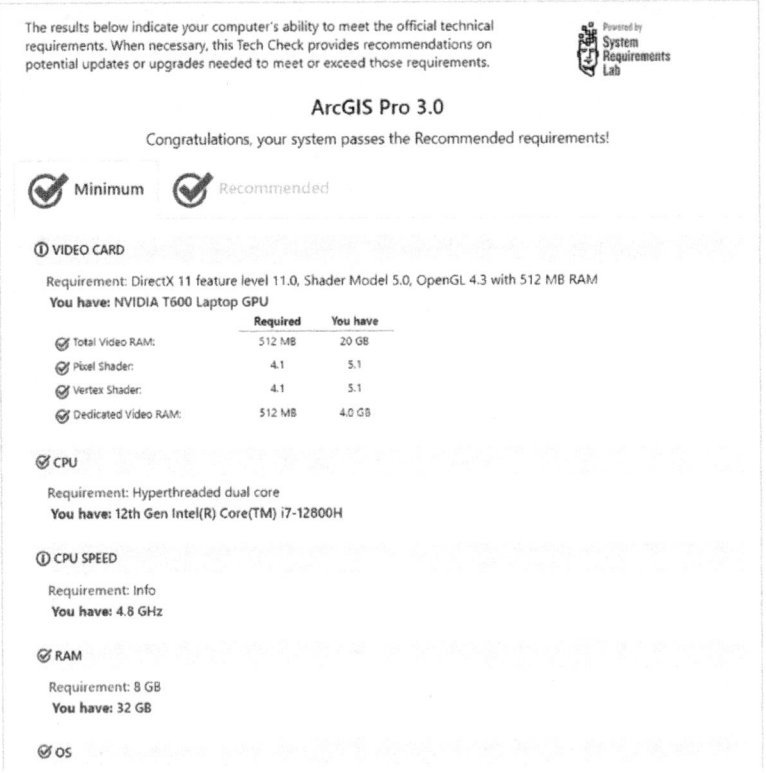

Figure 1.3 – System requirements results report

Notice that you can see whether your system meets the minimum and recommended specifications for running ArcGIS Pro. If your system just meets the minimum specifications, you can expect ArcGIS Pro to run slowly and require the application to be restarted much more frequently.

If your system fails, you will need to upgrade the components or software that the reports indicate are below the required specifications. This might be as simple as updating your drivers or web browser. It might require you to purchase new hardware if your CPU, RAM, or video card fails to meet the minimum requirements.

If your system meets or exceeds the system requirements, you may download and install ArcGIS Pro. Please refer to the installation instructions located at `https://pro.arcgis.com/en/pro-app/latest/get-started/download-arcgis-pro.htm`.

How it works...

ArcGIS Pro has very specific requirements that must be met in order to run effectively. In this recipe, you downloaded and used the tool provided by Esri to verify whether your system met or exceeded those requirements. This tool checks for both hardware and software dependencies needed to successfully run ArcGIS Pro and provides you with a detailed report so that you know without a doubt whether your computer has the horsepower required.

Determining your ArcGIS Pro license level

ArcGIS Pro has three different license levels: **Basic**, **Standard**, and **Advanced**. The license level determines the level of functionality available to the user. Basic has the least functionality, Advanced has the most, and Standard is somewhere in between.

In general, Basic allows you to visualize GIS data, produce maps, perform simple data edits, and perform basic GIS data analysis. Standard builds on the capabilities of the Basic level, with more advanced editing and data validation tools. Advanced expands the capabilities of both Basic and Standard by adding more data analysis tools. For a complete list of the capabilities of each license level of ArcGIS Pro, go to `https://pro.arcgis.com/en/pro-app/latest/get-started/license-levels.htm`.

It is important for you to know which license level you are using so that you know what capabilities are available to you. Some recipes in this book will require a Standard or Advanced license. In this recipe, you will learn how you can determine your ArcGIS Pro license level.

Getting ready

You will need to make sure that you have successfully installed ArcGIS Pro 2.0 or higher. If you have not installed ArcGIS Pro yet, please refer to `https://pro.arcgis.com/en/pro-app/latest/get-started/download-arcgis-pro.htm` for installation instructions.

ArcGIS Pro Capabilities and Terminology

How to do it...

You will now work through the steps needed to determine which license level of ArcGIS Pro has been assigned to you:

1. Go to the Windows **Start** button, which is normally located in the lower-left corner of your screen.
2. Scroll down to the **ArcGIS** program group and click **ArcGIS Pro**:

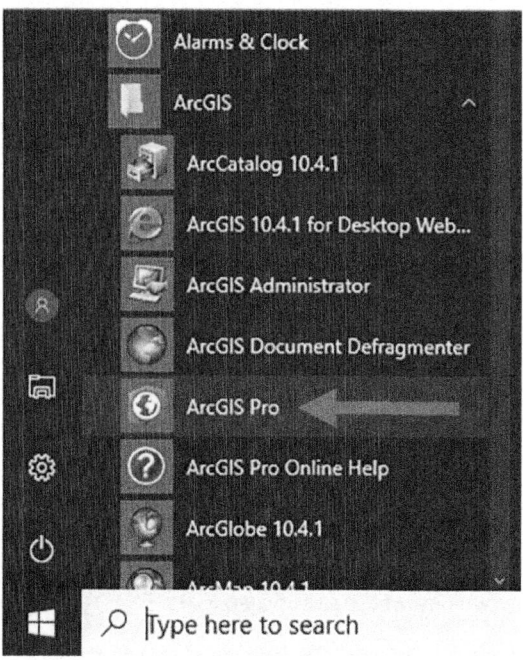

Figure 1.4 – Windows Start menu

> **Note**
> Depending on the version of Windows you are running, your Windows **Start** button and display might be a bit different. The screenshot was from Windows 10 Pro. You may also need to click **All Programs** to see the ArcGIS program group. If you right-click the **ArcGIS Pro** icon, you can select to add it to your taskbar at the bottom of your display. This makes starting ArcGIS Pro faster and easier.

3. The ArcGIS Pro start window will appear. Click **Settings**, located in the left panel of the start window. This will open the **About ArcGIS Pro** window.

4. In the **About ArcGIS Pro** window, select **Licensing**, located on the left side of the window. This will display your ArcGIS Pro licensing information:

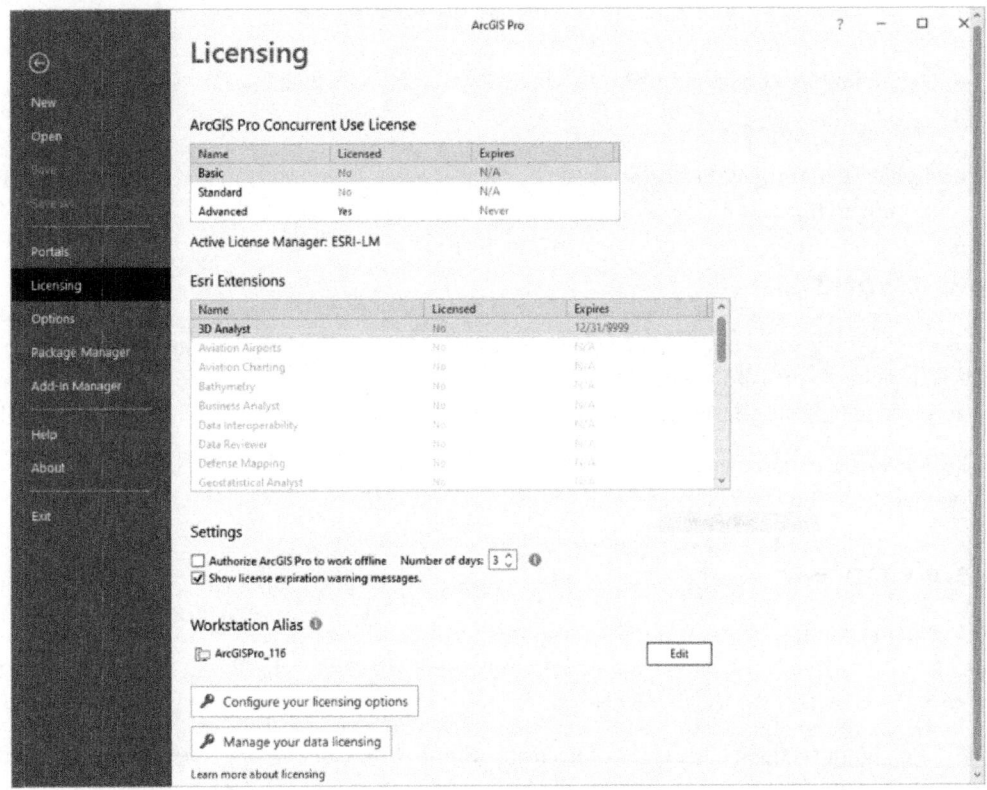

Figure 1.5 – Licensing in ArcGIS Pro

> **Tip**
> If you already have an ArcGIS Pro project open, you can access this same information from the project pane.

5. At the top of the **Licensing** window, you will see which license level you have been assigned.

As you can see in the preceding screenshot, I have access to the **Advanced** license level. From here, you can also see which extensions you have access to. Extensions are add-ons for ArcGIS Pro that provide additional functionality. These focus on a specific use, such as 3D analysis or network analysis. I have a license of 3D Analyst assigned to me.

Now, answer the following questions about the licenses you have available:

Question: Which license level have you been assigned: Basic, Standard, or Advanced?

Answer:

Question: Do you have access to any extensions, and if so, which ones?

Answer:

Once you have determined the license level you have access to, you can close ArcGIS Pro if you are not continuing to the next recipe. If you are continuing, keep ArcGIS Pro open.

How it works...

In this recipe, you learned how to determine which license level of ArcGIS Pro you had access to. You did this by going to the **About ArcGIS Pro** window from the ArcGIS Pro start window. From there, you accessed your license-level information by clicking the **Licensing** option located on the left side.

The license level is important as it determines what functionality you have access to within the program. As you will see in this book, some recipes will require you to have access to higher license levels.

There's more...

The **Licensing** window allows you to do more than just see which license level and extensions have been assigned to you. It also allows you to determine what type of ArcGIS Pro licensing you want to use, check out a license for use in the field, and more.

ArcGIS Pro supports three basic types of licensing: **Single Use**, **Concurrent Use**, and **Named User**. Named User is the default and requires you to have a username and password in your organization's ArcGIS Online or ArcGIS Portal. Single Use licenses are the traditional software licenses, where the software is licensed to a single computer. Concurrent Use licenses are sometimes referred to as network licenses. These make use of a license manager that is typically installed on a server and controls how many users can be running the software at once. To learn more about the types of licenses supported in ArcGIS Pro, go to `https://pro.arcgis.com/en/pro-app/latest/get-started/licensing-arcgis-pro.htm`.

Opening an existing ArcGIS Pro project

ArcGIS Pro makes use of project files that have an `.aprx` file extension. Projects store 2D maps, 3D scenes, database connections, folder connections, custom toolboxes, and more. When you start ArcGIS Pro, you must open a project.

In this recipe, you will learn how to open an existing project. Later in this chapter, you will learn how to create a new project.

Getting ready

To complete this recipe, you will need to make sure you have downloaded and installed the data associated with this book. The instructions to do so are present in the *Download the recipe data* section of the preface. If you followed the installation instructions, the data and projects should be located in C:\Student\ArcGISProCookbook. You will also need access to the internet. You will need to have ArcGIS Pro 3.0 or later installed as well. The recipe can be completed with any license level of ArcGIS Pro: Basic, Standard, or Advanced.

How to do it...

Now that you know which licenses are available to you, it is time to open a project in ArcGIS Pro:

1. If you closed ArcGIS Pro at the end of the last recipe, open ArcGIS Pro. If you still have ArcGIS Pro open, click the back arrow located in the top-left corner of the **About ArcGIS Pro** window.

2. In the **ArcGIS Pro** start window, click **Open another project**, as shown in the following screenshot:

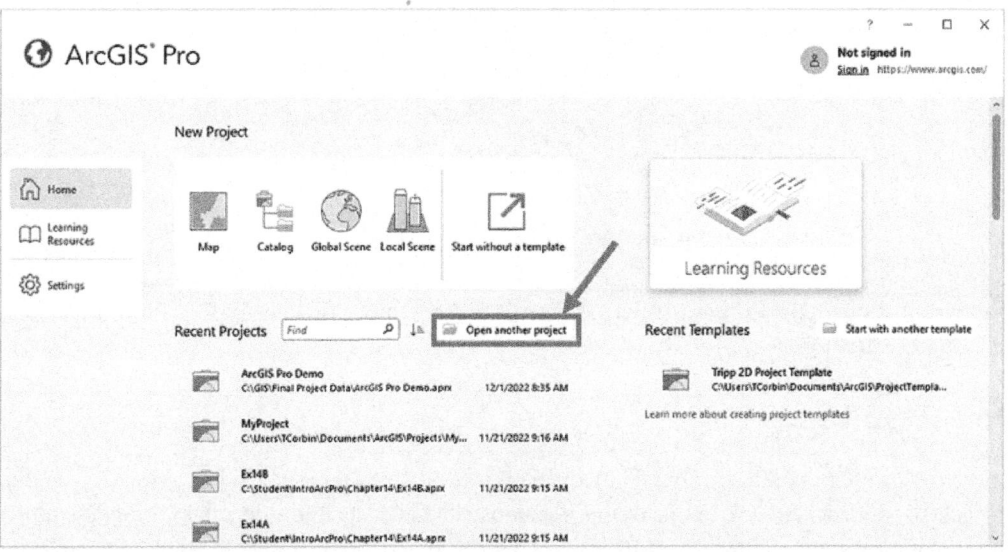

Figure 1.6 – ArcGIS Pro start window

3. The **Open Project** window should open. In the left panel, expand the tree under **Computer** by clicking on the small arrowhead next to **Computer**.
4. Select **This PC** from the tree that appears.
5. In the right panel, scroll down until you see **Local Disk (C:)** and double-click on it.
6. Scroll down until you see the Student folder and double-click on it.
7. Double-click on the ArcGISPro3Cookbook folder that contains your exercise data.

8. Double-click on the `Chapter1` folder and select the *Chapter 1 Ex 1.aprx* file. Then, click **OK** to open the project.

   If the project opened successfully, ArcGIS Pro should look similar to the following screenshot:

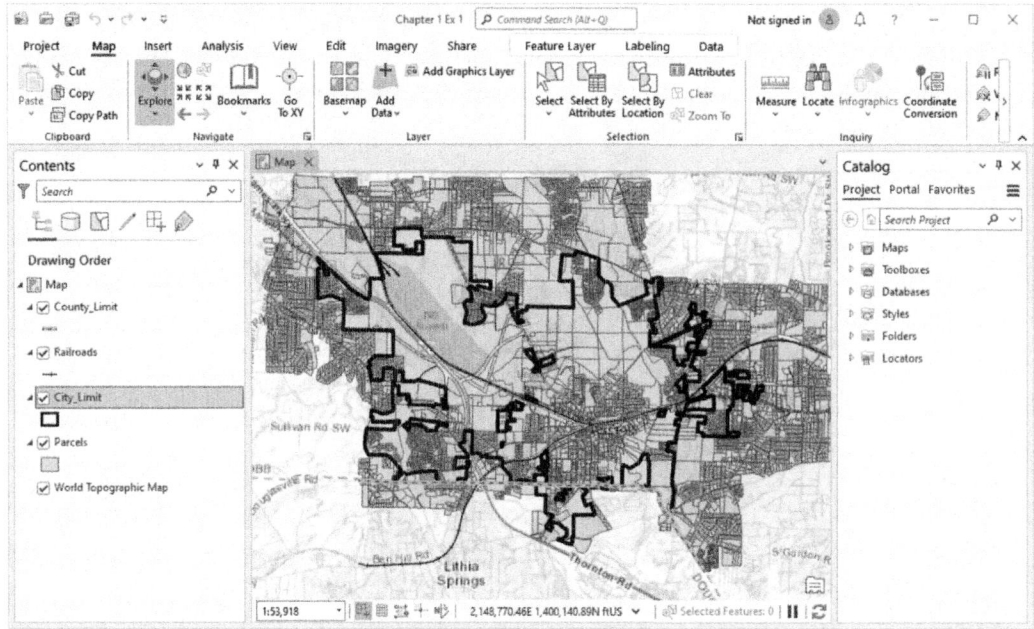

Figure 1.7 – ArcGIS Pro project opened

You have just opened your first ArcGIS Pro project. Remember that an ArcGIS Pro project has a `.aprx` file extension.

How it works...

Before you can start working in ArcGIS Pro, you must first open a project. To open a project, you must navigate to where it is stored. In this recipe, if you installed the data in the default location, the project was stored on your computer's `C:\` drive in a series of folders, so the full path was `C:\Student\ArcGISPro3Cookbook\Chapter1`. You were able to access this project by navigating to that location. You can save projects to your local computer or on a network server. It is also possible to save projects to external and flash drives. You may encounter issues if you do save and try to access projects stored on these devices because of slow data transfer rates.

Opening an existing ArcGIS Pro project   11

There's more...

You are not required to close ArcGIS Pro if you want to open another project. ArcGIS Pro does not allow you to open a project if you already have a project open. It will close the current project and open the one you select. To open another project, follow these steps:

1. Click the **Open** button located on the **Quick Access** toolbar at the top of the ArcGIS Pro interface, as indicated in the following screenshot:

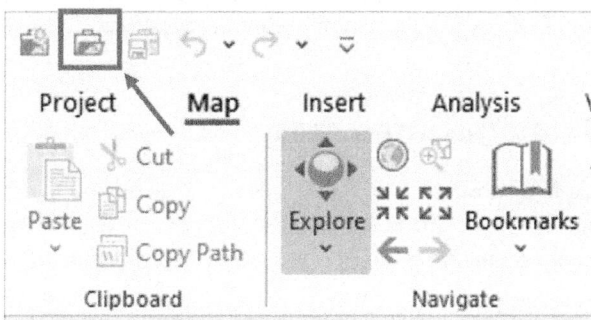

Figure 1.8 – Open button on the Quick Access toolbar

2. In the **Open Project** window that appears, navigate to `C:\Student\ArcGISProCookbook\Chapter1` using the same method you deployed to open the current project.

3. Select the *Chapter 1 Ex 1A.aprx* file and click **OK**.

   The project you originally opened will close and a new project should be open, which looks similar to this:

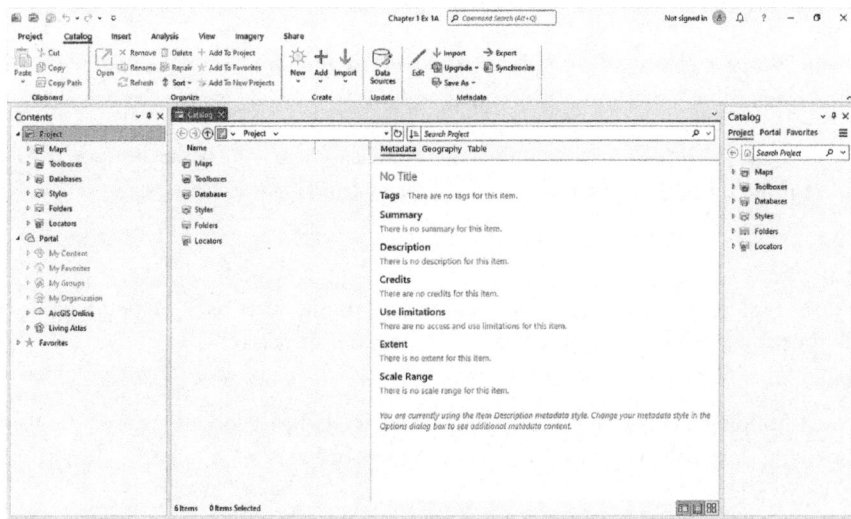

Figure 1.9 – Chapter 1 Ex 1A project open

You have now opened two projects in ArcGIS Pro using two different methods. You will find while using ArcGIS Pro that there are usually at least two ways to accomplish any task. Often, there are more.

4. If you are not continuing to the next recipe, close ArcGIS Pro without saving the project.

> **Tip**
> You can alternatively use *Ctrl + O* as a shortcut to open new projects. *Ctrl + S* works as a shortcut to save a project, and *Ctrl + N* is a shortcut to create a new project.

Opening and navigating a map

Now that you know how to open an existing project, it is time to learn how to open a map stored within the project. Projects can contain 2D maps, 3D scenes, data connections, layouts, styles, toolboxes, and more. However, 2D maps are still the primary canvas that GIS professionals work with.

In this recipe, you will learn how to open an existing 2D map. Once open, you will learn how to navigate within the map.

Getting ready

If you have successfully completed all the previous recipes, you should not need to do anything more to continue with this recipe. If you have not completed the other recipes in this chapter, you will need to do so before starting this one.

How to do it...

Now, you will continue to expand your skills. You will start to familiarize yourself with the ArcGIS Pro interface by opening a map and using common tools to navigate within the map:

1. Start ArcGIS Pro if you closed it at the end of the last recipe and open the `Chapter 1 Ex 1A` project located in `C:\Student\ArcGISProCookbook\Chapter1`.

> **Tip**
> If you don't remember how to open the project, you should refer back to the previous recipe or click the project name in the list of recently opened projects.

2. In the **Catalog** pane located on the right of the ArcGIS Pro interface, expand the `Maps` folder by clicking the small arrowhead to the left of the word **Maps**.

> **Tip**
> If the **Catalog** pane is not open, check that it is not set to **Auto-Hide**. If it is, you will see a small tab located on the right side, named **Catalog**. Simply click the tab to make the **Catalog** pane open. If you have closed the **Catalog** pane, click the **View** tab in the ribbon. Then, click the small arrowhead located below **Catalog** and select **Catalog Pane**.

3. Right-click on **Map** and select **Open**, as shown in the following screenshot:

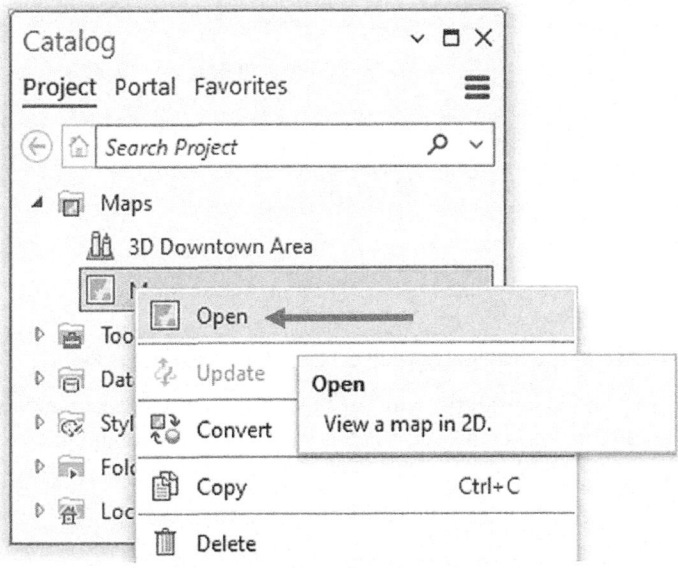

Figure 1.10 – Opening a map from the Catalog pane

You have just opened an existing project map. A project can contain multiple maps that can be either 2D or 3D. 3D maps are referred to as scenes. Now, you will learn how to navigate within maps.

4. Click the **Map** tab in the ribbon.
5. Click the **Explore** tool located on the **Map** tab in the **Navigate** tool group.

> **Note**
> The **Explore** tool is a jack of all trades. It allows you to pan, zoom, and access data about features in your map. For those who have used ArcMap, the **Explore** tool replaces the **Identify**, **Zoom in**, **Zoom out**, and **Pan** tools.

6. Move your mouse pointer into the map and roll the scroll wheel on your mouse away from you to zoom in on the map. Stop whenever you are zoomed in to the desired scale.

7. Now, roll the scroll wheel back toward you in the opposite direction to zoom out. Stop whenever you are zoomed out to the desired scale.
8. To return to the full extent of the map, click the **Full Extent** button located in the **Navigate** group on the **Map** tab in the ribbon, as shown in the following screenshot:

Figure 1.11 – Full Extent button

9. Click **Bookmarks** in the **Navigate** group on the **Map** tab in the ribbon.
10. Select **Washington Park** from the window that appears. This will zoom you to a predefined area in your map that focuses on Washington Park. Your map should now look similar to this:

Figure 1.12 – Map zoomed to Washington Park

Your display may look different depending on your monitor and resolution settings. So, do not be concerned if it does not match the preceding screenshot exactly.

11. Now, you want to zoom in closer to the block just to the north of Washington Park. Select the **Explore** tool again.

12. Hold your *Shift* key down and click near the intersection of Mulberry ST SW and Sweetwater ST SW. Continue holding down the *Shift* key and drag your mouse pointer to the southeast until you reach the intersection of Alabama St SW and Joe Jerkins Blvd SW. As you are dragging your mouse, you should see a dashed rectangular box appear on the map. This represents the area you want to zoom in on. Once you have created a box that looks similar to the following screenshot, release both the *Shift* key and your mouse button:

Figure 1.13 – Zoom area

13. Ensure the **Explore** tool is still active in the **Map** tab, as shown in the following screenshot:

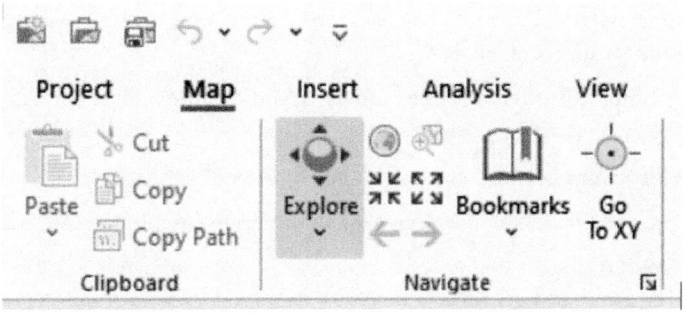

Figure 1.14 – Explore tool active in the Map tab

14. On the map, click the northwesternmost parcel in the block you just zoomed to.

    Question: What happens when you click on this parcel?

    Answer:

    If you look closely at the information window that appears, you will notice it shows information for **City Limits** and not the parcel. You need to adjust the settings for the **Explore** tool.

15. Close the information window by clicking the small **X** symbol located in the upper-right corner.
16. Click the arrowhead located below the **Explore** tool, as illustrated in the following screenshot:

Figure 1.15 – Clicking the arrowhead below the Explore tool

17. Select **Visible Layers**.
18. Click the same parcel once again.
19. Look at the top panel of the pop-up window. You should see two layers indicating **City Limits** and **Parcels**. Under each layer, you should see one feature presented.
20. In the pop-up window, click the feature located under the **City Limits** layer listed as `<Null>` to reveal information or attributes about **City Limits**.
21. Next, click on the feature located under the **Parcels** layer listed as **Sweetwater**. This will display attribute information for the parcel you clicked.
22. Close the pop-up window once you are done reviewing the attributes associated with the features in the location you clicked.
23. Try clicking other features in the map and using some of the other options associated with the **Explore** tool to see how they work.
24. Click the **Full Extent** button to return to the full extent of the map.
25. Save your project by clicking the **Save Project** button located in the **Quick Access** toolbar.
26. Close ArcGIS Pro.

So, you have just learned how to perform some basic navigation functions in a map using various tools including zoom extents, bookmarks, and the **Explore** tool. You also got to retrieve information about features in the map using the **Explore** tool, showing how versatile the tool is.

How it works...

In this recipe, you began exploring the contents of a project by opening an existing map that was contained in a project that you opened in the last recipe. Once you opened the map from the **Catalog** pane, you began to navigate within it using the **Explore** tool. You used the **Explore** tool to first zoom in and out within the map. Then, you used the **Full Extent** button to return to the full extent of the map view. Next, you used a bookmark to zoom in on **Washington Park** on the map. From there, you zoomed in on an even more specific area, using a combination of the **Explore** tool and the *Shift* key, along with your mouse. Once you zoomed in on a block of parcels you were interested in, you used the **Explore** tool to retrieve information about a specific parcel.

Creating a project with a template

We have mentioned several times that you must first open a project before you are able to access the functionality included in ArcGIS Pro. You have begun to experience this in the previous recipes where you opened and worked with existing projects. These were already configured and contained several project items including maps and database connections. How were these projects created?

In this recipe, you will create a new project using one of the four project templates included with ArcGIS Pro. You will see the structure created by ArcGIS Pro when a new project is created.

Getting ready

To complete this recipe, all you need to do is ensure that you have ArcGIS Pro installed, access to a license, and the data for the book downloaded and installed. This recipe does not even require you to have completed the previous recipes, though it might help provide a better understanding of what you are doing.

How to do it...

Now is the time for you to create your first new project using one of the Esri-provided project templates:

1. Start ArcGIS Pro.

2. When the ArcGIS Pro start window appears, select the **Catalog** template located under **New Project** near the top of the window, as illustrated in the following screenshot:

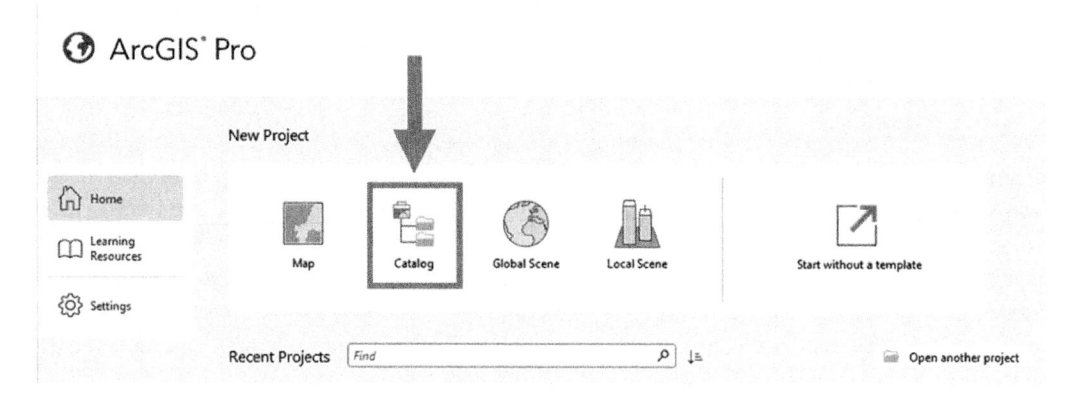

Figure 1.16 – Selecting your project template

3. In the **Name** cell, name your new project `your name_Chapter1NewProject` (for example, `Tripp_Chapter1Newproject`).

4. Click the **Browse** button located next to the **Location** cell. It looks like a small file folder with a blue arrow.

5. Under **Computer** in the left panel of the **New Project Location** window, click on **This PC**.

6. In the right panel of the **New Project Location** window, scroll down and double-click on the **Local Disk (C:\)** drive.

7. Scroll down and double-click the `Student` folder.

8. Double-click the `ArcGISPro3Cookbook` folder.

Creating a project with a template   19

9. Select the `MyProjects` folder and click **OK**. Do not double-click the folder:

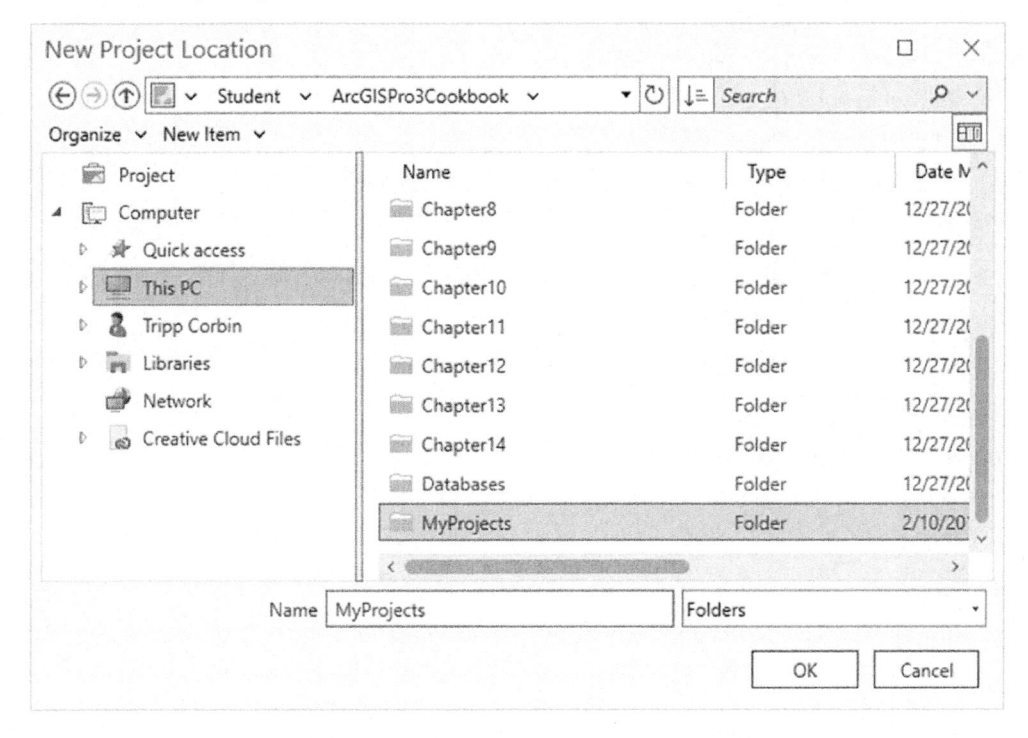

Figure 1.17 – Navigating to the MyProjects folder

10. Verify that your **Create a New Project** window looks like the following screenshot, except for your name being in place of `Tripp`. Then, click **OK**:

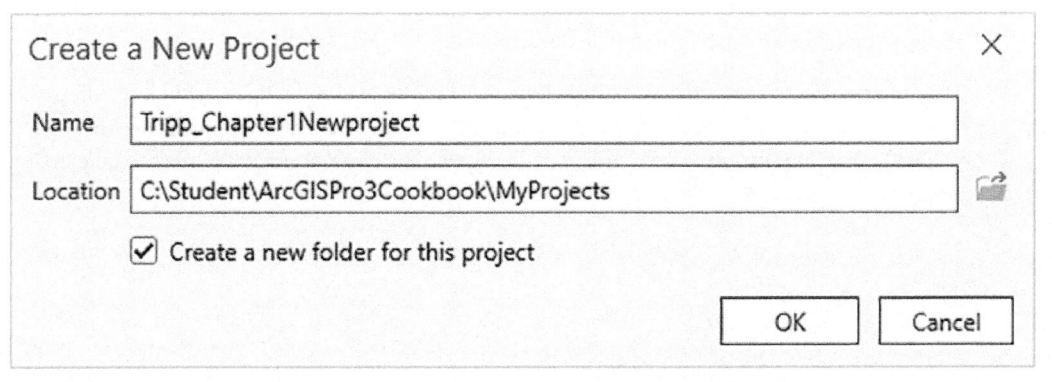

Figure 1.18 – Creating new project parameters

11. You have just created your first new project in ArcGIS Pro using the **Blank** template. This created a new bare-bones project. This project contains the minimum number of items you will find in a project. You can now begin to explore your new project.

12. To begin exploring the new project you just created, expand the `Toolboxes` folder in the **Contents** pane and answer the following question:

    Question: What do you see under the `Toolboxes` folder?

    Answer:

13. Next, expand the `Databases` folder on the **Contents** pane to see what is in that folder and answer the following question:

    Question: What do you see in the `Databases` folder?

    Answer:

14. Expand and look at the other folders included in the new project you created. See what they contain and what might be missing and answer the following question:

    Question: What do you think is missing from your new project?

    Answer:

    You should have noticed that your new project contains a single custom toolbox and file geodatabase that has the same name as your project. ArcGIS Pro automatically created these when you created the new project. You should have also noticed that several styles and locators were also automatically connected to your new project by default.

    There are four basic types of geodatabase: personal, file, enterprise (also called **SDE**, which stands for **Spatial Database Engine**), and mobile. ArcGIS Pro supports file, enterprise, and mobile geodatabases. It does not support the personal geodatabase. To learn more about geodatabases, go to `https://pro.arcgis.com/en/pro-app/latest/help/data/geodatabases/overview/what-is-a-geodatabase-.htm`.

15. This new project is missing several key elements that you will need to perform any GIS work. First, it is missing a map. Second, if you expand the geodatabase that is connected to the project, you will notice it is empty. Therefore, you also need data. Now, you will connect to an existing geodatabase and then add a map.

16. Right-click the Databases folder in the **Contents** pane and select **Add Database**, as illustrated in the following screenshot:

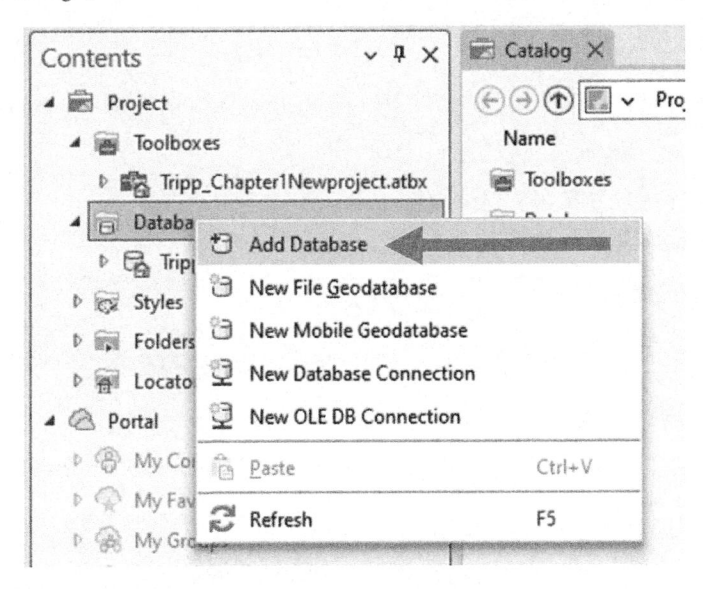

Figure 1.19 – Adding a geodatabase connection in the Contents pane

17. The **Select Existing Geodatabase** window should now be open. Click **This PC** under **Computer** in the area located on the left of the window.
18. Scroll down and double-click on **Local Disk (C:)** in the panel on the right side of the window.
19. Scroll down and double-click the Student folder.
20. Then, double-click ArcGISPro3Cookbook and Databases. You may need to scroll down to see the Databases folder.
21. Select the Trippville_GIS.gdb geodatabase and click **OK**.
22. The new database you just connected to should now appear in the Databases folder in the **Contents** pane. This database and all the data it contains are now accessible to you in the project you created. This database is one you will use in other new projects you may create. So, you will enable it so that this database will automatically be added to any new project you create.

22   ArcGIS Pro Capabilities and Terminology

23. In the **Contents** pane, right-click on the `Trippville_GIS.gdb` geodatabase you just added to the project. Then, select **Add to New Projects** from the menu:

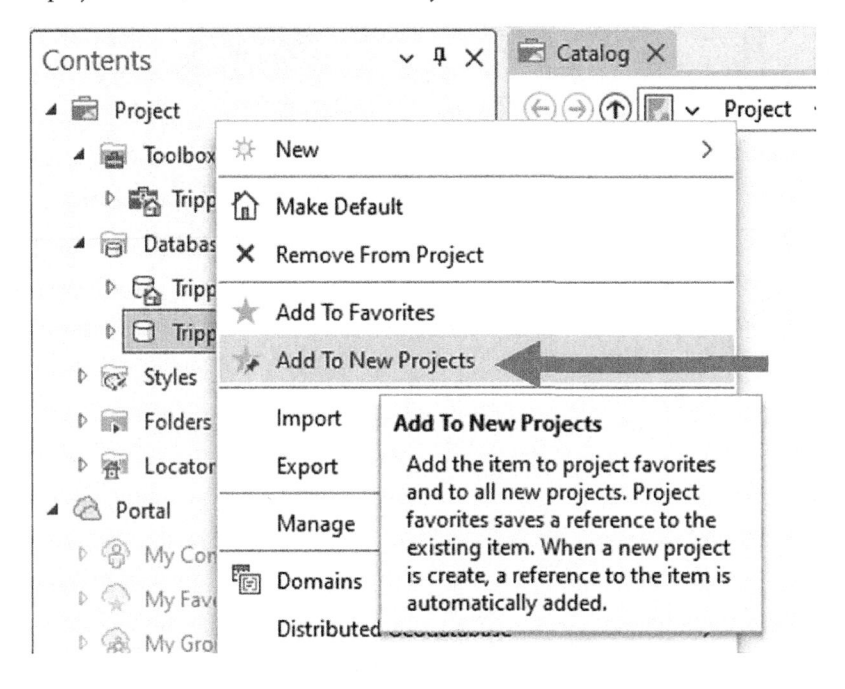

Figure 1.20 – Selecting the Add to New Projects option

24. Now, any new project you create will automatically include a connection to the `Trippville_GIS` geodatabase.

25. Next, you will add a new map to your project. You will use an existing map file so that the new map will already have layers displayed and preconfigured. You will learn how to add layers and configure them in *Chapter 2, Adding and Configuring Layers to a Map*.

26. Click on the **Insert** tab in the ribbon, as illustrated in the following screenshot:

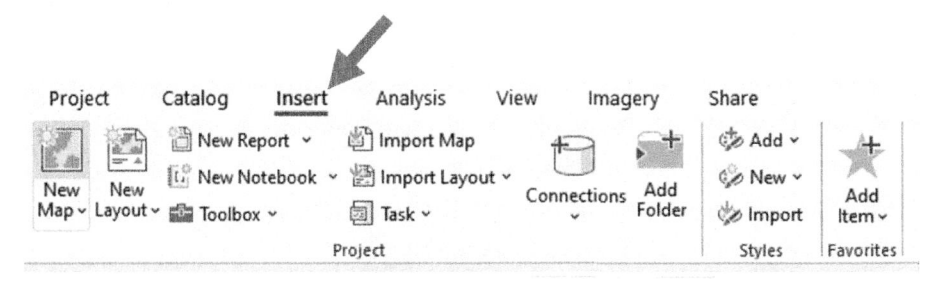

Figure 1.21 – Insert tab

Creating a project with a template 23

27. Select **Import Map** from the ribbon in the **Project** group. The **Import** window should appear.
28. Using the same process you have done in the past, navigate to `C:\Student\ArcGISPro3Cookbook\Chapter1`.
29. Select `NewMap.mapx` and click **OK**, as shown in the following screenshot:

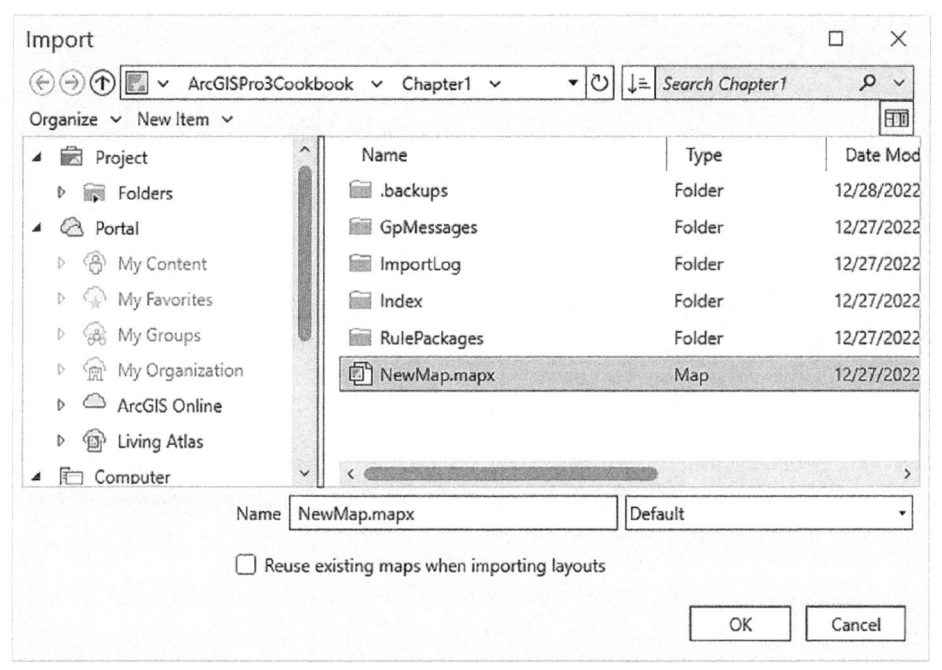

Figure 1.22 – Selecting a map file to import

> **Information**
> Map files are used to share a map created in a project with others for use in their projects or to add them to other projects you may have. Map files have a `.mapx` file extension and require the user to have access to data sources referenced in the map. They are created from the **Share** tab in the ribbon. To learn more about how to create a map file, go to `https://pro.arcgis.com/en/pro-app/latest/help/sharing/overview/save-a-map-file.htm`.

30. A new map should open in your project containing several layers. All these layers are already configured with specific symbology and other property settings. This illustrates the power of using map files.

31. Save your project using the *Ctrl + S* shortcut keys and close ArcGIS Pro.

You have now successfully created your first new project and added several components to it, such as a database connection and a map using a map file.

How it works...

In this recipe, you created your first new project. You used the **Catalog** template, which creates a very basic project for ArcGIS Pro. Projects created with the **Catalog** template include connections to a custom toolbox and geodatabase that have the same name as the project, along with several styles and locators. ArcGIS Pro automatically creates the toolbox and geodatabase when you create a new project. They are both empty and are intended for you to save project-specific items to them. However, in most cases, your organization will already have an established GIS database.

Once you created your project, you added a connection to an existing geodatabase, `Trippville_GIS`. This database contains GIS data for the City of Trippville that you used in other recipes. This is actually the primary database that you will use in the remainder of this book. So, you added it to your favorites and set it to automatically be added to all new projects you create.

After you established a connection to the primary geodatabase, you imported a new map. This new map was based on an existing map file. The new map was added with layers already added and their properties configured, saving you a lot of effort.

There's more...

In the last recipe, you created a new project using the **Catalog** template. As you saw, this template automatically created the project structure including a project geodatabase, custom toolbox, and more. The **Catalog** template is not the only one included with ArcGIS Pro. There are several others you can use when creating a new project. Let's take a quick look at them.

ArcGIS Pro stock project templates

As you may have noticed when you created your new project, ArcGIS Pro includes three other project templates: **Global Scene**, **Local Scene**, and **Map**. Each of these will create slightly different projects:

Name	Description	Project Items Created
Map	Creates a new project that automatically includes a new 2D map	- Project toolbox - Project file geodatabase - 2D map
Local Scene	Creates a new project that automatically includes a new local 3D map	- Project toolbox - Project file geodatabase - 3D local scene
Global Scene	Creates a new project that automatically includes a new global 3D map, similar to what you experience with Google Earth the first time you open it	- Project toolbox - Project file geodatabase - 3D global scene

Table 1.1 – Esri default project templates

In addition, connections to standard styles, locators, and any favorites that are set to **Add to New Projects** are also added to items created, based on the selected new project template.

Using the skills that you have learned in the previous recipes, try creating a new project using a template other than **Blank**. See how they differ from one another.

Creating a project without a template

In the last recipe, you created a new project using the **Catalog** template. This template automatically created several items associated with the project including a project geodatabase, folder, and toolbox. You may encounter situations where these items are not needed. You may just need to review some data and perform a quick edit or simple analysis. Can ArcGIS Pro create a project without those automatically created items? Yes, that is possible in ArcGIS Pro 3.X.

ArcGIS Pro 3.X allows you to create a project without a template. When you do this, ArcGIS Pro creates project-specific items such as databases, toolboxes, and folders temporarily. When ArcGIS Pro is closed, another project is opened, or a new project is created, these temporary files are deleted.

In this recipe, you will create a new project without using a template. You will see the structure created by ArcGIS Pro and how it differs from the project you created in the previous recipe.

Getting ready

To complete this recipe, all you need to do is ensure that you have ArcGIS Pro installed, access to a license, and the data for the book downloaded and installed. This recipe also requires that you have completed the previous recipe, *Creating a project with a template*.

How to do it...

Now, you will create a new project without using a template:

1. Start ArcGIS Pro.
2. When the ArcGIS Pro start window appears, select **Start without a template** located under **New Project** near the top of the window, as illustrated in the following screenshot:

Figure 1.23 – Starting a project without a template

3. Notice that, this time, you did not have to provide a location or name used to create and save the new project. ArcGIS Pro just opened with a blank project. Unlike the project you created in the previous recipe, the **Contents** pane of this new project is empty. Next, you will explore this project to see what ArcGIS Pro did create.
4. In the **Catalog** pane located on the right side of ArcGIS Pro, expand `Folders` so that you can see its contents. Then answer the following question:

   Question: Which folder do you see connected to this project?

   Answer:

5. Right-click on the folder and select **Show in File Explorer** so that you can see where this folder was created on your computer:

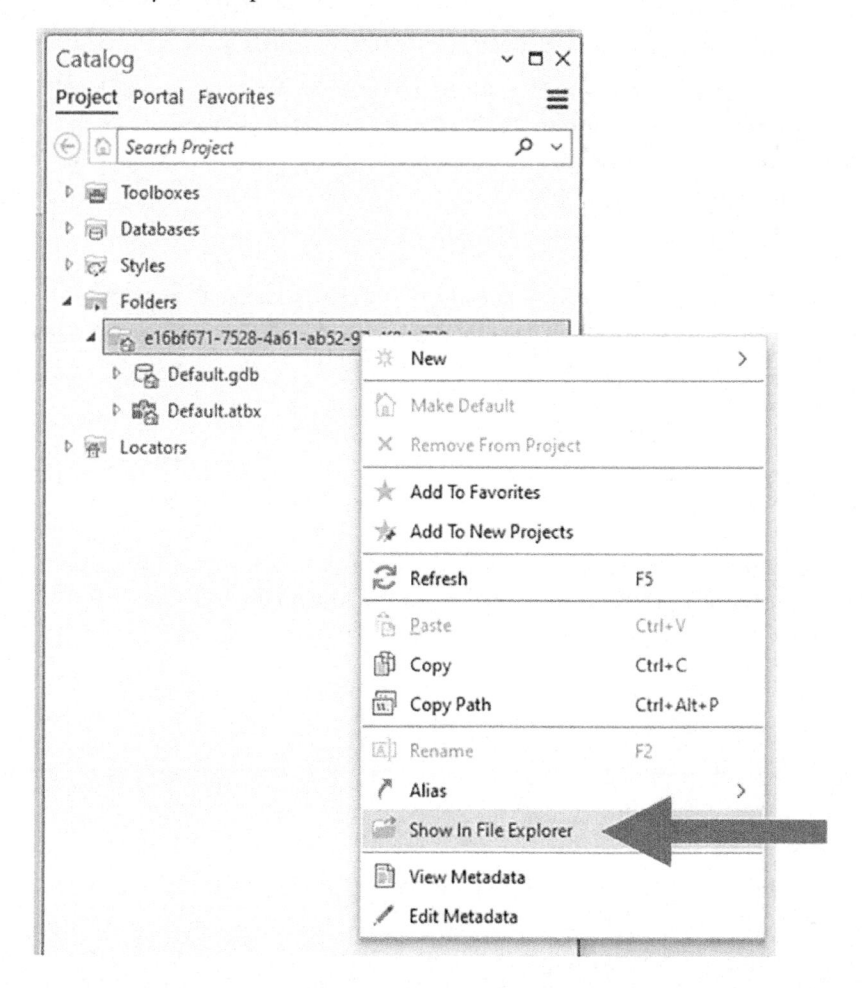

Figure 1.24 – Examining the results

6. When the **File Explorer** window opens, look in the address bar at the top of the window under the ribbon to see the location of this project and the files that were created. Answer the following questions:

   Question: Where is the project stored on your computer?

   Answer:

   Question: What files did ArcGIS Pro create with this new project?

   Answer:

7. Close the **File Explorer** window and return to ArcGIS Pro.
8. In the **Catalog** pane, expand Toolboxes so that you can see the contents.
9. Next, expand Databases so that you can see the contents.
10. You should notice that the contents of the toolboxes and databases match the files you saw in **File Explorer**. You should also see the Trippville_GIS geodatabase in the Databases folder. This connection appears in this new project because you set it to **Add to New Projects** in the previous recipe.
11. Close ArcGIS Pro without saving the project.

You have now created a new project without using one of the Esri default templates. You should now understand how a project created without a template differs from one that was created using a template.

How it works...

In this recipe, you created a new project without using a template. ArcGIS Pro still created several project items, including a geodatabase and toolbox. However, these were only temporary and were deleted from your system once you closed ArcGIS Pro. The Trippville_GIS.gdb geodatabase connection was also added to this new project. This database will not be deleted because it was not one of the temporary files or folders that were created by ArcGIS Pro.

2
Adding and Configuring Layers

One of the primary capabilities of any geographic information systems is the ability to create maps that allow multiple data layers to be overlaid on one another. This allows you to visualize the spatial relationships between features in one or more layers.

ArcGIS Pro allows you to use data from multiple sources and formats as layers in a map. This includes data from other Esri ArcGIS applications and other sources as well, such as AutoCAD and web services. This support for multiple data sources and formats expands your ability to visualize and analyze data, regardless of where it comes from and how it was created.

In this chapter, you will learn how to add new layers to a map using data from different sources. You will then learn how to configure various settings for the layer, such as symbology, labeling, and more.

We will cover the following recipes:

- Adding a layer from a geodatabase
- Adding a layer from ArcGIS Online
- Plotting X, Y points from a table
- Geocoding addresses

Adding a layer from a geodatabase

A geodatabase is the primary data storage format for the ArcGIS platform, which includes ArcGIS Pro. So much of the data that you will visualize, edit, and analyze using ArcGIS Pro will come from a geodatabase. There are several types of geodatabases, including personal, file, enterprise, and mobile.

A geodatabase stores related features as **feature classes**. A feature class is a collection of features that share the same geometry (point, line, polygon, annotation, or multipatch), attribute table, and coordinate system. A feature class can then be added as a layer to a map so that you can see both the spatial and tabular data. To take a deeper dive into the geodatabase format, go to `https://pro.arcgis.com/en/pro-app/latest/help/data/geodatabases/overview/what-is-a-geodatabase-.htm`.

Adding and Configuring Layers

In this recipe, you will act as a GIS analyst for the City of Trippville. The city manager has asked you to create a simple map showing some basic information about the city, including city limits, roads, railroads, and points of interest.

Getting ready

This recipe requires the sample data to be installed on your computer. It is recommended that you complete the recipes in *Chapter 1, ArcGIS Pro Capabilities and Terminology*, before starting this recipe. This will ensure you have a better foundational understanding of navigating within a map. You can complete this recipe with any ArcGIS Pro licensing level.

How to do it...

In this recipe, you will add the required layers to a map using different methods and then configure them.

Opening ArcGIS Pro and a project

To start, you must launch the ArcGIS Pro application and open a project. Follow these steps:

1. Start ArcGIS Pro by clicking on the Start menu button. Then if you are running Windows 10, expand the **ArcGIS** Program Group and select **ArcGIS Pro** as shown in the following screenshot. If you are running Windows 11, you will need to click on the **All apps** button before you expand the **ArcGIS** program group.

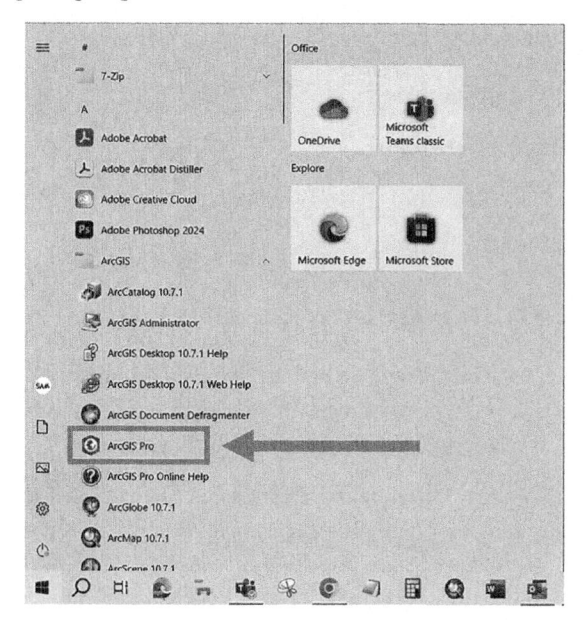

Figure 2.1 – Starting ArcGIS Pro from the Start button

Adding a layer from a geodatabase   31

2. In the **ArcGIS Pro Start** window, click on the **Open another project** button, as illustrated here:

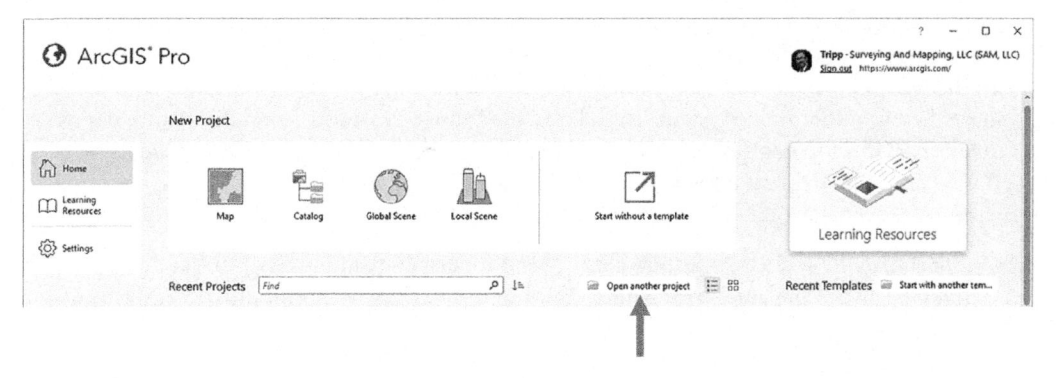

Figure 2.2 – Open another project

3. In the **Open Project** window, expand the **Computer** option in the left panel. Then, in the right panel, scroll down and double-click on the C: drive. It might be labeled as **Local Disk**, **Local Drive**, or **OS**.

4. Double-click on the `Student` folder, followed by the `ArcGISPro3Cookbook` and *Chapter 2* folders.

5. Double-click on the `AddingLayers` folder and select the `AddingLayers.aprx` project file. Then, click the **OK** button to open the project.

The project you selected should open with a single map. This map will contain a basemap but no other layers.

Adding layers to a map using the Add Data button

You will now begin adding the requested layers to this map. You will start with the city limits:

1. Activate the **Map** tab in the ribbon.
2. Click on the **Add Data** button.
3. In the **Add Data** window that just opened, expand the `Databases` folder located under **Project** in the left panel.
4. Double-click on the `Trippville_GIS.gdb` geodatabase in the right panel of the window.
5. Double-click on the `Base` feature dataset.

> **Information**
>
> Feature datasets are organizational units in a geodatabase. They act in a similar way as folders on your computer. They allow you to store related feature classes in a common container within the geodatabase so that you can easily find them. All feature classes stored within a feature dataset share the same coordinate system. This allows the feature classes stored in the feature dataset to take part in a topology and geometric network. Feature datasets only exist in geodatabases. You will not find them in other GIS data formats, such as shapefiles.

6. Select the `City_Limit` feature class. This is a polygon feature class. You can tell this by the icon located to the left of the feature class name. Click on the **OK** button to add this feature class to your map as a layer.

    Your map should look similar to what's shown in the following screenshot. Upon adding the new layer, your map should have automatically zoomed into the area covered by the new layer. ArcGIS Pro assigns random colors when a new layer is added, so your `City_Limit` layer might be a different color:

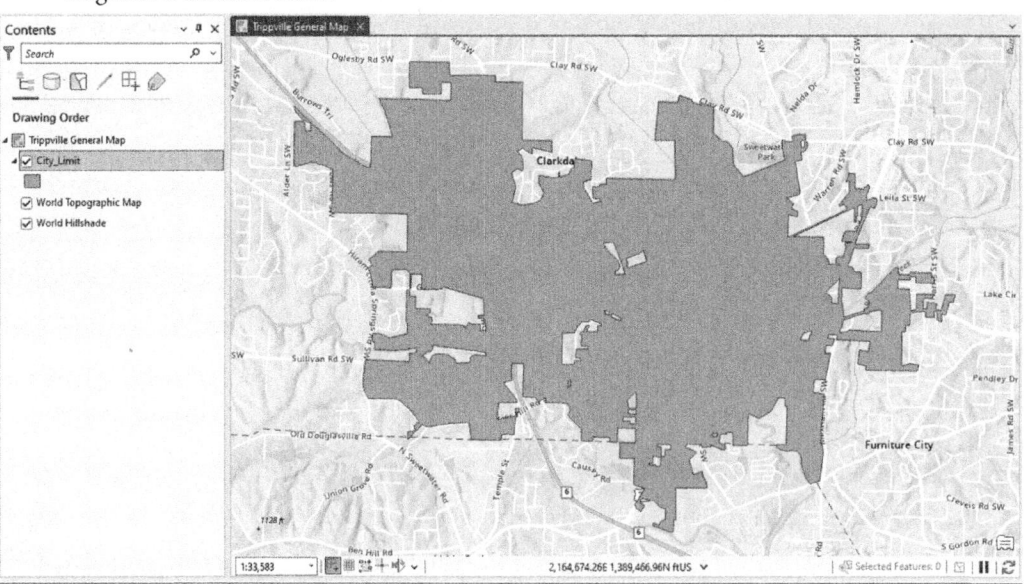

Figure 2.3 – Map with the City_Limit layer added

Adding a new layer from the Catalog pane

Next, you will add the street centerlines to represent the roads in the city. You will use a different method to add this layer to the map:

1. In the **Catalog** pane, which is normally docked to the right of the ArcGIS Pro interface, expand the `Databases` folder.
2. Next, expand the `Trippville_GIS.gdb` geodatabase and then expand the `Base` feature dataset.
3. Next, right-click on the `Street_Centerlines` feature class and select **Add To Current Map** from the menu that appears, as illustrated here:

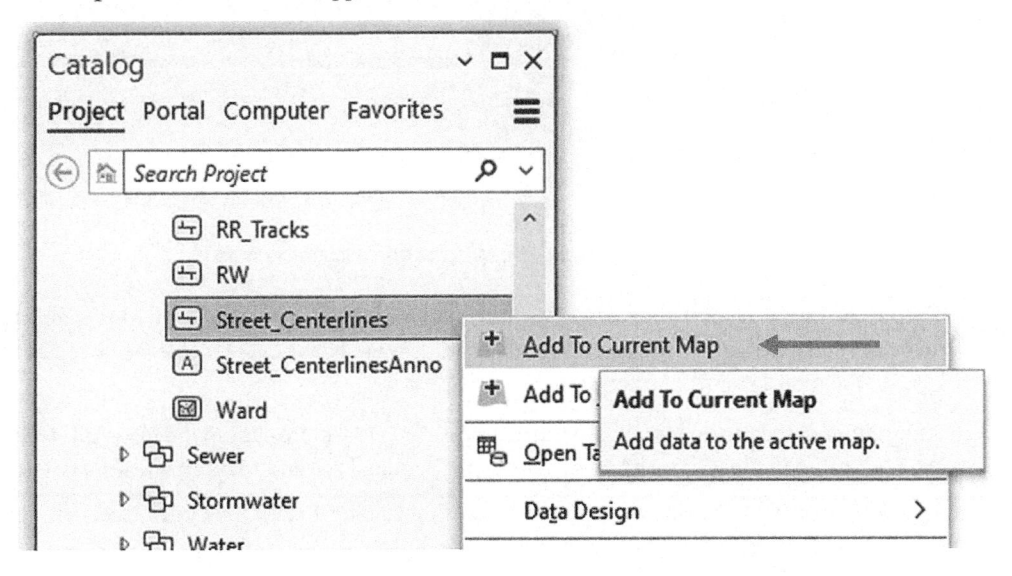

Figure 2.4 – Adding a feature class to a map from the Catalog pane

`Street_Centerlines` should now be visible on the map you are creating as a new layer. It should appear above the `City_Limit` layer in the **Contents** pane.

Dragging and dropping from the Catalog pane to add a new layer

There is another way you can add a feature class to a map as a layer from the **Catalog** pane: you can drag and drop the feature class from the **Catalog** pane into the map. You will use this method to add the `RR_Tracks` feature class to your map:

1. In the **Catalog** pane, locate the `RR_Tracks` feature class in the `Trippville_GIS.gdb` geodatabase and the `Base` feature dataset.

2. Click on the `RR_Tracks` feature class and then drag and drop it onto the map view area, as shown in the following screenshot:

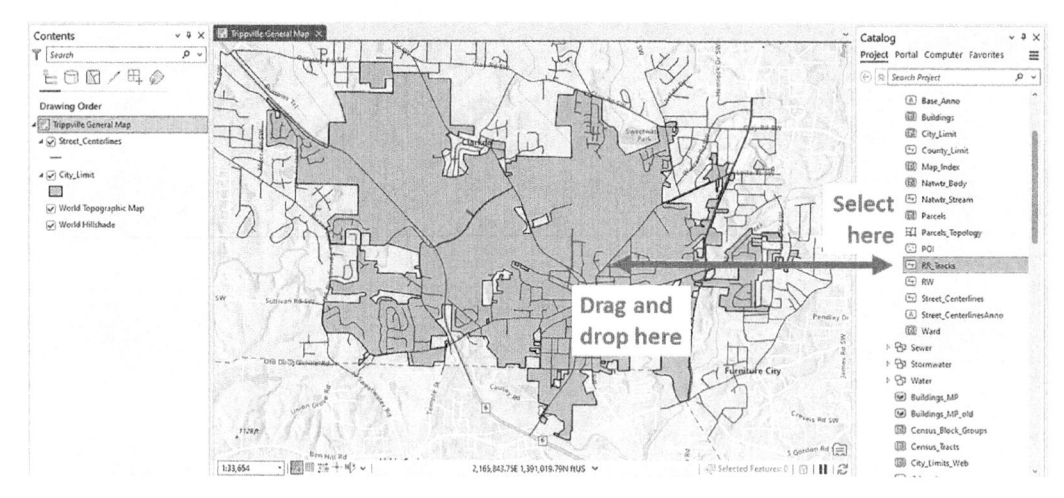

Figure 2.5 – Dragging and dropping to add a layer to a map

The `RR_Tracks` layer will be added to your map as the `Railroads` layer. This is another method you haven't used to add a new layer to your map. Now, you will add the last required layer to the map – points of interest.

3. Using either of the methods you have learned, add the `POI` feature class to the map. This feature class is located in the `Trippville_GIS.gdb` geodatabase and the `Base` feature dataset.

4. Save your project by clicking on the **Save Project** button located on the **Quick Access** toolbar in the top-left corner of the ArcGIS Pro interface:

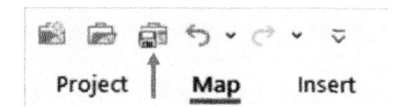

Figure 2.6 – The Save Project button on the Quick Access toolbar

The `POI` feature class will appear as the **Points of Interest** layer in your map. Points of interest is an alias that is applied to the POI feature class when it is added to a map automatically. An alias is a more descriptive name that can be created as part of the properties of a feature class or database field name. The newly added layer should be located at the top of the layer list. Now that the layers have been added to the map, you need to configure them so that you can distinguish one layer from another.

> **Note**
> ArcGIS Pro will automatically order layers based on the layer's geometry type. It will put points on top of lines, lines on top of polygons, and polygons on top of rasters as you add them.

Changing basic symbology settings

Now that you've added multiple layers to your map, you need to change the symbology so that each layer is easily distinguishable from the others. ArcGIS Pro allows you to symbolize layers using several methods, depending on the layer's purpose in the map and the data associated with that layer.

In this section, you will explore how to change simple symbology settings such as color, line type, and fill patterns, depending on the type of feature:

1. Click on the small symbol patch located below the `Railroads` layer, as shown in the following screenshot. This will open the **Symbology** pane:

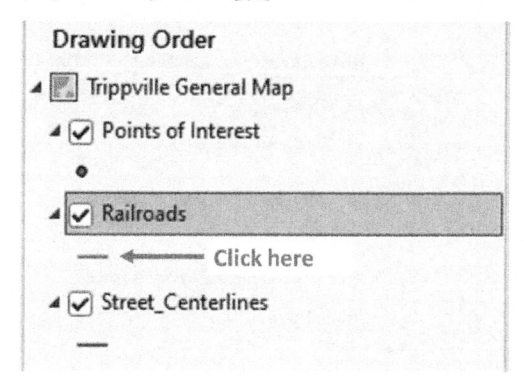

Figure 2.7 – Clicking on the symbol patch to open the Symbology pane

2. In the **Symbology** pane, click on the **Gallery** tab located near the top, just above the search cell. You should see many predefined symbology styles.
3. Scroll through the various predefined symbols and select the **Railroad** symbol. It should be located in the **ArcGIS 2D** style.
4. Click on the symbol patch located below the `City_Limit` layer in the **Contents** pane.
5. In the **Symbology** pane, verify that the **Gallery** tab is still active. Then, select the `Black Outline (2 Points)` option located in the **ArcGIS 2D** style. The symbology for the `City_Limit` layer should change. It should now be displayed as a hollow polygon with a black outline.
6. In the **Symbology** pane, click on the **Properties** tab.
7. Change **Outline width** from `2 pt` to `4 pt` using the small up arrow. Then, click **Apply**.
8. Close the **Symbology** pane.
9. Save your project by clicking on the **Save Project** button located in the **Quick Access** toolbar on the top left of the ArcGIS Pro interface.

Using unique attribute values for symbology

You have just adjusted basic symbology settings for a layer that impacts all features contained in that layer. Now, you will set up a symbology that is based on unique attribute values contained in the layer's attribute table:

1. Select the `Street_Centerlines` layer in the **Contents** pane.
2. Click on the **Feature Layer** tab in the ribbon.
3. Click on the small drop-down arrow located below the **Symbology** button and select **Unique Values** from the menu that appears, as illustrated here:

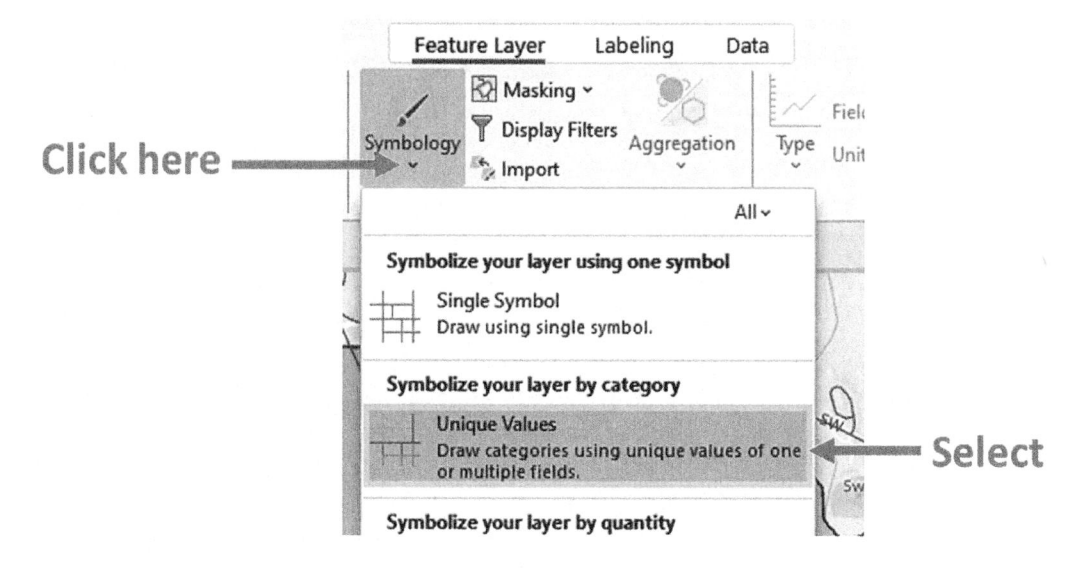

Figure 2.8 – Selecting Unique Values for symbology

4. In the **Symbology** pane, set **Field 1** to `RD_Class` using the drop-down arrow. You should see that classes have been added for city, county, and highway in the lower section of the **Symbology** pane.
5. Set **Color scheme** to Basic Random using the drop-down arrow.

> **Tip**
> There are two ways to see the names of the included color ramps. The first is to hover your mouse pointer over the color ramp; its name should be displayed as a small pop-up window. Second, you can check the box that says **Show names**. This will display the name of each color ramp above the graphic representation of that ramp.

6. In the cell located just above the three symbol classes where it says RD_Class, type Owner and press *Enter*. Watch what happens in the **Contents** pane under the Street_Centerlines layer.
7. Close the **Symbology** pane and save your project.

Your map should now look as follows:

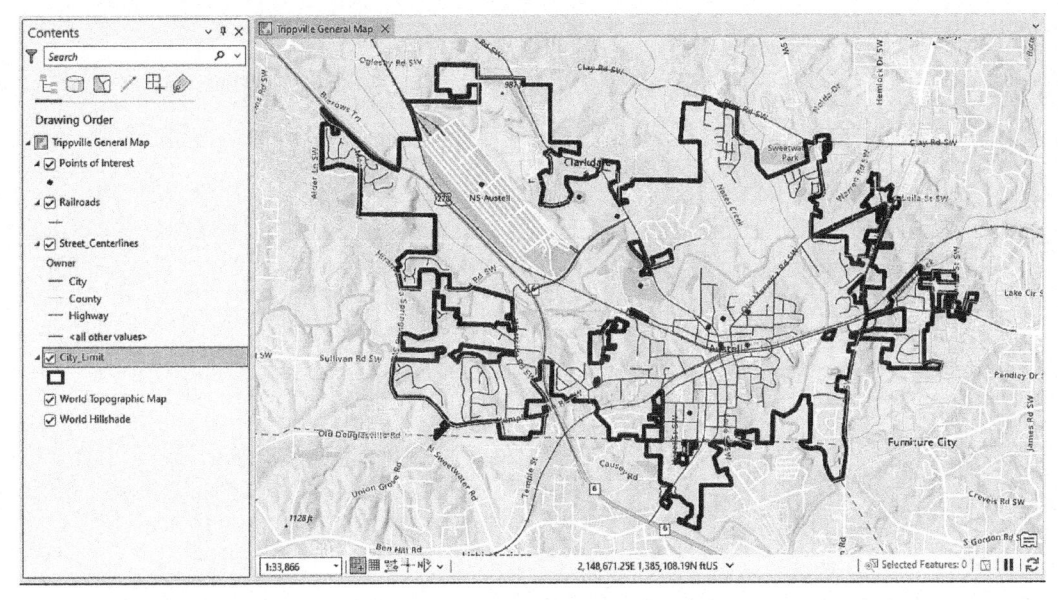

Figure 2.9 – Map with new layers added and some symbology configured

You have just configured the symbology for three of the four layers you've added to your map using two different methods. That leaves the Points of Interest layer.

Importing symbology settings from a layer file

You will use an existing layer file to not only update the symbology for the Points of Interest layer but also to apply label settings:

1. Select the Points of Interest layer in the **Contents** pane.
2. Activate the **Feature Layer** tab in the ribbon. Then, click on the **Import** button.
3. In the **Import Symbology** tool that opens, **Input Layer** should automatically be set to Points of Interest because that was the selected layer. Click the **Browse** button located to the right of **Symbology Layer**.
4. In the **Symbology Layer** window that opens, click on **Folders** beneath **Project** in the left panel.
5. Double-click on the Adding Layers folder in the right panel of the window.

38 Adding and Configuring Layers

6. Select the `Points of Interest.lyrx` layer file and click **OK**. This should return you to the **Import Symbology** tool. The symbology layer should now be set to the layer file you just selected.

7. Verify that your **Import Symbology** tool matches what's shown in the following screenshot and click **OK**:

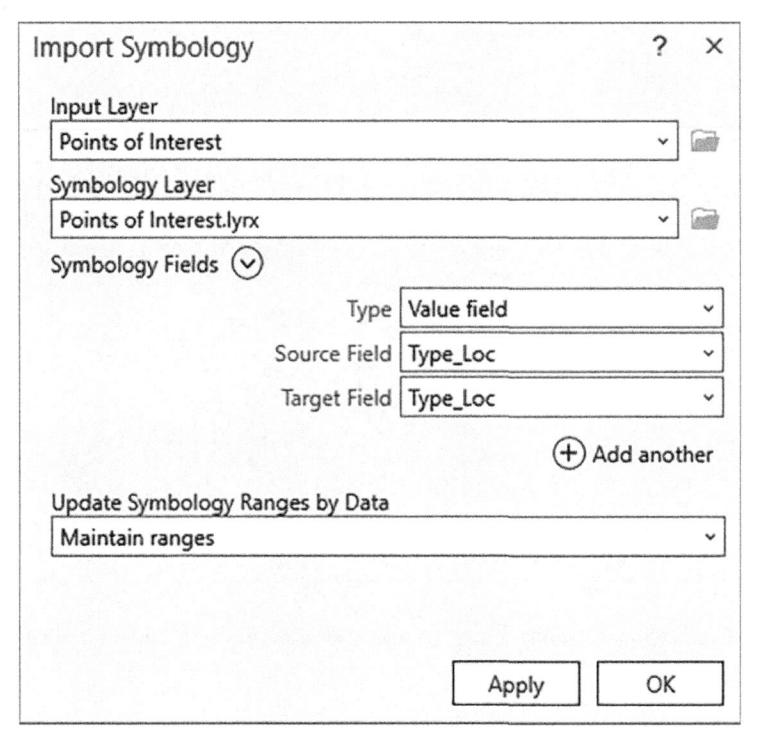

Figure 2.10 – Import Symbology with completed parameters

The symbology for the `Points of Interest` layer should now be set up to display based on the location type for each feature. By importing the settings from the layer, you did more than just update the symbology, as you will see next.

8. Activate the **Map** tab in the ribbon and select the **Explore** tool.

9. Zoom the map into the `Points of Interest` features grouped in the center of town, as shown here:

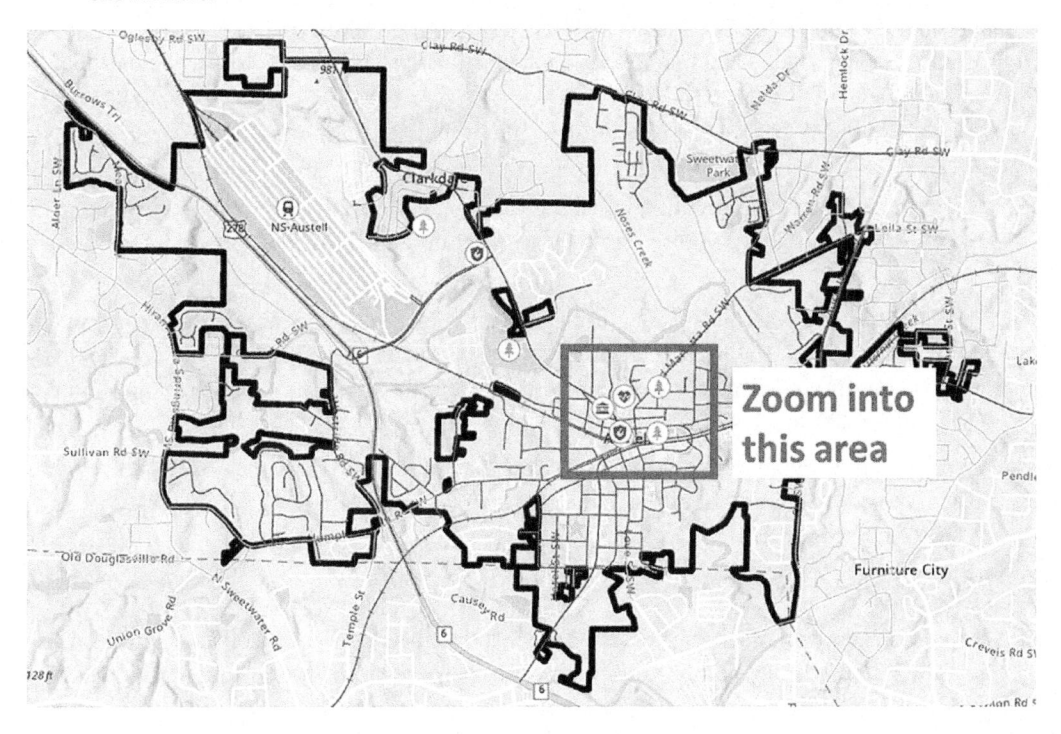

Figure 2.11 – Zoom into this area

As you zoom in to this area, text labels should appear showing the name of each point of interest. The label settings for this layer were also applied when you imported the symbology from the layer file.

You will now manually configure labels for the `Street_Centerlines` layer:

1. Select the `Street_Centerlines` layer in the **Contents** pane.
2. Activate the **Labeling** tab in the ribbon. Then, click on the **Label** button on the far-left end of the ribbon.

The text should appear on your map just above each street centerline feature. If you look at the text, you may notice it is incomplete. It does not show the full name of each road segment. It is missing the suffix that identifies if it is a street, circle, court, avenue, or lane. Also, some labels overlap with other features. You will now adjust some settings to see whether you can improve how the labels are displayed:

1. Set the **Field** option located in the **Label Class** group on the **Labeling** tab to `Label_Name` using the drop-down arrow. You should see the labels for the `Street_Centerlines` features change so that they now include the suffix.

> **Information**
>
> Labels are dynamic text that ArcGIS Pro will automatically generate and display based on values found in the attribute table of the layer. You can also build expressions using various programming languages, including Arcade, Python, VBScript, and JScript. To learn more about labeling in ArcGIS Pro, go to https://pro.arcgis.com/en/pro-app/latest/help/mapping/text/labeling-basics.htm.

2. Next, in the **Label Placement** group on the **Labeling** tab, select **North American Streets**. The labels should shift when you select this placement option so that they are cleaner and easier to read.

3. In the **Visibility Range** group, set **Out Beyond** to <Current> using the drop-down arrow. This will be populated with the current zoom scale of your map. The exact value will depend on the size of your monitor, the area you are zoomed into, and what panes are open, but it should be between 1:3200 and 1:4000.

4. Activate the **Map** tab in the ribbon and select the **Explore** tool. Use the scroll wheel on your mouse to zoom in and out of the map. Watch what happens to the labels for the Street_Centerlines layer.

5. Save your project and close ArcGIS Pro.

As you zoom in and out of the map, you should see that the labels turn on and off automatically based on your view scale. Setting visibility scales such as this helps reduce clutter within a map, making it more readable.

How it works...

In this recipe, you added four feature classes from a geodatabase to a map as new layers. You did this using three methods:

- The first method was using the **Add Data** button on the **Map** tab in the ribbon. This method had you click on the **Add Data** button, which opened the **Add Data** window. Next, you navigated to the Trippville_GIS.gdb geodatabase and expanded the Base feature dataset. From there, you selected the City_Limit layer and clicked **OK**. This created a new layer in your map that references back to the City_Limits feature class.

- The next method was adding the Street_Centerlines feature class to the map from the **Catalog** pane. To do this, you expanded the Databases folder to access the Trippville_GIS.gdb geodatabase that was connected to the project. Then, you expanded the Base feature dataset. You located the Street_Centerlines feature class and right-clicked on it. Lastly, you selected the **Add To Current Map** option from the menu that appeared.

- The last method you used to add a feature class to a map as a layer also involved using the **Catalog** pane. You were able to simply select the `RR_Tracks` feature class in the **Catalog** pane and drag and drop it into the map. This created a new `Railroads` layer in the map.

Once you added the new layers to your map, you had to configure them by setting up their symbology and labeling. You did this using the **Feature Layer** tab in the ribbon and by importing the settings from an existing layer file or manually setting them up.

Adding a layer from ArcGIS Online

One of the powerful capabilities of ArcGIS Pro is its integration with Esri's cloud solution, ArcGIS Online. ArcGIS Online allows users to share GIS content, including maps, data, and applications, with others. ArcGIS Pro's integration with this cloud solution allows users to access data and maps quickly and easily so that they can display, edit, and analyze them. It does this by utilizing web services.

ArcGIS Online contains a vast array of data that can help enrich your maps. A good example of this is Esri's catalog of basemaps. At the time of writing, ArcGIS Online hosts around 30 basemaps that you can use in ArcGIS Pro as backdrops for your maps and data. But that just scratches the surface. ArcGIS Online is not just limited to data published by Esri. Many organizations have published and shared data and maps that you can also access and use. To learn more about ArcGIS Online and its capabilities, go to `https://www.esri.com/en-us/arcgis/products/arcgis-online/overview`.

In this recipe, the city of Trippville's water system superintendent has asked you to review data that a consultant working for the city has created. The consultant has published the data via ArcGIS Online so that you can view it. You must add the data that the consultant published to ArcGIS Online to a map in ArcGIS Pro.

Getting ready

This recipe will require you to have an ArcGIS Online Named User account or higher in addition to your ArcGIS Pro license. If your organization does not have an ArcGIS Online Named User account that can be assigned to you, you can sign up for a free trial at `https://www.esri.com/en-us/arcgis/products/arcgis-online/trial`. As with all the recipes in this book, you must also have the sample data installed. You will also need to ensure you have a connection to the internet.

You don't need to have completed the previous recipe to complete this one. However, it is recommended that you complete the recipes in *Chapter 1* to ensure you have a foundational understanding of terminology and the ArcGIS Pro interface.

How to do it...

Follow these steps to learn how to add layers that are stored in ArcGIS Online to a map:

1. Start ArcGIS Pro from the Windows **Start** menu or the shortcut on your taskbar or desktop.
2. In the **ArcGIS Pro Start** window, click on the **Open another project** button, as shown here:

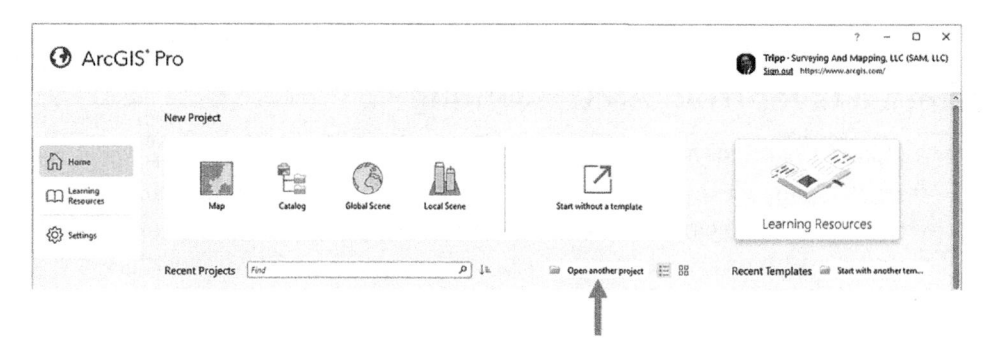

Figure 2.12 – Open another project

3. In the **Open Project** window, expand the **Computer** option in the left panel. Then, in the right panel, scroll down and double-click on the C: drive. It may be labeled **Local Disk**, **Local Drive**, or **OS**.
4. Double-click on the `Student` folder, followed by the `ArcGISPro3Cookbook` and *Chapter 2* folders.
5. Double-click on the `AGOL Layers` folder and select the `AGOL Layers` project file. Then, click the **OK** button to open the project.

   The project should open in ArcGIS Pro and display a map containing the `City_Limit`, `Parcels`, `World Topographic Map`, and `World Hillshade` layers. The `World Topographic Map` and `World Hillshade` layers are from ArcGIS Online as part of the `Topographic` basemap.

Verifying and changing the ArcGIS Online-hosted basemap

Now, you must verify where the `Topographic` basemap is coming from and then change the basemap to another one to help verify the consultants' water data:

1. In the **Contents** pane, click on the **List by Source** button, which resembles a short cylinder, so that you can see where the various layers in the map are sourced.
2. Locate the `World Topographic Map` and `World Hillshade` layers in the **Contents** pane. Note where the sources for these layers are.

As shown in the following screenshot, both of these layers reference two different web services. `World Topographic Map` points to a web map that is hosted via `arcgis.com`, whereas `World Hillshade` points to a feature service that is hosted via `arcgisonline.com`. Both URLs are part of ArcGIS Online:

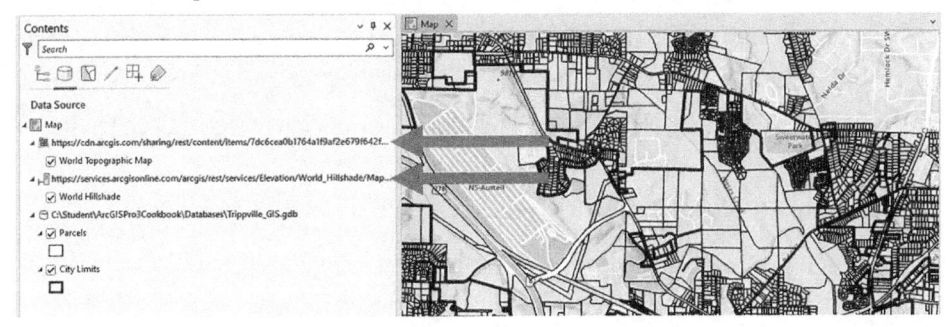

Figure 2.13 – Basemap sources from ArcGIS Online

As basemaps, you cannot change their display settings, edit their data, or use them for analysis. Basemaps just help provide context for the rest of your layers. You can, however, change the basemap being used in your map. You will do that now.

To help verify the water system data created by the consultant, you will switch to the `Imagery` basemap.

3. Active the **Map** tab in the ribbon. Then, click on the **Basemap** button to display the basemap gallery.
4. Select the **Imagery** option from the gallery that appears. This will change the basemap that's used in the map to one that shows aerial photography provided by Esri, as shown in the following screenshot:

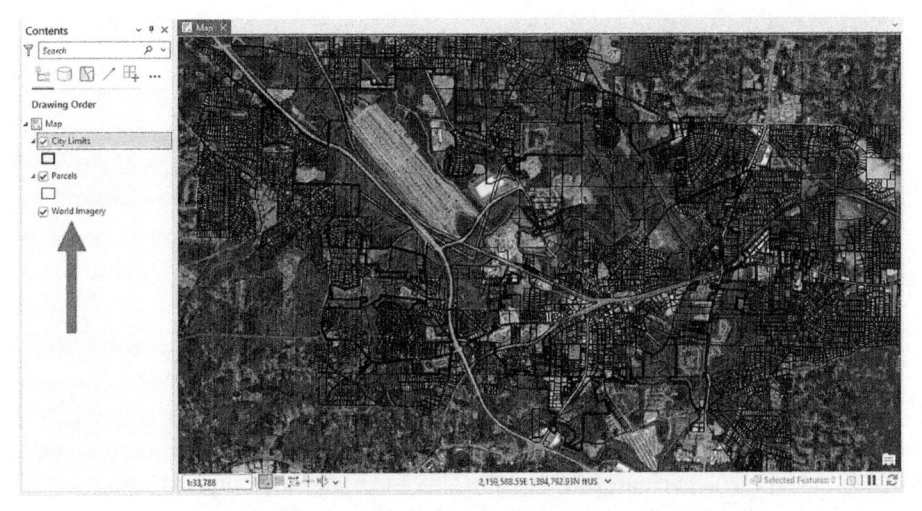

Figure 2.14 – Basemap changed to World Imagery from ArcGIS Online

As you saw in the gallery, Esri provides a wide range of basemaps you can use in your maps. These are available to all ArcGIS Pro users as part of your license.

Adding other layers from ArcGIS Online

Basemaps are just the start of the data you can access via ArcGIS Online. Next, you will add layers for the city of Trippville that were created by the city's consultant and hosted in ArcGIS Online:

1. In the **Catalog** pane, select the **Portal** tab located near the top of the pane and to the right of the **Project** tab.

2. Click on the **ArcGIS Online** button, which resembles a cloud, as shown here:

Figure 2.15 – The ArcGIS Online button in the Catalog pane

3. Verify that the **Content** option is selected. Then, in the **Search ArcGIS Online** cell, type `Trippville Water System Map 2023` and press *Enter* to run the search.

4. Right-click the `Trippville Water System Map 2023` feature service under **Search Results** and select **Add To Current Map**, as shown in the following screenshot:

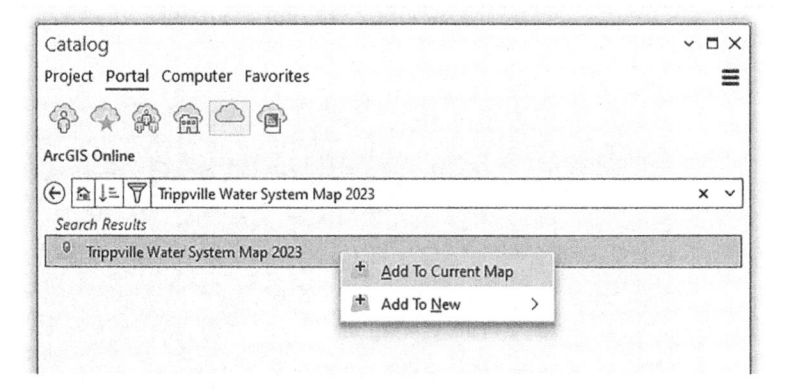

Figure 2.16 – Adding an ArcGIS Online feature service as a layer to the current map

You should now see a new group layer that's been added to your map called `Trippville Water System Map 2023`. This contains the data that was created and shared by the consultant. This is hosted in the ArcGIS Online cloud solution from Esri and is displayed on your computer.

Changing the symbology of an ArcGIS Online hosted layer

To help you evaluate the data, you will adjust the symbology for the `Water Lines` layer located within the group layer you just added:

1. Expand the `Trippville Water System Map 2023` group layer in the **Contents** pane so that you can see the layers that are included.
2. Select the `Water Lines` layer. Then, activate the **Feature Layer** tab in the ribbon.
3. Click on the **Import** button located in the **Drawing** group.
4. The **Input Layer** parameter should automatically be set to `Water Lines` because you have it selected in the **Contents** pane. Set the **Symbology Layer** parameter by clicking on the **Browse** button.
5. In the **Symbology Layer** window, click on the **Folders** option located below **Project** in the left panel of the window.
6. In the right panel, double-click on `AGOL Layers` to open it. Then, select the `Water Lines by Size and Material.lyrx` layer file and click **OK**.
7. Verify that your **Import Symbology** tool matches what's shown in the following screenshot and click **OK**:

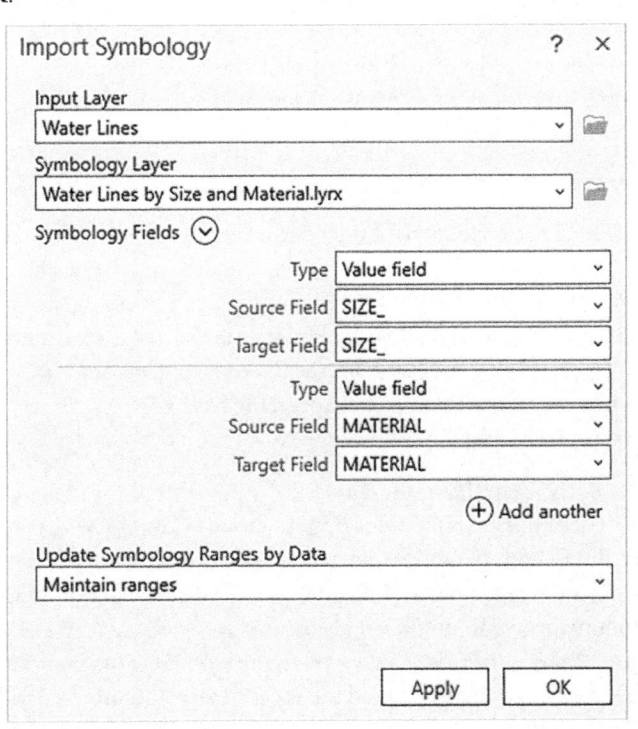

Figure 2.17 – The Import Symbology tool with completed parameters

When the tool finishes running, the symbology for the `Water Lines` layer will change and show the pipe size and material for each section of the water line. Symbolizing the water lines in this fashion will allow you or others to easily assess the completeness of the data and find possible errors.

Even though data is hosted in ArcGIS Online, you can interact with it just like you can data stored within your system or network, depending on the type of web service and your permissions.

> **Information**
>
> ArcGIS Online allows users to publish and share many types of content. From a map perspective, two basic types of web services can be published and shared – a Map Service and a Feature Service.
>
> **Map Service**: A rasterized version of the published map that allows data to be displayed, queried, and printed. It does not allow data editing.
>
> **Feature Service**: A vector version of the published map that allows all the functionality of a map service, plus it allows data editing.
>
> There is a lot more to web services and layers. To learn more, go to `https://pro.arcgis.com/en/pro-app/latest/help/sharing/overview/introduction-to-sharing-web-layers.htm`.

Save your project and close ArcGIS Pro.

You have just added a new group layer to your map that references data stored in ArcGIS Online and changed how one of the individual layers within the group layer was displayed.

How it works...

In this recipe, you got to experience some of the basic integrations between ArcGIS Pro and ArcGIS Online by changing the basemap included in the map and adding new layers that reference data stored in ArcGIS Online. You started by verifying the source for the `Topography` basemap included in the map. You did this by clicking on the **List by Source** button in the **Contents** pane. That allowed you to see that the `Topography` basemap source referenced different URLs – that is, `arcgis.com` and `arcgisonline.com`. Then, you changed the basemap to `World Imagery`, which is also hosted by Esri in ArcGIS Online.

Next, you added data to the map that represented the water system for the city of Trippville, which was also stored in ArcGIS Online. You did this by going to the **Catalog** pane and selecting the **Portal** tab. Next, you selected the **ArcGIS Online** button and performed a search for the data you needed. Then, you right-clicked on the returned result and selected **Add To Current Map**. This created a new group layer that contained several individual layers, including `Water Tank`, `Water Meters`, `Fire Hydrants`, and `Water Lines`. Lastly, you imported a layer file that changed the symbology for the `Water Lines` layer so that you could see the pipe size and material for each pipe segment.

Plotting X, Y points from a table

It is not unusual to get data from outside sources that is nothing more than a table with some information that includes X and Y data. This may come from a surveyor, someone who collected data with their smartphone, or some other source. The data might be a spreadsheet, a text file, a CSV file, or even a database table.

If the data includes coordinates for the location, you can turn these into points within a map. This is called an **event layer**. The coordinates can be in any known coordinate system if they are all the same, meaning that all the coordinates for all the records in the table must be listed in the same coordinate system.

In this recipe, you will plot the locations of crimes from a standalone database table. This table contains several records, each of which has a latitude and longitude coordinate. You will use that information to plot the location.

Getting ready

While not required, it is recommended that you complete the recipes in *Chapter 1* before starting this one to ensure you have basic skills in using ArcGIS Pro and understand the terminology associated with the application. This recipe does not require any previous recipes to have been completed. The sample data must be installed before you continue. This recipe can be completed with all license levels of ArcGIS Pro.

How to do it...

You will start by working through the process required to plot the locations of events listed in a standalone table found in a geodatabase:

1. Start ArcGIS Pro and open the `Plot XY from Table.aprx` project located in `C:\Student\ArcGISPro3Cookbook\Chapter2`. Click the **Open another project** button and navigate to `C:\Student\ArcGISPro3Cookbook\Chapter2`. Select the desired project file (`.aprx`) and click **OK**.

   The project will open with the Trippville map being displayed. This map contains two layers – `City Limit` and `Parcels`. Next, you will add the standalone table that contains the data you need to plot and examine.

2. In the **Catalog** pane, expand the `Databases` folder to reveal its contents.
3. Expand the `Trippville_GIS.gdb` geodatabase so that you can see what it contains.

48   Adding and Configuring Layers

4. Scroll through the contents of the `Trippville_GIS.gdb` geodatabase until you see the `Crimes_2023` standalone table. Right-click on the table and select **Add To Current Map**, as shown in the following screenshot:

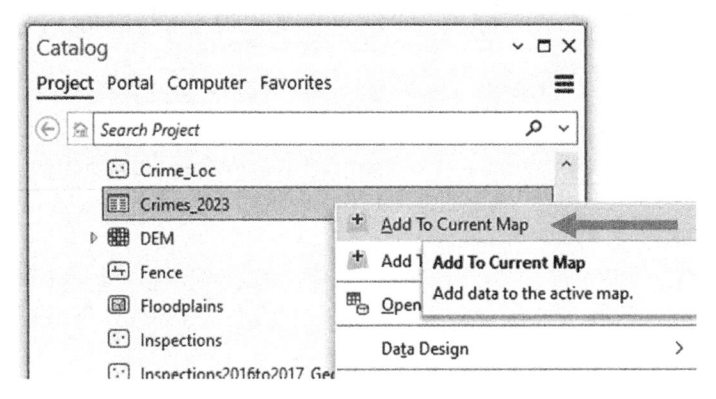

Figure 2.18 – Adding a table to a map from the Catalog pane

When added successfully, the table should appear at the bottom of the **Contents** pane.

> **Information**
>
> A standalone table is any table that is not directly associated with spatial data. This means it does not initially include a spatial component that is automatically displayed in a map. A standalone table might contain fields that hold coordinates, addresses, or other location information that can be used to display spatial information but requires additional steps to do so. Tables that are directly associated with spatial data are called attribute tables.

5. In the **Contents** pane, right-click on the `Crimes_2023` table you just added and select **Open** from the menu that appears. Take a moment to review the fields and data contained in the table so that you know what information is available for you to work with.

   You should see six different fields. The most important are the two at the end – **Lat** and **Long**. These are the coordinates that identify the location of each crime. You will use those to create points on your map that show the location.

6. Close the `Crimes_2023` table by clicking the small **X** in the tab at the top of the table.

7. Right-click the `Crimes_2023` table in the **Contents** pane and select **Create Points From Table**, then **XY Table To Point**, as shown in the following screenshot:

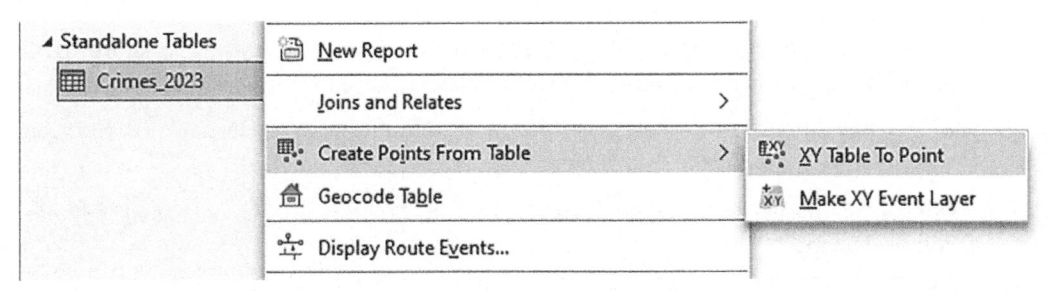

Figure 2.19 – Accessing the XY Table To Point tool

8. The **XY Table to Point** tool window should open. It will automatically populate with the required variables. Verify that yours looks as follows. If it does, click **OK**. If not, make the appropriate adjustments:

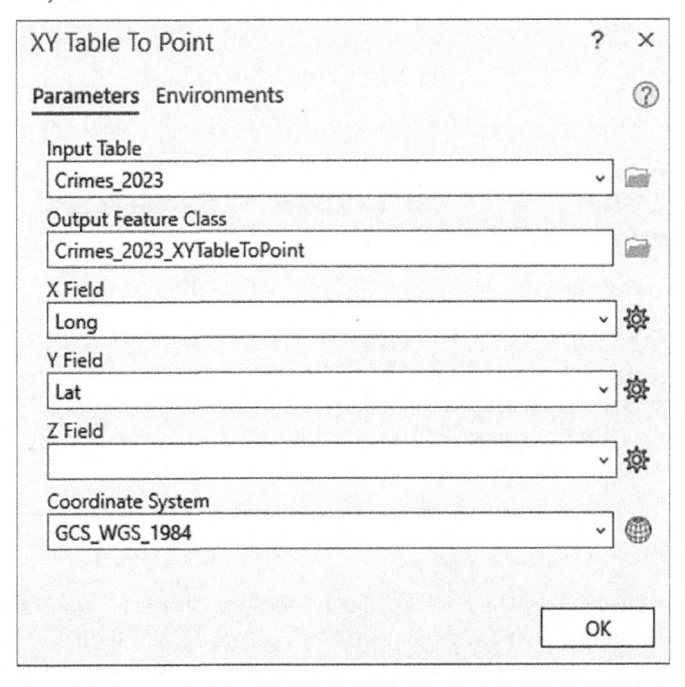

Figure 2.20 – The XY Table To Point tool with completed parameters

A new layer will appear in your map named `Crimes_2023_XYTableToPoint`. This is the results of the **XY Table to Point** tool. It generated a new point layer based on the latitude and longitude coordinates that were in the standalone `Crimes_2023` table. But it did more than that. You will explore the results later to see exactly what the tool created.

9. Right-click on the `Crimes_2023_XYTableToPoint` layer and select **Attribute Table** from the menu that appears. This will open the attribute table for the new layer you created.

10. Take a moment to review the data contained in the attribute table you just opened. Compare it to the standalone `Crimes_2023` table.

    As you can see, the data in the two tables is very similar. The attribute table contains an additional field called **Shape**.

11. In the **Catalog** pane, look at the contents of the `Trippville_GIS.gdb` geodatabase. If needed, expand it so that you can see its contents.

12. Scroll down and see whether you can locate a new feature class named `Crimes_2023_XYTableToPoint`.

13. In the **Contents** pane, double-click on the `Crimes_2023_XYTableToPoint` layer to open its properties.

14. Select **Source** in the left panel of the **Layer Properties** window.

15. Examine the values in the right panel for the `Database` and `Name` parameters.

    As you can see, the new layer references back to the new feature class you saw in the `Trippville_GIS.gdb` geodatabase. The **XY Table To Point** tool not only created a new layer but also created a new feature class in the geodatabase.

16. Close the **Layer Properties** window by clicking on the **OK** button.

17. Close all the tables you might have open.

18. Save your project and close ArcGIS Pro.

You have successfully created a new layer using data stored in a standalone table. This same process will work for data stored in other files, such as Excel spreadsheets, text files, or CSV files.

How it works...

In this recipe, you created points on a map showing the location of crimes using a standalone database table. You did this by adding the table to your map by right-clicking it in the database and selecting **Add To Current Map**.

Once you added the table to the map, you viewed the table to ensure it included the coordinate values for each record. This was as simple as opening the table and looking at the data it contained.

Lastly, you right-clicked the table in the **Contents** pane and selected **XY Table to Point**. This opened the **XY Table to Point** tool window displaying all the tool parameters. Because your table used several default fields, it automatically populated the required variables. Once this tool ran, a new point layer appeared in the map showing the actual location of the crimes from the table.

Geocoding addresses

In the previous recipe, you saw how data can be in something other than a traditional GIS format. It can be a standalone table that contains X and Y coordinates, along with other information. This can then be turned into points on a map. X and Y coordinates are not the only way we can identify a location.

Another even more common way to show a location is with a postal address. This is how postal carriers know where mail needed to be delivered well before the days of global navigation satellite systems such as GPS. ArcGIS Pro can also use an address to identify a location. This is called **geocoding**.

Simply put, geocoding is the process of converting an address or series of addresses into a location on a map or in a GIS. In this recipe, you will learn how to geocode addresses within ArcGIS Pro. This will include determining reference data in your GIS, creating an address locator, and geocoding an Inspections spreadsheet.

In this recipe, the city manager for the city of Trippville has a spreadsheet containing all the inspections recently completed by inspectors from various departments in the city. They want you to show where these inspections occurred on a map. The spreadsheet includes the address where each inspection occurred, so you will need to geocode this data.

Getting ready

To complete this recipe, you will need a spreadsheet application, such as Microsoft Excel or similar, that can open a `.xlsx` file. You also need to have the sample data installed. It is recommended that you complete the recipes in *Chapter 1* to ensure you understand the basic terminology associated with ArcGIS Pro and how to navigate the interface. This recipe can be completed with any license level of ArcGIS Pro (Basic, Standard, or Advanced).

How to do it...

You will now work through the process needed to bring in data from a spreadsheet and then geocode it so that those locations will be displayed on a map:

1. Before you start ArcGIS Pro, examine the spreadsheet containing the inspections data you will be using. Open Windows **File Explorer**. It should be on your Windows taskbar at the bottom of your screen and have an icon that resembles an old paper file folder in a holder.

2. Select **This PC** or **My Computer** in the left panel of the **File Explorer** window. Then, navigate to `C:\Student\ArcGISPro3Cookbook\Chapter2`.

3. Locate the `Inspections.xlsx` file and double-click on it to open it.

Tip

You will need an application such as Microsoft Excel or similar installed for this to work. If you do not have Microsoft Excel or a similar application, you can download and install Open Office for free. This is an open source application that has comparable functionality with Microsoft Office. You can download it from `https://www.openoffice.org/download/index.html`.

4. Review the data contained in the spreadsheet you just opened. Note what columns it contains and the data in each row:

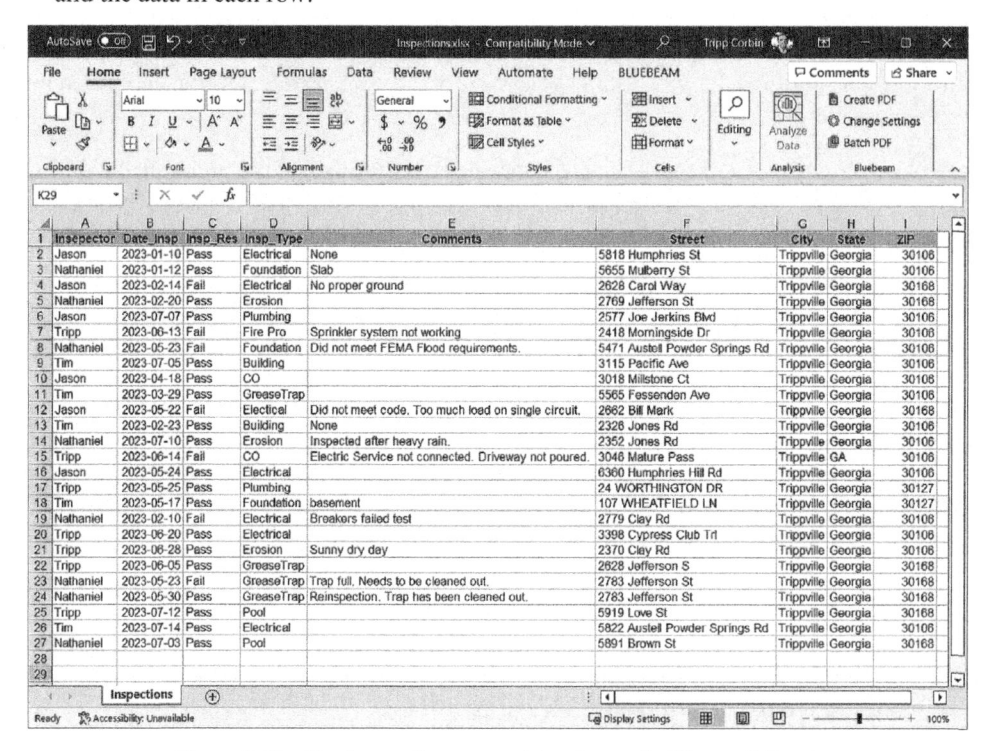

Figure 2.21 – The Inspections spreadsheet open in Microsoft Excel

This spreadsheet, provided by the city manager, represents an export from a permitting and inspections system used by the city. That system cannot display data on a map, but the city manager wants to see where inspections have been completed within the city. As you can see, the spreadsheet does include the address where the inspection took place. You will use that to geocode the location of each inspection onto a map.

Launching ArcGIS Pro and adding an Excel spreadsheet

Follow these steps to add the spreadsheet you just examined to an ArcGIS Pro project so that you can start using that data:

1. Start ArcGIS Pro and open the `Geocoding.aprx` project located in `C:\Student\ArcGISPro3Cookbook\Chapter2`. Click on the **Open another project** button and navigate to `C:\Student\ArcGISPro3Cookbook\Chapter2`. Select the desired project file (`.aprx`) and click **OK**.
2. In the **Catalog** pane, expand the **Folders** option so that you can see its contents. Then, expand the `Chapter2` folder.
3. Locate the `Inspections.xlsx` file and expand it so that you can see the individual sheets included in the file. Right-click on the **Inspections$** sheet and select **Add To Current Map**, as shown in the following screenshot:

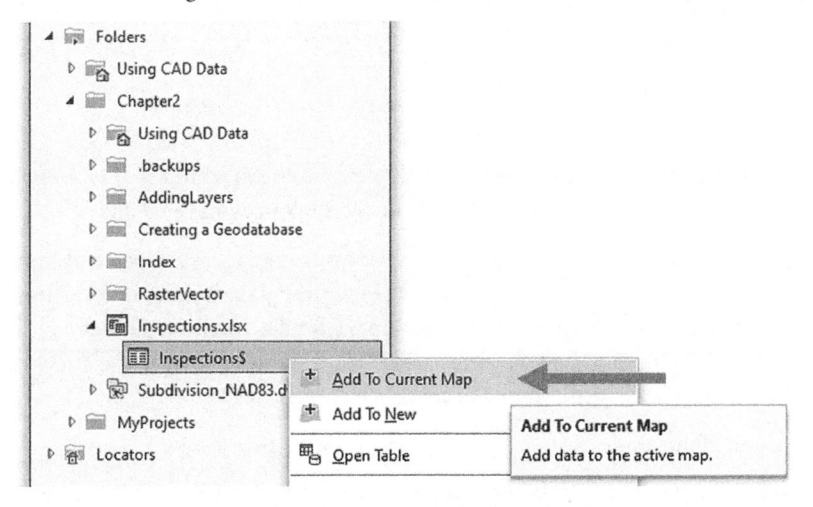

Figure 2.22 – Adding a spreadsheet to the current map from the Catalog pane

You have just added the spreadsheet you examined to your map so that you can geocode it.

Creating an address locator

Before you can geocode the data in the spreadsheet, you will need to create a locator. Geocoding in ArcGIS Pro requires three components if you are using just your own data. You will need the table you want to geocode, the reference data in your GIS that contains address information, and a locator.

Reference data is typically road centerlines, address points, or parcel polygons. Address points typically provide the greatest level of accuracy. This is followed by parcel polygons. The least accurate is road centerlines.

54 Adding and Configuring Layers

The locator is the translator between your source data and the data you are trying to geocode. It provides basic settings and options required to geocode. Several styles of locators are dependent on your reference data and how the address information is formatted.

So, before you can create your locator, you must first identify your source data. This will be a GIS layer that already contains address information:

1. Since there is no address point layer in the `Trippville_GIS.gdb` geodatabase, you know that will not be your reference data. You do have parcel polygons, so right-click the `Parcels` layer in the **Contents** pane and select **Attribute Table**:

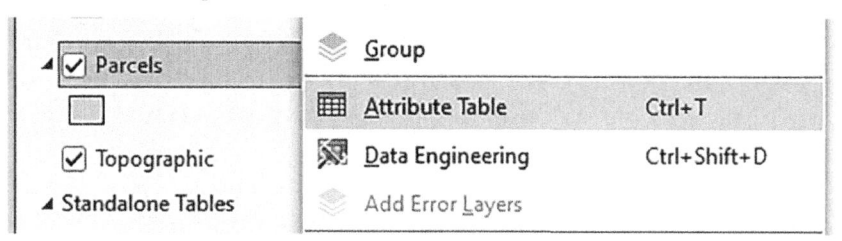

Figure 2.23 – Opening the Parcels layer attribute table

2. Examine the attribute table for the `Parcels` layer to determine whether it contains address data. Review the field names contained in the table, along with the data itself.

   The `Parcels` layer does include some address information. It has the `street name`, `street number`, and `street suffix` fields. However, the data is incomplete. Also, it is missing other information that is typically included in a complete address. Therefore, the `parcels` layer is not well-suited to be your reference layer. It's time to look at the street centerlines to see whether they have more complete address information.

3. Close the attribute table for the `Parcels` layer you just reviewed.

4. Open the attribute table for the `Street Centerlines` layer using the same process you did for the `Parcels` layer.

5. Review the `Street Centerlines` attribute table to see whether it includes the data needed to identify a complete address. You may need to scroll over to see the entire table.

   The `Street Centerlines` layer does have more complete address information. It has address ranges for both the left and right sides of the road. It also has fields for the road name and type. In addition, it has fields for the ZIP code and city on the left and right sides of each road. This is enough information to create complete addresses. So, the `Street Centerlines` layer is the best choice as your reference layer, even if it might not produce the most accurate data points. Now, you are ready to create your locator.

6. Close the `Street Centerlines` attribute table.

7. Click the **Analysis** tab in the ribbon. Then, click the **Tools** button to open the **Geoprocessing** window.

8. Click the **Toolboxes** tab at the top of the **Geoprocessing** pane. Then, scroll down and locate the **Geocoding Tools** toolbox.

9. Expand the **Geocoding Tools** toolbox and click on the **Create Locator** tool, as shown in the following screenshot:

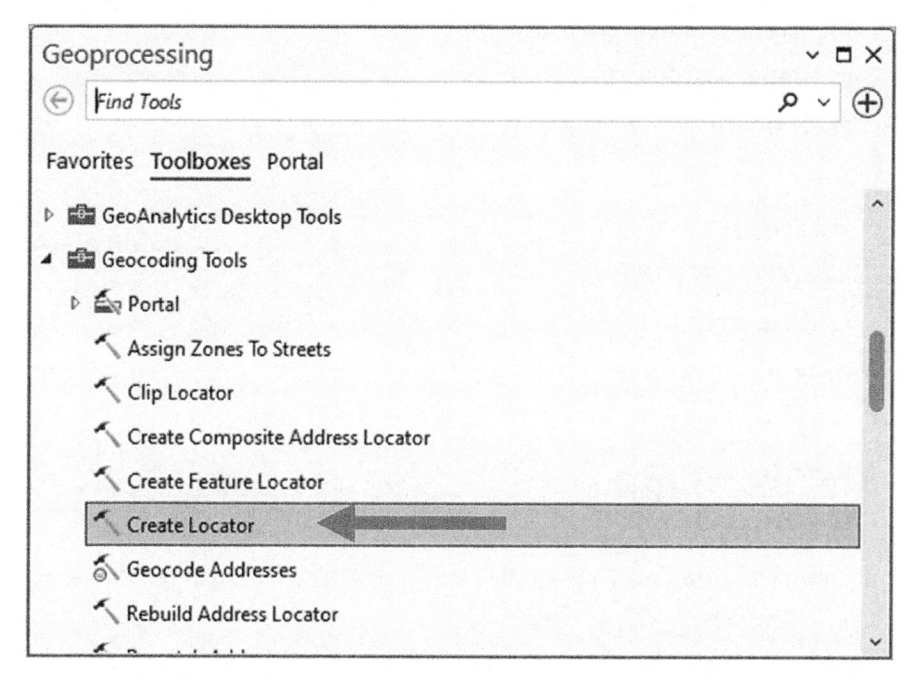

Figure 2.24 – Accessing the Create Locator tool in the Geoprocessing pane

10. In the **Create Locator** tool, set **Country or Region** to `United States` using the drop-down arrow.

11. Set **Primary Table(s)** to `Street Centerlines` using the drop-down arrow. Then set **Role** to `Street Address`, also using the drop-down arrow. You should see **Field Mapping** options appear when you do this.

> **Information**
>
> Field mapping allows you to identify which fields in the primary table you want the locator to consider when you geocode your data. Those with an asterisk (*) before the field name are required. Those without an asterisk are not required by the locator to work but might improve your geocoding results if you have them. To learn more about the **Create Locator** tool, go to `https://pro.arcgis.com/en/pro-app/latest/tool-reference/geocoding/create-locator.htm`.

12. Using the drop-down arrows, set the following field mapping values:

    - **Left House Address From**: L_F_ADD
    - **Left House Address To**: L_T_ADD
    - **Right House Address From**: R_F_ADD
    - **Right House Address To**: R_T_ADD
    - **Prefix Direction**: PREFIXDIR
    - **Street Name**: NAME
    - **Left City**: GEONAME_L
    - **Right City**: GEONAME_R
    - **Left State**: STATE_L
    - **Right State**: STATE_R
    - **Left Zip**: POSTAL_L
    - **Right Zip**: POSTAL_R
    - **Country**: COUNTRY

13. For **Output Locator**, click on the **Browse** button. Then, navigate to C:\Student\ArcGISPro3Cookbook\MyProjects.

14. Name the new locator Trippville_Locator, as shown in the following screenshot, and click **Save**:

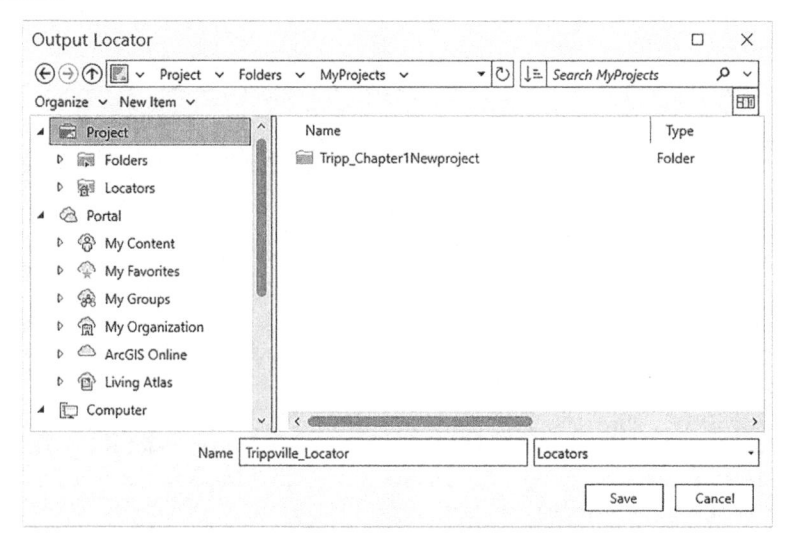

Figure 2.25 – Naming the new locator and creating it

15. Set **Language Code** to English using the drop-down arrow. Then, click the **Run** button.

> **Information**
> If the tool completes but has warnings, you can ignore them for the sake of this recipe. In a real-world situation, you would want to investigate the warnings and determine whether they will affect your ability to use the locator successfully. If they negatively affect the use of the locator, you will need to fix the issues causing the warning. In this case, most of the issues involve street centerlines located outside the city limits for Trippville so that they do not impact the use of the locator within the city.

16. Close the **Geoprocessing** pane once the tool completes.

Adjusting address locator settings

Now that you created the address locator, you will need to adjust some of the property settings to improve the final results when you geocode the spreadsheet of addresses:

1. In the **Catalog** pane, expand the **Locators** option to reveal its contents. You should see the new locator you just created. This means it is now available for you to use in your project.
2. In the **Catalog** pane, right-click on the Trippville_Locator locator you just created and select **Properties** from the menu that appears.
3. In the **Locator Properties** window, select **Geocoding options** from the left panel.
4. In the right panel of the **Locator Properties** window, expand **Display Options**.
5. Set the **Side Offset** and **End Offset** parameters to 35.

> **Information**
> **Side Offset** is how far off the centerline the geocoding process will create a new point along the edge of the centerline. **End Offset** works similarly but is at the end of the line segment. Since we know most of the street rights-of-way are between 50 and 60 feet, 35 should put the point on or close to the parcel it belongs on.

6. Verify that your **Locator Properties** window looks like what's shown in the following screenshot and click **OK**:

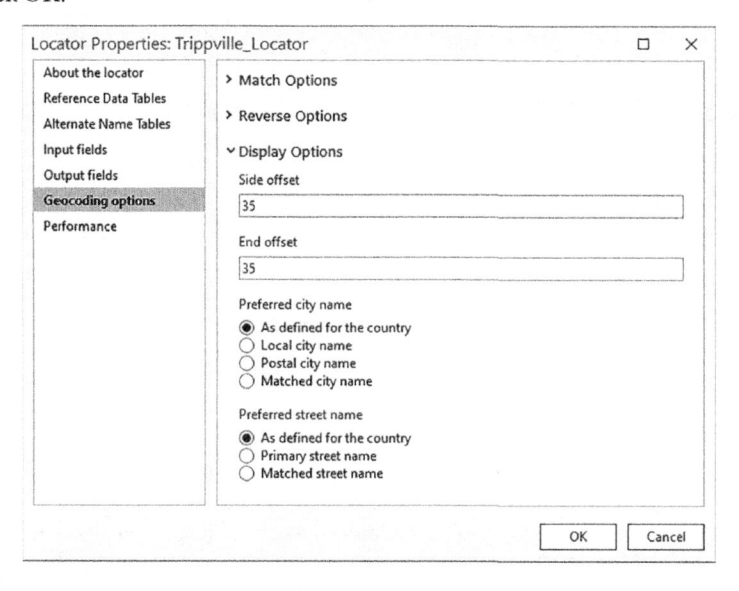

Figure 2.26 – Locator Properties with adjusted settings

Geocoding the spreadsheet data

You are now ready to geocode the address locations shown in the spreadsheet you reviewed at the start of the recipe. This will create points in the map at the address locations:

1. In the **Contents** pane, right-click on the `Inspections$` table and select **Geocode Table** from the menu that appears, as illustrated in the following screenshot:

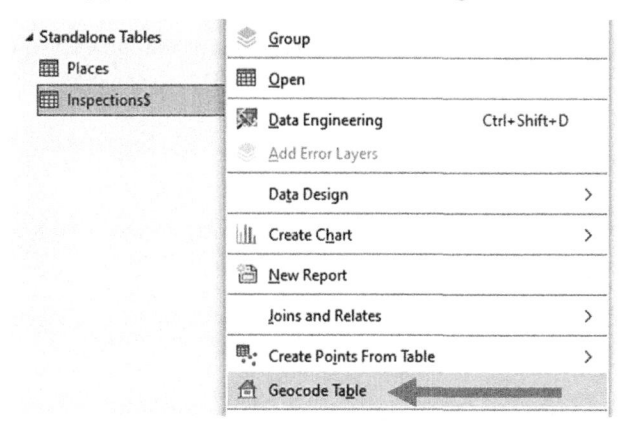

Figure 2.27 – Selecting Geocode Table from the Contents pane

2. The **Geocode Table** tool will open. Initially, it will show you a list of steps the tool requires to complete the process. Scroll down and click **Start**.
3. Set **Input Locator** to `Trippville_Locator` using the drop-down arrow. Then, click **Next**.
4. Verify that **Input Table** is set to `Inspections$`, which it should be because you right-clicked on that table to start this process.
5. Scroll down in the tool and verify that it is set to `More than one field`. Then, click **Next**.
6. Verify that the input address fields are set as follows:

    - **Address or Place**: `Street`
    - **City**: `City`
    - **State**: `State`
    - **ZIP**: `ZIP`
    - **All others**: `<None>`

7. Click **Next**.
8. In **Step 4** of the **Geocode Table** tool, set the **Output** by clicking the **Browse** button. Then, select the **Databases** option under **Project** in the left panel of the **Output** window.
9. Double-click the `Trippville_GIS.gdb` geodatabase in the right-hand panel of the **Output** window.
10. In the **Name** field, type `Inspections_Geocoded` and click **Save**.
11. Verify that **Add output to map after completion** is checked.
12. Set **Preferred Location Type** to `Address location` using the drop-down arrow and set **Output fields** to `All`. Then, click **Next**.
13. Under **Step 6: Limit by Category**, set **Category** to `Address` by clicking the checkbox. Then, click **Finish**. This might be listed as **Step 5 of 5** at the bottom of the tool. This isn't a cause for concern.

14. Review your input in the **Geocode Table** tool and ensure it matches what's shown in the following screenshot. Then, click **Run**:

Geocode Table	
Guided Workflow Complete — Review your inputs below and run the tool.	
Input Table	Inspections$
Input Locator	Trippville_Locator
Input Address Fields	More than one field
Locator Field	Data Field
Address or Place	Street
Address2	<None>
Address3	<None>
Neighborhood	<None>
City	City
County	<None>
State	State
ZIP	ZIP
ZIP4	<None>
Country	<None>
Output	C:\Student\ArcGISPro3Cookbook\Databases\Trippville_GIS.gdb\Inspections_Geocoded
☑ Add output to map after completion	
Preferred Location Type	Address Location
Output Fields	All
Category	Address

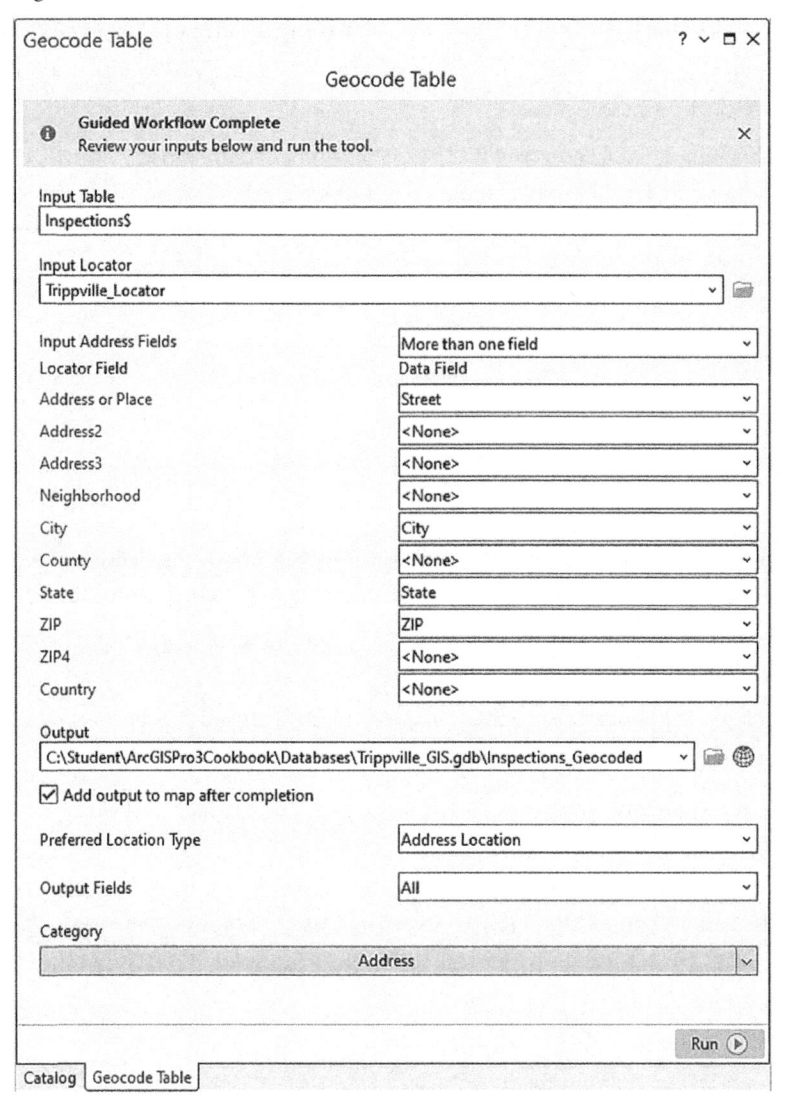

Figure 2.28 – The Geocode Table tool with completed parameters

When the tool completes, a new window will appear showing you the results. Luckily, you don't have any unmatched locations. A new layer should appear on your map showing the results of your geocoding efforts.

15. In the **Geocoding Complete** window, select **No** when you're asked to start the rematch process and close the **Geocode Table** tool.

16. Now, examine your results more closely. In the **Contents** pane, right-click on the `Inspections_Geocoded` layer and select **Zoom to Layer**. Your map should look similar to the following:

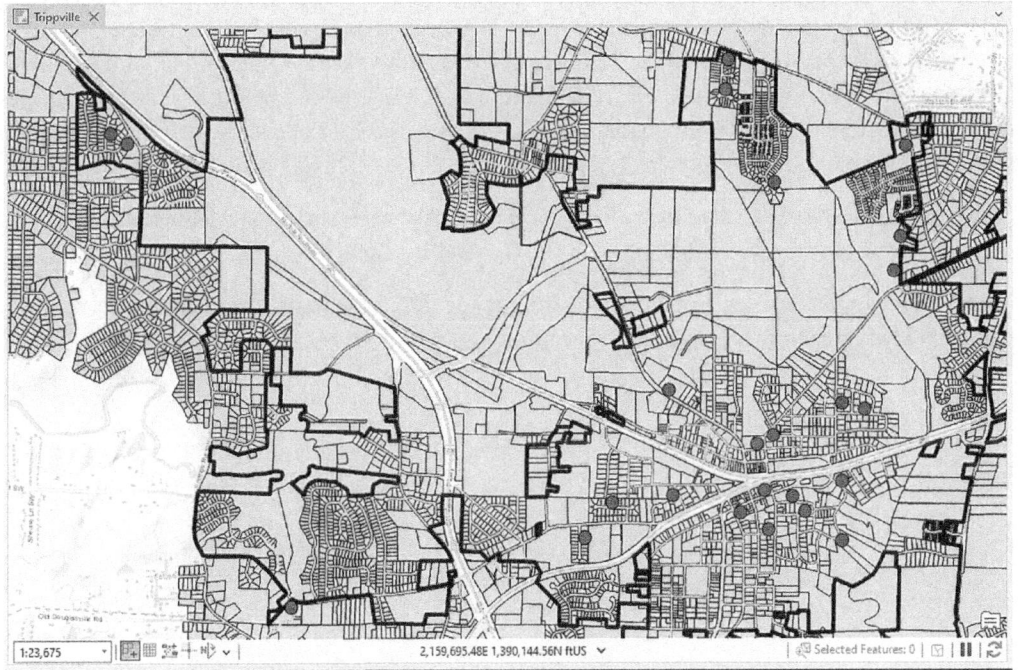

Figure 2.29 – Map zoomed into the geocoding results

17. Using the skills you have learned in this chapter, open the attribute table for the `Inspections_Geocoded` layer. Examine the table to see what fields and data it contains.

    When you examine the attribute table, you should recognize several of the fields as coming from the `Inspections$` table you geocoded. You will also see several fields that were created by the geocoding process. This includes `Status`, `Score`, `Match_type`, and many more. The `Status`, `Score`, and `Match_type` values are calculated from the **Geocode Table** tool. The `Status` field will contain one of three values – **M = Match**, **T = Tie**, or **U = Unmatched**. The `Score` field indicates how well the address in the input table (`Inspections$`) matched the reference data (`Street Centerlines`) used by the locator. It will be between 1 and 100, with 100 being a perfect match. `Match_type` is how the location was geocoded. **A** is automatic and **M** is manual.

18. Close the attribute table once you've finished reviewing it.
19. Save your project and close ArcGIS Pro.

You have successfully geocoded the data in an Excel spreadsheet so that it is displayed on a map.

How it works...

In this recipe, you geocoded a spreadsheet of inspections that had been exported from a permitting and inspections system that was external to your GIS. To do this, you had to determine a reference layer. This reference layer was a GIS layer that contained address data. You examined the `Parcels` and `Street Centerline` layers to determine which one would be the best reference layer. You determined that the `Street Centerlines` layer was best due to the completeness of the address information.

Then, you created a new locator that allowed you to use the centerlines for reference. You used the **Create Locator** geoprocessing tool found in the **Geocoding Tools** toolbox to do this.

Once the locator was created, you were able to geocode the Inspections spreadsheet. This created a new point layer in your map showing the locations of the inspections.

3
Linking Data Together

GIS data includes more than just what you might see on a map. Each layer has additional information linked to it that is stored in an attribute table. In addition, not all data you display in a map is stored in a traditional GIS format. Some data may be stored in standalone tables or even spreadsheets. This data can also be displayed on a map if it includes a mailing address or x and y coordinates.

However, there is a lot of data out there in various databases that may not have x and y coordinates, or an address, or even be part of our GIS, but we need to be able to use that information to perform queries, display information, or conduct analysis in the GIS. This data may come from other systems, such as tax appraisal, permitting, inspections, work order, and asset management systems. If we want to use data stored in these systems, we must be able to link it to our GIS data. ArcGIS Pro provides a couple of methods to do this: joins and relates.

At other times, we may need to transfer or link data together that is in our GIS. ArcGIS Pro supports several tools that allow you to do this, including performing a spatial join and creating relationship classes within a geodatabase.

In this chapter, you will learn the steps needed to link data in ArcGIS Pro to increase the capabilities of your datasets. You will learn how to create and use joins and relates. You will perform a spatial join to combine attributes between two layers based on a spatial relationship. Lastly, you will learn how to create and configure a relationship class within a geodatabase. You will start with a simple relationship class by creating an annotation feature class that is linked to another layer. Then, you will create a relationship class manually that will link two feature classes together.

In this chapter, we will cover the following recipes:

- Joining two tables
- Labeling features using a joined table
- Querying data in a joined table
- Creating and using a relate
- Joining features spatially

- Creating feature-linked annotation
- Creating and using a relationship class using existing data

Joining two tables

A join is one of two basic methods that can be used to link data in ArcGIS Pro. Joins link two datasets together to create a single virtual dataset within a single map in a project. This allows you to use the joined data to query, label, and symbolize using the information from both joined datasets.

In this recipe, you will join the `Parcels` layer to a table that contains a list of owner names. You will learn about the requirements needed to join two tables and how to complete the join.

Getting ready

For this recipe, you will need to ensure that you have installed the book data and have ArcGIS Pro installed. An ArcGIS Pro **Basic** license level will be sufficient for this recipe.

How to do it...

You will now get an opportunity to create a join in ArcGIS Pro between a `Parcels` layer and a standalone table containing the property owners. Then, you will examine the results of the join you create.

To begin, you will need to start ArcGIS Pro.

1. In the **ArcGIS Pro** start window, click on **Open another project**.
2. Expand **Computer** in the left panel of the **Open Project** window and select **This PC**.
3. In the right panel of the **Open Project** window, scroll down and double-click on **Local Disk (C:)**.
4. Continue to navigate to `C:\Student\ArcGISPro3Cookbook\Chapter3\Joining Data`.

5. Select the `Joining Data.aprx` file and click **OK**, as illustrated in the following screenshot:

Figure 3.1 – Opening the Joining Data project

The project will open with a single map named `City of Thomaston`. The map contains two layers, `CityLimits` and `Parcels`, plus a `Topographic` basemap. First, you will review the information contained in the attribute table for the `Parcels` layer.

6. In the **Contents** pane, right-click on the `Parcels` layer and select **Attribute Table** from the menu that appears, as shown in the following screenshot:

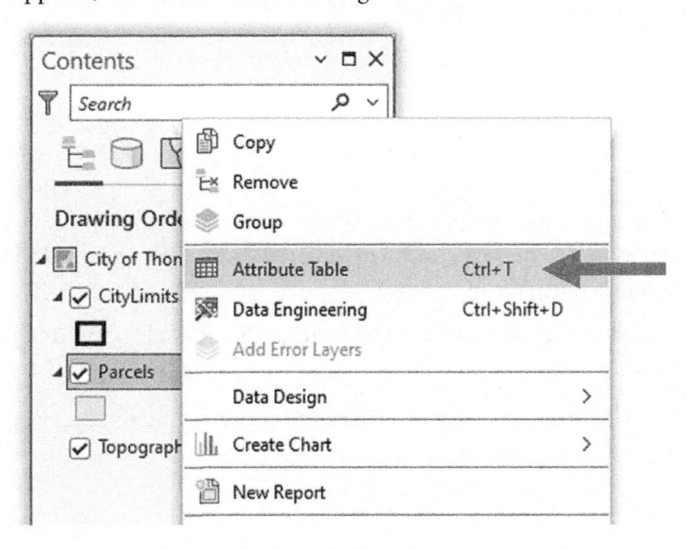

Figure 3.2 – Selecting the Attribute Table option from the menu

The attribute table for the `Parcels` layer should open in a window located at the bottom of the map view. The attribute table contains additional information about parcels you cannot see on the map. Each row in the table represents a parcel you see on the map:

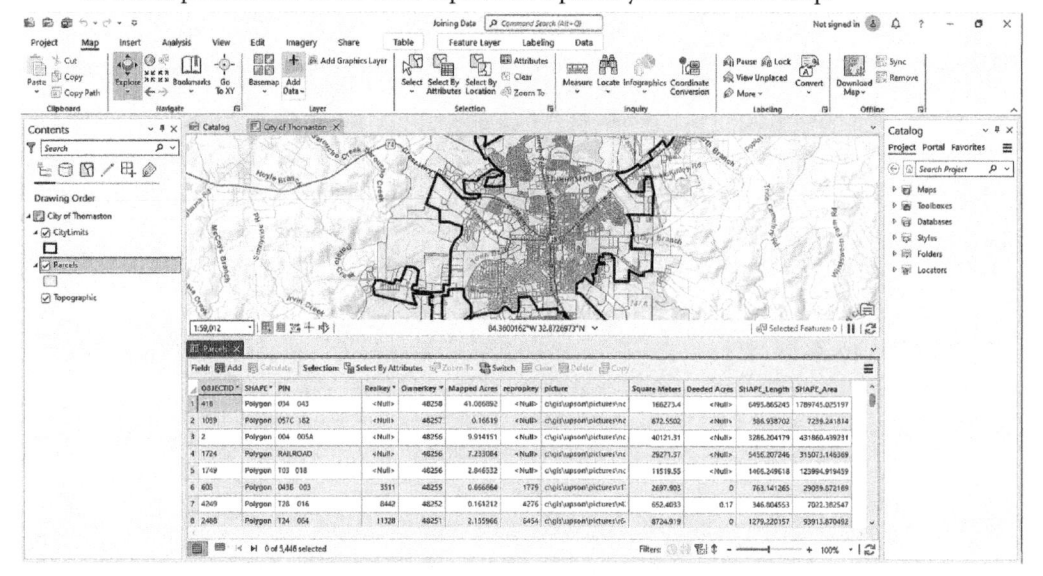

Figure 3.3 – Parcels attribute table open in ArcGIS Pro

7. Review the fields contained in the **Parcels** attribute table. Fields are the columns that you see. The text at the top of each column is the name of the field the column represents. Now, answer the following question.

   Question: What fields are contained in the **Parcels** attribute table?

   Answer:

   As you can see, the table contains information about the parcel, such as the `Parcel Identification Number (PIN)`, `RealKey`, `OwnerKey`, and more. It does not, however, contain any information about who owns the parcels. However, you do know that information is available in the **Computer Assisted Mass Appraisal (CAMA)** system located in the tax appraisal office for the county.

   They have provided you with a table of all landowners in the county. You now need to join it to your `Parcels` layer.

8. In the **Catalog** pane, normally located on the right side of the interface, expand the `Databases` folder by clicking on the small arrowhead to the left.

9. Expand the `Thomaston.gbd` geodatabase so that you can see its contents. You should see at least three feature classes (`CityLimits`, `Parcels`, and `Street_CL`) and a couple of standalone tables (**Owners** and **Parcels_Sales**).

10. Right-click on the **Owners** table and select **Add to Current Map**, as shown in the following screenshot:

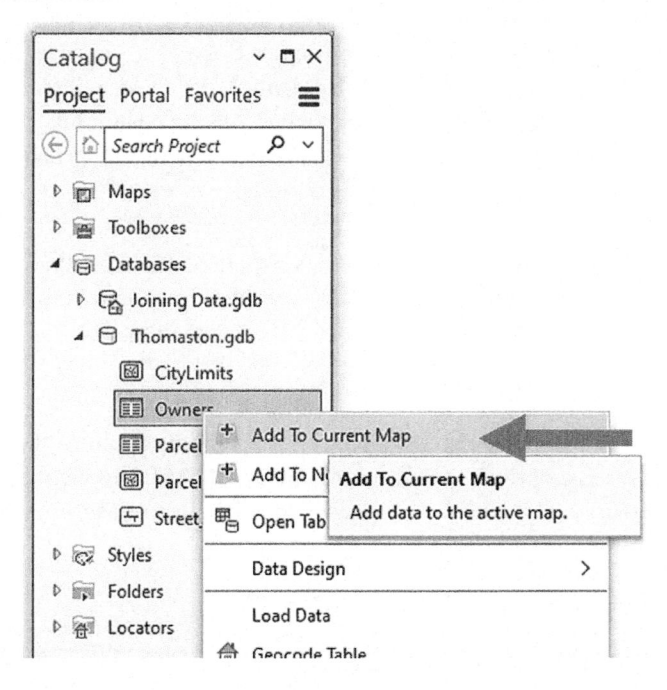

Figure 3.4 – Adding the Owners standalone table to the current map

The newly added table should appear in your **Contents** pane at the bottom of the list under the heading of **Standalone Tables**. Now, you will look at the data contained within this newly added table.

11. In the **Contents** pane, right-click on the **Owners** table you just added and select **Open**. The **Owners** table should open in the same area as the **Parcels** attribute table.

12. Review the **Owners** table, paying close attention to what fields it contains. Then, answer the following question:

Question: What fields does the **Owners** table contain?

Answer:

As you can see, this table does have information about all the landowners in the county. However, it does not have any information about the parcels they own. In order to join the owner information to the parcels, you must identify a key field. A key field is a field in each table that contains data common to the two tables you are attempting to link. Now, you will identify a key field in each table, which you will use to link them together.

> **Tip**
> Key fields are not required to have the same names in both tables. The names can be different. However, they must be the same data type and contain the same data values. By the same data values, that means they must be exactly the same. This includes capitalization and the number of spaces if the key fields contain text. This is why it is often best to use numeric values if possible.

13. Click on the tab at the top that says **Owners**. Hold your mouse button down and begin to drag it downward. A series of icons should appear in the middle of the table view. Drag your mouse to the icon that has a filled beige area located on the right side, as shown in the next screenshot. Then, release your mouse:

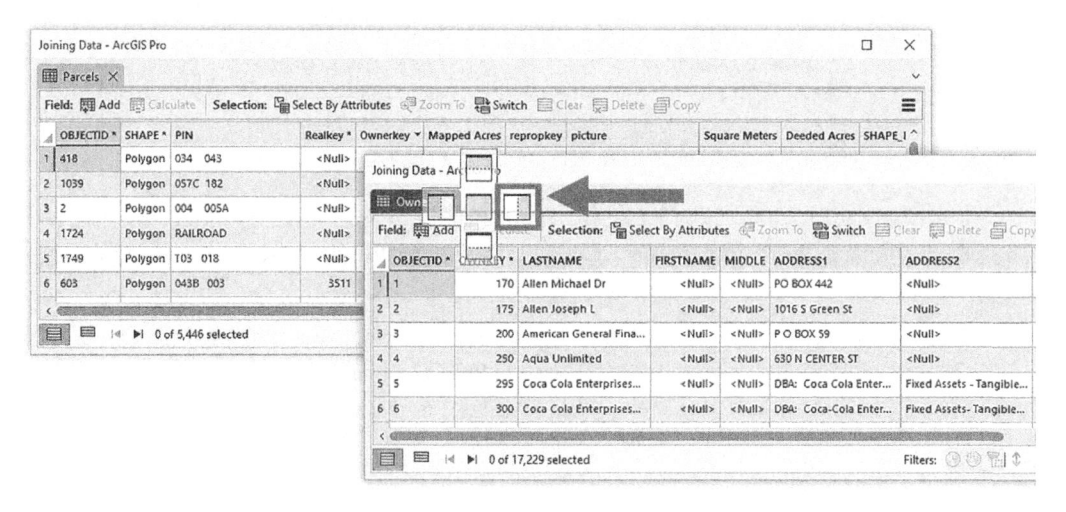

Figure 3.5 – Arranging the tables so that you can view both

The **Parcels** attribute table and the **Owners** table should now appear side by side, as shown in the following screenshot. This allows you to see the data in both tables so that you can evaluate them together:

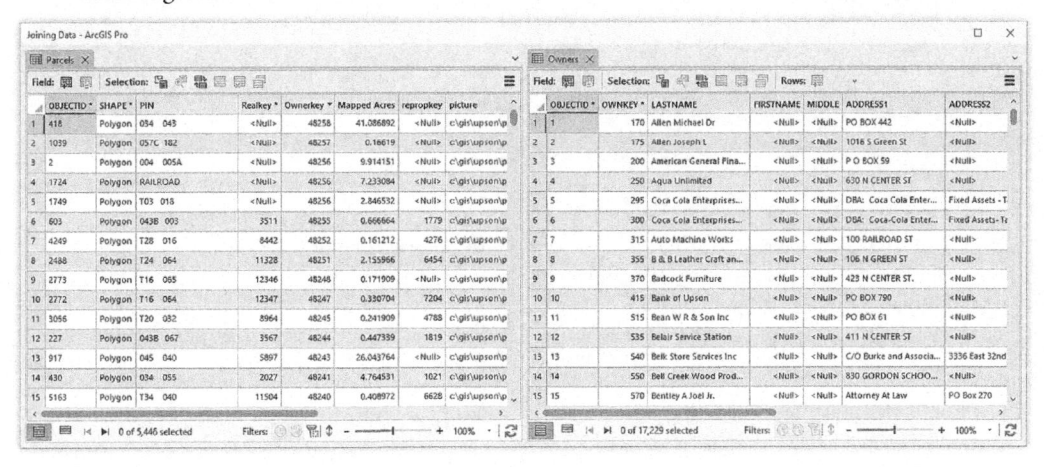

Figure 3.6 – Two tables docked side by side

14. In the **Parcels** table, right-click on the `OwnerKey` field and select **Sort Descending**.
15. In the **Owners** table, right-click on the `OWNKEY` field and select **Sort Descending**.
16. Review and compare the values for these fields in each table, then answer the following question:

    Question: Do the `OwnerKey` and `OWNKEY` fields appear to contain the same values?

    Answer:

    Once you have verified these two fields contain like values, you need to verify the field types are the same. When you create or add a field to a database table, you must define the type of data it will contain. There are several field types that a database table can contain, including, but not limited to, text, integer, single, and date. In order to join two tables, the field types between the two tables must match.

70　Linking Data Together

17. Right-click on the `OwnerKey` field in the **Parcels** table and select **Fields** from the menu that appears:

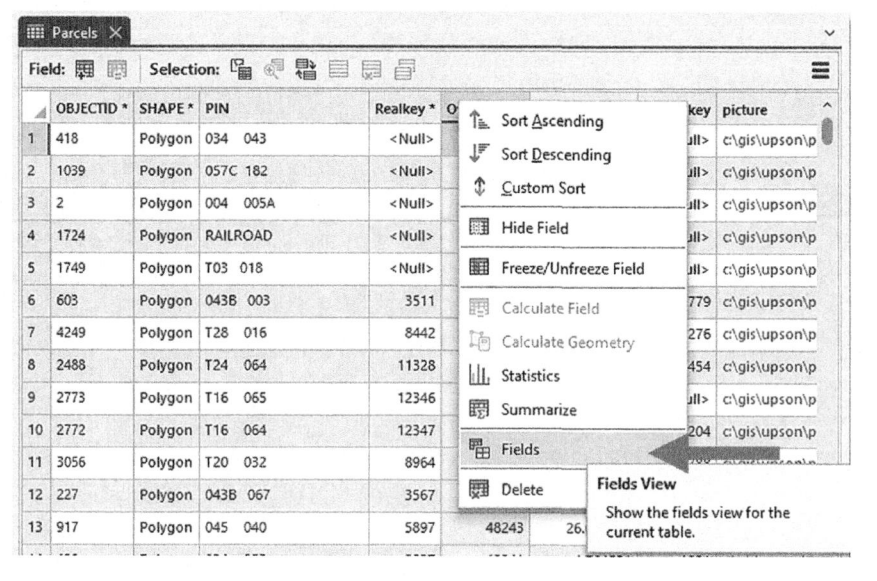

Figure 3.7 – Opening the field properties using the menu

18. Perform the same operation on the **OWNKEY** field in the **Owners** table.

    Your table view window should now look like the following screenshot:

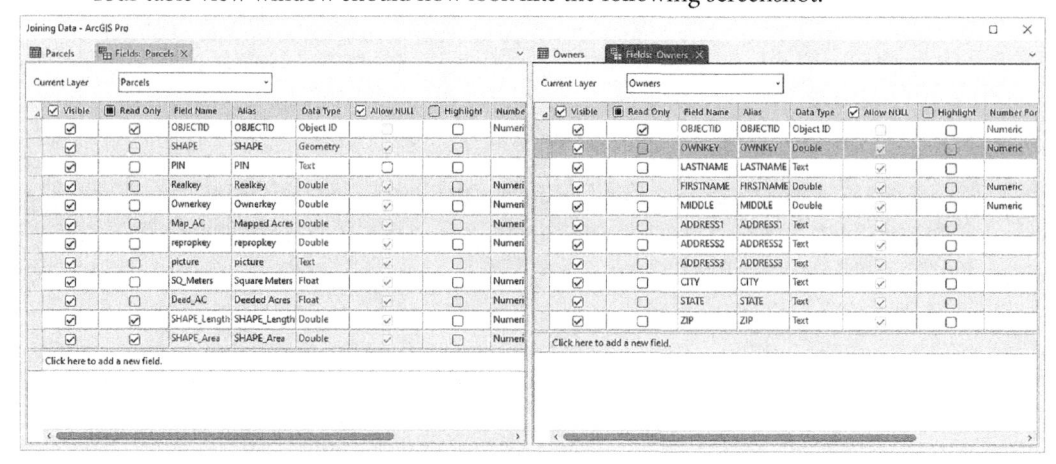

Figure 3.8 – Reviewing the field properties for the two tables

19. Verify the data type for the `OwnerKey` field in the **Parcels** table and the `OWNKEY` field in the **Owners** table are the same.

As you have now determined, the fields are both the same field type and contain similar data values. This means these two fields can be used to link these two tables together. Fields that meet these two requirements are often referred to as key fields. In this case, both fields have the word *Key* in their name to help identify them as such, but that will not always be the case. Now is the time to link these two fields using a join.

20. Close the two open tables and the **Fields** tabs by clicking on the small **X** symbol.

> **Tip**
> It is not required to close tables before joining them. You can leave them open if desired. Closing them removes clutter from your desktop and reduces computer resources being used.

Creating a join

Now that you have identified the key fields you need to join the two tables together, you are ready to create a join.

1. In the **Contents** pane, right-click on the `Parcels` layer. Then, go to **Joins and Relates** in the menu that appears.
2. Select **Add Join**, as illustrated in the following screenshot:

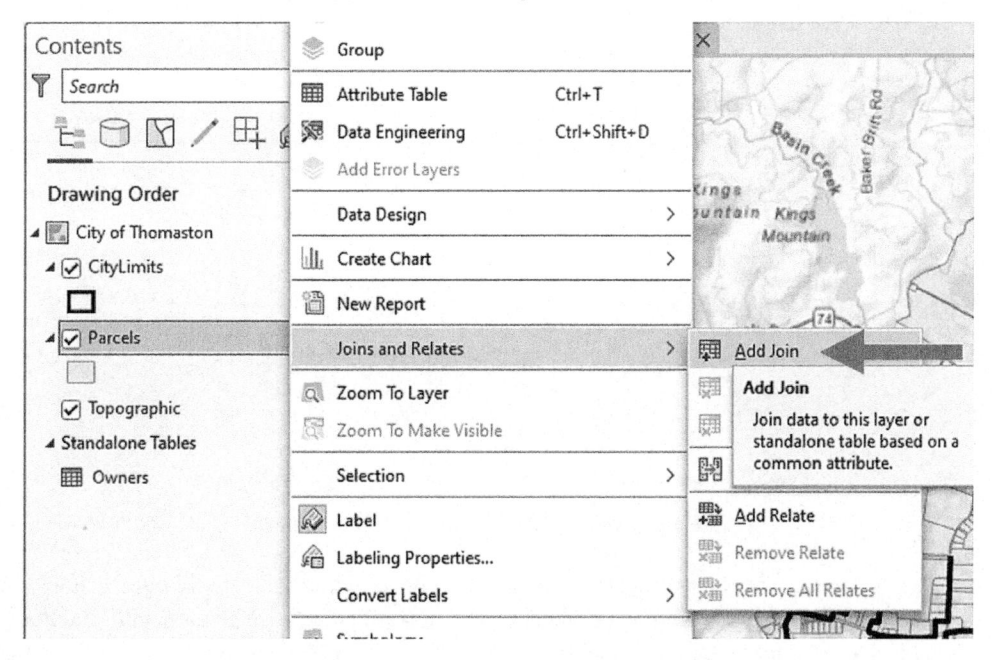

Figure 3.9 – Starting to create a new join in ArcGIS Pro

3. The **Add Join** tool window should now be open. Verify the **Input Table** parameter is set to **Parcels**.

4. For the **Input Join** parameter, click the drop-down arrow at the far right of the cell and select **OwnerKey**.

5. Verify that the **Join Table** parameter is set to **Owners** and the **Join Table Field** parameter is set to **OWNKEY**. If they are not, use the drop-down arrow in each cell to select the correct values.

6. Verify that **Keep All Input Records** is enabled. It should be enabled by default unless you have changed this setting in a previous ArcGIS Pro session. In older versions of ArcGIS Pro 3.x, this may be **Keep All Target Features**.

> **Tip**
> If you need to get a better understanding of the parameters associated with a tool you are using, you can click on the **?** symbol located in the upper-right corner of the tool window. This will open help documentation for the tool.

7. Click the **Validate Join** button. This will provide a quick report that can help you get an idea of the result of the join you are about to perform:

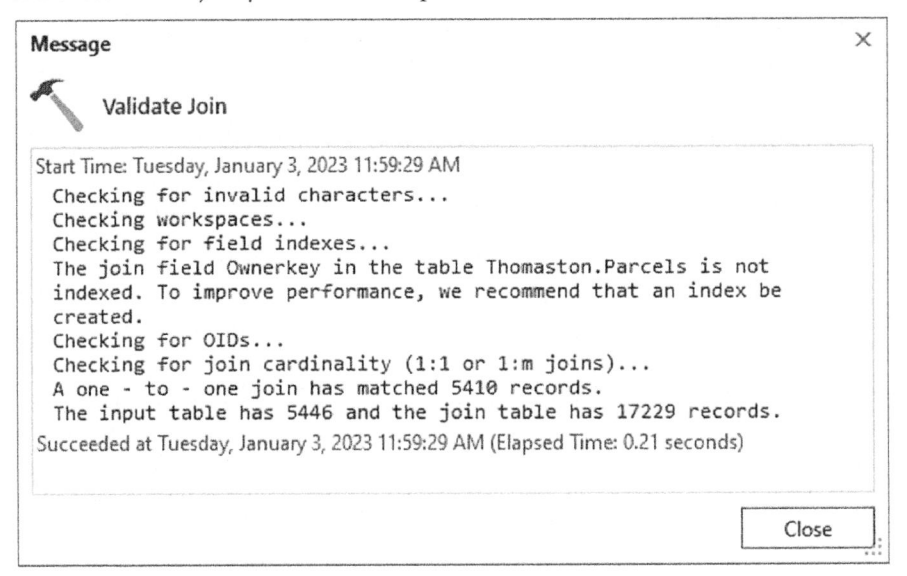

```
Message                                                                    ×

    Validate Join

Start Time: Tuesday, January 3, 2023 11:59:29 AM
   Checking for invalid characters...
   Checking workspaces...
   Checking for field indexes...
   The join field Ownerkey in the table Thomaston.Parcels is not
   indexed. To improve performance, we recommend that an index be
   created.
   Checking for OIDs...
   Checking for join cardinality (1:1 or 1:m joins)...
   A one - to - one join has matched 5410 records.
   The input table has 5446 and the join table has 17229 records.
Succeeded at Tuesday, January 3, 2023 11:59:29 AM (Elapsed Time: 0.21 seconds)

                                                                    Close
```

Figure 3.10 – Validating the join to ensure it will work

As you can see, when you validate the join, the report provides some useful information. First, it indicated you could improve performance by indexing the OwnerKey field in the **Parcels** table. Next, it checks for cardinality between records in both tables. Cardinality is how many records in one table match records in the other tables. It then tells you how many records match using a one-to-one match.

Joining two tables 73

8. After reviewing the validation report, click the **Close** button.
9. Verify the **Add Join** tool looks like the following screenshot and click **OK**:

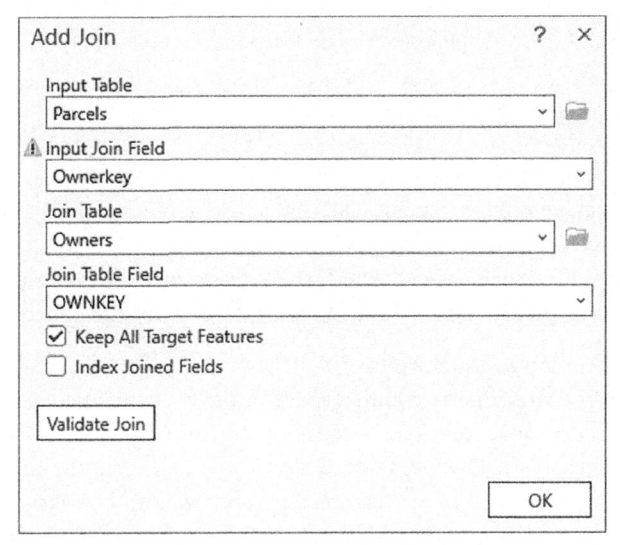

Figure 3.11 – Filling out the parameters for the Add Join tool

10. Once the **Add Join** tool finishes, right-click on the **Parcels** layer in the **Contents** pane and select **Attribute Table** from the menu.
11. Using the scroll bar at the bottom of the **Parcels** table, scroll through the table fields. Look to see if any new fields appear in the table:

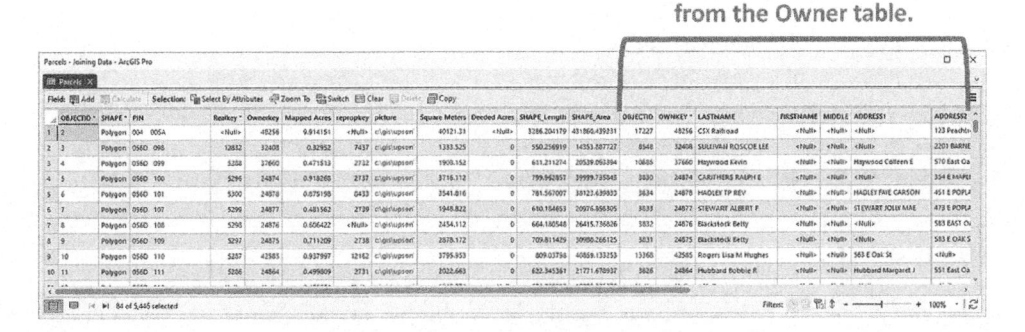

Figure 3.12 – Results of successful join in ArcGIS Pro

As you can see, the fields that were on the **Owners** table have now been added to the **Parcels** attribute table. Now that those fields have been added, you can use that information to locate all parcels owned by a specific owner or label parcels with the owner. Let's put that to the test. You will start with labeling the parcels with fields from the joined table in the next recipe.

12. Close the **Parcels** table by clicking on the small **X** symbol in the tab that shows the table name.
13. Click on the **Project** tab in the ribbon, and click on **Save**, located in the panel on the left side.

If you are continuing to the next recipe, keep ArcGIS Pro open. If you are not, you may close ArcGIS Pro.

How it works...

In this recipe, you joined the **Owners** table to the **Parcels** table. ArcGIS Pro links these two tables together, based on a key field in each table. The key fields are not required to have the same name but must have the exact same values and be the same data type. ArcGIS Pro then compares the two tables, and where the values are identical, it adds those fields and values to the primary table in the map. Where the values are not identical, it leaves the field values for the joined fields blank or null.

A join only exists in the map in which you create it. It will not be applied to other maps within the ArcGIS Pro project. It is also not permanently applied to the feature class or table. A join works best when you want to link data that comes from other systems or datasets that will not be maintained in your GIS directly, such as the **Owners** table, which is maintained in the county's CAMA system. You just get an updated download of the data on a regular basis and replace the old table with the new one so that when you open the map that contains the join, you always see the most current available information.

Labeling features using a joined table

Now that you have learned how to join tables and seen how it links them to provide more information, what can you do with that information? In short, you can do anything with the joined information that you can do with normal attributes in the layer's attribute table.

In this recipe, you will learn how you label features using the data from the joined table. You will label each parcel with its PIN and owner name.

Getting ready

You will need to complete the previous recipe before you can begin this one.

How to do it...

You will now create text labels for parcels using data coming from the **Owners** table you just joined to the Parcels layer. This will allow you to see the power joining data can provide.

1. If you closed ArcGIS Pro after completing the previous recipe, open the Joining Data.aprx project by following the same instructions as those shown at the beginning of the previous recipe.

Labeling features using a joined table   75

2. Click on **Bookmarks** on the **Map** tab in the ribbon. Select the **Labeling** bookmark, as shown in the following screenshot. This will zoom you into an area located in the center of the city:

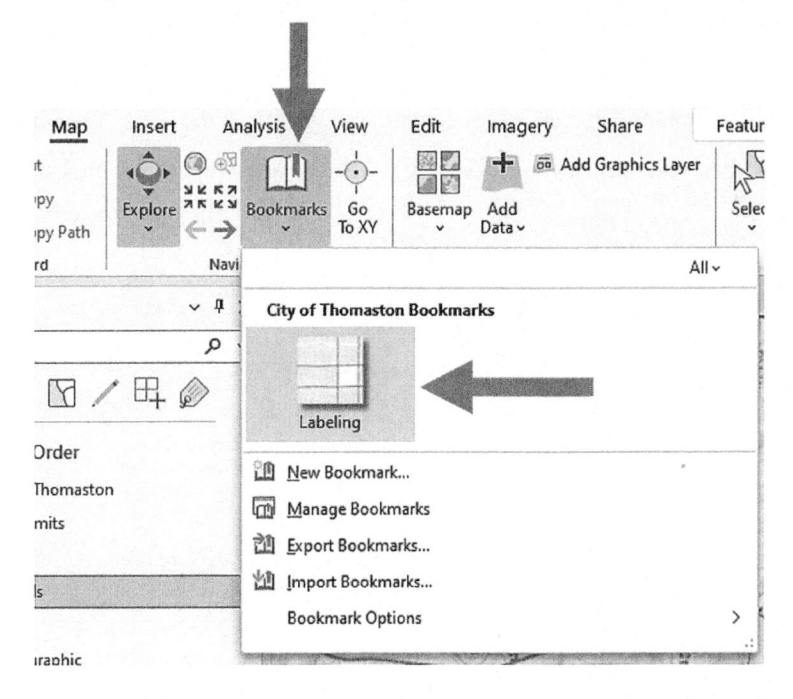

Figure 3.13 – Accessing the Labeling bookmark to zoom your map to the proper location

3. Select the `Parcels` layer in the **Contents** pane so that layer contextual tabs appear in the ribbon. Then, click on the **Labeling** tab in the ribbon:

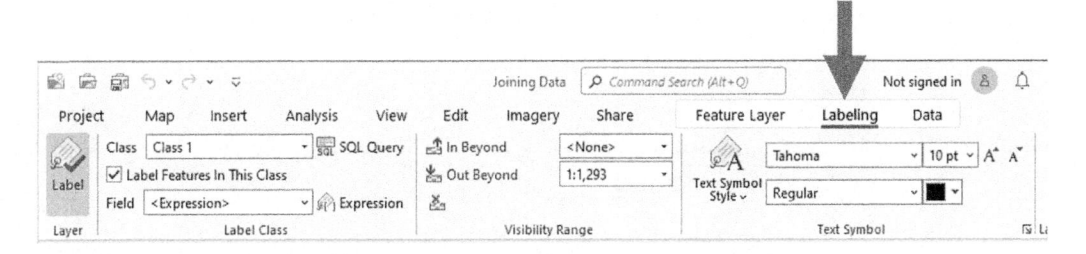

Figure 3.14 – The Labeling tab in the ArcGIS Pro ribbon

4. Click on the **Label** button located in the **Layer** group on the **Labeling** tab.

Text will appear on the map showing the PIN for each parcel. The values being displayed are being pulled from the attribute table for the Parcels layer. If you look in the **Label Class** group on the **Labeling** tab, you will see a setting for **Field**. It is set to Parcels.PIN. This means that the values being displayed are those found in the PIN field from the original **Parcels** table. You will now build an expression using the Arcade language, which will label each parcel with its PIN from the **Parcels** table and with its owner from the joined **Owners** table.

5. Click on the **Expression** button located to the right of the **Field** cell in the **Label Class** group. The **Label Class** pane should appear on the right of the interface. This is where you will build your labeling expression:

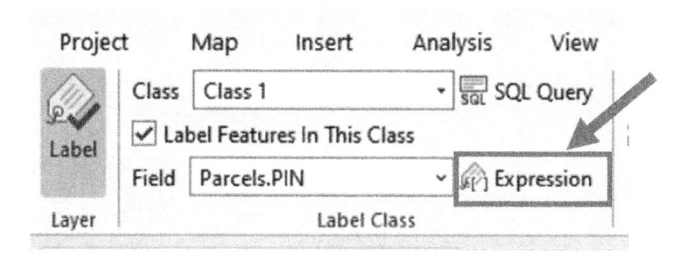

Figure 3.15 – The Expression button on the Labeling tab

6. Click in the **Expression** cell where you see $feature['Parcels.PIN']. This is where you will create your new labeling expression.

7. Type "Pin: " + just before the existing value shown in the **Expression** area, as illustrated in the next screenshot:

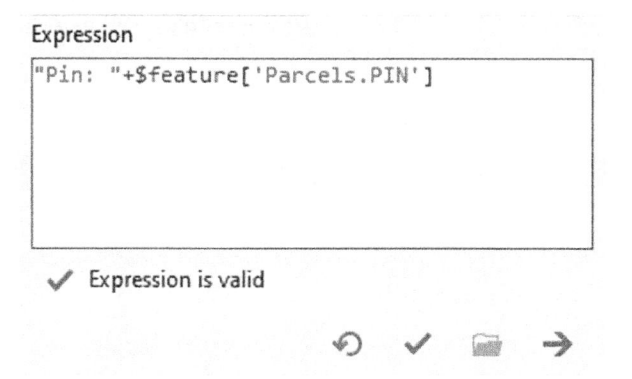

Figure 3.16 – Creating a custom labeling expression

8. Then, click the **Validate** button, which looks like a green checkmark located below the **Expression** area to ensure you have typed the expression correctly. If the result says the expression is valid, continue to the next step. If the result says the expression is not valid, double-check your expression to ensure you have the correct syntax.

9. You will continue to expand your expression. At the end of the expression you have created so far, type +Textformatting.Newline+"Owner: "+ $feature['Owners.LASTNAME'].

   Your full labeling expression should now look like this:

   ```
   "Pin: "+$feature['Parcels.PIN']+Textformatting.Newline+"Owner: "+ $feature['Owners.LASTNAME']
   ```

10. Click on the **Validate** button once again to verify your syntax. If it says your expression is valid, click **Apply**. If it does not, verify your expression. If you continue to have issues, read the following tip.

Tip

If you are having issues creating an expression, you can load one that has already been created. Just click on the **Import** button located next to the **Validate** button and navigate to `C:\Student\ArcGISProCookbook\Chapter3\Joining Data`. Select the `PIN_Owner_Label_Expression.lxp` file. Then, click **OK** to import the expression.

Your map should now look similar to the following screenshot. You should see that each parcel is labeled with its PIN value and the owner's last name. The label includes descriptive text so that each value is understood:

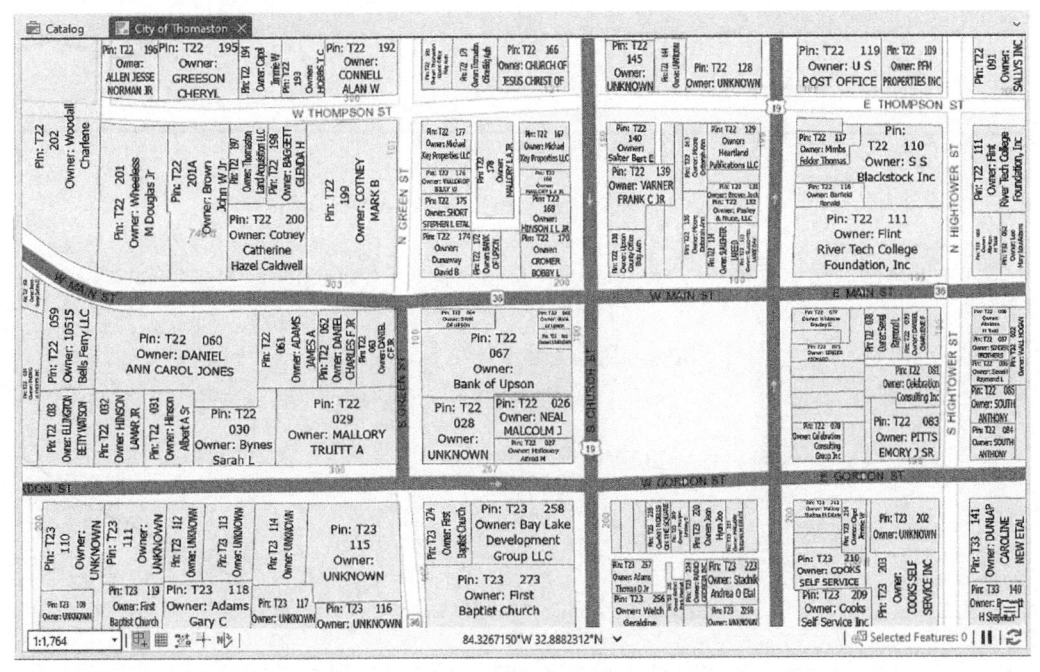

Figure 3.17 – Results of custom label expression

11. Save your project by clicking on the **Save** button in the **Quick Access Toolbar**.

You just created a labeling expression using Esri's Arcade expression language that labeled each parcel using the `PIN` field found in the original **Parcels** attribute table and the `LastName` field from the **Owners** table. The expression you built separated the PIN and owner's names so that they were on two separate lines.

ArcGIS Pro supports several languages for writing labeling expressions. In addition to Arcade, you can write expressions using Python, VBScript, and JScript. Labeling expressions written for ArcMap and saved to expression files (`.exp`) are not compatible with ArcGIS Pro.

Adjusting label placement

Now, you will adjust some placement settings for the labels to make them fit better and be easier to read.

1. In the **Label Class** pane where you just created your expression, click on the **Position** tab, as shown in the following screenshot:

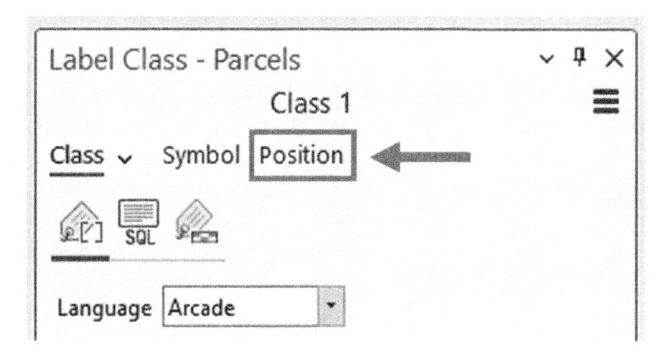

Figure 3.18 – The Position tab in the Label Class pane

2. As illustrated in the following screenshot, click on the **Positions** button in the **Label Class** pane:

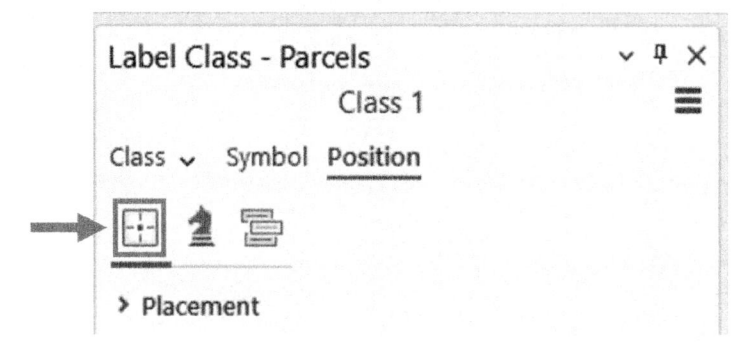

Figure 3.19 – The Positions button in the Label Class pane

3. Expand the **Placement** option and select **Land parcel placement** from the drop-down menu.
4. Locate where it says **Horizontal in Polygon**, click on the small drop-down arrow to the right, and select **Straight in polygon** from the list that appears:

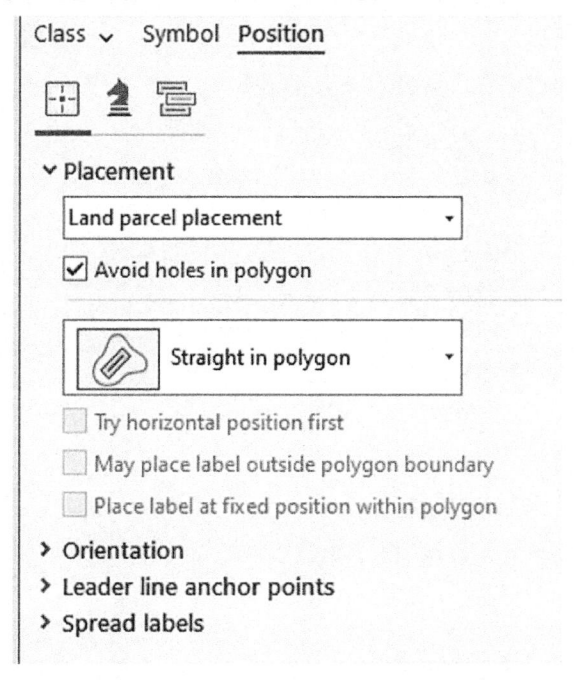

Figure 3.20 – Configuring placement options for your labels

5. Click on the **Fitting Strategy** button, which resembles a knight (horse head) from chess.
6. Expand **Reduce size** by clicking on the small arrowhead.
7. Click on the box located to the left of **Reduce font size** to enable this function.
8. Set the following parameters:

    - **Font size reduction**

        - **Lower limit**: 4.0

    - **Step interval**: 1.0

    - **Font width compression**

        - **Lower limit**: 60.0

    - **Step interval**: 5.0

Your **Label Class** pane should now look like the following. The settings should automatically be applied as you make changes:

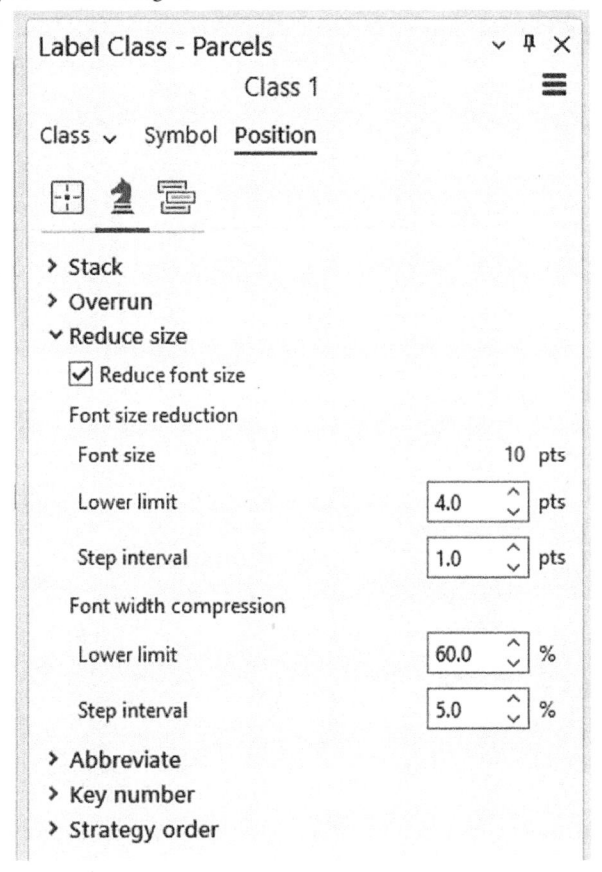

Figure 3.21 – Parameters for label placement complete

9. Save your project.
10. Go back to the **Labeling** tab in the ribbon.
11. In the **Visibility Range** group, click on the drop-down arrow for **Out Beyond** and select <**Current**>. This will mean that the labels you just configured will only appear when you zoom in to the current scale of your map. If you zoom out to show a larger area, the labels will not be displayed:

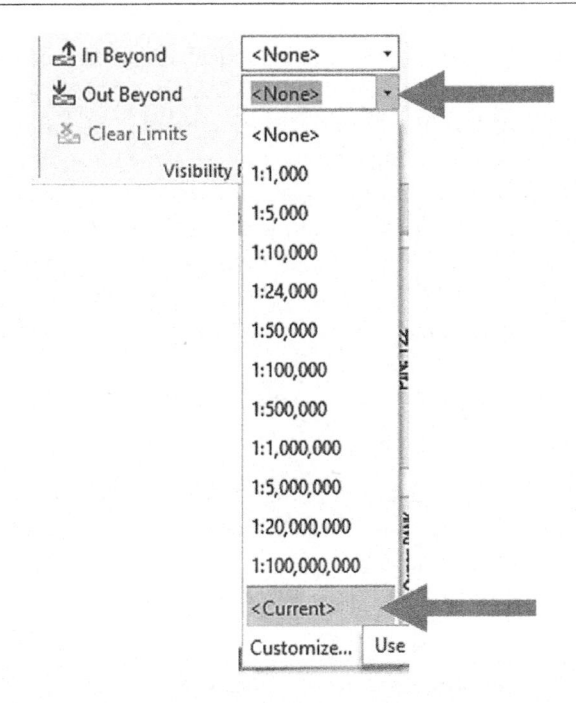

Figure 3.22 – Setting zoom scales for labels

12. Close the **Label Class** pane.
13. Save your project and close ArcGIS Pro.

You have now been able to see some of the power a join can offer by providing access to additional data. Using that data for labeling is just one example of how you can use additional data. In the next recipe, you will experience how joins can be used in queries to locate data.

How it works...

In this recipe, you labeled the parcels in the map with values from the PIN field in the **Parcels** attribute table and the LASTNAME field in the **Owners** table, which you had joined to the parcels in the last recipe. You did this using Esri's new Arcade expression language.

In the first part of your expression, you added descriptive text that allowed users to know what the labeled values were. You did this in Arcade by using double quotes. This tells ArcGIS Pro to simply display the text values found inside the double quotes. For example, your expression contained "PIN: ". This produced a label that showed up in the map displaying the text PIN: followed by a space.

This was then followed up with a reference to a database field. In Arcade, you did this using the + sign to indicate another part of the expression, followed by `$feature['Parcels.PIN]`. This displays the value found in the `PIN` field within the **Parcels** attribute table. The + sign must be used between each part of the expression. You then displayed the owner's name on another line in the label using `Textformatting.Newline` in your expression. Once you set the owner name to be displayed on a separate line, you add more descriptive text by including `"Owner:  "` in your expression. Lastly, you displayed the owner's name by calling the `LASTNAME` field in the joined **Owners** table by including `$feature['Owners.LASTNAME']` at the end of your expression. This produced a label for each parcel that looked like this on the map:

- **PIN**: T22 028
- **Owner**: Unknown

You then adjusted several configuration settings for the labels in the **Label Class** pane to make the labels fit in the parcels better and be easier to read. You did this by setting the placement option to **Land parcel placement**. This automatically applies some settings that are preconfigured in ArcGIS Pro for placing text within a parcel. This cleaned up several of the overlapping labels but left many parcels without a label.

So, next, you set the positioning to be **Straight in a polygon** as opposed to the default **Horizontal in a polygon**. This allowed ArcGIS Pro to fit labels diagonally within the polygon if needed, first adjusting the size of the font. It also allowed ArcGIS Pro to label a few more parcels, but still not all of them.

Next, you moved over to the **Fitting Strategy** button to adjust the font size for your labels. You enabled **Reduce font size**. This allowed ArcGIS Pro to automatically resize the font within the parameters you designated to help fit the label within the parcels. By making this adjustment, you were able to get all the parcels in the view labeled.

Querying data in a joined table

Labeling is not the only thing you can do with a joined table. You can also use the joined information to perform queries and analysis. In this recipe, you will perform a query to locate all the parcels owned by the city of Thomaston. You will then export that information to a spreadsheet using a geoprocessing tool.

Getting ready

You must have completed the *Joining two tables* recipe from this chapter before you can perform this recipe. You will also need to have Microsoft Excel or a similar application installed that will open a spreadsheet.

How to do it...

You will now go through the steps to create queries using data that is the result of a join. This illustrates another use of joined data.

1. If you closed ArcGIS Pro after completing the previous recipe, open the `Joining Data.aprx` project by following the same instructions as shown at the beginning of the previous recipe. Otherwise, continue with this recipe.
2. Click on the **Map** tab in the ribbon.
3. Click on the **Full Extent** button in the **Navigate** group on the **Map** tab. It looks like a small globe.
4. Click on the **Select by Attributes** tool in the **Selection** group on the **Map** tab in the ribbon.
5. Set the **Input Rows** parameter to **Parcels** using the drop-down arrow.
6. Ensure the **Selection type** value is set to `New selection`.
7. Under **Expression** in the cell to the right of the word **Where**, click in the **Select a field** cell and select `LASTNAME`, as shown in the following screenshot:

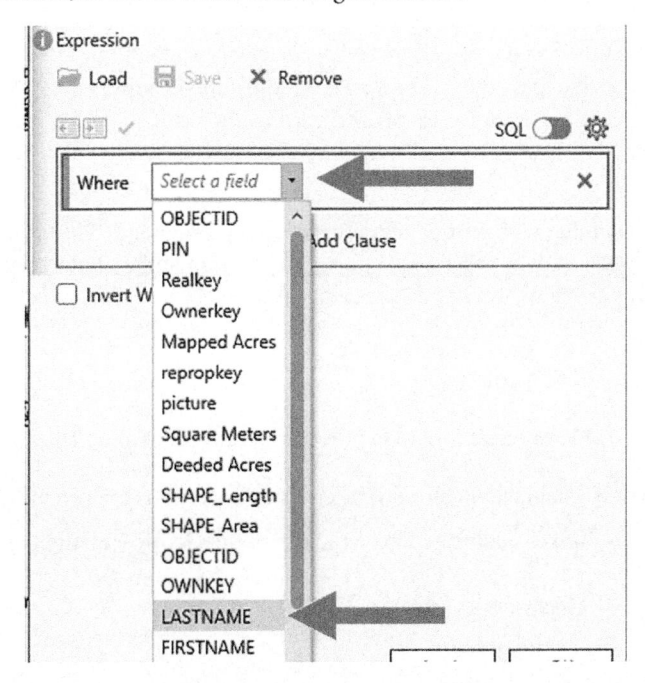

Figure 3.23 – Selecting the field to query by

Once you have selected the `LASTNAME` field, the `is equal to` operator should appear along with a blank cell that allows you to enter the specific value you are looking for.

84   Linking Data Together

8. Click in the blank cell and either type `City of Thomaston` or pick it from the list of values shown. Your expression should now look like the following:

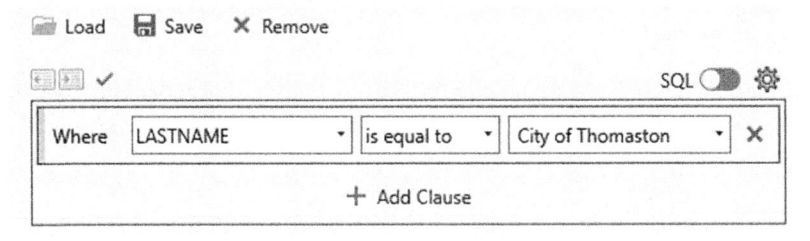

Figure 3.24 – Completed query

9. Once you have verified your expression matches the previous screenshot, click **Apply** in the **Select by Attributes** tool to run your query.

10. Look under the map view on the right side. It should say **Selected Features: 84** if your query was successful. Close the **Select by Attributes** window.

You have just selected all the parcels owned by the city of Thomaston using a **Select by Attributes** query. You were able to do that using data from the joined **Owners** table. Next, you will export those selected records to an Excel spreadsheet.

> **Tip**
> If you are already familiar with writing **Structured Query Language** (**SQL**) WHERE clauses, you can click on the small **SQL** button located under **Expression** and manually type the query you just built. The proper syntax would be `Owners.LASTNAME = 'City of Thomaston'`.

Exporting to a spreadsheet

You will now use the **Table to Excel** tool to export the selected data to an Excel spreadsheet.

1. Select the **Analysis** tab in the ribbon and click on the **Tools** button to open the **Geoprocessing** pane.

2. Click on the **Toolboxes** tab located near the top of the **Geoprocessing** pane:

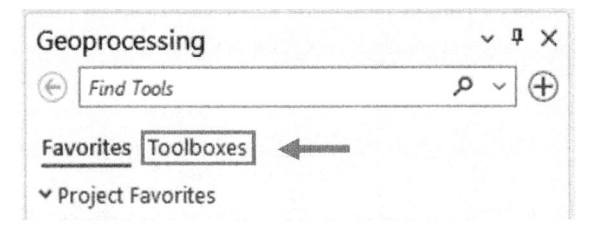

Figure 3.25 – The Toolboxes tab in the Geoprocessing pane

3. Expand the **Conversion Tools** toolbox by clicking on the small arrowhead.
4. Expand the **Excel** toolset using the same method. Then, double-click on the **Table to Excel** Python script.

> **Note**
> In the toolboxes, different icons indicate different types of tools. Hammers indicate system tools. Scrolls indicate Python scripts. A series of connected squares indicate models created with `ModelBuilder`.

5. Set the **Input Table** parameter to **Parcels** using the drop-down arrow. A warning should appear that says `the input has a selection. Records to be processed: 84`. This appears because of the selection query you performed earlier.
6. For **Output Excel File (xls or xlsx)**, click on the **Browse** button located at the end of the cell.
7. In the **Output Excel File (xls or xlsx)** window that appears, navigate to the `MyProjects` folder located in `C:\Student\ArcGISPro3Cookbook\`.
8. In the **Name** cell, type `Parcels owned by Thomaston`, as shown in the following screenshot. Then, click **Save** to return to the **Geoprocessing** pane and the **Table To Excel** tool:

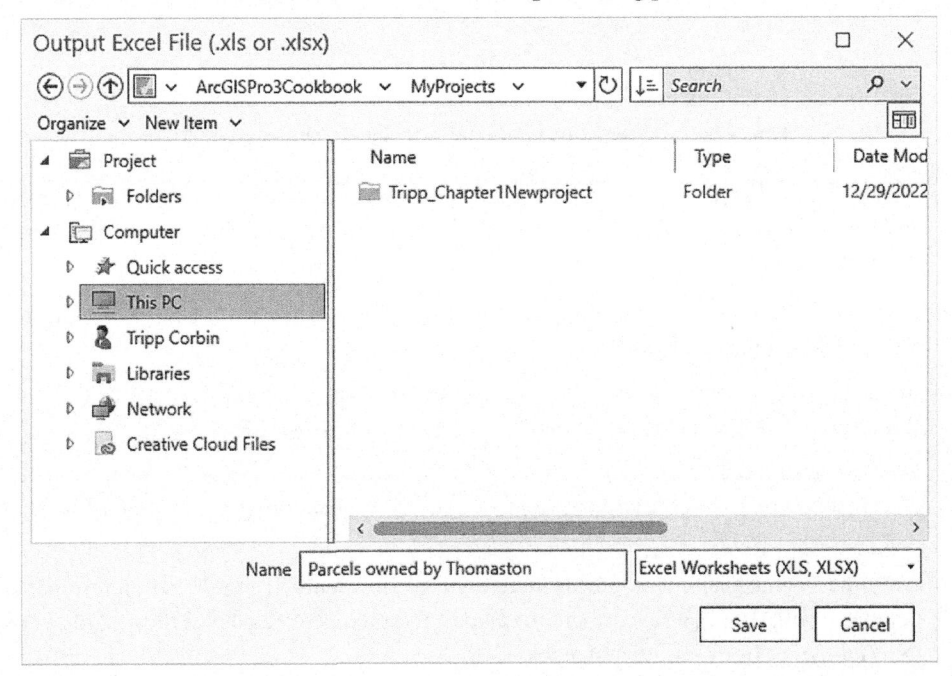

Figure 3.26 – Setting the output for the Table to Excel script

9. Verify the **Table To Excel** tool looks like the following screenshot and click **Run** to execute the tool:

```
Geoprocessing                                          ⌄  ▢  ✕
  ←                    Table To Excel                       ⊕

  Parameters  Environments                                  ?

  Input Table ⌄
  ┌─────────────────────────────────────────────────┐  ┌──┐
  │ Parcels                                      ⌄  │  │📁│
  └─────────────────────────────────────────────────┘  └──┘
  ⓘ The input has a selection. Records to be processed: 84   ↻
  ┌─────────────────────────────────────────────────┐  ┌──┐
  │                                              ⌄  │  │📁│
  └─────────────────────────────────────────────────┘  └──┘
  Output Excel File (.xls or .xlsx)
  ┌─────────────────────────────────────────────────┐  ┌──┐
  │ C:\Student\ArcGISPro3Cookbook\MyProjects\Parcels owned by Thomaston.xls │  │📁│
  └─────────────────────────────────────────────────┘  └──┘
  ☐ Use field alias as column header
  ☐ Use domain and subtype description

                                              ▶ Run   ⌄
```

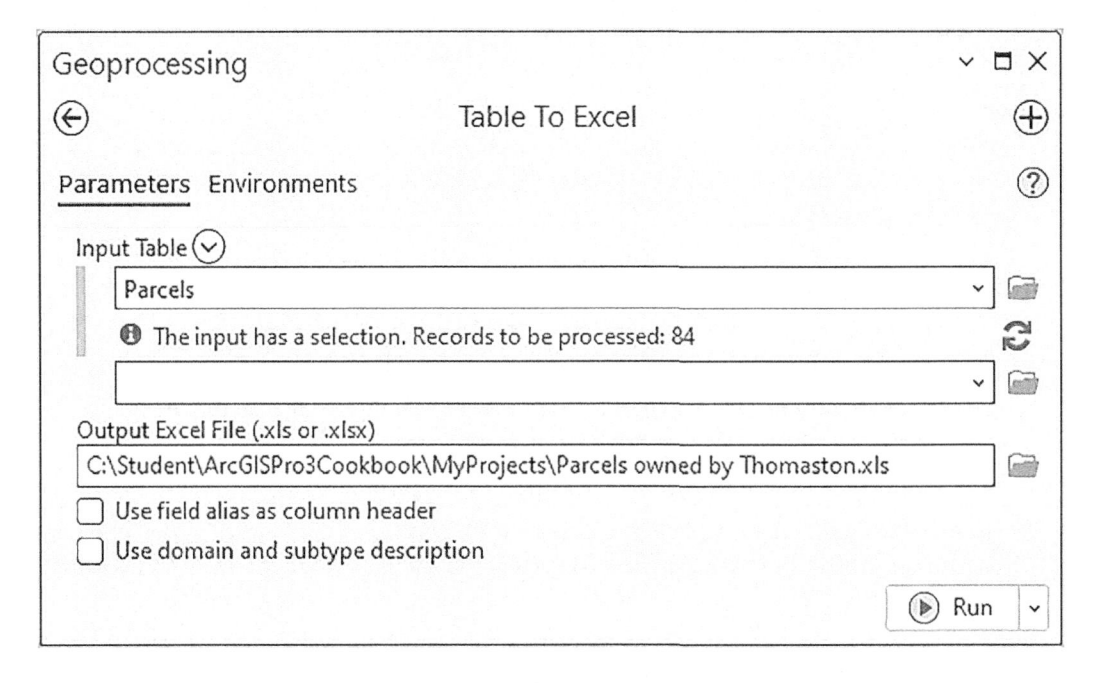

Figure 3.27 – Parameters for the Table to Excel script completed.

A status bar will appear at the bottom of the **Geoprocessing** pane showing the tool is running and letting you know when it completes. Depending on the specifications of your computer, this tool could complete very quickly or may take several minutes.

10. Once the tool completes, open Windows **File Explorer**. It is normally located on your taskbar near the **Start** button and looks like a file folder.

11. In File Explorer, navigate to `C:\Student\ArcGISPro3Cookbook\MyProjects`. You should see the Excel spreadsheet you just exported from ArcGIS Pro.

12. Double-click on the `Parcels owned by Thomaston.xls` spreadsheet to open it. This does require you to have Microsoft Excel or another application such as OpenOffice installed that can read this file.

13. Once the spreadsheet opens, take time to review its contents. It should have 85 rows that are populated with data. These represent the 84 parcels you had selected plus the first row showing the field names from the attribute table.

14. Close the spreadsheet once you are done reviewing it and return to ArcGIS Pro.

15. Back in ArcGIS Pro, select the **Map** tab in the ribbon. Then, click on the **Clear** button located in the **Selection** group. This will deselect all selected features:

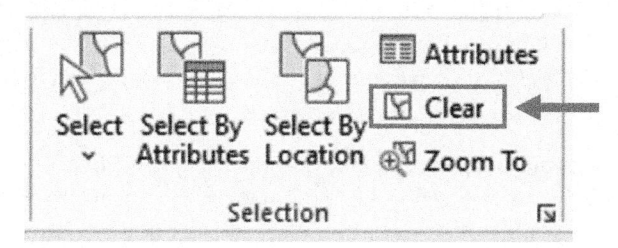

Figure 3.28 – Clear button on the Map tab in the ribbon

16. Save your project and close ArcGIS Pro.

You have seen another use of joined data. You were able to use the data that originally came from the **Owners** standalone table that was joined to the `Parcels` layer to select all parcels owned by the city of Thomaston. You then exported that selection set to an Excel spreadsheet for others to use.

How it works...

In this recipe, you queried all the parcels owned by the city of Thomaston and then exported the selected list to a spreadsheet. To query the parcels owned by the city, you created a SQL `WHERE` clause using the query builder. This allows you to create SQL expressions without having to know SQL.

SQL is the standard language used by databases to communicate. It allows you to perform many functions within a relational database. `WHERE` clauses allow you to select records (or features within a GIS) that meet specific criteria. In the recipe you just completed, the criterion was where the parcel's owner equaled the city of Thomaston, thus why it is called a `WHERE` clause. As with all languages, SQL has a specific syntax that must be followed to work. The ArcGIS Pro query builder you used knows the syntax and automatically creates the proper expression based on the parameters you provide. You used the **Create Feature Class** geoprocessing tool to generate a new shapefile. After you created the new shapefile, you added attribute fields to the table, which will store additional information about the street signs.

After you selected the parcels owned by the city of Thomaston, you exported them to a spreadsheet using the **Table to Excel** Python script tool found in the **Conversion Tools** toolbox and Excel toolset. Python is a scripting language that has been heavily integrated into the ArcGIS platform. Python scripts can be created to assist with automating and streamlining workflows. To learn more about using Python with ArcGIS, we recommend *Python for ArcGIS Pro* by Silas Toms and Bill Parker from Packt Publishing.

Creating and using a relate

A join is just one of the basic methods you can use in ArcGIS Pro to link data together. Another method is a relate. A relate links two tables together, but unlike a join, which adds information to the primary table, the two tables remain separate when related. This allows you to see all related records in the linked table.

A relate works best when you have one record in your primary table which matches multiple records in the linked table or when you have multiple records in the primary table that match multiple records in the linked table. In these situations, a join would not work as well because it would have multiple records that match.

In this recipe, you will create a relate between the **Parcels** table and a land sales table. The land sales table contains all parcel sales that have occurred over the last several years. This table also comes from the county's CAMA system, like the **Owners** table did. Once you relate the two tables, you will see how you can view all sales for a selected parcel.

Getting ready

You will need to ensure that ArcGIS Pro and the book data are installed. This recipe is not based on previous recipes, so you are not required to have completed any previous recipes before starting this one.

How to do it...

You will now link the `Parcels` layer to a standalone table containing records of sales over the last several years using a relate. You will see the steps required to create a relate and see how it differs from a join.

1. Start ArcGIS Pro and open the `Creating and using a Relate.aprx` project, located in `C:\Student\ArcGISPro3Cookbook\Chapter3\Creating and using a Relate`.
2. In the **Catalog** pane located on the right of the ArcGIS Pro interface, expand **Databases** by clicking on the small arrowhead.
3. Expand the `Thomaston.gdb` geodatabase so that you can see its contents.

4. Right-click on the **Parcels_Sales** table and select **Add to Current Map**. The table should then appear at the bottom of the list in the **Contents** pane:

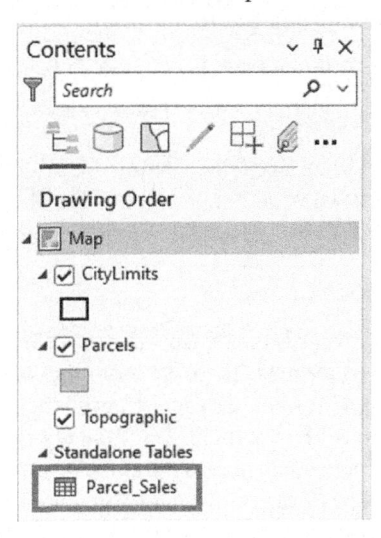

Figure 3.29 – The Parcels_Sales standalone table in the Contents pane

5. Right-click on the `Parcels` layer and select **Attribute Table** from the menu that appears. The attribute table for the `Parcels` layer should now be open, allowing you to see the fields and data it contains.
6. Right-click on the **Parcels_Sales** table and select **Open** from the menu that appears. The table should open, allowing you to see its contents.
7. Using the skills you learned in the previous recipe, try to determine the key field in each table that you will use to link the two tables together. Remember that key fields must contain the same data values and be the same field data type. The field names do not need to be the same.

    Question: What is the key field for each table that can be used to link them?

    Answer:

    After reviewing each table, the key field should be fairly easy to pick out. `Realkey` is the only field in each table that has the same data values and the same data type.

8. In the **Parcels** attribute table, right-click on the `Realkey` field and select **Sort Descending**.
9. In the **Parcels** attribute table, locate the records with a `Realkey` value of `15812`. Then, answer the following question:

    Question: How many records in the **Parcels** attribute table have a `Realkey` value of 15812?

    Answer:

10. In the **Parcels_Sales** table, right-click on the **REALKEY** field and select **Sort Descending**.

11. In the **Parcels_Sales** table, locate the rows with a **REALKEY** value of 15812. You will need to scroll down until you see that value, or you could use the **Select by Attribute** tool to select the records. Then, answer the following question:

Question: How many records in the **Parcels_Sales** table have a REALKEY value of 15812?

Answer:

You should now know that there is one parcel and two sales records that have a realkey value of 15812. This means that you have a one-to-many cardinality, which works well with a relate. Now, you need to configure the relate to link the `Parcels` layer to the `Parcels_Sales` table.

> **Note**
> Cardinality refers to how many records in one table match records in another table, based on a comparison of values found in a key field. There are four types of cardinality: one-to-one, one-to-many, many-to-one, and many-to-many. This relationship between records in the two tables can impact how you are able to work with them. Traditionally, if your data had a cardinality of one-to-one or many-to-many, you would always want to use a join to link the data. If your data had a cardinality of one-to-many or many-to-many, you would always want to use a relate to link the data. This is not the case with ArcGIS Pro. ArcGIS Pro offers more flexibility, so you can choose either a join or a relate depending on how you need to use the data.

Creating a relate

With the key fields and cardinality determined, you are ready to create a relate that will link the two tables together.

1. Right-click on the **Parcels** layer in the **Contents** pane. Go to **Joins and Relates** and select **Add Relate**:

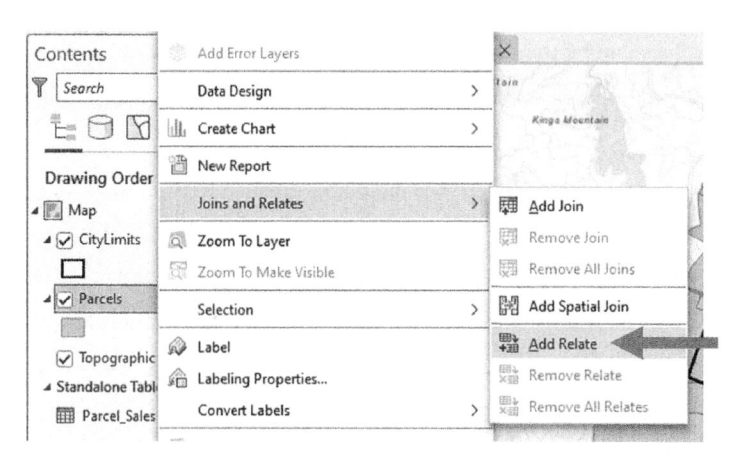

Figure 3.30 – Selecting Add Relate from the menu

2. Ensure **Layer Name or Table View** is set to **Parcels**. Since you right-clicked on the Parcels layer to add the relate, this should be automatically populated.
3. Set the **Input Relate Field** to **Realkey** using the drop-down arrow.
4. The **Relate Table** and **Output Relate Field** parameters should automatically populate because you have no other standalone tables in your map.
5. Change the **Relate Name** value to **Parcel Sales**.
6. Verify that your **Add Relate** tool looks like the following screenshot. If it does, click **OK**:

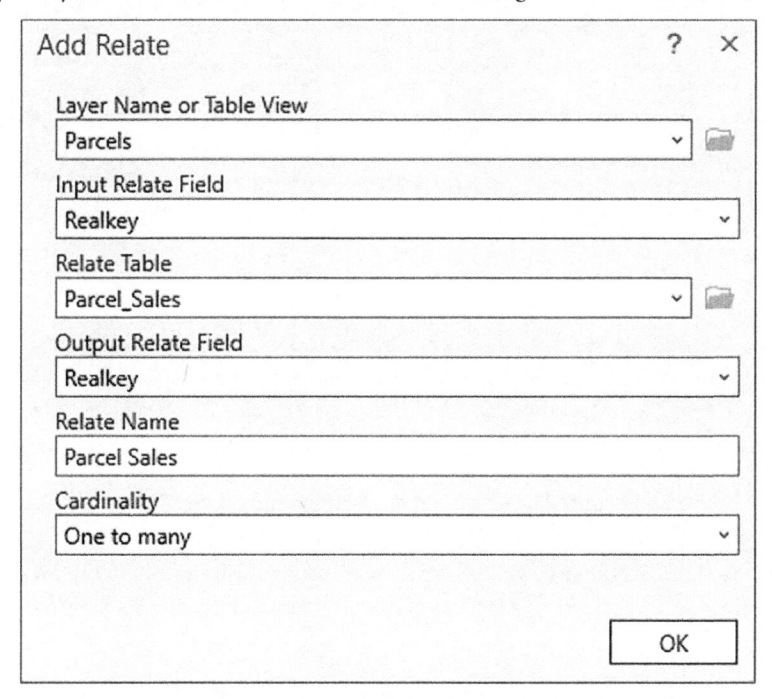

Figure 3.31 – Add Relate parameters complete

7. Once the **Add Relate** tool completes, save your project.

Verifying the relate works

Now that you have established a relate between the two tables, you need to verify it worked and you can access the linked data as expected. You will use the **Select** tool to do this.

1. Select the **Map** tab in the ribbon and activate the **Explore** tool.

2. In the map, move your mouse pointer to the approximate center of the city and zoom in by pushing your scroll wheel away from you until the map looks like the following screenshot:

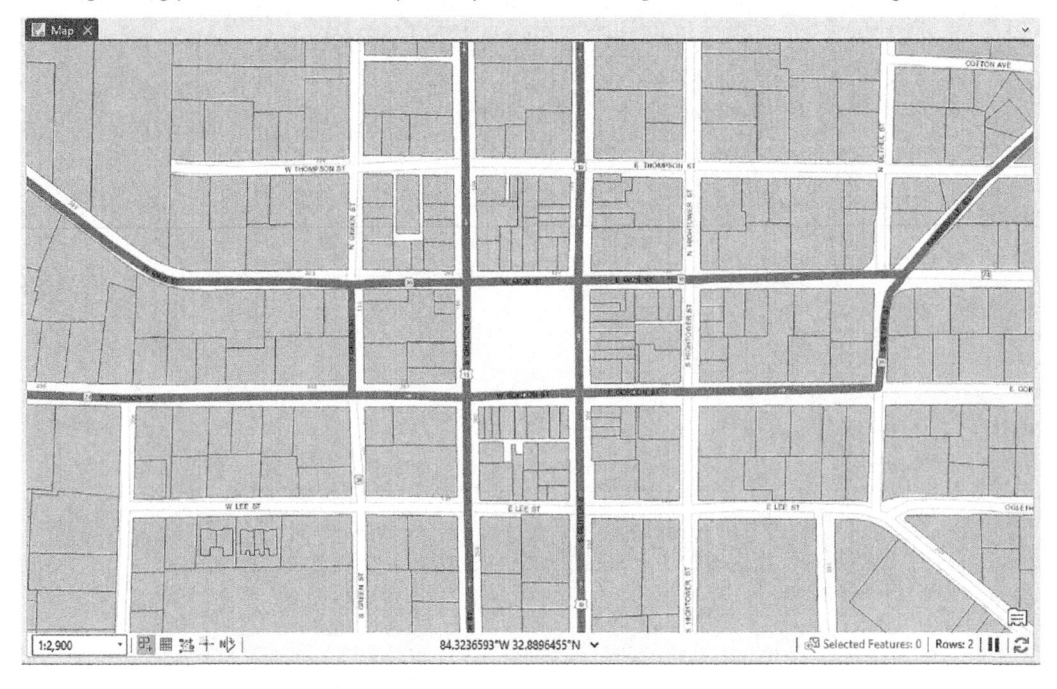

Figure 3.32 – Map zoomed in on the correct location

3. If you closed the `Parcels` layer attribute table, open it by right-clicking on the layer in the **Contents** pane and selecting **Attribute table**. If you still have the table open, proceed to the next step.

4. Click on the **List by Selection** button on the top of the **Contents** pane, as illustrated in the following screenshot:

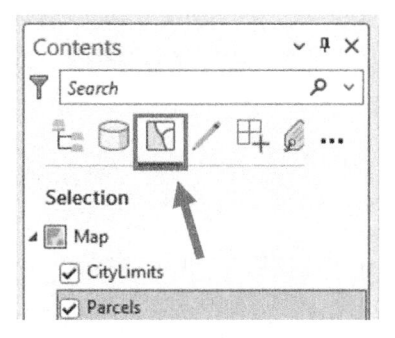

Figure 3.33 – List by Selection button in the Contents pane

5. Right-click on the `Parcels` layer and select **Make this the only selectable layer** from the menu that appears.
6. In the **Map** tab in the ribbon, select the **Select** tool.
7. Click on the parcel in the map indicated in the following screenshot. The parcel will be highlighted in light blue when selected:

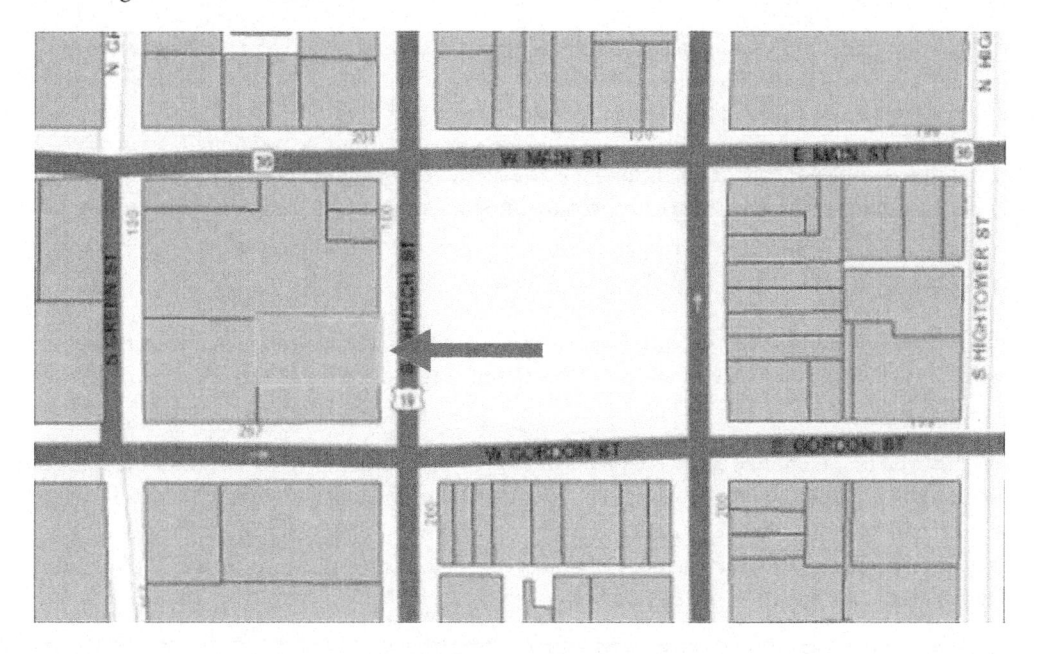

Figure 3.34 – Selecting a parcel

8. Click on the **Data** tab. Then, click on the **Related Data** button and select **Parcels_Sales** from the option that appears, as illustrated in the following screenshot:

Figure 3.35 – Selecting the Parcel_Sales relate to access sales data

The **Parcels_Sales** table should now open with two records selected. These are the two times the selected parcel was sold. The **Parcels** attribute table and **Parcels_Sales** tables are linked but remain separate. It should also be mentioned that both relates and joins only exist in projects where you create them. Neither of these creates permanent links between data in the actual databases themselves. We will explore how to do that in a later recipe in this chapter, *Creating and using a relationship class using existing data*.

9. Continue selecting other parcels to see how many times they have been sold. As you select other parcels, you will need to click on the **Related Data** button to refresh the `Parcel_Sales` table. When you are done, save your project.

You have just created a relate between the `Parcels` layer and a standalone table that contains sales records. This then allowed you to see how many times a selected parcel had been sold.

How it works...

In this recipe, you linked the `Parcels` layer to a standalone table containing sales information using a relate. This allowed you to access all recorded sales for each selected parcel, further enriching your GIS data.

To do this, you first identified a key field in each table. These key fields had to contain the same data values and be the same data type. Once you identified the key fields, you then needed to see how many records in the **Parcel** table matched records in the **Parcel_Sales** table. This relationship is called cardinality. You discovered that each parcel potentially had several matching sales records, meaning the two tables had a one-to-many cardinality.

Due to the established cardinality, you used a relate to link these two tables together. This linked them but kept them separate so that when you selected a parcel in the map, you could then access the related records in the sales table quickly.

There's more...

As was mentioned earlier in this recipe, cardinality was a major consideration when choosing whether you should use a join or a relate to link tables. Historically, you should always use a join if you have a one-to-one or a many-to-one cardinality. However, if you have a one-to-many or a many-to-many cardinality, you should use a relate; otherwise, you will miss data because the result would be a joined table that contained `null` values or just the values from the first record it finds that matches.

ArcGIS Pro does not work that way. You can join two tables regardless of the cardinality, as long as they are in the same database. If you join two tables in ArcGIS Pro that have a one-to-many or a many-to-many cardinality, you see all the matched records. ArcGIS Pro does this by creating virtual records in the table view for each match, as shown ahead:

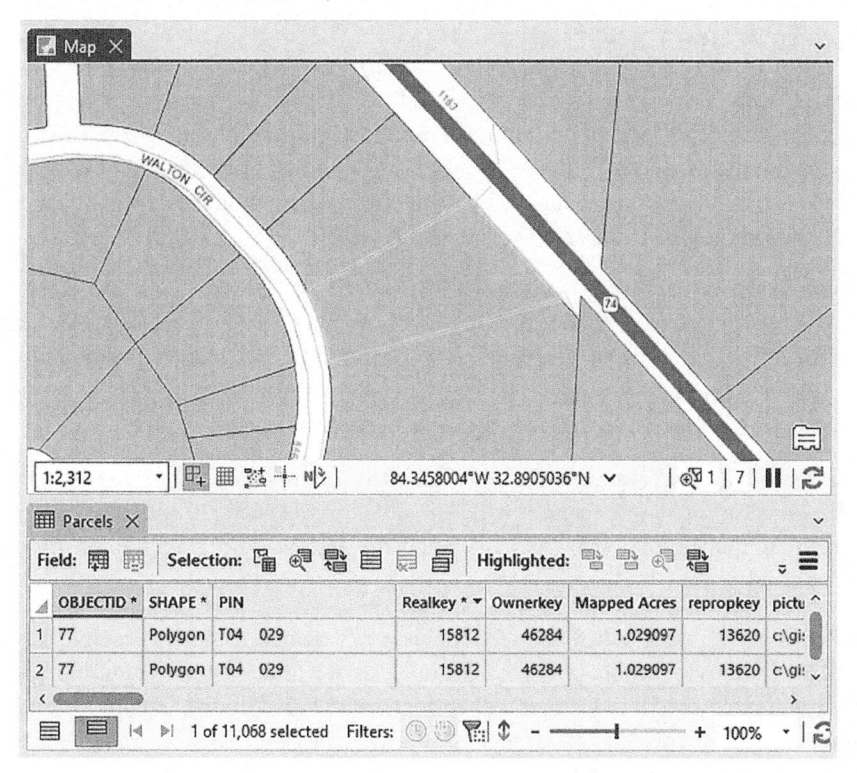

Figure 3.36 – Results of what happens when you create a join when cardinality is one to many

As you can see, a single parcel is selected but two records are shown in the attribute table. This is because the **Parcels_Sales** table was joined to the **Parcels** attribute table instead of being related. As you saw in the recipe, there is a one-to-many cardinality between these tables. So, there are two sales associated with the selected parcel in the **Parcel_Sales** table, and both are displayed.

In ArcGIS Pro, this method of joining tables with one-to-many or many-to-many cardinality has other advantages. You are able to label features with values from each record in the joined table. This is something relates do not allow. However, remember that this only works with tables in the same database. If the tables are in two different databases or are different formats, such as linking a spreadsheet to a geodatabase feature class, then the normal rules about when to use a join and when to use a relate still apply.

If you wish, use the skills you have learned in this and the previous recipe to join the **Parcel_Sales** table to the `Parcels` layer. Then, explore how this join works by selecting several parcels while looking at the attribute table.

Joining features spatially

In the previous recipes, you have seen how you can link external data to layers or other tables using a join or a relate. However, what if you want to transfer data from one layer to another but there is no key field to use to link the data? Maybe the two layers in question overlap one another, are next to one another, or share some other spatial relationship; surely there should be some way to link or join the two layers together based on a spatial relationship.

You can join two layers together based on a spatial relationship. This is called a spatial join. A spatial join creates a new feature class that adds the attributes from the joined feature class to the target feature class based on a spatial relationship you define when you run the tool. It is not required that the target and joined feature classes be the same type. You can spatially join lines with polygons, points with lines, or points with polygons, as well as those of the same feature type.

When you run the **Spatial Join** tool, you can specify the join operation. This determines how joins between target features and join features will be handled in the output feature class if multiple join features are found that have the same spatial relationship with a single target feature. The options for this include the following:

- JOIN_ONE_TO_ONE: When this option is chosen, if a single feature in the target layer matches multiple features in the join layer because they have the same spatial relationship, the attributes from all the matched join features will be aggregated based on the field map merger rule you configure. For example, if you set the target layer to be road centerlines and a road centerline crosses two wetland polygons that are in your join layer, the attributes from the two wetland polygons will be aggregated and added to the resulting new line feature class. If one wetland polygon has an area value of 2 acres and the other has an area value of 5 acres, and a Mean merge rule is specified, the aggregated value in the output feature class will be 3.5.

- JOIN_ONE_TO_MANY: When this option is chosen, if a single feature in the target layer is found to match multiple features in the join layer because they have the same spatial relationship, the resulting output feature class will contain multiple records representing the single target feature and each match it has with features in the join layer. Using the same example as previously, if a single road centerline crosses two wetland polygons, the output feature class will contain two road centerlines: one will have the attributes from one of the wetland polygons it crossed and the second will have the attributes from the other.

Joining features spatially | 97

In this recipe, you have been asked to assign each sewer line and sewer manhole the name of the watershed it is located within. This is for an annual report the sewer superintendent must submit to the state every year. You have the boundary of the existing watersheds in your area, and you have layers for the sewer lines and sewer manholes, so you will use the **Spatial Join** geoprocessing tool to assign the name of each watershed to the sewer features that are inside it.

Getting ready

This recipe can be completed with any license level of ArcGIS Pro. Completion of other recipes is not required, though it is recommended that you complete the recipes in *Chapter 1, ArcGIS Pro Capabilities and Terminology*. This will ensure that you have the basic skills required to successfully complete this recipe. Of course, you must install the sample data before you begin.

How to do it...

You will now work through the steps required to create a spatial join between two layers.

1. Start ArcGIS Pro and open the `Spatial Join.aprx` project located in `C:\Student\ArcGISPro3Cookbook\Chapter3\Spatial Join\`.

2. Right-click on the `Sewer Manholes` layer in the **Contents** pane and click **Attribute Table** from the menu.

3. Explore the attribute table for the `Sewer Manholes` layer and answer the following question:

    Question: What fields do you see in the attribute table for the `Sewer Manholes` layer?

    Answer:

    You will notice that it does not indicate which watershed each manhole is located in. So, you need to add this attribute to each manhole. Luckily, you do have a watershed layer that shows the boundary of each watershed and its name. You will be able to use a spatial join to add this information to the manholes. First, you will explore the attribute table for the `Watersheds` layer. Then, you will use the **Spatial Join** tool.

4. Right-click on the `Watersheds` layer and select **Attribute Table**. Then, explore the available fields in the table and answer the following question:

    Question: What fields do you see in the attribute table for the `Watersheds` layer?

    Answer:

98   Linking Data Together

5. Click on the **Analysis** tab in the ribbon and select **Spatial Join** from the **Tools** group to open this tool in the **Geoprocessing** pane:

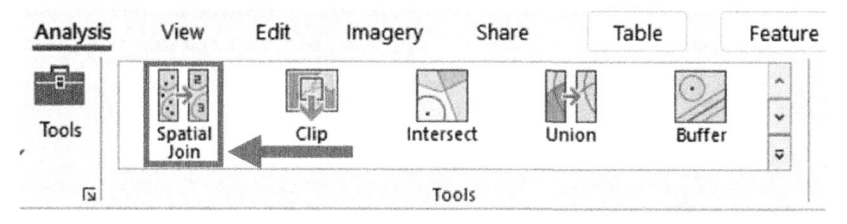

Figure 3.37 – Selecting the Spatial Join tool from the Analysis tab in the ribbon

6. In the **Spatial Join** tool, set the **Target Features** parameter to **Sewer Manholes** using the drop-down arrow.
7. Continuing in the **Spatial Join** tool, set the **Join Features** parameter to **Watersheds** using the drop-down arrow.
8. For **Output Feature Class**, click on the **Browse** button, which resembles a file folder. Then, select **Databases** in the left panel of the **Output Features Class** window.
9. In the right-hand panel of the **Output Feature Class** window, double-click on the Spatial Join.gdb geodatabase.
10. In the **Name** cell located near the bottom of the window, type Sewer_Manholes_Watersheds, as shown next, and click **Save**. You should be returned to the **Spatial Join** tool in the **Geoprocessing** pane:

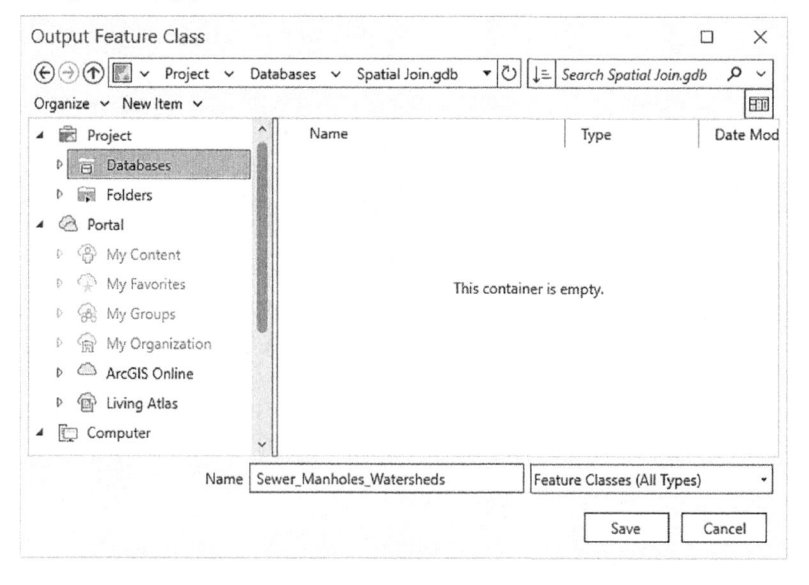

Figure 3.38 – Setting the location and name of the output for the Spatial Join tool

11. In the **Spatial Join** tool, set the **Join Operation** to **Join one to one** and ensure the box next to **Keep All Target Features** is enabled.
12. For the **Match Option** parameter, set it to **Within** using the drop-down arrow. This is the spatial relationship the tool will use to join the watershed name to each manhole.
13. Ensure that **Search Radius** is blank. Because each manhole should fall within a watershed boundary, you do not need to specify a radius.
14. Verify your **Spatial Join** tool looks like the following screenshot and then click **Run**:

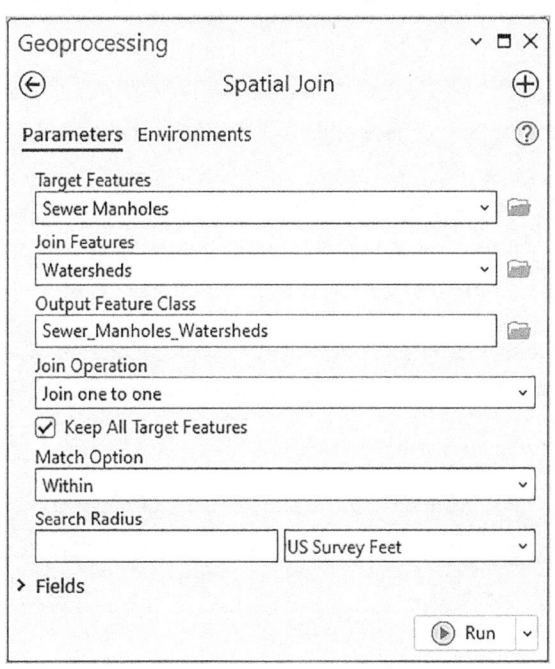

Figure 3.39 – Parameters complete for the Spatial Join tool

15. When the **Spatial Join** tool completes, close the **Geoprocessing** pane.

Verifying the spatial join

With the spatial join completed, you will check the results to verify they contain the data you need them to.

1. A new layer has appeared in the **Contents** pane and the map called `Sewer_Manholes_Watersheds`. This is the result of the **Spatial Join** tool you just ran. Right-click on this new layer and select **Attribute Table**.

2. Review the attribute table for the new layer and answer the following questions:

   Question: What fields does the attribute table contain for the new layer you created?

   Answer:

   Question: How does the new table compare to the attribute tables for the `Sewer Manholes` and `Watersheds` layers?

   Answer:

   As you should hopefully see, the new layer contains the combined attributes of both the `Sewer Manholes` and `Watersheds` layers. The output was a new point layer because the target was the `Sewer Manholes` layer, which was a point layer.

3. Close any attribute tables you have open in ArcGIS Pro and save your project.

4. Now, using the same process, perform a spatial join between the `Sewer Lines` layer and the `Watersheds` layer. However, this time, set the **Match Option** parameter to `Intersect` and the **Output Feature Class** parameter to `Sewer_Lines_Watersheds`.

5. Your **Spatial Join** tool should look like this. If it does, click **Run**:

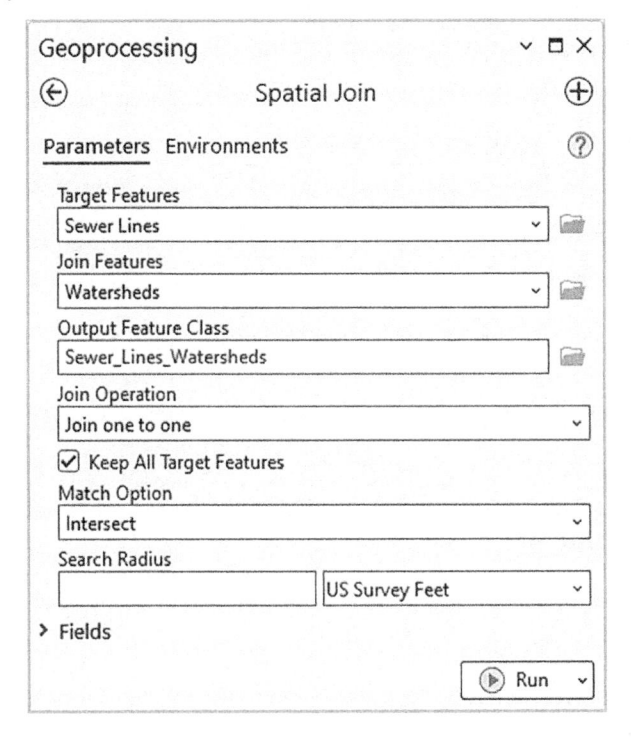

Figure 3.40 – Spatial join of Sewer Lines layer with Watersheds layer

6. Once the tool completes, a new layer will appear on your map like before. Close the **Geoprocessing** pane and save your project.

   This new layer is a line layer. This is because your **Target Features** parameter was set to the `Sewer_Lines` layer, which referenced a line feature class. The output feature type of a spatial join will always be the same as the target features used in the tool.

7. Close ArcGIS Pro.

You have now seen how a spatial join can be used to merge data in two different layers into a single layer or feature class. This can increase the types of analysis you can perform on the data or how you can display it. Remember – when you perform a spatial join, it does create a new data layer. This means that your original data is still intact for you to use as well.

How it works...

In this recipe, you merged two layers together so that the result contained the attributes of the two input layers using a spatial join. Unlike the joins and relates you performed in previous recipes that linked features together based on common attribute values in key fields, a spatial join uses spatial relationships to merge data.

The spatial join required you first to identify target features. This is the layer that determines what feature type (point, line, or polygon) the result will be, and the attributes from the **Join Features** parameter are appended. Next, you identified join features. This represents the attributes that will be added to the resulting features in the output based on the spatial relationship you select. Lastly, you had to select what spatial relationship must exist between the target and join features. In this recipe, you used two: **within** and **intersect**. However, there are many options you can use with this tool, including but not limited to **within a distance of**, **contains**, **are identical to**, and **completely within**. The result is a new feature class that contains data from both the target and join feature inputs.

Creating feature-linked annotation

You have now seen how to link data together based on a common key field using a join or relate, as well as based on a spatial relationship using a spatial join. These methods are extremely useful. However, they do have limitations. Joins and relates only exist in a single map and do not transfer easily to other maps or projects. Spatial joins create new feature classes or tables while still leaving the original data unaltered. So, is there a way to permanently link two tables or two feature classes together?

Yes, if the data is all in the same geodatabase. Geodatabses allow you to create a relationship class that permanently links data together. That link carries over into any map, scene, or project that uses that data. You can create a relationship class between two feature classes, two standalone tables, or a feature class and a standalone table.

Relationships not only link data together but also impact the behavior or contents of the linked features. For example, if a sewer-line feature class is linked to a manhole feature class via a relationship class, the relationship class can be configured to allow information such as an attribute value from one feature class to also pass to another, or if I delete a feature in one feature class, it deletes the connected feature in the other feature class. This makes relationship classes very powerful for updating and maintaining data.

One of the easiest relationship classes to create is called a feature-linked annotation. In this recipe, you will create a feature-linked annotation and then explore its behavior.

Getting ready

To complete this recipe, you will need a **Standard** or **Advanced** license of ArcGIS Pro. A **Basic** license does not support this level of functionality. If you are not sure what license you have, you can check it by doing the following. Open ArcGIS Pro, and on the left of the ArcGIS Pro start window, click on **Settings**. Then, select **Licensing** from the list on the left. This will display the license level available to you and any extensions you may have access to as well.

If you only have a **Basic** license, check with your account administrator to see if they can assign you a **Standard** or **Advanced** license at least temporarily. If a higher license is not available, you can request a trial license from Esri by going to their website, at `https://www.esri.com/en-us/arcgis/products/arcgis-pro/trial`. This will provide you with access to an **Advanced** license for ArcGIS Pro and more.

How to do it...

You will now see how to create a feature-linked annotation using labels.

1. You will begin by starting ArcGIS Pro. Then, open the `FeatureLinkedAnno.aprx` file located in `C:\Student\ArcGISPro3Cookbook\Chapter3\FeatureLinkedAnno`.

2. The project should open with a single map called **Trippville**. Click on the **Bookmarks** button located on the **Map** tab. Select **Labeling Area** to zoom to the middle of Trippville:

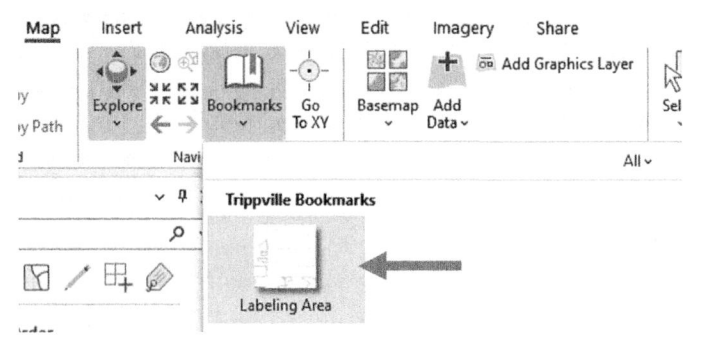

Figure 3.41 – Accessing the Labeling Area bookmark

Creating feature-linked annotation   103

3. In the **Contents** pane, select the `Street Centerlines` layer.
4. From the ribbon, select the **Labeling** tab, which appears in the ribbon.
5. In the **Map** group, select the **More** dropdown. Ensure **Use Maplex Labeling Engine** is not checked, as shown in the following screenshot. If it is, click on it to deactivate. If you get a warning about switching the labeling engine, click **Yes**:

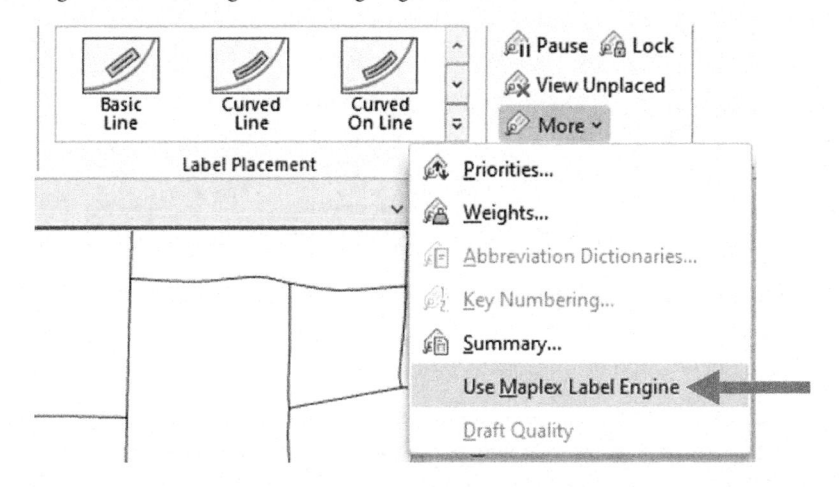

Figure 3.42 – Maplex Label Engine

6. Click on the **Label** button located on the far-left side of the **Labeling** tab to turn on labels for the `Street Centerlines` layer:

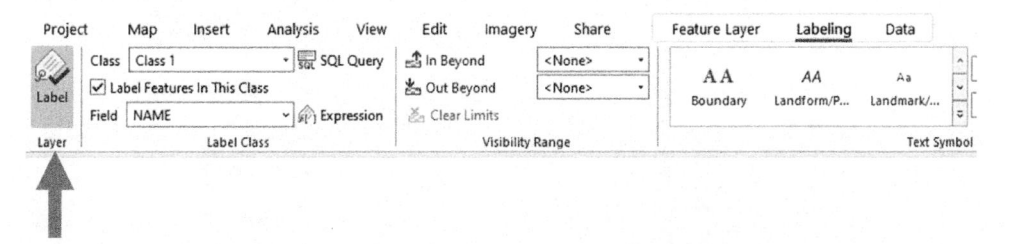

Figure 3.43 – Turning on labels from the Labeling tab

Text should appear on the map along the street centerlines. This text is labels pulled from values found in the attribute table. In this case, they are coming from the `Name` field. This field only has a partial name. You will now adjust the labeling settings so that the full name is displayed and the text follows the centerline geometry more closely.

104　　Linking Data Together

7. In the **Label Class** group, set the **Field** parameter to ST_NAME using the drop-down arrow. This field in the attribute table contains the full name of each street. You should see the text in the map update when you make this change.

8. In the **Label Placement** group, select the **Curved Line** option, as illustrated in the following screenshot. The text should now follow the street centerline geometry much more closely:

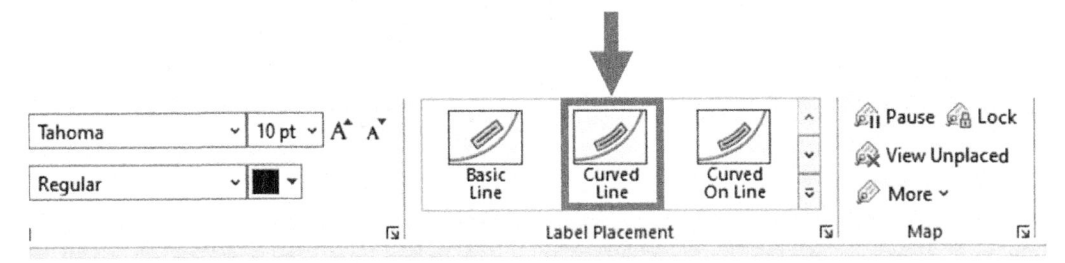

Figure 3.44 – Setting label placement option from the Labeling tab

9. In the **Text Symbol** group, set the font to **Tahoma**, size to **10**, type to **Regular**, and color to **Black**.

The street centerlines should now be labeled with their full name. The names should also follow the geometry of each street so that it is easy to identify the name of each street. In general, your map should now look similar to the following screenshot. Yours might look slightly different depending on the size of your monitor and resolution settings:

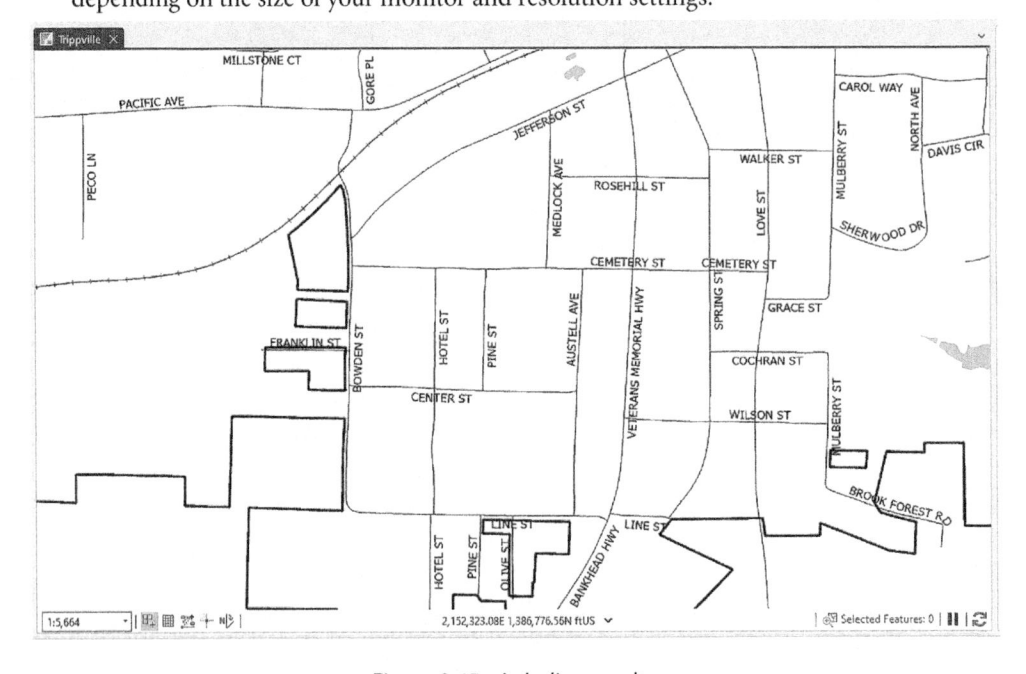

Figure 3.45 – Labeling results

10. Save your project.

Converting labels to annotation

With your streets labeled, you will convert those labels to annotation features that will be linked to the centerline feature they reference.

1. Activate the **Map** tab in the ribbon.
2. Click on the **Convert Labels To Annotation** button located in the **Labeling** group, as shown in the following screenshot. This will open the tool in the **Geoprocessing** pane:

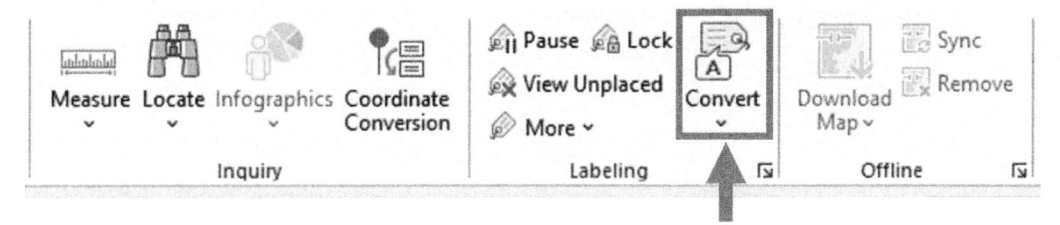

Figure 3.46 – Convert Labels to Annotation button on the Map tab

> **Note**
> ArcGIS Pro supports two types of text display in a map or scene: labels and annotation. Labels are dynamic text that is displayed and placed based on a specific layer, attributes, and placement settings configured by the user. Labels are extremely useful because they automatically adjust as your map view changes in size or location. Annotation is fixed text that can be individually placed and modified. This provides much greater control over the text, allowing you to put it exactly where and how you want it. Its big disadvantage is that it is more labor-intensive to maintain and will not work at all scales. Feature-linked annotation is the best of both worlds. It provides the control of annotation with some of the dynamic behavior of labels, as you will see in this recipe.

3. In the **Convert Labels to Annotation** tool, verify the **Input Map** parameter is set to **Trippville**. Because this is the only map in the project, it should be set by default.
4. Set the **Conversion Scale** parameter to **1:5000** using the drop-down arrow.
5. Set the **Convert** parameter to **Single Layer** using the drop-down arrow. Then, select the `Street Centerlines` layer as the **Feature Layer** parameter.
6. Click on the **Browse** button to the right of the **Output Geodatabase** setting. In the **Output Geodatabase** window, click on the `Databases` folder in the left panel.

7. Double-click on the `Trippville_GIS.gdb` database and then select the **Base** feature dataset. Verify your **Output Geodatabase** window looks like the following screenshot and click **OK**:

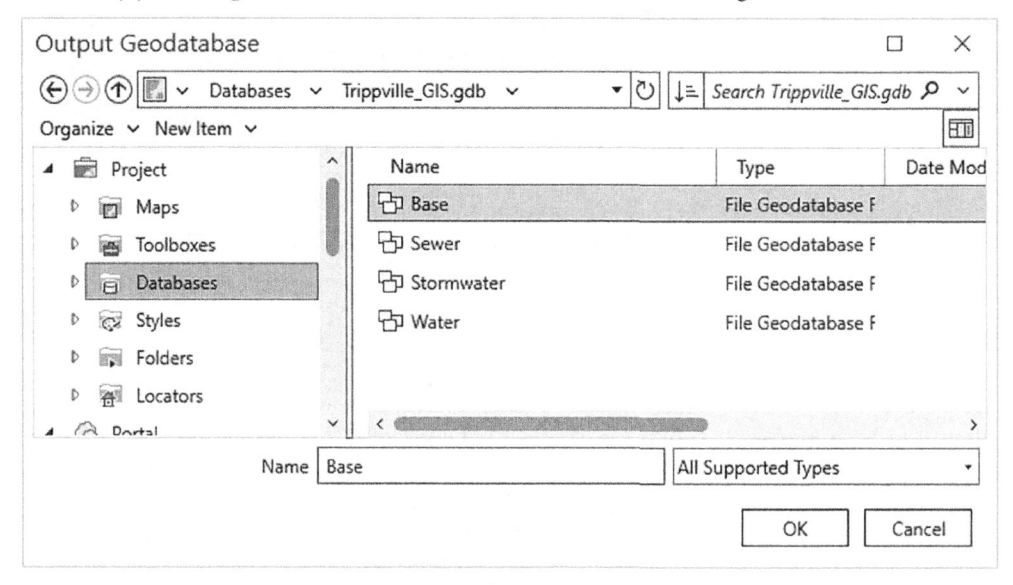

Figure 3.47 – Selecting the output location for the Convert Labels to Annotation tool

8. Set the **Extents** parameter to use the extents of the `Street Centerlines` layer using the drop-down arrow. Once you select the `Street Centerlines` layer, the numbers for the extents will update, and the **Extent** value will revert to **As Specified Below**.

9. Enable the **Create Feature Linked Annotation** parameter. Two other options will appear: **Create annotation when new features are added** and **Update annotation when feature's shape is modified**. Ensure these two parameters are also enabled. These parameters allow the relationship class to push changes to the annotation layer this tool will create when updates or changes are made to the `Street Centerlines` layer.

10. Change the **Output Layer** value to `StreetNames`.

Creating feature-linked annotation   107

11. Verify your **Convert Labels to Annotation** tool matches the following screenshot and click **Run**:

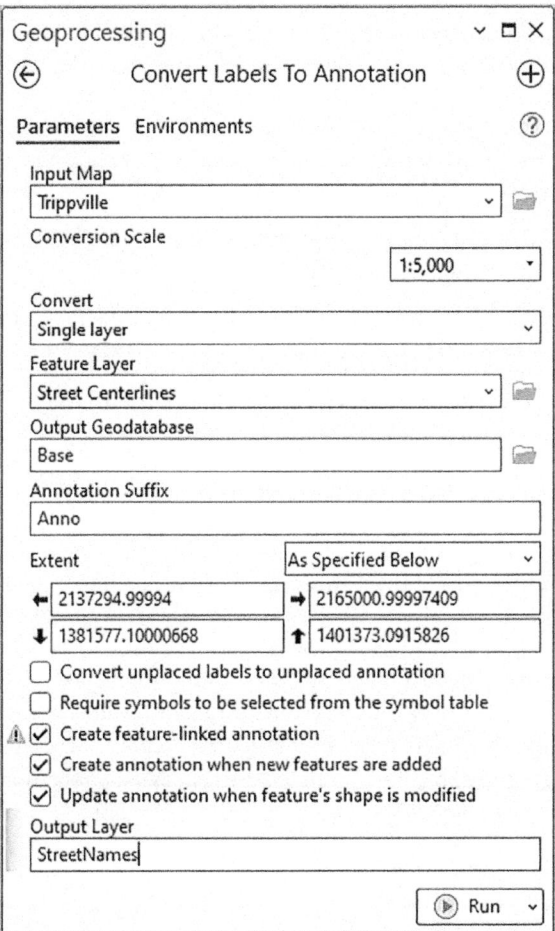

Figure 3.48 – Convert Labels to Annotation parameters complete

12. Close the **Geoprocessing** pane once the **Convert Labels to Annotation** tool is complete.

You should now see a new layer has been added to the map. This new layer is the new annotation feature class you created with the **Convert Labels to Annotation** tool. Like other layers, it has the same basic capabilities. You can turn it on and off, move it up or down in the draw order, change the symbology, and open its attribute table. You will examine the new layer to gain a better understanding of what you created.

Verifying the results

With the labels now converted to feature-linked annotation, you will verify and examine the resulting new layer. You will look closely at the attribute table and see how the annotation is linked to the centerline features.

1. In the **Contents** pane, expand the `StreetNames` layer by clicking on the small arrow located next to the layer name.
2. Right-click on the `Street_CenterlinesAnno` layer and select **Properties** from the bottom of the context menu that appears.
3. Select the **Source** option from the list on the left panel in the **Layer Properties** window, then answer the following questions:

   Question: What is the data type and feature type for the new layer?

   Answer:

   Question: What database is the feature class that the layer references named and where is it located?

   Answer:

   Question: What is the name of the feature class you created that is referenced by the new layer?

   Answer:

   As you will see when you answer the preceding questions, the **Convert to Annotation** tool created not only a new layer in your map but also a new feature class within the geodatabase. It is an annotation feature class that is a specialized version of a standard `polygon` feature class; basically, it acts like a textbox. As with all feature classes, an annotation feature class is another form of vector data. As such, it has both a spatial component and an attribute component.

4. Click **OK** to close the **Layer Properties** window.
5. Right-click on the `Street_CenterlinesAnno` layer in the **Contents** pane. Select **Attribute Table** from the context menu that appears.
6. Take a moment to explore the attribute table for the new layer you just created, then answer the following question:

   Question: What attribute fields are in the table for the `Street_CenterlineAnno` layer?

   Answer:

7. Click on the **List by Selection** button located near the top of the **Contents** pane. It looks like three polygons, one of which has a blue fill.
8. Set the `Street Centerlines` and `Street_CenterlinesAnno` layers as selectable and all others as not selectable as illustrated in the following screenshot. Those with a checkmark in the box are selectable layers, and those without a checkmark are not. To change the status from selectable or not selectable, simply click on the box next to the layer name:

Creating feature-linked annotation   109

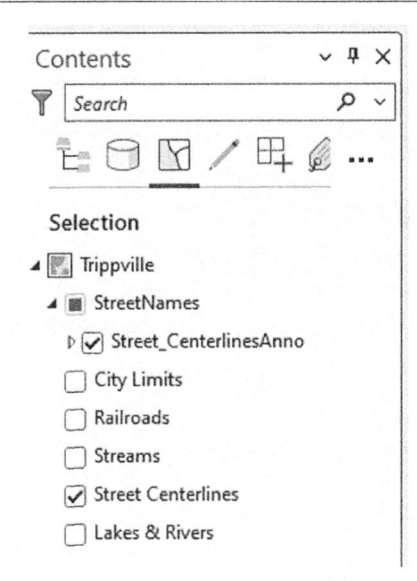

Figure 3.49 – Setting the Street_CenterlinesAnno and Street Centerlines Layers as selectable

9. Click on the **Select** tool in the **Map** tab in the ribbon. If you click on the drop-down arrow, select **Rectangle**.

10. Click just to the southeast of **PINE ST** on the map. Continue to hold your mouse button down and move your pointer just to the northwest of PINE ST, as shown in the following screenshot, so that you select both the street centerline and the road name:

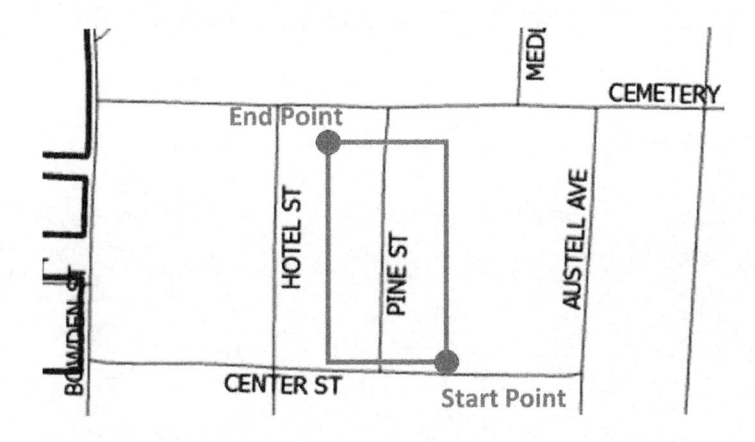

Figure 3.50 – Selecting PINE ST in the map

11. Right-click on the `Street Centerlines` layer in the **Contents** pane and select **Attribute Table** from the context menu that appears.

12. At the bottom of the table view, click on the **Show Selected Records** button to only display attributes for the street centerline you have selected.
13. Look at the attribute values for the selected street. Pay special attention to the `ObjectID` value.
14. Ensure the `Street Centerlines` layer is still selected in the **Contents** pane, then click on the **Data** tab in the ribbon.
15. Select the **Related Data** button in the **Relationship** group on the **Data** tab in the ribbon. Choose the `Street_CenterlinesAnno` option that appears. This should open the attribute table for the `Street_CenterlinesAnno` layer and select the related annotation record.
16. Review the attribute values for the selected record, then answer the following question:

    Question: Which fields for the `Street_CenterlinesAnno` layer contain similar values to those for the street centerline you have selected?

    Answer:

    Your comparison of the values for the selected street centerline and the related annotation should have revealed that the `ObjectID` value for the street centerline matched the `FeatureID` value for the annotation, and the `ST_NAME` value for the street centerline matched the `TextString` value for the annotation.

    The two feature classes are linked to one another in the relationship class using the `ObjectID` value of the street centerlines and the `FeatureID` value of the annotation as the key field, similar to how you used a key field to create both a join and a relate in previous recipes. As you will see next, the `ST_NAME` field and `TextString` fields are also linked as part of the relationship class.

17. Ensure that you still have the **Pine St** centerline selected, and then click on the **Edit** tab in the ribbon.
18. Click on the **Attributes** button located in the **Selection** group on the **Edit** tab, as shown in the following screenshot:

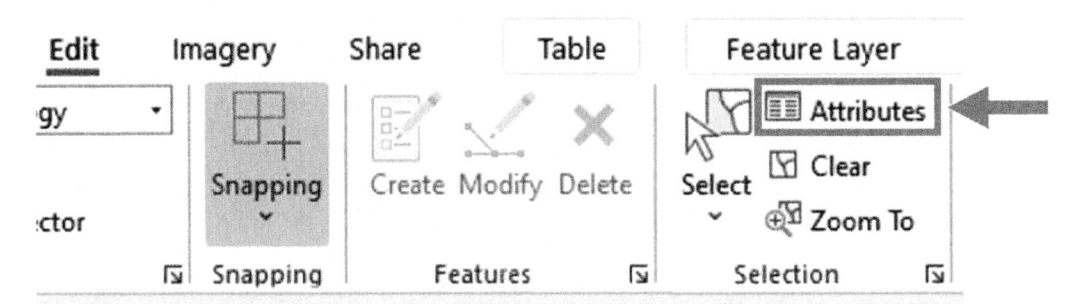

Figure 3.51 – Attributes button on the Edit tab

19. In the **Attributes** window, select **Street Centerlines** in the top panel. This will display attributes for the selected street centerline segment in the lower panel of the **Attributes** window.
20. Click in the cell located to the right of the `ST_NAME` field. Then, type `OAK AVE`. Press the *Enter* key when done and click **Apply**.

> **Tip**
> If ArcGIS Pro does not allow you to type a new value, this might be due to a couple of settings. First, in the **Contents** pane, verify the `Street Centerline` layer is set as editable on the **List by Editing** button. Second, if you see an **Edit** button on the **Edit** tab, make sure it is highlighted in blue, which indicates you have enabled an edit session.

Notice what happens to the annotation for the street name on the map. It should have also changed because the `ST_NAME` field is linked to the `TextString` field for the annotation layer. Changing the `ST_NAME` value in the `Street Centerlines` layer automatically updates the value for the annotation layer because of the relationship class you created. For feature-linked annotation, this is a one-way relationship. This means the change is only pushed from the `Street Centerlines` layer to the `Street_CenterlinesAnno` layer, but not from the `Street_CenterlinesAnno` layer to the `Street Centerlines` layer.

21. Close the **Attributes** window.
22. In the **Edit** tab in the ribbon, click the **Clear** button to deselect both the annotation and street centerline segment.
23. In the **Contents** pane, click on the **List by Selection** button. Then, right-click on the `Street Centerlines` layer and select **Make this the only selectable layer**.
24. Return to the **Edit** tab in the ribbon and choose the **Select** tool.
25. Click on the street centerline for the newly renamed Oak Ave that was formerly Pine St. Then, press your *Delete* key or click on the **Delete** button on the **Edit** tab.
26. If you get a warning message asking if you are sure you want to delete this data, click **Yes**.

Notice that the annotation is also deleted when you delete the street centerline for Oak Ave. The relationship class not only links fields together but also creates a link between the features in each feature class so that by deleting the centerline, it also deletes all related annotation features. As with the field relationship you explored earlier, this is only a one-way relationship. This means that if you were to delete the annotation, it would not delete the related street centerline.

27. On the **Edit** tab in the **Manage Edits** group, select **Discard** to discard all edits you have made.
28. In the **Catalog** pane, expand the `Databases` folder so that you see the two geodatabases connected to the project.
29. Expand the `Trippville_GIS.gdb` geodatabase to see its contents.

30. Now, expand the **Base** feature dataset and examine its contents. You should see the two feature classes that were referenced by the two layers you have been working with, `Street_Centerlines` and `Street_CenterlinesAnno`. You should also see the relationship class you created when you created the feature-linked annotation layer, `Anno_80_85`. Again, the name of your relationship class may have different numbers.

31. Save your project and close ArcGIS Pro.

Because feature classes and relationship classes exist as objects in your geodatabase, they may be used in other maps, scenes, or projects. This provides much greater flexibility than the joins and relates you performed in past recipes.

How it works...

As you saw in this recipe, feature-linked annotation links text in one layer to features in another layer. It does this using a relationship class. As with joins and relates, the relationship class relies on a key field from the attribute tables for both feature classes to establish a basic link between text features in the annotation layer and the feature that the text references in the other feature class. In the case of feature-linked annotation, the key field was automatically determined, and the relationship class was configured when you converted the labels for the street centerlines to annotation using the **Convert to Annotation** tool.

Feature-linked annotation takes advantage of the additional functionality the relationship class allows. In addition to relating two feature classes together so that you can locate related features or records, it also allows you to push changes from a feature to an annotation. This is called a composite relationship. In a composite relationship, the existence or values of the origin feature controls or changes the existence or values of the destination feature or record. You saw this demonstrated when you changed the street name for the centerline and deleted the centerline, which caused changes to the street annotation layer.

Creating and using a relationship class using existing data

You have seen the power of a relationship class in the previous recipe. You saw how linking two feature classes together allowed you to not only access information about linked features but also control some behavior. However, this was all automatically set up by the **Convert to Annotation** tool. How do you create a relationship class that would link that existing data together?

In this recipe, you will create a relationship class between a feature class and a standalone table. This will be between the same `Parcels` layer and sales table you related in the *Creating and using a relate* recipe earlier in this chapter. However, once you establish the relationship class, the link becomes permanent, unlike a relate, which is limited to the map in which it was created.

Getting ready

To complete this recipe, you will need a Standard or Advanced license of ArcGIS Pro. A Basic license does not support this level of functionality. If you are not sure what license you have, you can check it by doing the following. Open ArcGIS Pro, and on the bottom left of the start window, where you would normally select a project to open, click on **About ArcGIS Pro**. Then, select **Licensing** from the list on the left. This will display the license level available to you and any extensions you may have access to as well.

If you only have a Basic license, check with your account administrator to see if they can assign you a Standard or Advanced license at least temporarily. If a higher license is not available, you can request a trial license from Esri by going to their website, at https://www.esri.com/en-us/arcgis/products/arcgis-pro/trial. This will provide you with access to an Advanced license for ArcGIS Pro and more.

How to do it...

You will now get to create a new relationship class from the beginning. You will get to work through the required steps to gain a better understanding of the effort this requires.

1. As normal, you will need to start ArcGIS Pro and then open the RelationshipClass.aprx file located in C:\Student\ArcGISPro3Cookbook\Chapter3\RelationshipClass\. The project should open with a single map called Thomaston. If you completed the previous *Creating and using a relate* recipe, it should look familiar.

2. In the **Contents** pane, you need to verify that you see the Parcels layer and the **Parcels_Sales** table. If not, use the skills you have learned to add them to the map from the Thomaston.gdb geodatabase, which is connected to the project.

    In order to create a relationship class between the Parcels layer and the **Parcels_Sales** table, you need to determine two things: a key field in each table and the cardinality. If you completed the *Creating and using a relate* recipe, you will be using the same key field and should already know the cardinality, so you can skip to *Creating a relationship class Step 1* and continue from there. If you did not complete that recipe, continue to the next step.

3. Right-click on the Parcels layer and select **Attribute Table** from the menu that appears.

4. Right-click on the **Parcels_Sales** table and select **Open**.

5. Click on the name of the **Parcels_Sales** table located at the top of the table view, and drag it until your mouse point is over the docking icon, as shown here:

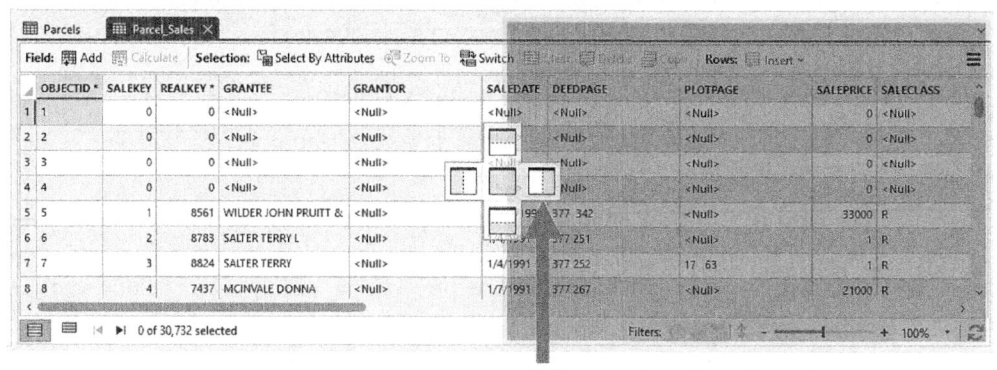

Figure 3.52 – Docking the Parcels and Parcels_Sales tables so that you can see both

The two tables should now appear side by side so that you can more easily assess each one and identify key fields:

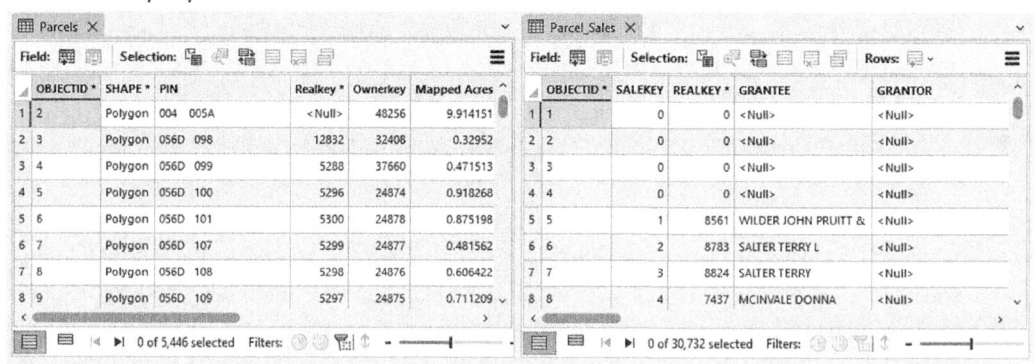

Figure 3.53 – The Parcels and Parcels_Sales table side by side

6. Review the two tables, looking for a field in each that contains values that match values found in the other. These will be the key fields that you will use to link these two tables together. Remember – the field names do not need to match. They just need to contain the same values and be the same data type.

7. After reviewing the two tables, answer the following question:

    Question: What is the key field you identified in each table?

    Answer:

    After reviewing each table, the key field should be easy to identify. The **Realkey** field is the only field in both tables that contains the same values and has the same field type. It just so happens in this example that they also have the same name, though the capitalization is different. Now that you have determined the key fields, it is time to assess the cardinality between the two tables.

8. In the **Parcels** attribute table, right-click on the **Realkey** field and select **Sort Descending**.
9. In the **Parcels** attribute table, locate the records with a **Realkey** value of 15812 and then answer the following question:

    Question: How many records contain a **Realkey** value of 15812?

    Answer:

10. In the **Parcels_Sales** table, right-click on the **REALKEY** field and select **Sort Descending**.
11. In the **Parcels_Sales** table, locate the rows with a **REALKEY** value of 15812. You will need to scroll down until you see that value, or you could use the **Select by Attribute** tool to select the records. Answer the following question:

    Question: How many records in the **Parcels_Sales** table contain a value of 15812 in the **REALKEY** field?

    Answer:

12. Close both tables once you have answered the question.
13. On the **Map** tab in the ribbon, click on the **Clear** button to deselect any records you may have selected.

Based on your review of the two tables, you should now know that 1 parcel and 2 sales records have a value of 15812. This means they have a cardinality of one to many. With that knowledge, you will be able to create and configure a new relationship class to link the `Parcels` layer to the `parcels_sales` table.

Creating a relationship class

You are now ready to create a relationship class in the geodatabase that will link these two tables together.

1. In the **Catalog** pane, expand the `Database` folder so that you can see the two databases that are connected to the project.

2. Expand the `Thomaston.gdb` geodatabase, then right-click on it. Go to **New** and select **Relationship Class**. The **Create Relationship Class** tool should open in the **Geoprocessing** pane:

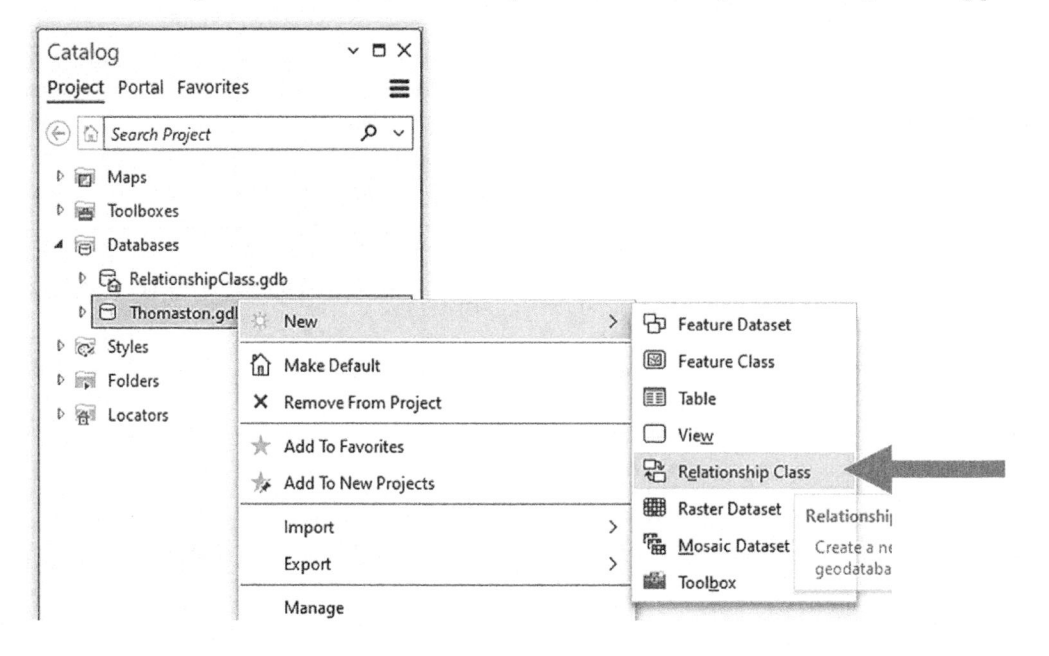

Figure 3.54 – Accessing the Create Relationship Class tool from the menu

3. Set the **Origin Table** parameter to `Parcels` using the drop-down arrow. This establishes the **Parcels** table as the primary table for the relationship.

4. Set the **Destination Table** parameter to `Parcels_Sales` using the drop-down arrow. This establishes it as the secondary or child table for the relationship.

5. The **Output Relationship Class** parameter should be automatically populated with `Parcels_Parcels_Sales`. You can change the name if you desire or accept the default.

6. Verify the **Relationship Type** parameter is set to `Simple`.

7. Both the **Forward Path Label** and **Backward Path Label** parameters should be automatically populated as well. Accept these default values.

8. Set the **Message Direction** parameter to **None (no messages propagated)** using the drop-down arrow, since we just want to be able to view related records between the tables. We are not trying to have one table control the information in the other.

9. Using the drop-down arrow, set the **Cardinality** parameter to **One to Many (1:M)**, since this was what you determined the cardinality to be previously.

10. Set the **Origin Primary Key** parameter to **Realkey** and set the **Origin Foreign Key** parameter to **REALKEY**.

11. Verify that your **Create Relationship Class** tool looks like the following screenshot. If it does, click **Run**:

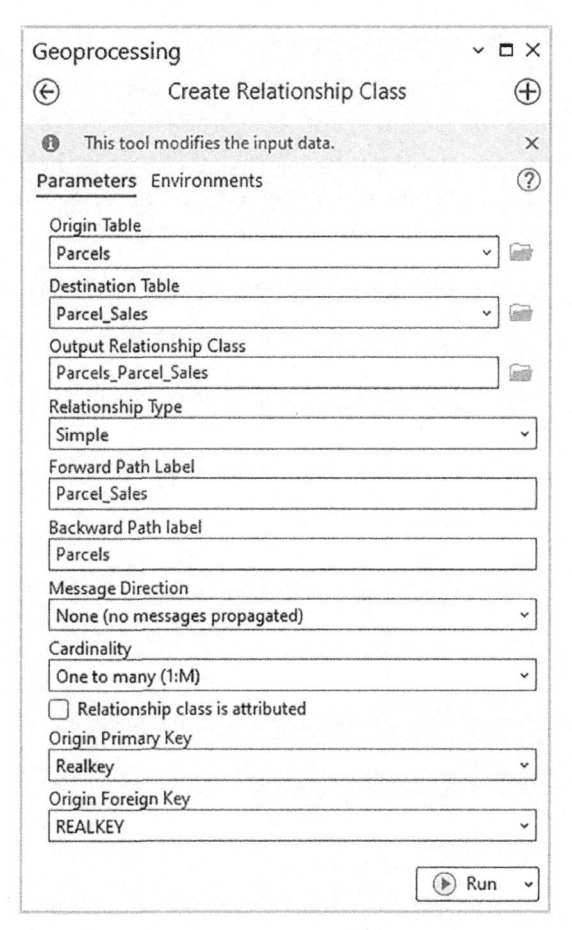

Figure 3.55 – The Create Relationship Class with parameters complete

12. Once the tool completes, close the **Geoprocessing** pane.

Verifying the relationship class properties

With the new relationship class created, you will verify it retained the properties you set up when creating it. You will also test to ensure it works as expected.

1. In the **Catalog** pane, expand the Thomaston.gdb geodatabase so that you can see its contents. You should see the new relationship class you just created, named Parcels_Parcel_Sales.
2. Right-click on the Parcels_Parcel_Sales relationship class in the **Catalog** pane and select **Properties**.

118 Linking Data Together

3. Take a moment to review the properties. They should reflect the settings you defined in the **Create Relationship Class** tool.

4. Click **OK** when done reviewing the properties.

5. Now, to see if the new relationship class you create works, select the `Parcels` layer in the **Contents** pane.

6. Using the **Explore** tool on the **Map** tab, zoom into the center of the city of Thomaston. Zoom in so that you can easily see the boundary of individual parcels. This should be somewhere close to a scale of 1:4800:

Figure 3.56 – Map zoomed to the required area

7. On the **Map** tab in the ribbon, activate the **Select** tool and click on a single parcel in the map. The selected parcel should have a blue highlighted border around it.

8. Click on the **Data** tab in the ribbon and select the **Related Data** button and the **Parcels_Sales** option that is displayed. The **Parcels_Sales** table should open, displaying any sales associated with the parcel you selected.

9. Select a couple more parcels and view the related sales associated with them. You may need to click on the **Related Data** button with each newly selected parcel to see the sales records.

10. Once you are done, save your project and close ArcGIS Pro.

You have now successfully created a relationship class that linked the `Parcels` layer to the `Parcel_Sales` standalone table. Unlike a relate or a join, a relationship class persists across any project or map that references the `Parcels` feature class. So, the sales data will always be available anytime the parcel data is used.

How it works....

In this recipe, you created a relationship class that linked the `Parcels` feature class to the `Parcels_Sales` standalone table. This permanently linked these two objects in the geodatabase so that you were able to select a parcel in the map and easily see each time it had been sold from the `Parcels_Sales` table.

This link was established using a key field you identified in each table. The key field used by a relationship class, such as a Relate or a Join, must contain the same exact values and be the same field type (text, long integer, short integer, and others). However, the field names are not required to be the same. ArcGIS Pro is then able to find records in both tables that have the same value and link those together, as you saw demonstrated in this recipe.

Unlike a Join or a Relate, a relationship class is part of your geodatabase. This means that anytime you add the `Parcels` feature class to a map in any project, you will have access to the data in the `Parcels_Sales` table automatically through the Related Data function.

Once you create a relationship class, you are not able to change how it is configured. If you wanted to change a simple relationship to a composite for example, you would have to create a completely new relationship class.

4
Editing Existing Spatial Features

GIS data should represent the real world as completely and accurately as possible so that any analysis we do is correct. Since the world is constantly changing, we need to be able to update our GIS to reflect these changes. ArcGIS Pro includes a wealth of tools for editing our existing data so that it can accurately reflect the real world.

Before we start looking at specific tools and methods that you can use to edit and update existing GIS data, you need to understand what types of data ArcGIS Pro can edit. While ArcGIS Pro supports the use of a wide range of data in maps and analysis, it does have limits regarding data formats it can edit. On the spatial side, this is generally limited to data stored in geodatabases, shapefiles, and ArcGIS web feature services. Another limiting factor will be your license level. ArcGIS Pro has three licensing levels:

- Basic
- Standard
- Advanced

Each of these has different levels of capability, with Basic having the lowest and Advanced having the highest. To learn more about the different capabilities of these three levels, go to `https://pro.arcgis.com/en/pro-app/latest/get-started/license-levels.htm`.

In this chapter, you will learn how to configure various editing options and use multiple tools to edit existing features you see on your map. You will reshape polygons, split lines, align features so that they follow other features, and combine features.

We will cover the following recipes:

- Configuring editing options
- Reshaping an existing feature
- Splitting a line feature
- Merging features
- Aligning features

Configuring editing options

Before you start editing data, ArcGIS Pro has several options you need to check out and configure that impact the overall editing process. You will review and update these options to ensure your edits are done smoothly and saved correctly. This will include setting the units of measure, verifying tolerances, and more.

In this recipe, you will configure and verify several editing options, including how and when to save, setting your units of measure, making newly added layers editable by default, and configuring snapping.

Getting ready

This recipe requires the sample data to be installed on your computer. It is recommended that you complete the recipes in *Chapter 1, ArcGIS Pro Capabilities and Terminology*, before starting this recipe. This will ensure you have a better foundational understanding of navigating within a map. You can complete this recipe with any ArcGIS Pro licensing level.

How to do it...

Follow these steps to learn how to access the editing options in ArcGIS Pro and how to configure them:

1. Start ArcGIS Pro by clicking on your **Start** menu button. Then, expand the **ArcGIS** program group and select **ArcGIS Pro**, as shown in the following screenshot:

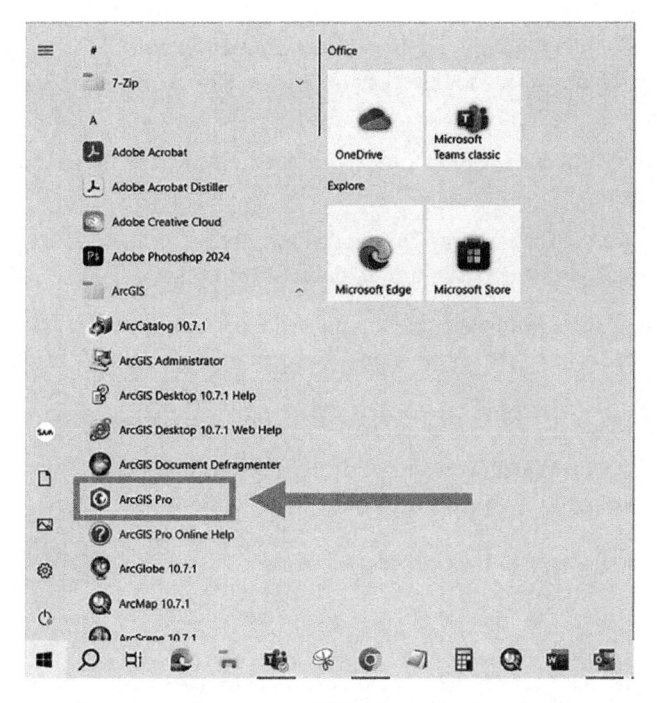

Figure 4.1 – Starting ArcGIS Pro from the Start button

> **Information**
> The preceding screenshot is based on Windows 10. If you are using Windows 11, it will work a little differently. First, you must click on the **Start** button within Windows 11 and go to **All Apps**. Then, scroll to **ArcGIS** and expand it. Lastly, select **ArcGIS Pro** from the list.

2. In the ArcGIS Pro Start window, click on the **Open another project** button, as shown here:

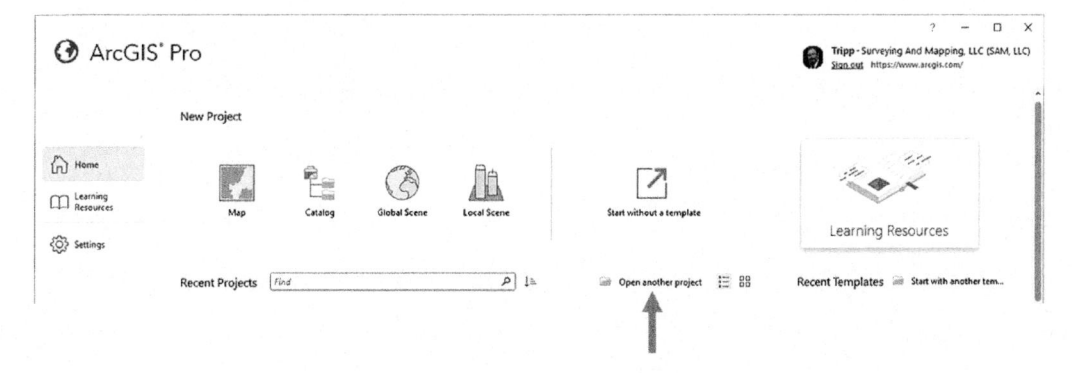

Figure 4.2 – Open another project

3. In the **Open Project** window, expand the **Computer** option in the left panel. Then, in the right panel, scroll down and double-click on the `C:` drive. It may be labeled **Local Disk**, **Local Drive**, or **OS**.

4. Double-click on the `Student` folder, followed by the `ArcGISPro3Cookbook` and *Chapter 4* folders.

5. Double-click on the `Editing` folder and select the `EditingExisting Data.aprx` project file. Then, click the **OK** button to open the project.

   The project will open with a single map named **City of Trippville**. The map contains three layers – `City Limits`, `Parcels`, and `Railroads`, plus the `Topographic` basemap.

Accessing and configuring ArcGIS Pro options

Now that you have a project open, you will begin setting various options that can impact how you edit data. You will start with the **Application** and **Project** options:

1. Click on the **Project** tab of the ribbon:

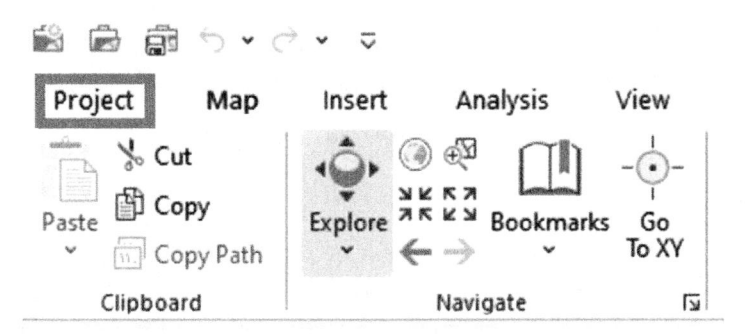

Figure 4.3 – Selecting the Project tab of the ribbon

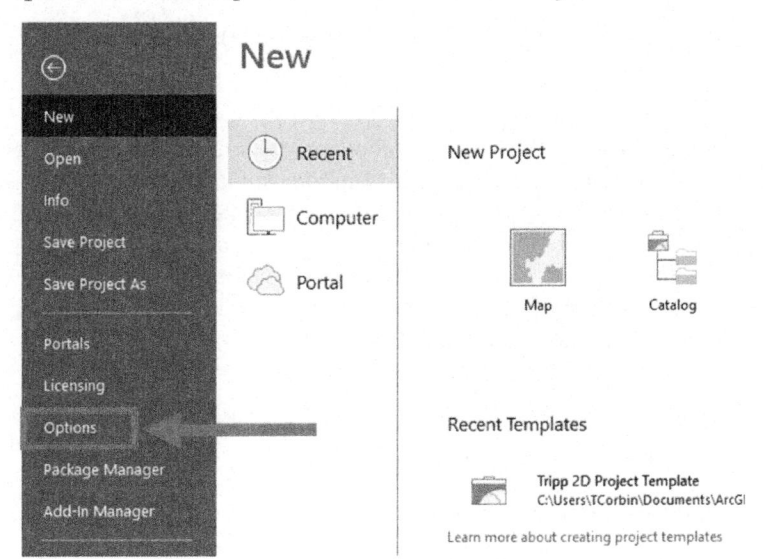

2. Select **Options** from the left panel, as shown in the following screenshot:

Figure 4.4 – Accessing Options from the Project tab

The **Options** window should now be open. You can now make changes to several settings in this window.

> **Information**
>
> ArcGIS Pro categorizes option settings into two types: **Project** and **Application**. Project settings are applied to the current project you have open, whereas application settings are applied to the application as a whole and will be applied regardless of what project is open. You should always verify these before editing to ensure they are set properly.

3. Select **Units** under **Project** in the panel on the left of the **Options** window. You should see several items appear in the right-hand panel, including **Distance**, **Angular**, **Area**, **Location**, **Direction**, and **Page Units**, along with others.

4. Click on the small arrowhead located to the left of **Distance Units** to expand the options settings.
5. Since Trippville is in the United States, select `Foot_US` as your distance unit.

> **Information**
>
> There is a difference between the US survey foot and the international foot. The difference starts after the sixth decimal place. Most places in the United States use the US survey foot as the standard for most measurements and local coordinate systems such as state planes. While only being slightly different, that difference does add up and can cause alignment and accuracy issues if it's not considered. For more information about the difference between the US survey foot and the international foot, go to `https://geodesy.noaa.gov/corbin/class_description/NGS_Survey_Foot/`.

6. Click on the small arrowhead located to the left of **Area Units** to expand those options.
7. Select `Square_Foot_US` from the list that is presented.
8. Select the **Editing** option from the panel on the left-hand side of the **Options** window, as illustrated in the following screenshot:

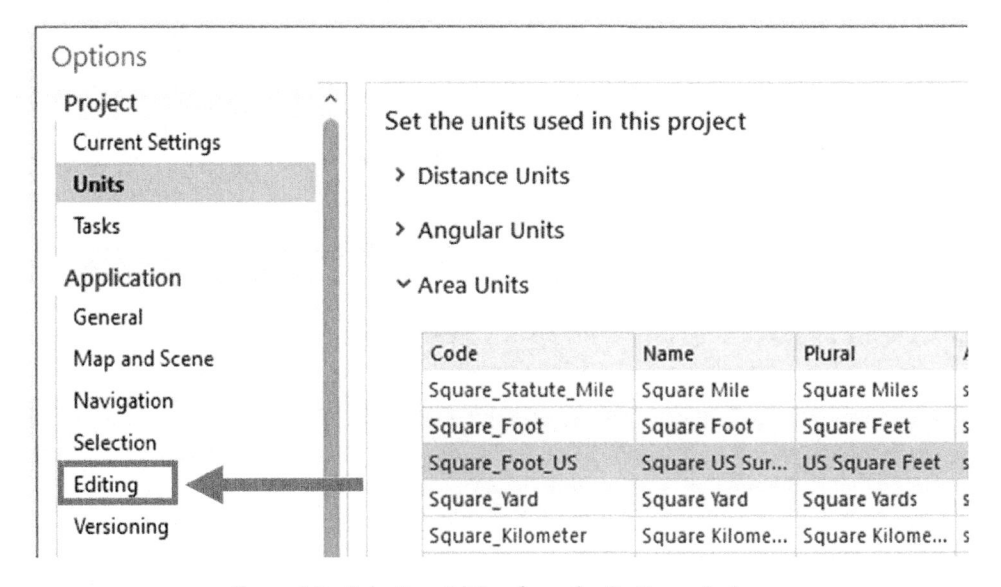

Figure 4.5 – Selecting Editing from the Options window

9. If necessary, expand **General** by clicking on the small arrowhead.
10. Take a moment to review all the options available under **General**.
11. Scroll down and ensure **Enable double-click as a shortcut for the Finish button** is checked.
12. Ensure **Show feature symbology in the sketch** is enabled.

13. If required, expand the **Session** options in the right-hand panel of the **Options** window:

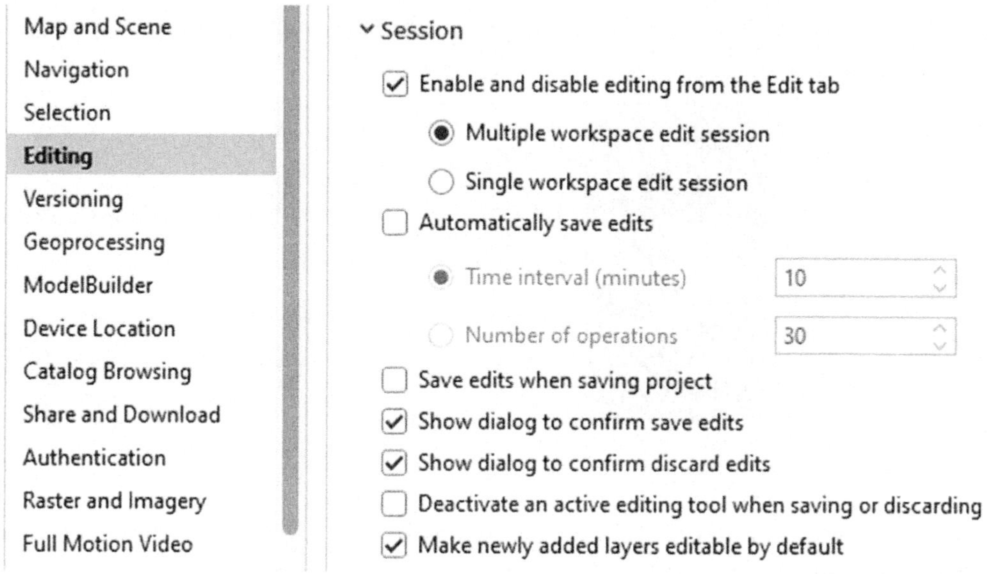

Figure 4.6 – The Sessions options expanded

14. Ensure **Show dialog to confirm save edits** and **Show dialog to confirm discard edits** are both enabled. Feel free to explore other options.
15. Click **OK** to apply these settings. The **Options** window should close automatically.

> **Tip**
>
> You may have noticed that ArcGIS Pro has an autosave feature for edits. This is something new for ArcGIS users and not just for ArcGIS Pro. Having to manually save edits has been the norm for those using ArcGIS desktop applications for decades. If autosave is enabled, it can reduce the need to remember to manually save.
>
> When editing, you need to ensure you save your edits frequently to avoid losing work if your computer crashes, locks up, or loses power. Unlike other applications, ArcGIS Pro does not create a restorable copy of your edits. This is because ArcGIS Pro never works directly with the data you see on a map or table. Instead, ArcGIS Pro references data and loads visible features into your computer's **Random Access Memory** (**RAM**). This allows ArcGIS Pro to work with large datasets that would cause other applications to slow or crash. This also means your edits are stored in your computer's RAM until they are saved. **So, make sure you save often or enable the autosave option!**

Now that you have configured the application and project options, you will configure your snapping options.

1. Click on the **back arrow**, which looks like an arrow inside a circle, in the **Project** tab to return to the project view, as shown in the following screenshot:

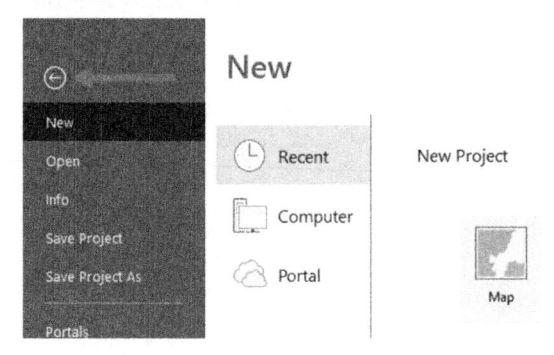

Figure 4.7 – The back button on the Project tab in ArcGIS Pro

With that, you have configured the basic editing options for ArcGIS Pro. However, there are some others that you need to set up as well.

Changing snapping settings

One of the options you'll need to set up is snapping. Snapping helps ensure that the features you edit or create are connected to other features. This helps ensure data integrity by reducing the chance of having free-floating features or introducing gaps in your data. Gaps can also be referred to as slivers. Follow these steps to configure your snapping tolerances:

1. Click on the **Edit** tab in the ribbon.
2. Click on the drop-down arrow located below **Snapping** and select **Snapping Settings…**, as shown in the following screenshot. The **Editor Settings** window should open:

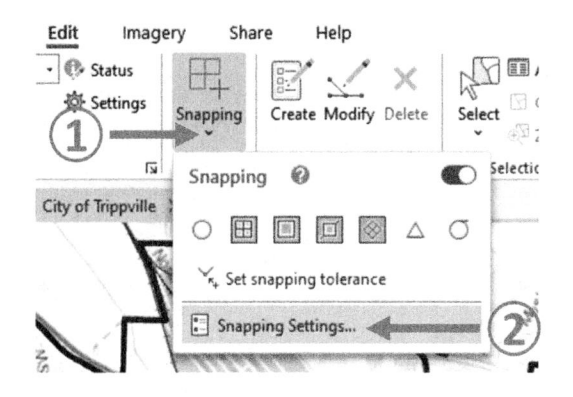

Figure 4.8 – Accessing Snapping Settings… in ArcGIS Pro

3. Ensure **Snapping** is selected in the left panel of the window. For **XY Tolerance**, set the value to `10` and change the units from **Pixels** to **Map units**. To do this, simply click on the drop-down arrow located to the right of **Pixels** and select **Map units** from the list that appears.

> **Tip**
> If you are working with data that contains Z coordinates or elevation information, you will also need to enable Z snap and set a Z tolerance, even when editing data in a 2D map.

4. Set **Snap tip color** to `red` by clicking on the drop-down arrow located to the right of the color block. You can choose any shade of red you like.
5. Verify that your **Snapping** settings window looks as follows and click **OK**:

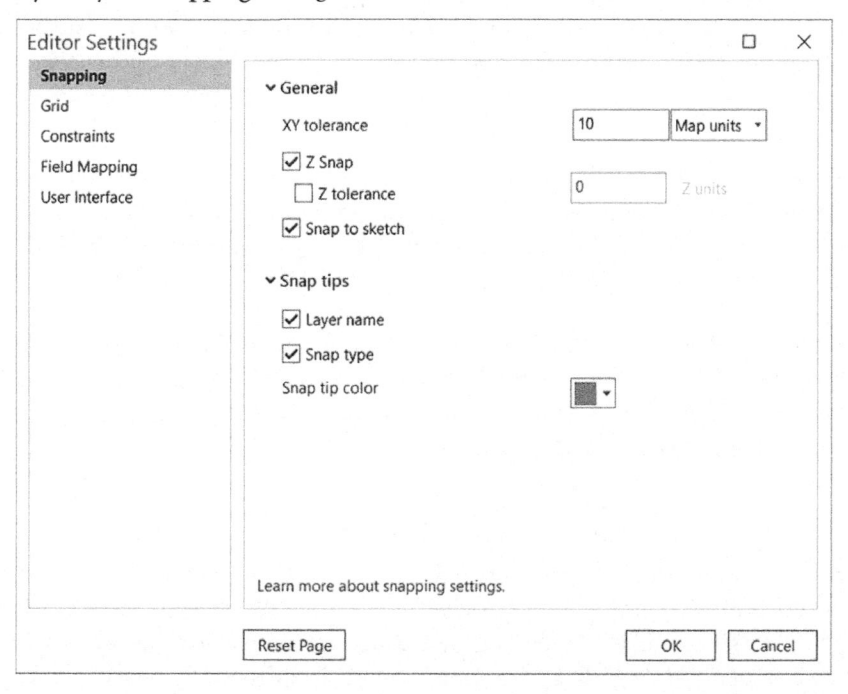

Figure 4.9 – Setting your snapping tolerance and other settings

You will notice that the **Editor Settings** window allows you to configure other settings that impact your editing experience. For this recipe, you will not make any more adjustments to the default settings. Feel free to examine other settings if you desire.

1. Click **OK** to close the **Editor Settings** window.
2. Save your project by clicking on the **Save Project** button located on the **Quick Access** toolbar in the top-left corner of the ArcGIS Pro interface.

3. If you are not continuing to the next recipe, close ArcGIS Pro. If you are continuing, you can leave the application open.

With that, you have configured several key options and settings that will help ensure edits to your GIS data will be performed effectively and correctly.

How it works...

In this recipe, you configured several settings for editing GIS data. You did this in two different locations:

- You started by changing **Editing Options**. To access this area, you clicked on the **Project** tab in the ribbon. Next, you selected **Options** from the left panel in the **Project** pane to open the **Options** window. In the **Options** window, you selected **Units** and set those to match the units used by your data. In this case, it was **US Survey Feet**. Next, you selected **Editing** in the left panel. Under the **Editing** options, you verified that several options were enabled, including double-clicking as a shortcut to finish, showing feature symbology in the sketch, and showing a dialog to confirm save edits.

- Then, you returned to the map and selected the **Edit** tab in the ribbon so that you could configure snapping options. On the **Edit** tab, you clicked on the drop-down arrow located below the **Snapping** button and selected **Snapping Settings…** to open the **Editor Settings** window. On that window, you set the snapping tolerance, including the distance and units.

It is important to remember that several of these settings are dependent on your location and how your data is configured. Next, you'll learn how to edit existing features.

Reshaping an existing feature

Now that you have the various editing options set, it is time to start editing data. Unlike the older ArcMap application that preceded ArcGIS Pro, there is no need to start and stop editing. You can start editing data at any time by default. You can change this by enabling the **Enable and disable editing** option from the **Edit** tab. Enabling this option will add a new **Edit** button to the **Edit** tab that turns the ability to edit data on and off. In this book, we assume you have not enabled that option and are using the default setting.

In this recipe, you will reshape the city limits boundary to reflect a recent annexation. You will need to ensure the `City Limits` layer is editable, snappable, and selectable. You will then draw a sketch using various construction tools that will represent a change to the city limits boundary.

Getting ready

To complete this recipe, you need to have completed the *Configuring editing options* recipe. You will also need to start with the `EditingExistingData.aprx` project open. If you closed ArcGIS

Pro after completing the previous recipe, follow *Steps 1 to 5* in the previous recipe to open the `EditingExistingData.aprx` project.

How to do it...

You will now use editing tools to make changes to the existing `City Limits` layer to reflect a recent annexation:

1. To begin, ensure the **Map** tab is active in the ribbon, then click on **Bookmarks** and select the `Reshape City Limits` bookmark. This will zoom you into the parcel that is being annexed into the city.

2. Click on the **List by Selection** button located at the top of the **Contents** pane. It should be the third button from the left and includes three polygons: one is blue and the other two are white.

3. Right-click on the `City Limits` layer and select **Make this the only selectable layer**, as illustrated in the following screenshot:

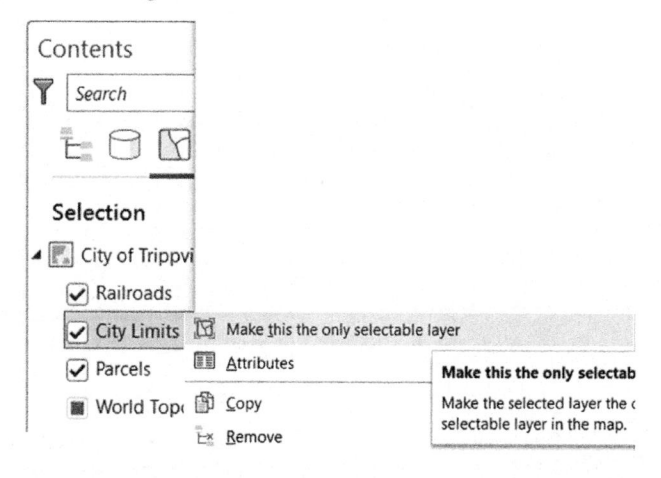

Figure 4.10 – Making the City Limits layer the only selectable layer

4. Click on the **List by Snapping** button in the **Contents** pane. It looks like four squares with a blue plus sign in the lower-right corner, as shown here:

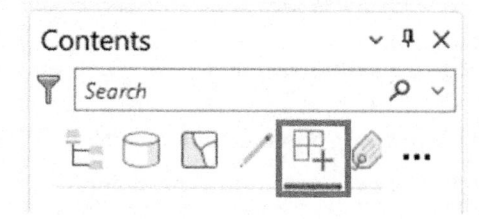

Figure 4.11 – The List by Snapping button in the Contents pane

5. Disable snapping to the `Railroads` layer by clicking on the box to the left of the layer name so that the check mark disappears.

6. Ensure snapping is enabled for the `City Limits` and `Parcels` layers. You know when snapping is enabled when the box next to the layer contains a check mark.

7. Click on the **List by Editing** button in the **Contents** pane. The icon looks like a pencil, as shown here:

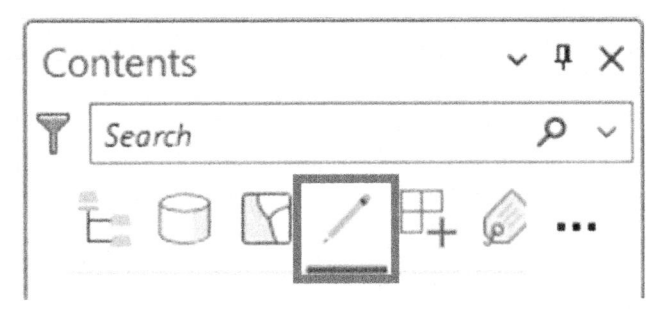

Figure 4.12 – The List by Editing button in the Contents pane

8. Right-click on the `City Limits` layer and select **Make this the only editable layer** in the menu that appears, as shown in the following screenshot:

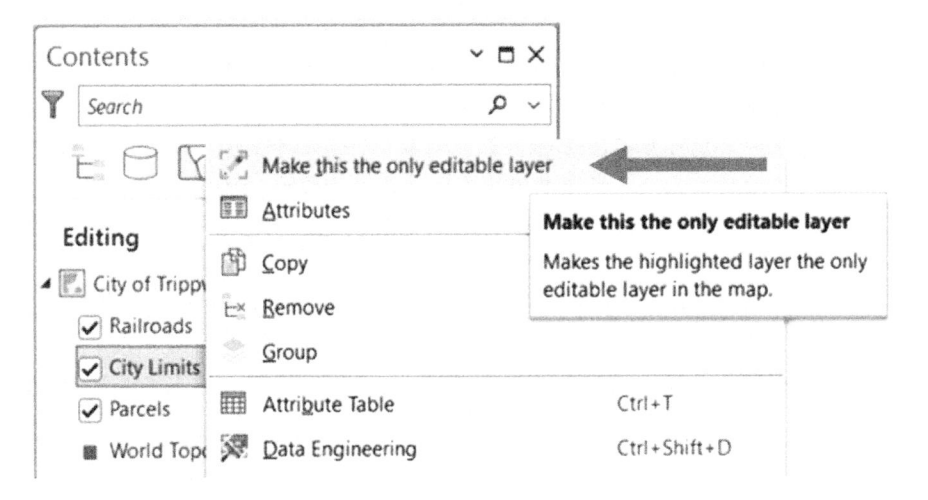

Figure 4.13 – Making the City Limits layer the only editable layer

9. Select the **Edit** tab in the ribbon to access the editing tools in ArcGIS Pro.

10. Click on the drop-down arrow located below the **Snapping** button on the **Edit** tab in the **Snapping** group.

11. Enable snapping on the `End`, `Vertex`, and `Edge` areas, and disable all other snapping options, as indicated in the following screenshot. To enable or disable a snapping option, simply click on it. If the icon is highlighted in blue, it is enabled. If it is not highlighted, it is disabled:

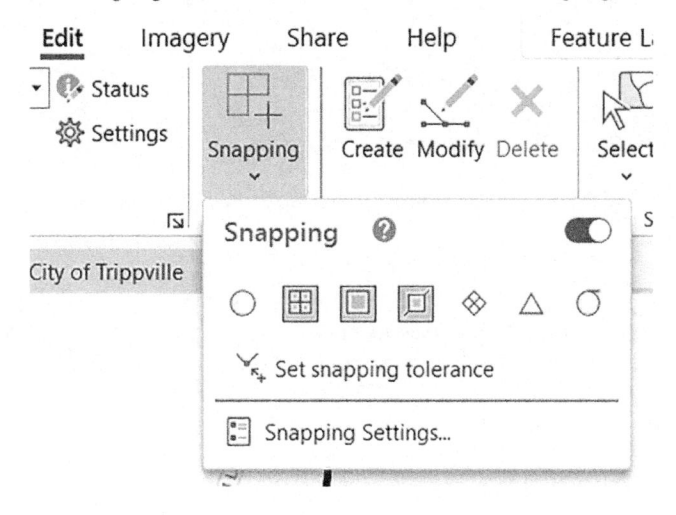

Figure 4.14 – Setting snapping locations

Selecting and reshaping a feature

Now that you have configured your data and map for editing, it is time for you to select the feature you need to edit:

1. Click on the **Select** tool on the **Edit** tab.
2. Click inside the city limits boundary to select it, as shown here:

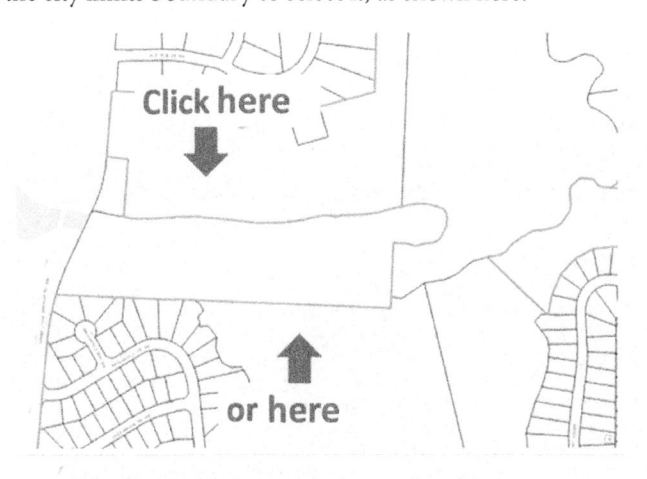

Figure 4.15 – Selecting the City Limits boundary using the Select tool

3. Click on the **Reshape** tool in the **Tools** group on the **Edit** tab:

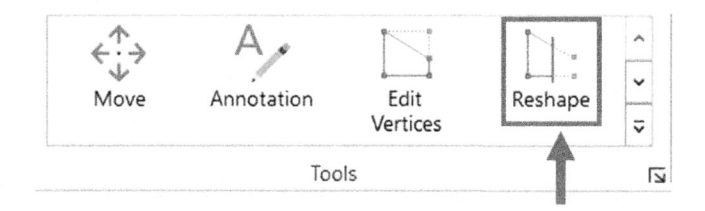

Figure 4.16 – Selecting the Reshape tool on the Edit tab

> **Tip**
> If these tools are grayed out, this means one of two things. First, it might indicate you do not have one or more layers set as editable or the map contains no editable layers. Secondly, it could indicate you have enabled the **Editing Session** option for **Enable and disable editing** from the **Edit** tab. You will see an **Edit** button on the **Edit** tab if you have done this. To allow editing if you see this button, click on it so that it is highlighted in blue. This indicates you have enabled the ability to edit data.

4. Click on the south-western corner of the parcel being annexed into the city, as shown in the following diagram:

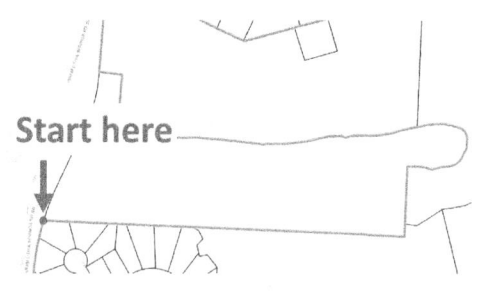

Figure 4.17 – The starting location to reshape the city limits boundary

5. Select the **Trace** tool from the **Edit** toolbar that appears at the bottom of the map view area, as shown here:

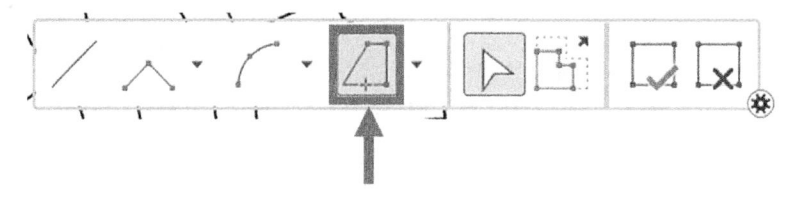

Figure 4.18 – Selecting the Trace tool from the Edit toolbar

Reshaping an existing feature   135

6. Click on the line that forms the western boundary of the parcel being annexed into the city. Then, move your mouse pointer along the boundary of the parcel, as shown in the following diagram. Double-click when you reach the intersection of the parcel's northern boundary with the existing city limits:

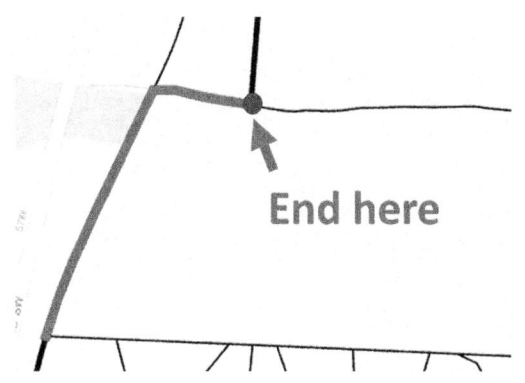

Figure 4.19 – Ending location for the Trace tool

7. Click on the **Clear** button located in the selection group in the **Edit** tab to deselect the city limits boundary you were just editing.

When you double-clicked to end your sketch, your city limits boundary should have been reshaped to include the newly annexed parcel, as shown in the following diagram. Using the **Trace** tool, ensure you followed the boundary of the existing parcel exactly. This means your data is clean and does not have gaps between the city limits boundary and the parcel:

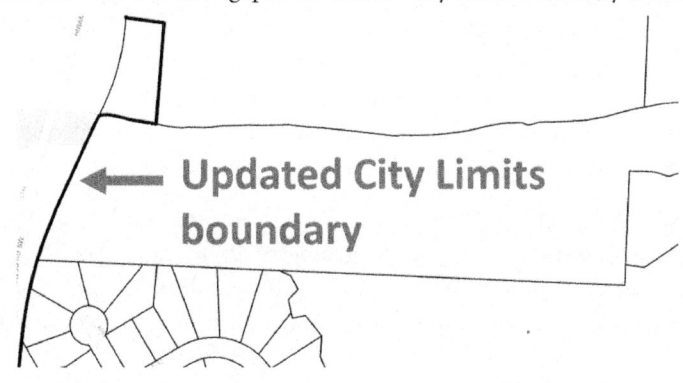

Figure 4.20 – The results of using the Reshape tool to update the city limits boundary

8. Click the **Save Edits** button in the **Manage Edits** group on the **Edit** tab. The icon looks like a floppy disk with a pencil. When you're asked to confirm whether you wish to save your edits, click **Yes**.

> **Tip**
>
> Remember that, by default, ArcGIS Pro does not autosave data edits. You must do that manually. Also, saving your project does not save your data edits by default. You can enable autosave for edits as well as have your edits saved when you save the project by going to the **Options** area of the **Project** pane.
>
> If you don't enable the autosave feature in ArcGIS Pro, you will want to make sure you save often; otherwise, you will lose any edits you have not saved if ArcGIS Pro or your computer crashes.

9. Close the **Modify Features** pane.
10. Save your project. If you are going to continue to the next recipe, keep ArcGIS Pro open. Otherwise, close ArcGIS Pro.

You have just edited an existing feature – the city limits boundary – using ArcGIS Pro.

How it works...

In this recipe, you reshaped the city limits boundary for the City of Trippville so that it included a newly annexed parcel. The first thing you needed to do to perform this edit was set which layer you wanted to have as selectable and editable. You did this through the lists in the **Contents** tab.

When editing data, remember that data can be selectable but not editable, editable but not selectable, both, or neither. If you need to edit a feature by first selecting it from the map, as you did in this recipe, you need the layer to be both selectable and editable. If you select the feature for editing using an attribute query, the layer does not need to be selectable. Setting a layer as selectable in the **Contents** pane only applies when performing an interactive selection in the map.

Using the **Trace** tool allowed you to edit the city limits boundary so that it followed the parcel boundary exactly. The **Trace** tool automatically copies the geometry of the feature you are tracing to the new shape so that it is an exact duplicate. This ensures your data is clean and accurate. It reduces errors such as overlaps and gaps, which can impact the results of analysis and measurements.

Splitting a line feature

Polygons are not the only type of feature you may need to split. Lines often need to be split as you add other features to the same layer to maintain the correct connections. Lines often form a network that's used to travel across, such as roads, railroads, sidewalks, and trails. Travel is not limited to transportation. Utilities also use linear features to move water, sewer, electricity, and communications from one point to another. Other systems, such as emergency dispatch systems, rely on linear data from GIS as well. So, we must maintain these layers correctly.

In this recipe, you will draw new sewer lines based on a scanned and georeferenced plot. Once you've drawn the new sewer lines, you will split them where they connect to manholes, as shown on the plot, so that you can get accurate measurements of each section of the sewer line between the manholes.

Getting ready

Before you begin this recipe, you will need to complete the first recipe in this chapter, *Configuring editing options*. It is also recommended that you complete all the recipes in *Chapter 1* before beginning this recipe to ensure you have a foundational understanding of ArcGIS Pro and its associated terminology.

How to do it...

Follow these steps to create some new sewer lines and split them at the appropriate locations:

1. If you closed ArcGIS Pro at the end of the last recipe, then launch the application and open the `EditingExistingData.aprx` project again. It should be in your list of recently opened projects.
2. To add the scanned plans showing the location of the new sewer pipes, you will need to add a folder connection to your project. In the **Catalog** pane, right-click on the `Folders` folder and select **Add Folder Connection**, as shown in the following screenshot:

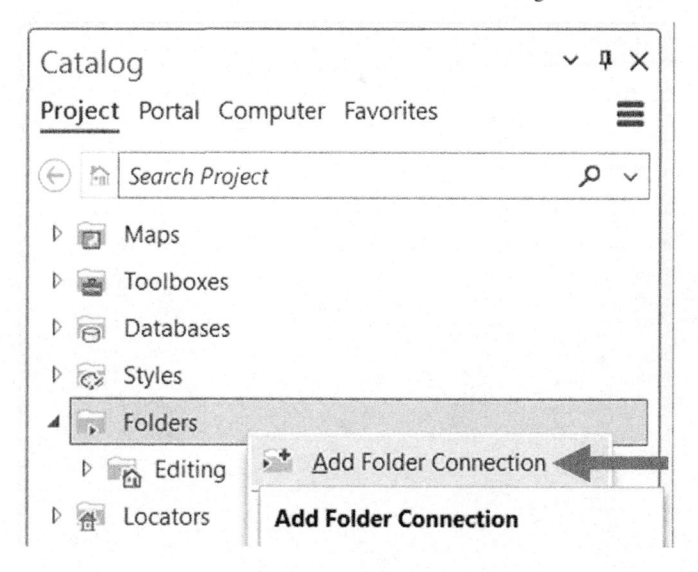

Figure 4.21 – Adding a new folder connection

3. In the left panel of the **Add Folder Connection** window, expand the **This PC** option located under **Computer**.

4. In the right panel, scroll down and double-click on Local Disk (C:)

5. Scroll down and double-click on the Student folder.

6. Scroll down, select the ArcGISPro3Cookbook folder, and click **OK**.

> **Tip**
> **Do not double-click on the folder name**. Double-clicking on the folder will open it instead of selecting it for a connection. If you do double-click on it, simply click on the **back** button located near the top of the window. It is also a horizontal arrow in a circle. It will take you back one folder and allow you to select the correct folder.

The new folder connection should appear in your list in the **Catalog** pane. You should now have at least two folder connections: Editing and ArcGISPro3Cookbook.

Adding new layers to the map and changing symbology

Now, it is time to use the new folder connection to add a new layer to the map you have open. Then, you will adjust the symbology for a layer so the layers are clearer:

1. Expand the ArcGISPro3Cookbook folder connection you just added so that you can see its contents.

2. Next, expand the Chapter4 folder so that you can see its contents.

3. Right-click on the Forrest Park Subdivision.jpg file and select **Add to Current Map**. If you're asked to calculate statistics for Forrest Park Subdivision.jpg, click **Yes**. The file should now appear as a new layer in the **Contents** pane.

4. In the **Contents** pane, right-click on the Forrest Park Subdivision.jpg layer you just added and select **Zoom To Layer** from the menu that appears:

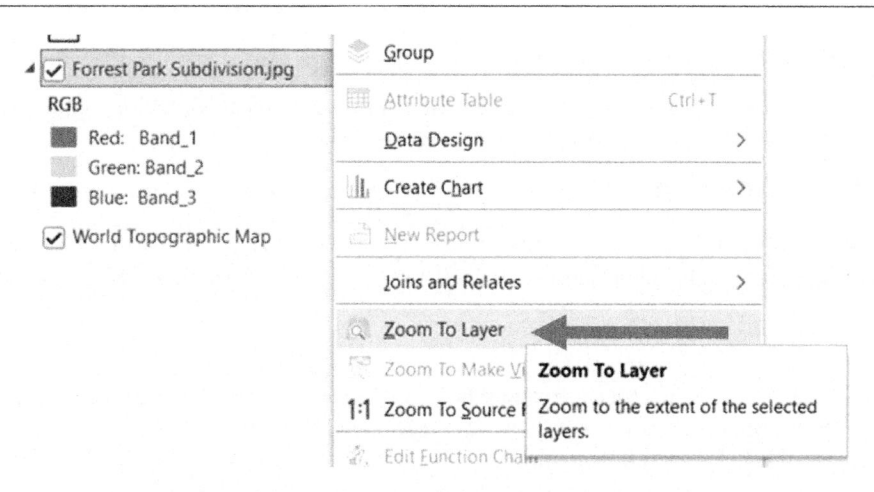

Figure 4.22 – Zooming into the selected layer

Your map will be zoomed into the area containing the subdivision that contains the new sewer lines you will need to digitize. However, you cannot see the layer you added because it is hidden by the `Parcels` layer. You need to adjust its symbology settings so that you can see the underlying layer you added.

5. Select the `Parcels` layer in the **Contents** pane. Then, select the **Feature Layer** tab that appears in the ribbon.
6. Click on the **Symbology** button on the **Feature Layer** tab to open the **Symbology** pane.
7. Click on the small square symbology patch located to the right of **Symbol**, as shown in the following screenshot:

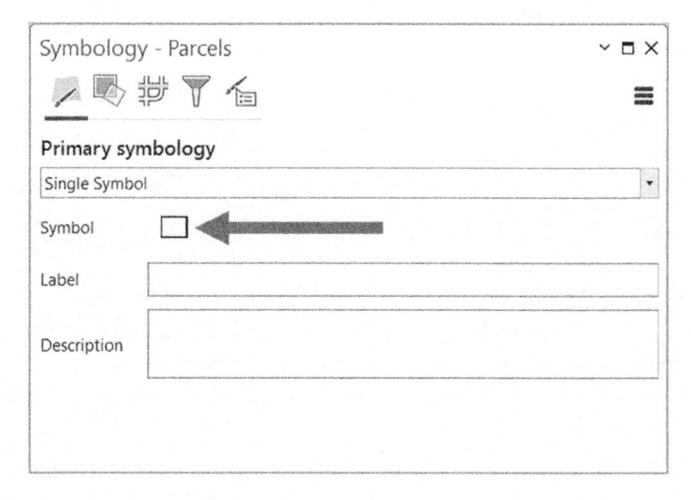

Figure 4.23 – Selecting the Parcels symbol patch to change

8. Select the **Properties** tab in the **Symbology** pane.

9. Click on the drop-down arrow for the **Color** option and select **No Color** to make the fill for the `Parcels` layer transparent. Then, click **Apply**.

10. Close the **Symbology** pane.

You should now see the `Forrest Park Subdivision.jpg` layer in the map beneath the `Parcels` layer. Setting the fill to **No Color** for the `Parcels` layer allows you to see what's underneath.

> **Tip**
> It is helpful to close panes such as the **Symbology** pane when you've finished using them. This limits the number of panes or windows you have open and the amount of screen space they take up. It also helps keep them from becoming lost in a stack of panes, which often happens if you don't manage the number of panes you have open. Another way to manage panes is to undock them and place them on a secondary monitor. This frees up screen space on your primary display while still making items easy to locate. It is recommended that you keep the **Contents** and **Catalog** panes open at all times.

Adding a new layer using a layer file

Next, you will draw the new sewer lines by tracing the ones shown on the new plot. Before you can draw the new sewer lines, however, you will need to add the sewer line layer:

1. Activate the **Map** tab in the ribbon. Then, select the **Add Data** button.

2. Expand the **Folders** option located beneath **Project** in the left panel of the **Add Data** window.

3. Select the `ArcGISPro3Cookbook` connection so that its contents appear in the right panel of the window.

4. Double-click on the `Chapter4` folder in the right panel of the **Add Data** window.

5. Select the `SewerLines.lyrx` layer file and click **OK**. The `Sewer Lines` layer should appear in the **Contents** pane at the top of the layer list as a thick green line.

> **Information**
> Layer files created with ArcGIS Pro have a `.lyrx` file extension. They allow you to add layers to a map that already have their properties predefined. This includes properties such as symbology, labels, scale ranges, definition queries, and others. Layer files allow you to establish consistent display settings for layers so that they look and behave the same way in multiple maps.

Creating new line features

Now that you have made the subdivision plot visible and added the `Sewer Lines` layer to the map, you are ready to start adding the new sewer line:

1. Activate the **Edit** tab in the ribbon and select the **Create** button. This should open the **Create Features** pane.
2. On the **Edit** tab, verify that **Snapping** is enabled. The **Snapping** button should be highlighted in blue if it is, as shown in the following screenshot:

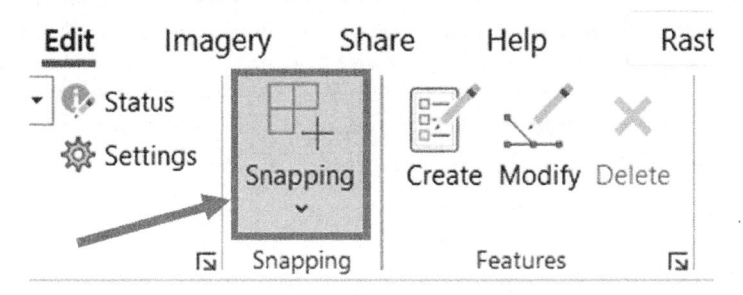

Figure 4.24 – Enabling Snapping

3. Click on the drop-down arrow located below the **Snapping** button and set the snapping location to **End**. Disable all others. Like enabling and disabling the snapping functionality, those locations that are highlighted in blue are enabled, as shown in the following screenshot:

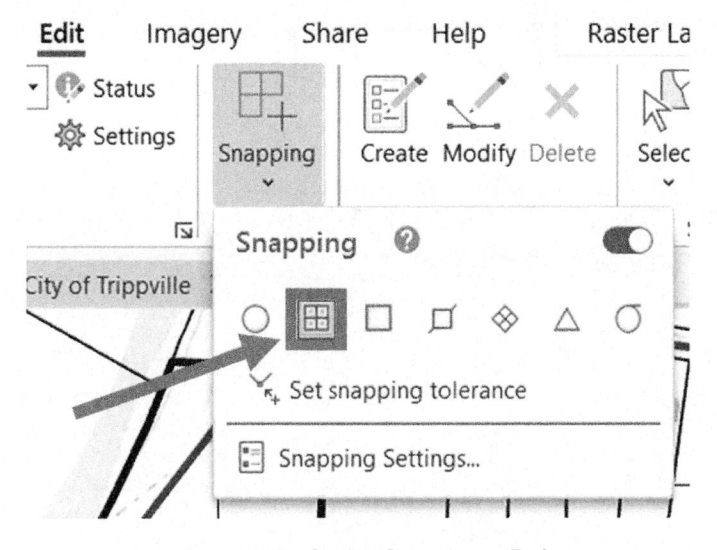

Figure 4.25 – Setting Snapping to End

4. Active the **Map** tab in the ribbon, then click on the **Bookmarks** button.

5. Select the `Create New Sewer Lines` bookmark to zoom your map into the area where you will be creating the new lines.

6. Activate the **Edit** tab in the ribbon.

7. In the **Create Features** pane, select the **Sewer Lines** feature template. The template is located below the layer of the same name and has a green line to the left of it.

8. Move your mouse pointer to the end of the existing sewer line located on the north-eastern side of the new subdivision. When you see the snapping indicator appear, click at the end of the existing line to start creating the new sewer line, as shown in the following diagram:

Figure 4.26 – Selecting the starting point for the new sewer line

9. Continue drawing the new sewer line by clicking on the location of each manhole, as shown in the following diagram. To finish drawing the new sewer line, you can either double-click on the last point or single-click and press *F2*:

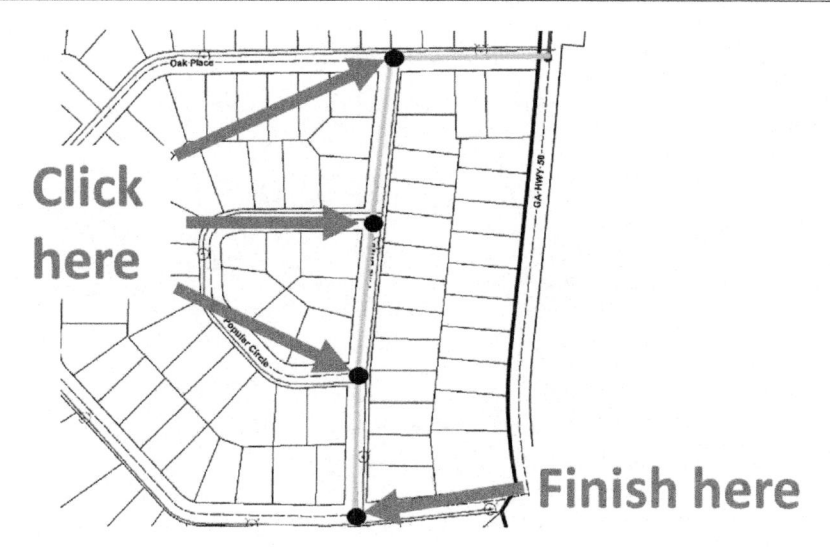

Figure 4.27 – Digitizing new sewer lines

With that, you have successfully drawn the new sewer line. Next, you need to split it at the manhole locations. This will allow you to determine the pipe distance between manholes and provide a more accurate data model of the pipes. First, though, save your edits to ensure you do not lose the work you have completed so far.

1. Click on the **Save** button located in the **Manage Edits** group on the **Edit** tab in the ribbon. Click **Yes** when you're asked to confirm whether you wish to save your edits.

> **Tip**
> Remember that saving your project does not necessarily save edits to your data. Saving your project only saves changes to the project, such as the addition of new layers, the creation of a new map, the addition of a layout, and so on. Data edits are saved separately unless you have adjusted the **Editing** options so that saving the project also saves data edits.

2. Close the **Create Features** pane.

Splitting line features

Now that you have created the new sewer lines, you need to split them where the manholes are located so that the new lines better represent real-world conditions:

1. On the **Edit** tab in the ribbon, click on the drop-down arrow located below the **Snapping** button on the **Edit** tab in the ribbon.

Editing Existing Spatial Features

2. Enable **Vertex** snapping and disable **End** snapping, as shown here:

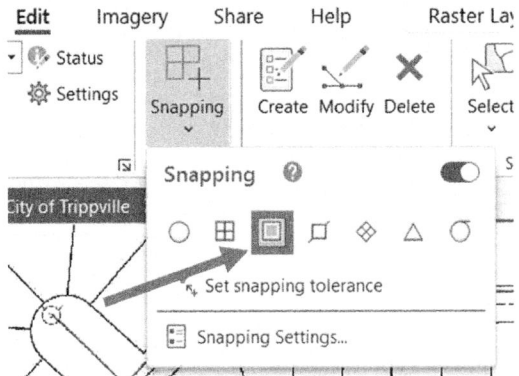

Figure 4.28 – Setting Snapping to Vertex

3. Ensure the new sewer line you drew is still selected. If it isn't, click on the **Select** button and then click on the sewer line you drew.
4. Select the **Split** tool in the **Tools** group on the **Edit** tab. You may need to click on the down arrows located on the right-hand side of the **Tools** group to find the **Split** tool. The **Modify Features** tab should open on the right-hand side of the interface.
5. Click on the sewer line of each manhole, as shown in the following diagram, to split the line at each manhole:

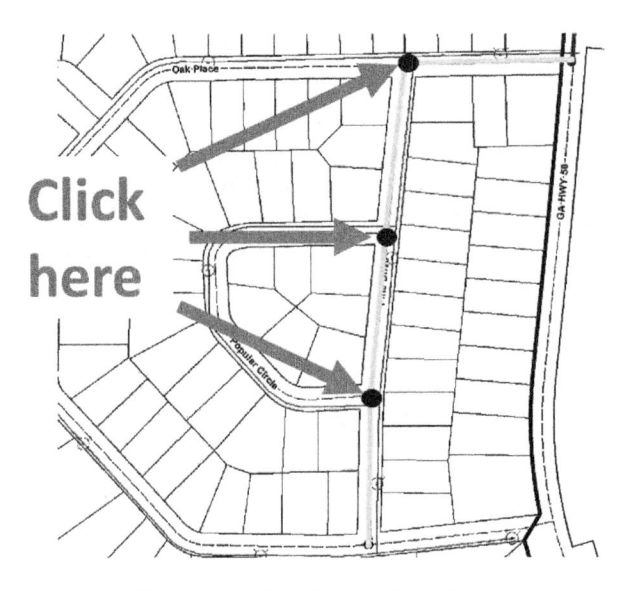

Figure 4.29 – Split line locations shown

In the **Modify Features** pane, you should now see four features listed; you saw only one before using the **Split Line** tool. This indicates that you created three new features using the tool. Now, you will verify that new features were created:

1. Click on the **Clear** button located in the **Selection** group on the **Edit** tab in the ribbon to deselect the sewer lines.
2. Activate the **Select** tool in the **Selection** group on the **Edit** tab in the ribbon.
3. Click on the southernmost sewer line segment to select it.
4. If you split the sewer line correctly, only the segment of the line from the last manhole to the manhole located at the intersection of **Popular Circle** and **Pine Drive** should be selected, as indicated in the following diagram:

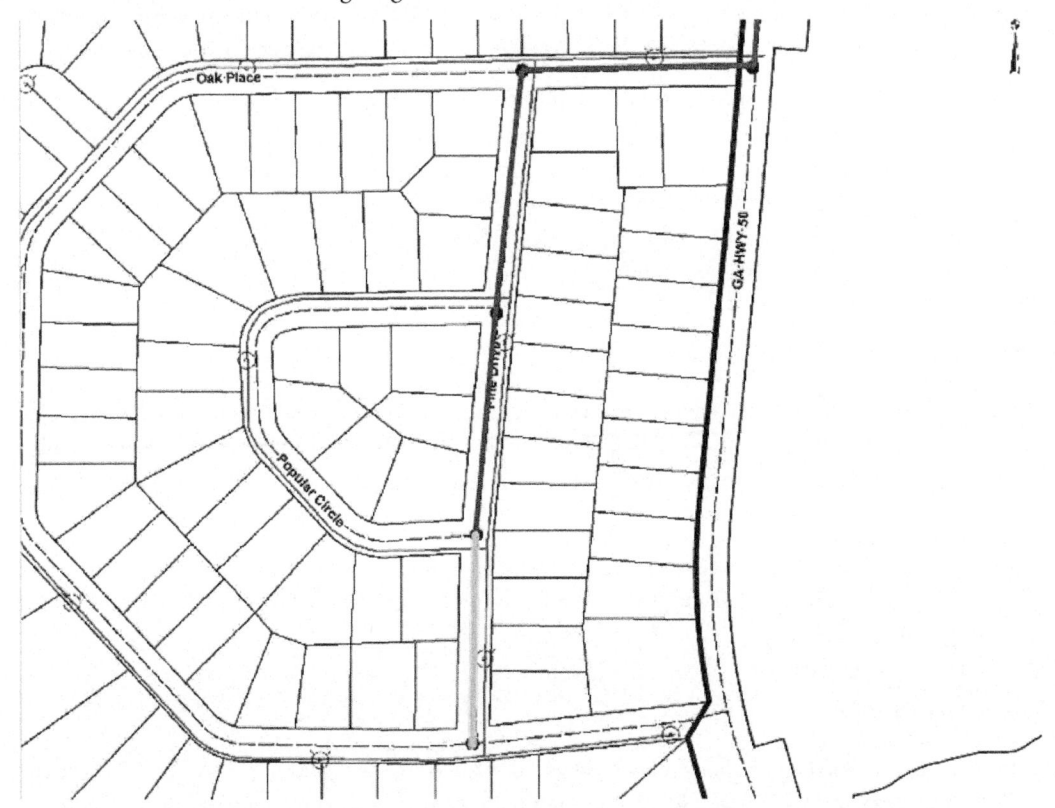

Figure 4.30 – The southern sewer line selected to verify the results of the Split tool

5. On the **Edit** tab, click the **Save** button. When you're asked if you're sure you want to save, click **Yes**.
6. Close the **Modify Features** pane.
7. Save your project using the **Save** button on the **Quick Access** toolbar. If you are not continuing to the next recipe, close ArcGIS Pro; otherwise, you can leave it open.

With that, you have successfully created and split new sewer lines based on a scanned plot.

How it works...

In this recipe, you created new sewer lines located in a new subdivision that is being built in the City of Trippville. You based these on a scanned plot of the new subdivision. Creating these new sewer lines required you to perform several steps so that you added them correctly.

First, you needed to configure several editing options. To do this, you went to the **Project** tab and opened the **Options** window. In the **Options** window, you set the units you needed to use. In this project, that was **US Survey Feet**. Then, you covered the specific settings for editing data. You verified that several of the defaults were still enabled. From there, you closed the **Options** window and returned to the map view. Then, you selected the **Editing** tab and opened the **Editor Settings** window using the drop-down arrow below the **Snapping** button. Here, you set your snapping tolerance and snap tip color.

With your editing options configured, you added two layers to your map. This included the scanned subdivision plot for the Forrest Park subdivision and the sewer lines. From there, you traced the new sewer lines using the **Sewer Lines** feature template based on the data you saw in the scanned plot. Once you drew the new sewer lines, you used the **Split** tool to break the lines at the locations of the manholes. This will allow you to get measurements between manholes, as well as associate other information with those specific segments.

Merging features

You now know how to split and reshape features, but not all edits will be limited to these. Sometimes, you need to combine two or more existing features. This might be done to simplify a layer for easier analysis or to reflect a change to the real-world features, such as someone buying two adjacent parcels to combine them into a single parcel.

In this recipe, you have found a road centerline, which is split into many unnecessary segments. This is causing issues when you're trying to calculate the total length of each road. You need to merge these segments.

Getting ready

This recipe is not based on previous recipes, so you don't need to have completed any previous recipes before starting this one. However, it is recommended that you complete the recipes from *Chapter 1* to ensure you have the basic skills needed to complete this recipe. This recipe can also be completed with any license level of ArcGIS Pro.

How to do it...

Follow these steps to merge multiple individual features into a single one:

1. If you closed ArcGIS Pro when you completed the last recipe, launch ArcGIS Pro and open the `EditingExistingData.aprx` project. This should be in your **Recent Projects** list. The project should open with the `City of Trippville` map that you used in previous recipes in this chapter.

2. Active the **Map** tab in the ribbon and click on the **Bookmarks** button.

3. Select the `Merge Street Centerlines` bookmark to zoom the map into the location containing the street centerline segments that need to be combined, as illustrated in the following screenshot:

Figure 4.31 – Zoomed map showing the location of the centerlines to be edited

Now that you've zoomed into the correct area, you should notice that the street centerline data is not included in the map. Next, you will add that layer so that you can edit it.

4. In the **Catalog** pane, expand the Databases folder, if necessary, by clicking on the small arrowhead next to the folder name. You should see two file geodatabases connected in the project – `Editing.gdb` and `Trippville_GIS.gdb`.

5. Expand the `Trippville_GIS.gdb` file geodatabase by clicking on the small arrowhead located to the left of its name.
6. Expand the `Base` feature dataset using the same method.

> **Information**
> A feature dataset is an item inside a geodatabase that can be used to group related feature classes for better organization and to make use of advanced geodatabase functionality such as topologies. All feature classes in a feature dataset share the same coordinate system.

7. Scroll down to locate the `Street_Centerlines` feature class and right-click on it. Then, select **Add to Current Map** from the menu that appears. A `Street_Centerlines` layer should now appear in the `Contents` pane.

Setting layers as selectable and editable

With the required layers added to the map, you need to set which ones are selectable and editable to make the editing process easier:

1. In the **Contents** pane, click on the **List by Selection** button. This icon for this button looks like three polygons with one of them having a blue fill.
2. Right-click on the `Street_Centerlines` layer and select **Make this the only selectable layer** from the menu that appears, as illustrated in the following screenshot:

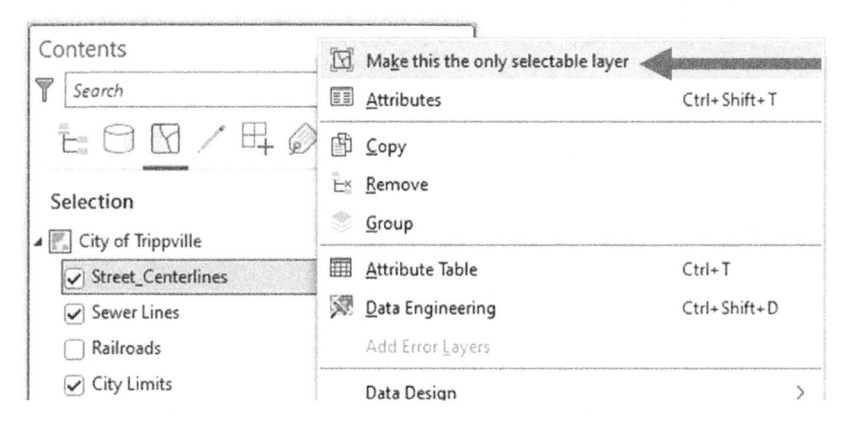

Figure 4.32 – Making Street_Centerlines the only selectable layer

3. In the **Contents** pane, click on the **List by Editing** button. The icon for this button looks like a pencil.
4. Once again, right-click on the `Street_Centerlines` layer and select **Make this the only editable layer**.

> **Note**
> Before you begin editing, you will need to turn off the World Topographic basemap or whatever basemap you are using. This will help reduce clutter and make editing the data easier.

5. In the **Contents** pane, click on the **List by Drawing Order** button. The icon for this button looks like a directory tree that you would see in Windows Explorer.
6. Turn off the Work Topographic Map layer by clicking on the box to the left so that it is empty. This layer is located at the very bottom of the layer list.

> **Tip**
> Depending on your settings, you might see a different basemap, such as Imagery, Streets, or Navigation, including custom basemaps from your organization. ArcGIS Pro provides access to many different basemaps published by ArcGIS Online. Turn off whatever basemap you might be using.

Selecting and merging features

Now that you have set which layers are selectable and which are editable, you are ready to merge features. Merging features simplifies your data, making it easier to maintain and analyze:

1. Active the **Edit** tab in the ribbon and click on the **Select** tool.
2. Click on the Street_Centerline segment, as shown in the following screenshot, using the **Select** tool:

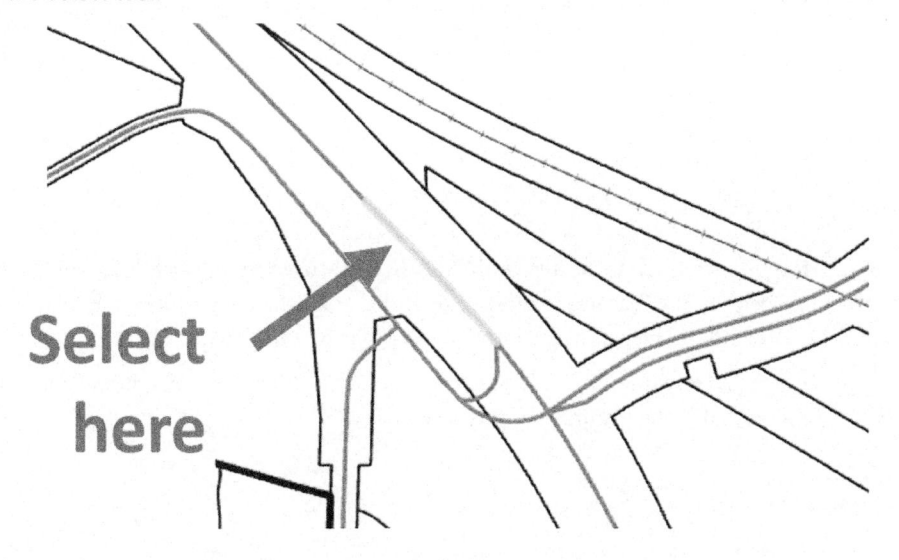

Figure 4.33 – Selecting the first centerline segment

3. Hold the *Shift* key down and click on the other two street centerline segments, as shown here:

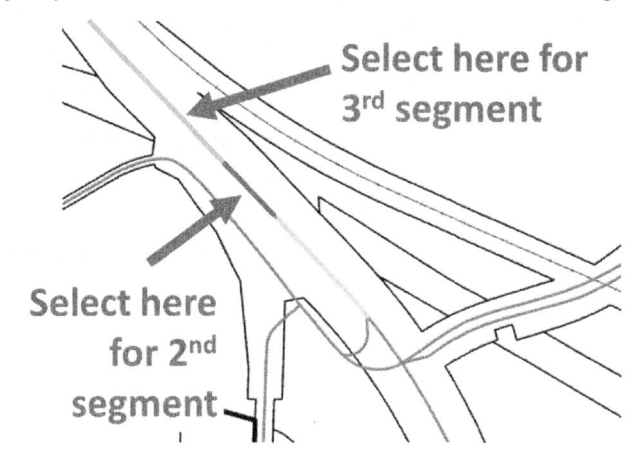

Figure 4.34 – Selecting additional centerline segments

You should now have three segments selected. You can verify this by clicking on the **List by Selection** button in the **Contents** pane. This will tell you how many features in each layer are selected. Now, you will merge the three selected features into a single feature.

4. Active the **Edit** tab in the ribbon and then select the **Merge** tool in the **Tools** group, as shown in the following screenshot. This will open the **Modify Features** pane:

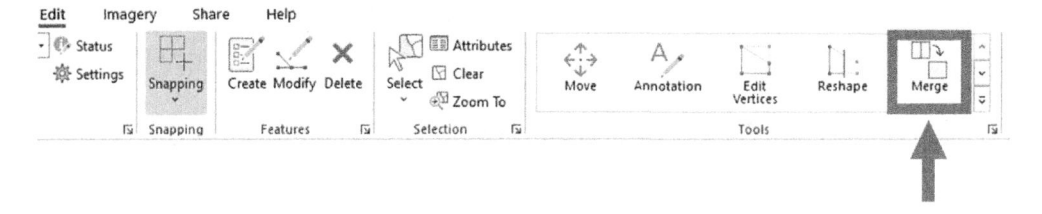

Figure 4.35 – Selecting the Merge tool

5. In the **Merge** tool, which is located in the **Modify Features** pane, verify that you have three segments selected in the **Features to merge** panel. One of the three segments will have **(preserved)** beside it. This will be the parent feature that the other two will be merged into.

6. Verify that the **Layer** parameter is set to `Street_Centerlines`. This should automatically be set because that is the feature you have selected.

7. Verify that your **Merge** tool looks as follows and click the **Merge** button at the bottom of the **Modify Features** pane to run the tool:

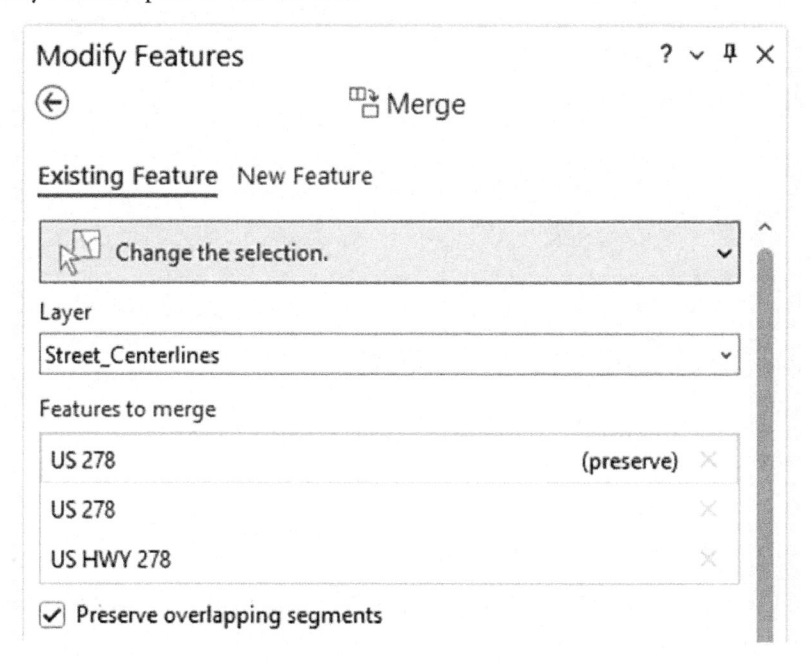

Figure 4.36 – The Merge tool with completed parameters

The three features you originally selected have now been combined into a single feature. This should be indicated in the **Merge** tool, with only one feature being shown in the **Features To Merge** panel of the tool.

Verifying your results

Now that you have used the **Merge** tool, you will verify the results just to make sure it worked as expected:

1. Close the **Modify Features** pane.
2. Click on the **List by Selection** button in the **Contents** pane.

3. Verify how many features are selected in the `Street_Centerlines` layer. You should see one to the right of the layer's name, as shown here:

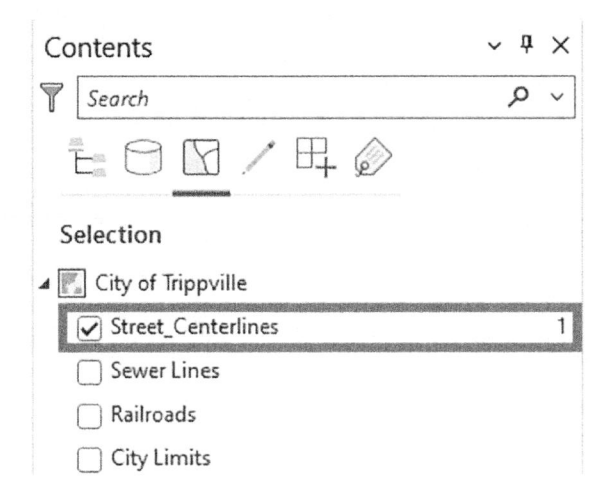

Figure 4.37 – List by Selection showing merge results

4. Click on the **Save** button in the **Manage Edits** group on the **Edit** tab to save the edit you just made. Click **Yes** to confirm that your edit is saved when asked.
5. You can save your project by clicking on the **Project** tab in the ribbon and then clicking on the **Save** option that is shown.
6. If you are not continuing to the next recipe, close ArcGIS Pro. If you are continuing to the next recipe, leave ArcGIS Pro and the project open.

With that, you have successfully combined three separate features into a single one.

How it works...

In this recipe, you merged three center-line segments into a single segment. This simplifies the layer by reducing the total number of features, which makes it easier to analyze and maintain. The **Merge** tool allows you to replace existing features or create a new feature while leaving the existing features in place. This option can be accessed by clicking on one of the two tabs located at the top of the tool – **Existing Feature** or **New Feature**. The **Existing Feature** option replaces the selected features with a new feature that has the attributes of one of the selected features. The **New Feature** option creates a brand-new feature and leaves the selected features in place as well. You can merge any type of feature if the features are in the same layer.

When using the **Existing Feature** option, as you did in this recipe, you get to select which selected feature the newly created feature will inherit. You can do this by clicking on the feature in the **Features To Merge** grid. The feature that's listed as **(preserved)** is the one that the new feature will inherit.

Aligning features

ArcGIS Pro can be a very valuable tool for performing analysis. However, the result of any analysis is only as good as the quality of the data used. The adage of garbage in, garbage out certainly applies.

So, our data must be as clean as possible. This means we need to remove unwanted gaps or overlaps within our data. ArcGIS Pro has several tools to help you clean up your data. One of those is the **Align Features** tool. This tool is available for all license levels.

In this recipe, you will locate an area where the city limits are supposed to follow the boundary of two parcels but do not. You will use the **Align Features** tool to fix this error in your data.

Getting ready

You will need to have completed the first recipe in this chapter, *Configuring editing options*, before you can complete this recipe. This recipe works at all licensing levels of ArcGIS Pro.

How to do it...

You will now work through the process required to plot the locations of events listed in a standalone table found in a geodatabase:

1. If you closed ArcGIS Pro when you completed the last recipe, launch ArcGIS Pro and open the `EditingExistingData.aprx` project. This should be in your **Recent Projects** list if you have completed the other recipes in this chapter. The project should open with the `City of Trippville` map that you used in previous recipes in this chapter.

2. In the **Contents** pane, turn off the `Sewer Lines` layer so that it is no longer visible. If you do not see the `Sewer Lines` layer in the **Contents** pane, that is because you haven't completed the *Splitting a line feature* recipe in this chapter and can ignore this task.

3. Active the **Map** tab in the ribbon and click on the **Bookmarks** button. Select the **Align Features** bookmark to zoom the map into the location that contains the `City Limits` and `Parcels` layer's boundaries that do not match.

    You should now see the area where the city limits boundary does not line up with the parcel boundaries. This creates a gap or sliver between the two that should not exist, as shown in the following diagram:

154   Editing Existing Spatial Features

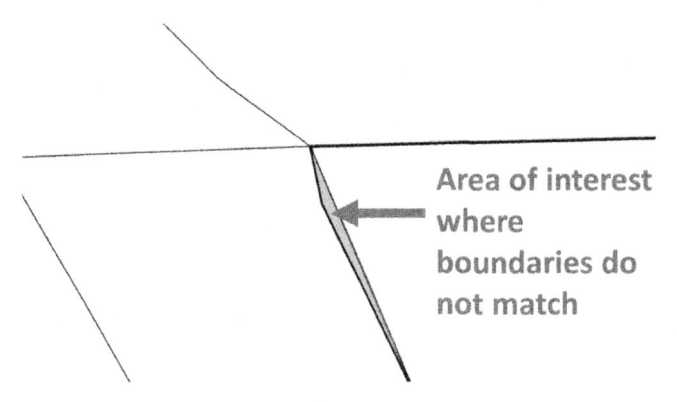

Area of interest where boundaries do not match

Figure 4.38 – Map zoomed in to show where City Limits and Parcels do not match

4. Turn off `City Limits` so that you can see the parcel boundaries by clicking on the checkbox next to the `City Limits` layer. As you can see, there are issues with the alignment of the parcel boundaries as well.

Determining if there's a gap or an overlap

Now that you can see the issue in the data, you need to determine whether there's a gap or an overlap between the features:

1. In the **Contents** pane, click on the **List by Selection** button located near the top of the pane. The icon looks like three polygons, with one of them being blue.
2. Right-click on the `Parcels` layer and select **Make this the only selectable layer** from the menu that appears.
3. Choose the **Select** tool in the **Selection** group on the **Map** tab of the ribbon.
4. Click on the western parcel, as indicated in the following diagram:

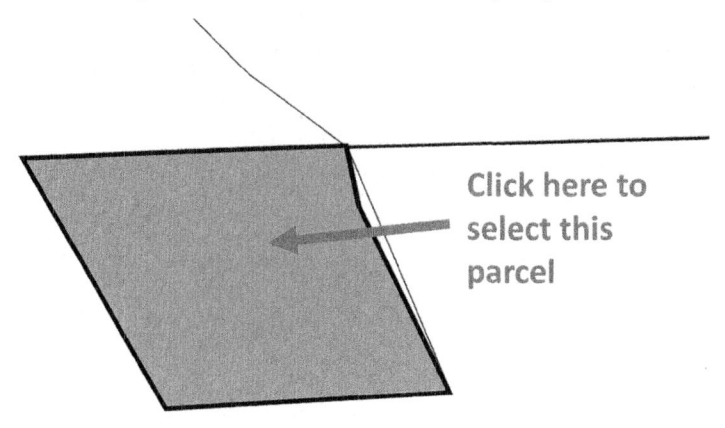

Click here to select this parcel

Figure 4.39 – Selecting the western parcel to determine whether there's a gap or an overlap

5. Review the boundary for the selected parcel; this should be highlighted in blue. Determine if it overlaps the parcel to the east or forms a gap.
6. To verify your determination of a gap or overlap, using the **Select** tool on the **Map** tab, select the parcel to the east, as shown in the following diagram:

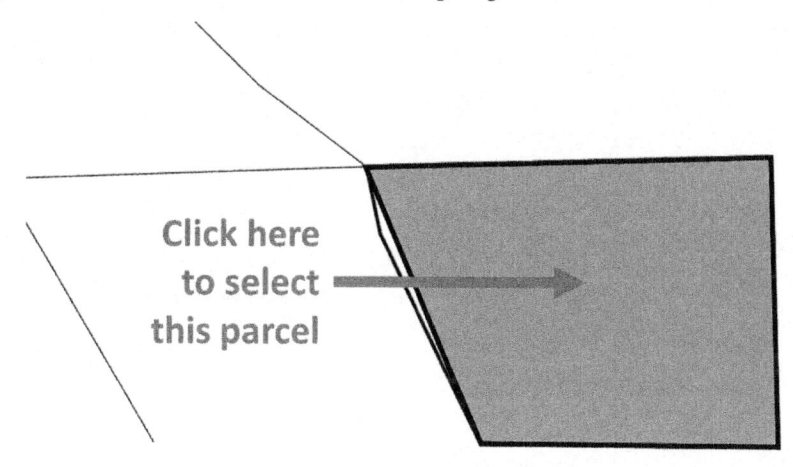

Figure 4.40 – Selecting the eastern parcel to verify whether there's a gap or an overlap

After selecting both parcels, you should see that they overlap with one another. Parcels should not overlap with one another. Landowners tend to get upset if this happens. So, you need to remove this overlap so that your data provides a more realistic representation of the parcel boundaries. The city limits boundary should also follow the common boundary formed between the two parcels, so you will need to ensure it is adjusted as well.

Fixing the overlap and alignment issues

Now, you must fix the issues you have determined using the **Align** tool from the **Edit** tab in the ribbon. As you will see, this provides a quick and efficient method for correcting issues like this in your data:

1. Click on the **Clear** button located in the **Selection** group on the **Map** tab of the ribbon. This will deselect all selected features.
2. Turn the `City Limits` layer back on so that you can see it on the map. To do this, click on the **List by Drawing Order** button at the top of the **Contents** pane. It is the first button on the left. Then, click on the box located to the left of the layer name.
3. Click on the **List by Selection** button at the top of the **Contents** pane.

156　Editing Existing Spatial Features

4. Click on the box next to the `City Limits` layer so that the `Parcels` and `City Limits` layers can both be selected, as shown in the following screenshot:

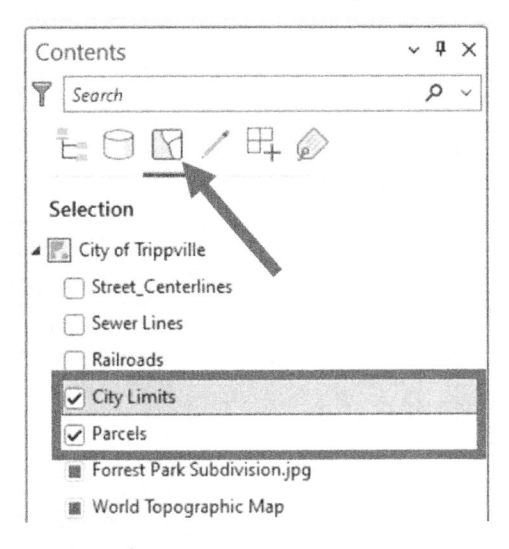

Figure 4.41 – Setting the City Limits and Parcels layers so that they can be selected

5. Click on the **List by Editing** button in the **Contents** pane. The icon resembles a pencil.
6. Ensure the `City Limits` and `Parcels` layers are the only layers that have been set as editable, as shown in the following screenshot:

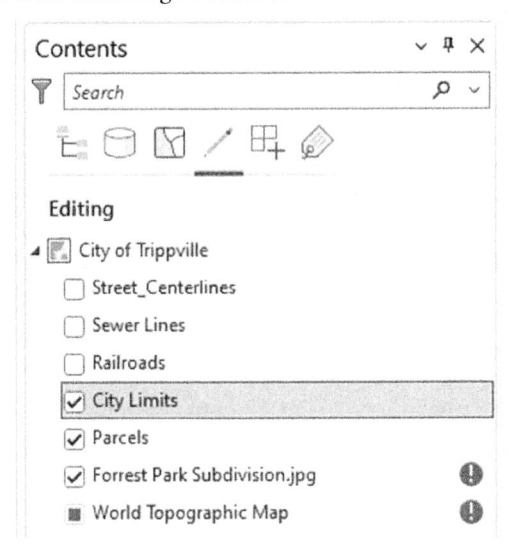

Figure 4.42 – Setting the City Limits and Parcels layers as editable

Aligning features | 157

> **Information**
> You may see red circles containing exclamation points to the right of the two layers. These indicate layers that are not editable. Layers may not be editable for several reasons. They might not be in an editable format, or you might not have permission to edit the layers. In the case of this recipe, the two non-editable layers are stored as raster data, which can't be edited by ArcGIS Pro.

7. Active the **Edit** tab in the ribbon. Then, click on the drop-down arrow located below the **Snapping** button and set **Endpoint**, **Edge**, and **Vertex** as the snapping locations, as shown in the following screenshot:

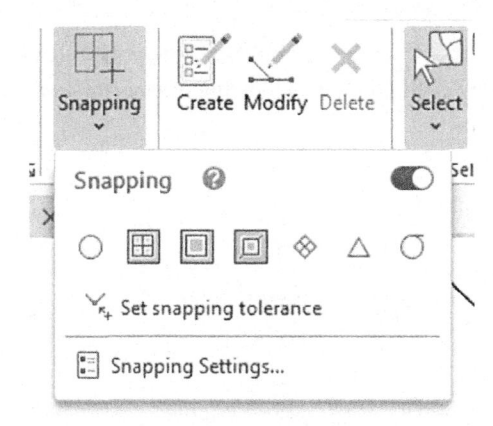

Figure 4.43 – Setting the snapping locations to Endpoint, Vertex, and Edge

8. Click on the **Select** tool on the **Edit** tab. Click on the starting location shown in the following diagram. While holding your mouse button down, drag your mouse to the northeast until you reach the approximate location of the *Finish here* point, as shown in the following diagram. This will create a selection box that will select the two parcels and the city limits:

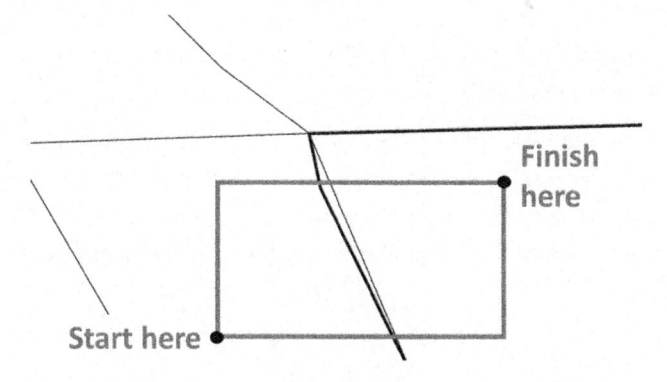

Figure 4.44 – Creating a selection rectangle for the City Limits and Parcels layers

You should see that both parcels and the city limits are selected. If you are not sure, you can go back to the **List by Selection** option in the **Contents** pane to verify this. It will show you how many features you have selected in each layer.

> **Tip**
> When editing, it is recommended to keep the **Contents** pane on the **List by Selection** option as much as possible. This allows you to track how many features are on the layers you have selected. This information can stop you from accidentally editing or deleting features you do not intend to. This is a lesson I have learned the hard way. Remember, you cannot undo an edit once you have saved the edits you have performed.

9. In the **Tools** group on the **Edit** tab, click on the arrowhead with a small line above it, located in the lower-right corner, as shown in the following screenshot, to access additional editing tools:

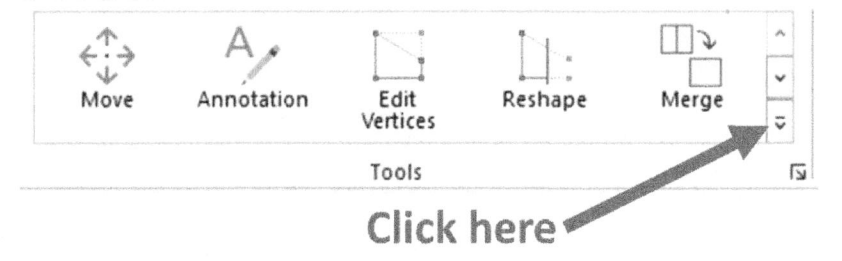

Figure 4.45 – Accessing additional editing tools

10. Select the **Align Features** tool from the list of tools presented. This will open the **Modify Features** pane.
11. Set the **Alignment Tolerance** parameter to 5 and **Units** to `FtUS` using the drop-down arrow.
12. Set the **Method** parameter to `Fit shapes to path`.
13. Set the **Area Side** parameter to `Both` and **End** to `Round` using the radial buttons.

Next, you will need to draw the alignment path. This is a line that represents how the selected features need to be moved so that they match.

14. At the top of the **Align Features** tool, select the **Draw the alignment path** option.
15. In the map area, draw your alignment line, as shown in the following diagram, snapping to the line that forms and contains the vertices for the city limits boundary. To end the alignment path line, double-click on the last point or press the *F2* key after you have clicked on the last point location:

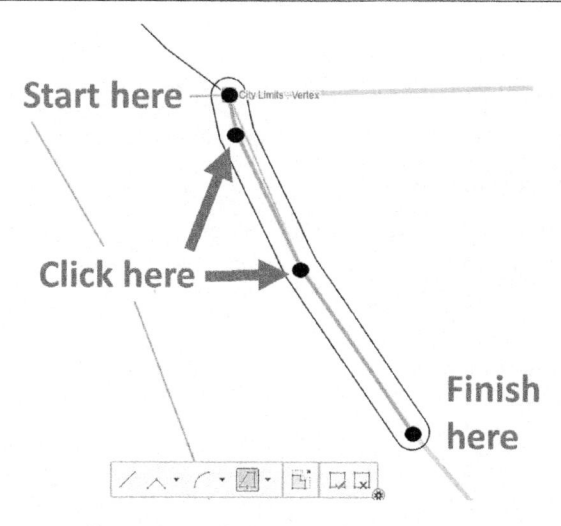

Figure 4.46 – Creating an alignment path

16. At the bottom of the **Align Features** tool, click the **Align** button to run the tool.

    If you traced the alignment path line correctly, all three features (one city limit and two parcel boundaries) should have been edited so that they all now line up properly and there is no longer any overlap, as shown in the following figure:

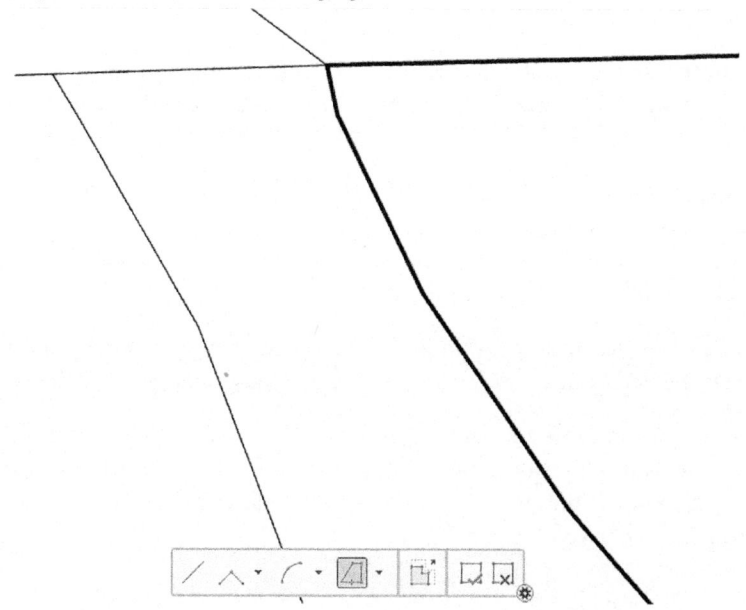

Figure 4.47 – Results of using the Align Features tool

This tool allowed you to correct and edit three features on two different layers at one time. This is much more efficient than editing each one individually.

17. Click the **Clear** button on the **Edit** tab to deselect all the selected features.
18. Close the **Modify Features** pane.
19. Click the **Save** button on the **Edit** tab to save the edits to the `City Limits` and `Parcels` layers you just performed.
20. Save your project and close ArcGIS Pro.

With that, you have successfully removed an overlap between multiple features on different layers using the **Align Features** tool.

How it works…

You've seen the **Align Features** tool in action, but how does it work? It adjusts line and polygon features so that they share coincident locations. This means the vertices of the lines and polygons that are adjusted end up being in the same locations.

Selecting the features tells the tool which ones will be adjusted. **Alignment Tolerance** provides a value to account for the distance difference between the features you wish to adjust. It will only adjust the portions of the features that are within the tolerance distance you specify. This prevents you from making undesired changes to your data.

Then, you can start drawing a sketch that represents the stitch line to which all selected items are adjusted. You can choose which side of the stitch line you wish to adjust and how to handle the end of the stitch line.

The **Align Feature** tool automatically defaults to using the **Trace Construction** tool. This allows you to easily follow an existing feature exactly because you are tracing its form. This is what you did in the recipe. However, you can use any of the construction tools, such as **Line** and **Perpendicular**, or any of the **Curve** tools to create the stitch line.

Once you've created the stitch line, ArcGIS Pro will automatically adjust the selected features so that they match the stitch line. It will either extend or trim line features and subtract or add to polygon features.

In this recipe, we learned how to use various editing tools to create and clean up GIS data so that it accurately represents real-world conditions. This included using tools such as **Split**, **Merge**, and **Align**. These are just a few of the commonly used tools contained in ArcGIS Pro that you can use to help keep your GIS data current.

5
Creating New Spatial Data

In *Chapter 4*, *Editing Existing Spatial Features*, you learned about various methods and tools in ArcGIS Pro that you can use to update or change existing features stored in your GIS. Those are extremely useful as you work to keep your data current to reflect changes to those existing features. However, you will often encounter situations where you need to add completely new features to your GIS.

ArcGIS Pro contains a host of tools you can use to create new features. The tools you can use will depend on several things. The first is your license level. As mentioned in *Chapter 4*, the license level (Basic, Standard, or Advanced) you use will determine what tools you can access and – in some cases –what those tools you can access do. The second thing that will impact which tools you can use to create new features is the type of features you are creating. There are different tools to create points versus lines versus polygons. Your data storage format and schema will also impact what tools might be available for use in creating new data.

In this chapter, you will learn how to use basic tools and methods to create new points, lines, and polygons. These tools and methods should work with all licensing levels of ArcGIS Pro.

The following recipes are included in this chapter:

- Creating new point features
- Creating new line features
- Creating new polygon features
- Creating a new polygon feature using the **Autocomplete Polygon** tool

Creating new point features

If you completed *Chapter 4*, you successfully edited features that already existed in GIS data. You realigned, reshaped, merged, and split those features. Now, it is time to look at how these features were created to begin with. You will start with the simplest of features: a point.

A point identifies an object at a single location. It is stored and located using a single coordinate pair. A coordinate pair consists of one X and one Y coordinate. It is also possible for a point to have a Z coordinate; this normally represents its elevation.

In this recipe, you will create several new point features. You will start by adding the manholes that are in the new subdivision you were looking at in the *Splitting a line feature* recipe of *Chapter 4*.

Getting ready

Before starting this recipe, you will need to have completed the *Configuring editing options* and *Splitting a line feature* recipes from *Chapter 4*. This recipe can be completed with all licensing levels of ArcGIS Pro.

How to do it...

You will use simple tools in ArcGIS Pro to create the sanitary sewer manholes that are part of the new Forrest Park Subdivision you referenced in the *Splitting a line feature* recipe of *Chapter 4*:

1. Launch ArcGIS Pro and open the `CreateNewFeatures.aprx` project file located in `C:\Student\ArcGISPro3Cookbook\Chapter5\CreateNew` by clicking on the **Open another project** button in the ArcGIS Pro **Start** window and navigating to the indicated folder.

   The project will open with a single map named `City of Trippville`. The map contains five layers – `City Limits`, `Parcels`, `Railroads`, `Sewer Lines`, and `Forrest Park Subdivision.jpg`, plus `World Topographic Map`. Now that you have a project open, you can start creating new points to represent sewer manholes.

2. In the **Contents** pane, ensure the `Forrest Park Subdivision.jpg` layer is visible. To do this, go to the **List by Drawing Order** button and make sure there is a checkmark in the box to the left of this layer.

3. In the **Catalog** pane, expand **Databases**. Then, expand the `Trippville_GIS.gdb` geodatabase so that you can see its contents.

4. Expand the `Sewer` feature dataset in the `Trippville_GIS.gdb` geodatabase. Then, right-click on the `Manhole` feature class and select **Add To Current Map**, as shown in the following screenshot:

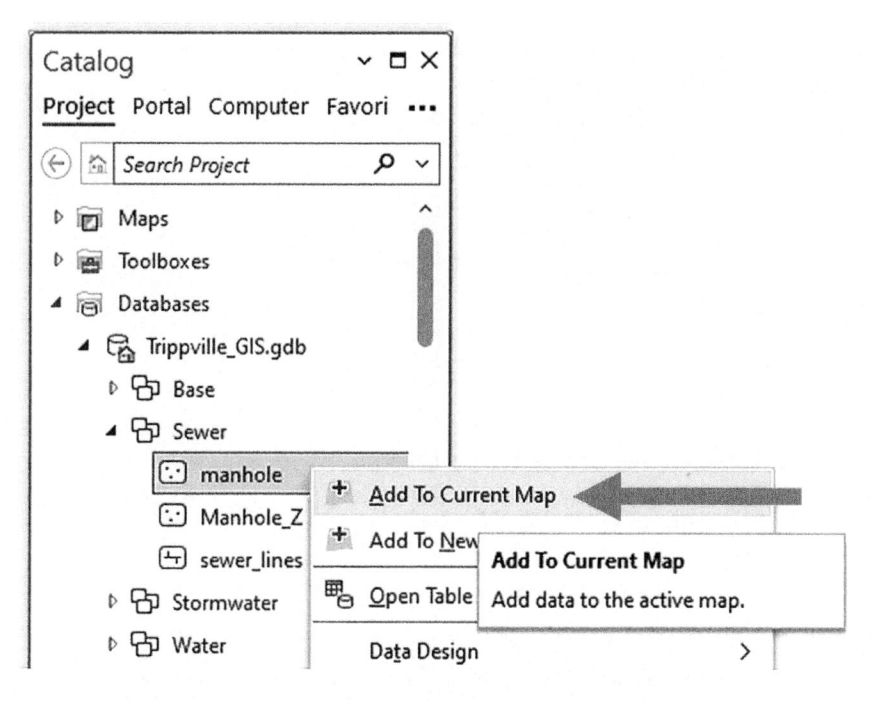

Figure 5.1 – Adding the Manhole feature class to the current map

5. Active the **Map** tab in the ribbon. Click on the **Bookmarks** button and select `Creating manholes` from the list presented. This will zoom your map into the location of the new Forrest Park Subdivision.

6. Select the `manhole` layer in the **Contents** pane. Next, click on the **Feature Layer** tab that appears in the ribbon.

7. Click on the **Symbology** button to open the **Symbology** pane.

164　Creating New Spatial Data

8. Click on the dot or point symbol patch located to the right of **Symbol**, as shown here, to access the **Format Point Symbol** options:

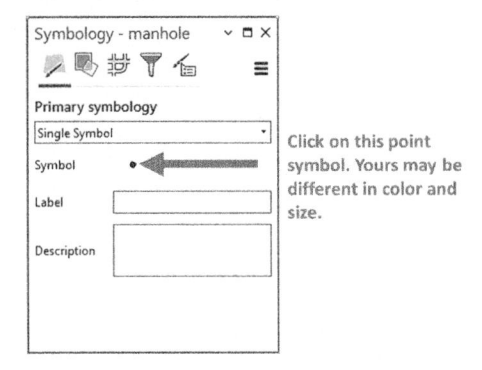

Figure 5.2 – Accessing the Format Point Symbol options

9. Click on the **Properties** tab located to the right of **Gallery** in the **Format Point Symbol** options.
10. Set **Color** to `Mars Red` (second column and third row) using the drop-down arrow.
11. Set **Size** to `12 pt`. Then, accept the default values for all other parameters.
12. Verify that your settings in the **Format Point Symbol** window match what's shown here and click **Apply**:

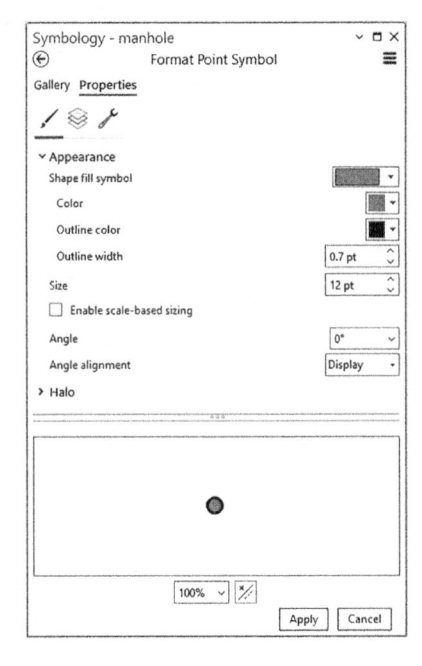

Figure 5.3 – Setting the point symbol parameters for the manhole layer

13. Close the **Symbology** pane and save your project.

Setting snapping options and editable layers

Now that you have added the manhole layer to the map and configured its symbology so that it stands out, it is time to configure some snapping options and ensure you have set the manhole layer as the only selectable and editable layer:

1. Active the **Edit** tab in the ribbon and click on the drop-down arrow located below the **Snapping** button.
2. Enable the **End Point snapping** option and disable all others, as shown in the following screenshot:

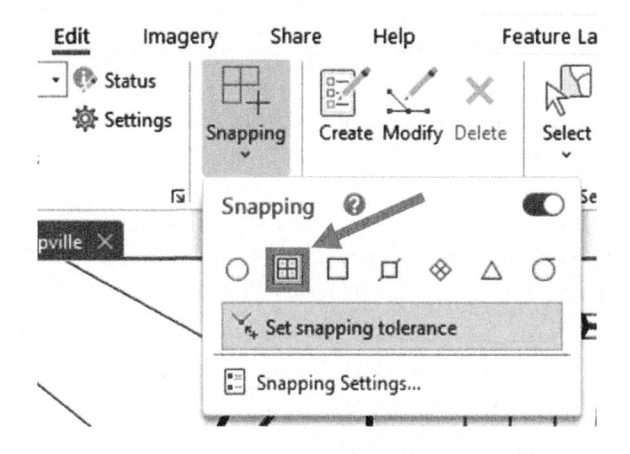

Figure 5.4 – Setting End Point snapping

3. Click on the **List by Snapping** button in the **Contents** pane. It has an icon that looks like four squares with a blue plus sign in the lower-right corner.
4. Right-click on the `Sewer Lines` layer and select **Make this the only snappable layer** for the menu that appears.
5. Click on the **List by Editing** button in the **Contents** pane. Its icon looks like a pencil.
6. Ensure editing is enabled for the `manhole` layer you recently added. If the box next to the layer name is checked, editing is enabled. If it is empty, editing is not enabled. Click on the box to enable editing for the layer.

Creating new point features

You are now ready to start creating new point features in the `manhole` layer:

1. Click on the **Create** button located on the **Editing** tab in the ribbon. This will open the **Create Features** pane on the right-hand side of the interface.

2. Select the `manhole` feature template. The feature template is located below the layer name and includes the symbol and the feature name.

3. Ensure the **Point** tool is selected below the feature template, as illustrated in the following screenshot:

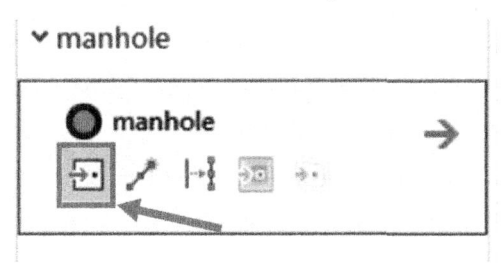

Figure 5.5 – Selecting the Point tool in the Create Features pane

4. Move your mouse into the map until you reach the first sewer line intersection located near the intersection of Oak Place and Pine Drive, shown in the Forrest Park Subdivision. Your mouse should automatically snap to the correct location once you get close. Once you see the indicator showing that you are snapping to the end point of those intersecting sewer lines, click the location with your mouse to create the new manhole at that location. The new manhole should appear at the location, as shown in the following diagram:

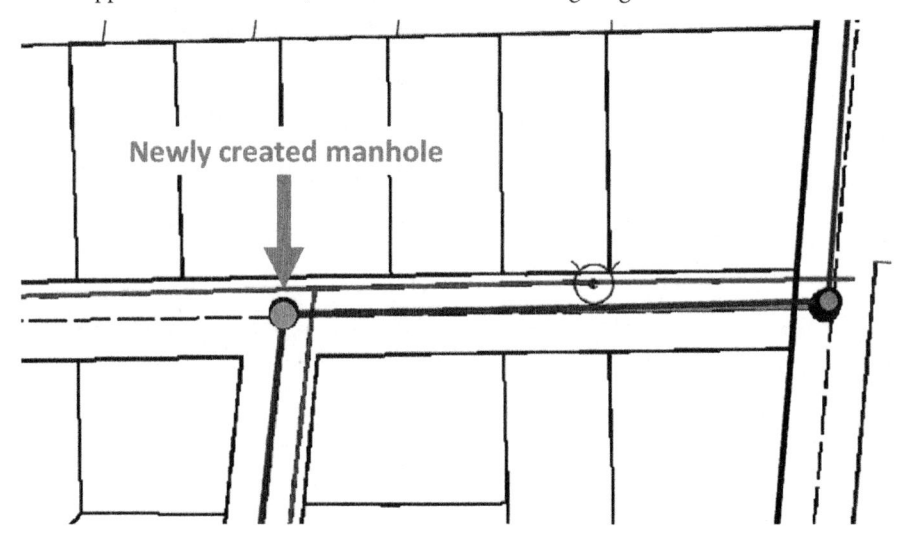

Figure 5.6 – The location of the newly created manhole feature

5. Now, click on the **Attributes** tool on the **Editing** tab. This will open the **Attribute** pane next to the **Create Features** pane.

6. In the grid located at the bottom of the **Attribute** pane, click on the cell located next to MANHOLE_ID, which should say <NULL>. Type 101.

7. Go down to the **Condition** field in the grid and click on the cell that says <NULL>. A drop-down menu should appear; select **Good** from the list, as shown here:

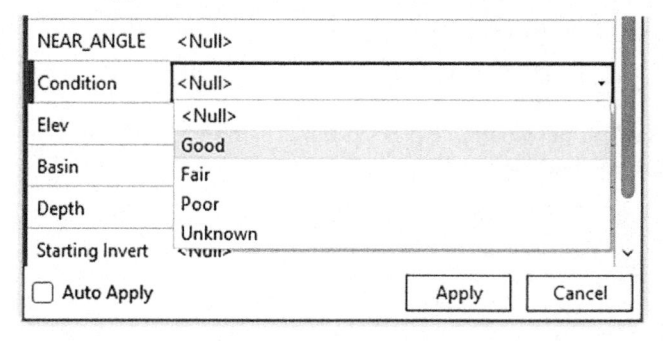

Figure 5.7 – Selecting Good from the provided picklist

> **Information**
> The **Condition** field has a domain assigned to it. A domain defines acceptable values that may be stored in the field. This can be a list of values, as you just saw, or a range of values if the field stores numbers.

8. Use the same process to add new manholes at the locations indicated in the following diagram. Increment the MANHOLE_ID value by one so that the next one will be 102. The condition for all new manholes you create should be **Good**:

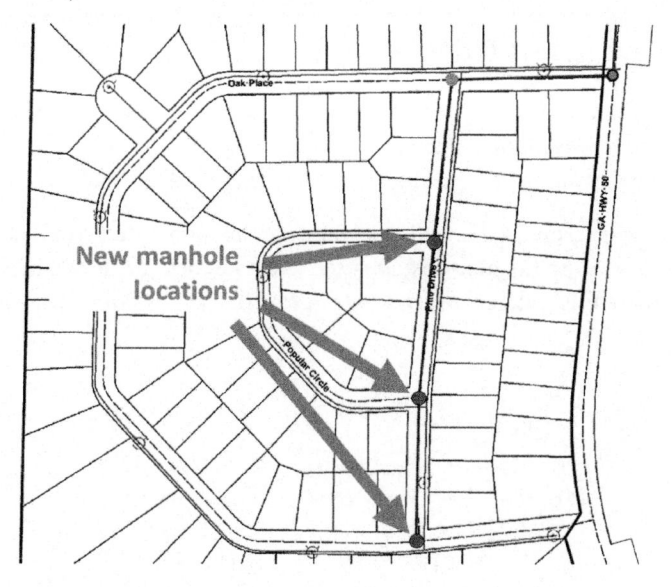

Figure 5.8 – The locations of three additional manholes

9. On the **Edit** tab in the ribbon, click on the **Save** button to save the new features you just created to the `Trippville_GIS.gdb` geodatabase.

10. Close the **Create Features** pane.

11. Save your project and close ArcGIS Pro.

You have just created four new manhole point features using the **Point** tool, as well as updated the attribute values for those new features.

How it works...

In this recipe, you created new manhole point features that were snapped to existing sewer lines based on the location shown in a scanned plan for a new subdivision. You then updated the key attributes associated with these new manholes, including `MANHOLE_ID` and `Condition`.

To accomplish this, you added the manhole feature class from the `Trippville_GIS.gdb` geodatabase and `Sewer feature` dataset to your map as a new layer. Then, you set the existing `Sewer Lines` layer as the only snappable layer and set the snapping location to **End Point** to ensure the new manholes would be connected to the sewer line features at their ends reflecting real-world conditions.

Next, you opened the **Create Features** pane by clicking on the **Create** button on the **Edit** tab in the ribbon. From there, you selected the manhole feature template and then the **Point** tool located below the template. From there, you digitized the four new manholes at the locations indicated in the scanned plan for the subdivision. Lastly, you updated the `MANHOLE_ID` and `Condition` attributes for each new manhole in the **Attribute** pane.

Creating new line features

Now, we will move on to creating line features. These are more complicated because they require multiple vertices. At a minimum, a line requires two vertices: a beginning and an ending. It is not uncommon for a line feature to have multiple vertices. This is called a polyline.

As far as ArcGIS is concerned, a line and a polyline are the same thing. They are stored together in the same feature classes and the tools that are used to create them are the same. So, you will see the terms *line* and *polyline* used interchangeably within ArcGIS. This is not true of all applications, such as AutoCAD.

Line features stored in a geodatabase feature class can also include curved segments. These segments are stored and created as arcs. Not all data storage formats support arcs. A shapefile is a good example of one that does not support arcs. Instead of using arcs, a shapefile typically uses multiple very short straight segments to simulate the arc. When displayed to scale, these short straight segments appear to be a curve. Geodatabases and CAD formats (DWG, DXF, and DGN) do support true arcs.

There are many construction techniques for creating line features. You can simply digitize the vertices by clicking with your mouse, you can specify exact measurements, or you can enter coordinates. Which method will work best will depend on the source data you have and its quality.

In this recipe, you will create several new line features that represent the street centerlines for the new Forrest Park Subdivision. You will use various construction techniques to do this.

Getting ready

To complete this recipe, you need to have completed the *Configuring editing options* and *Splitting a line feature* recipes of *Chapter 4*. This recipe can be completed with any license level of ArcGIS Pro (Basic, Standard, or Advanced).

How to do it...

In this recipe, you are going to start by creating new centerlines for the roads in the new Forrest Park Subdivision. You will get the opportunity to use several tools and methods to do this so that you can gain some experience with the capabilities available in ArcGIS Pro:

1. Launch ArcGIS Pro and open the `CreateNewFeatures.aprx` project. This project is located in `C:\Student\ArcGISPro3Cookbook\Chapter5\CreateNew` if you installed the data in the default location. If you completed the *Creating new point features* recipe in this chapter, this project should be in your list of **Recent Projects**.

2. Once the new project opens, activate the **Map** tab. Then, click on the **Bookmarks** button and select the `Create Manholes` bookmark to zoom your view into the Forrest Park Subdivision.

3. In the **Contents** pane, select the **List by Drawing Order** button at the top. The icon for this button resembles a folder directory tree.

170　Creating New Spatial Data

4. Turn off the `manhole` and `Sewer Lines` layers in the **Contents** pane so they are not displayed in the `City of Trippville` map, as illustrated in the following screenshot:

Figure 5.9 – The City of Trippville map with the manhole and Sewer Lines layers turned off

5. If needed, turn on the `Forrest Park Subdivision.jpg` layer in the **Contents** pane so that it is visible on the map.
6. Activate the **Map** tab in the ribbon. Then, click on the **Add Data** button in the **Layer** group.
7. Select the **Databases** folder under **Project** in the panel on the left-hand side of the **Add Data** window.
8. Double-click on the `Trippville_GIS.gdb` geodatabase in the panel on the right-hand side of the **Add Data** window.
9. Double-click on the `Base` feature dataset. Feature datasets are symbolized by an icon that looks like three overlapping squares.
10. Scroll down and select the `Street_Centerlines` feature class using a single click, then click **OK** to add `Street_Centerlines` as a layer to your map.
11. Select the newly added `Street_Centerlines` layer in the **Contents** pane. Then, click on the **Feature Layer** tab in the ribbon.
12. Click on the **Symbology** button in the **Feature Layer** tab in the ribbon. This should open the **Symbology** pane.

13. Click on the small line patch located to the right of **Symbol** in the **Symbology** pane, as shown in the following screenshot:

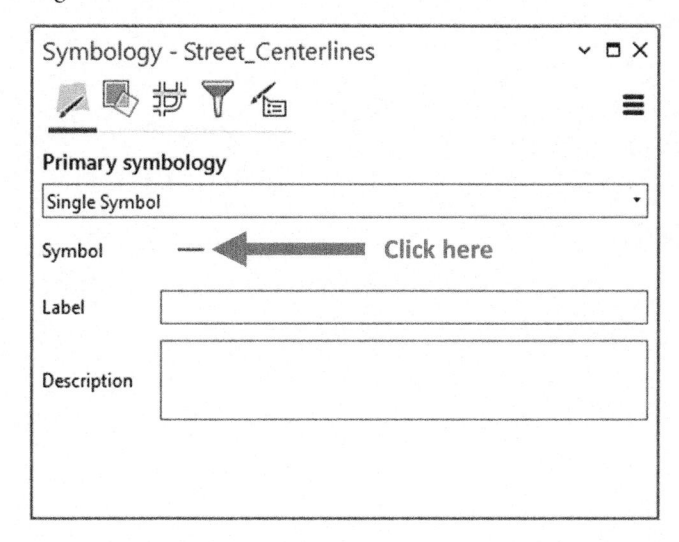

Figure 5.10 – Accessing the symbol properties for the Street_Centerlines layer

14. The **Format Line Symbol** options should appear in the **Symbology** pane. This allows you to adjust the display properties of the layer. Click on the **Properties** tab in the **Symbology** pane if needed. It is located next to the **Gallery** tab in the pane, as shown here:

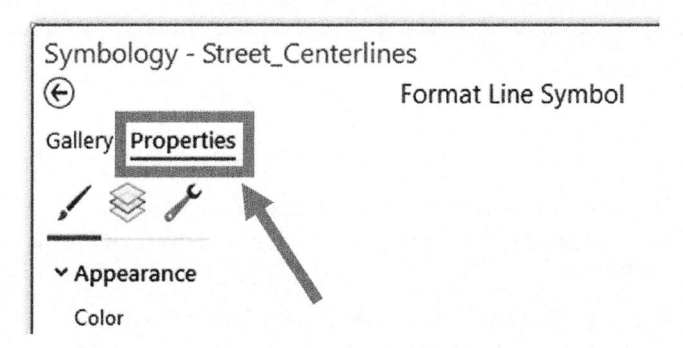

Figure 5.11 – The Properties tab in the Symbology pane

15. Ensure the **Symbol** button is selected in the **Symbology** pane. It looks like a paintbrush.
16. Set the **Color** parameter to `Ultra Blue` using the drop-down arrow. It is the tenth column and fourth row. If you would prefer to use a different color, then feel free to do so.
17. Set **Line width** to `3.0` and click **Apply** at the bottom of the pane.
18. Close the **Symbology** pane.

Your map should now look similar to the map shown here. The newly added `Street_Centerlines` layer should be visible and easy to see over the Forrest Park Subdivision image. This will help you when you start creating the new centerline features:

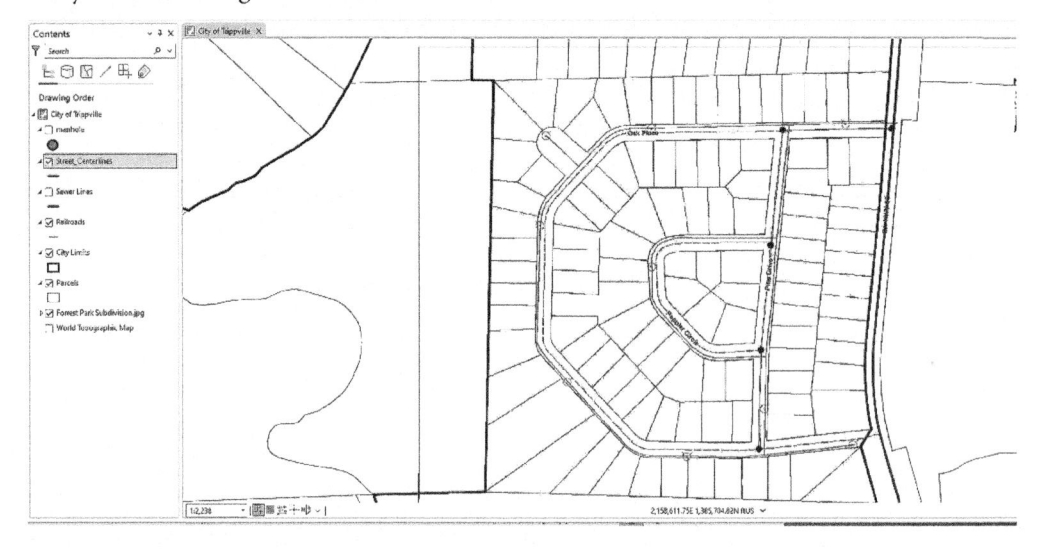

Figure 5.12 – The map with updated symbology for Street_Centerlines

Setting up editable layers and snapping settings

Now that you have added the layer you wish to edit and configured its symbology so you can easily distinguish its features, it is time to verify and configure a few settings that will ensure you correctly create new centerline features:

1. Click on the **List by Editing** button in the **Contents** pane. Then, right-click on the `Street_Centerlines` layer and select **Make this the only editable layer** from the menu that appears.
2. Click on the **List by Snapping** button in the **Contents** pane. Set the `Street_Centerline` layer as the only snappable layer by right clicking on the layer and selecting **Make this the only snappable layer**.
3. Click on the **List by Selection** button in the **Contents** pane and set the `Street_Centerlines` layer as the only selectable layer.

With that, you have configured your map so that you can successfully create the new street centerline features. You have ensured that the `Street_Centerlines` layer is selectable, snappable, and editable.

Creating new line features by tracing the source document

Now, you are ready to begin creating the new centerlines using the Forrest Park Subdivision image as a guide. The street centerline is shown as a dashed black line. You will use various methods to trace this line:

Creating new line features 173

1. Activate the **Edit** tab in the ribbon so that you can access the various editing tools.
2. Click on the **Create** button in the **Features** group on the **Edit** tab in the ribbon. This will open the **Create Features** pane, as shown in the following screenshot, so that you can access the feature template for the `Street_Centerlines` layer:

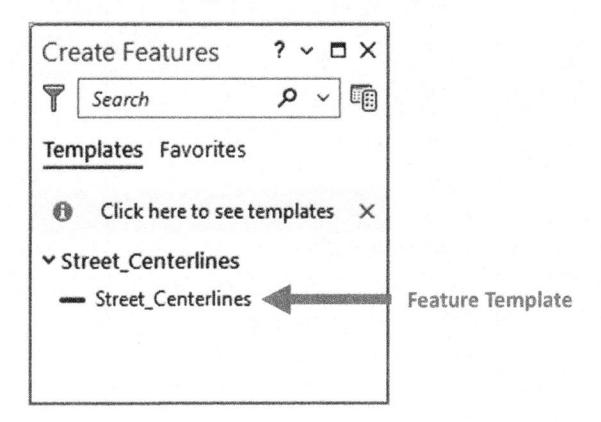

Figure 5.13 – The Street_Centerlines feature template displayed in the Create Features pane

3. Select the `Street_Centerline` feature template in the **Create Features** pane.
4. Using the scroll wheel on your mouse, zoom and pan to the southern intersection of Oak Place and GA HWY 50 until your map looks similar to the following:

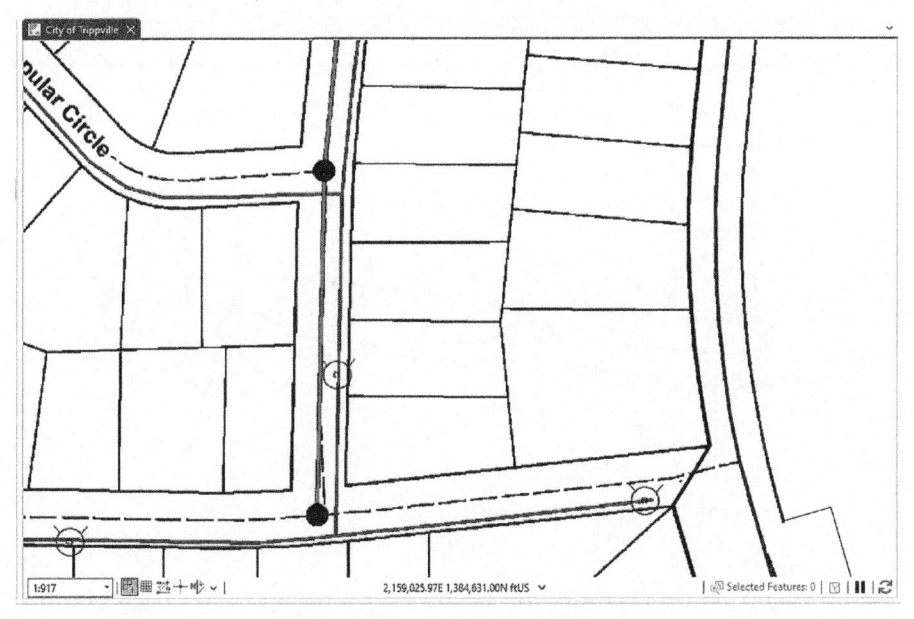

Figure 5.14 – Mapped zoomed into the identified area

5. Select the `Street_Centerlines` feature template in the **Create Features** pane. A series of tool icons should appear below the template. Select the **Line** tool; this should be the first one on the left:

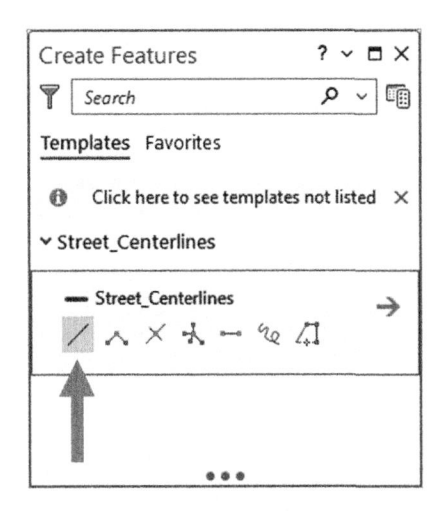

Figure 5.15 – The Line tool below the Street_Centerlines feature template

6. Click on the approximate intersection between Oak Place and GA HWY 50 to start drawing the new line. Your mouse should snap to the location when you get close. Then, trace the centerline shown on the scanned image by clicking on the locations, as illustrated in the following diagram:

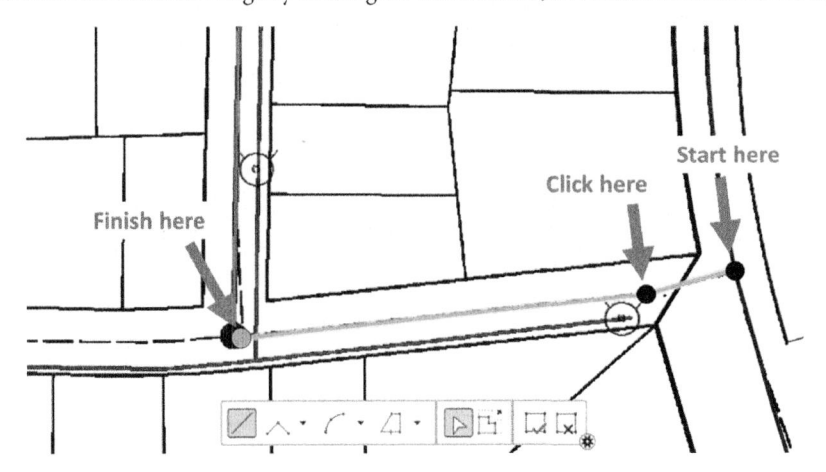

Figure 5.16 – Creating the first street centerline segment

You have just created your first new line feature representing a segment of a street centerline between two intersections. You will continue to draw the remainder of the centerline segments that represent this road in the new Forrest Park Subdivision:

1. Ensure the **Edit** tab is still active in the ribbon and click on the **Save** button to save your new `Steet_Centerlines` feature. When you're asked to save all edits, click **Yes**.

2. Press and hold the scroll wheel down on your mouse to pan to the right so that you can see the next segment from where you ended the last one, including the curve located to the west of your end point:

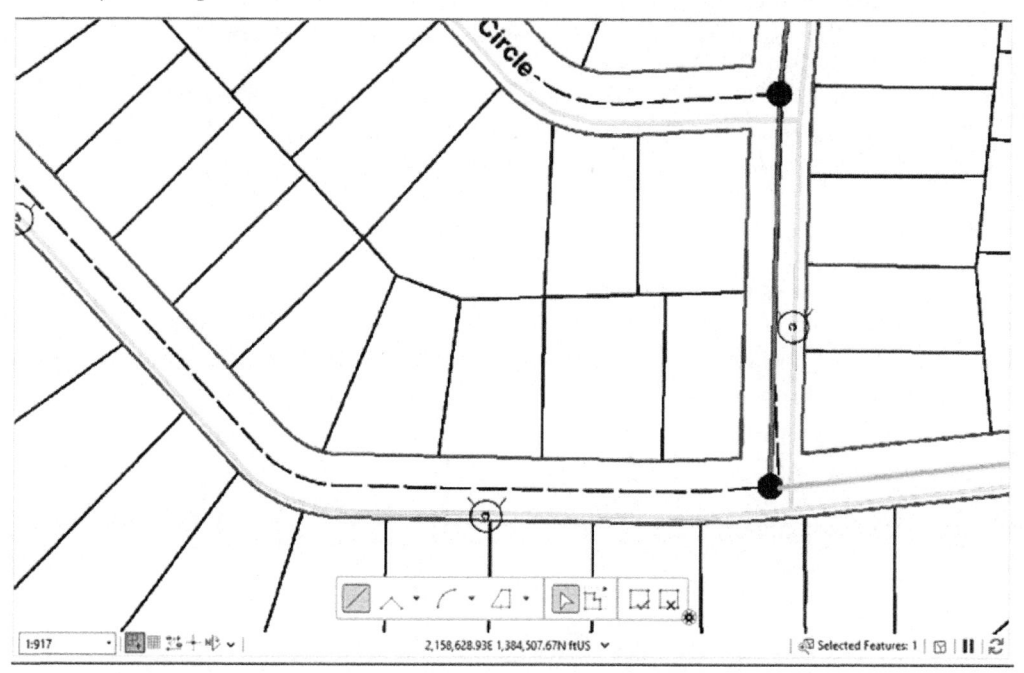

Figure 5.17 – Zoom area to draw the next segment

3. In the **Create Features** pane, select the **Line** tool located below the `Street_Centerlines` feature template again.

4. Move your mouse pointer so that it is over the end of the line you drew previously. When the snapping tip appears, click on that location to start a new line segment.

5. Move your mouse pointer along the street centerline shown in the Forrest Park Subdivision.jpg scanned image. Then, click on a point just before it starts to curve to the northwest, as shown here:

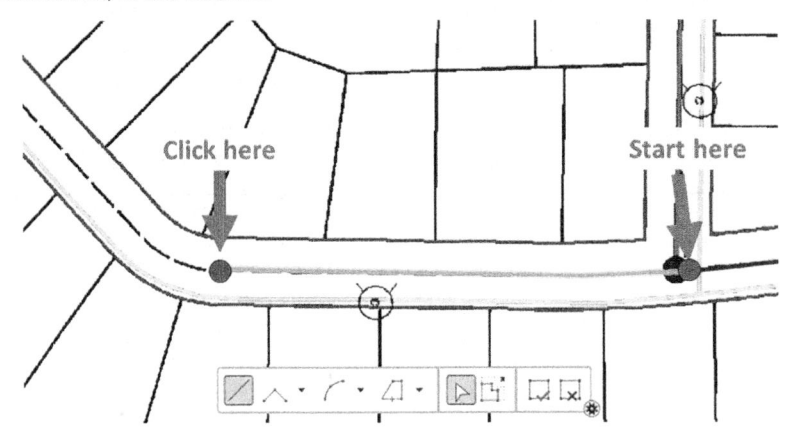

Figure 5.18 – Drawing the first part of the second segment

6. Select the **Arc Segment** tool from the **Edit** toolbar located at the bottom of the **Map** view, as illustrated in the following screenshot:

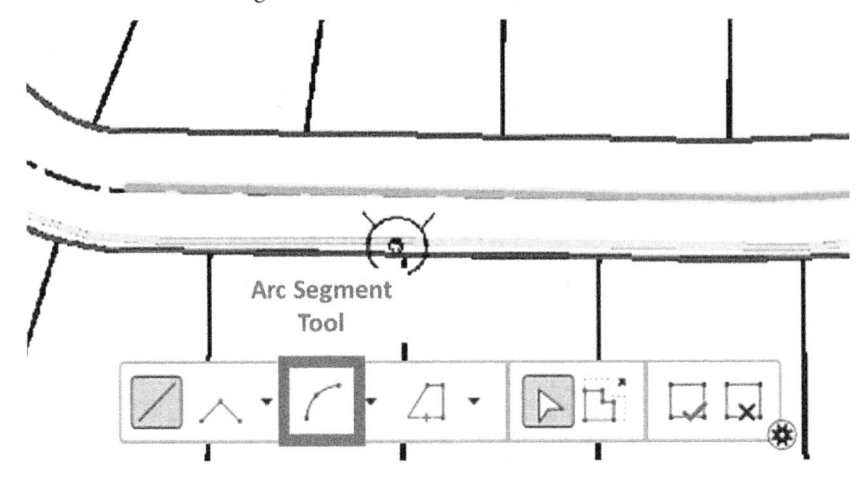

Figure 5.19 – The Arc Segment tool on the Edit toolbar

7. Using the **Arc Segment** tool, click on the locations indicated in the following diagram. The starting point is the same as the last point you clicked on before changing to the **Arc Segment** tool. Then, click on the remaining two locations. The location in the middle of the arc sets the radius. The third point sets the end of the arc:

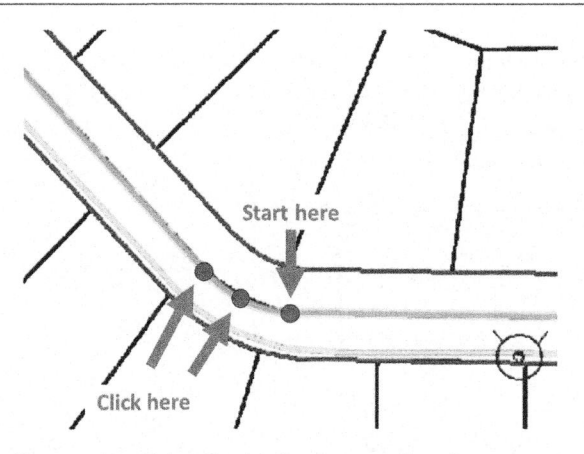

Figure 5.20 – Using the Arc Segment tool to draw a curve

> **Tip**
> When using the **Arc Segment** tool, you can specify a radius by pressing the *R* key. This will open a window where you can enter the specific radius measurement you want to use.

8. Select the **Line** tool from the **Edit** toolbar. It is the first tool on the left of the **Edit** toolbar. Remember that the **Edit** toolbar is located at the bottom of the map view area.

9. If needed, press down on the scroll wheel on your mouse and pan your map until you can see the next curve in the road, as shown in the following diagram. Click the location where you believe the road starts to curve to create a straight-line segment from the end of the first curve to the start of the next curve:

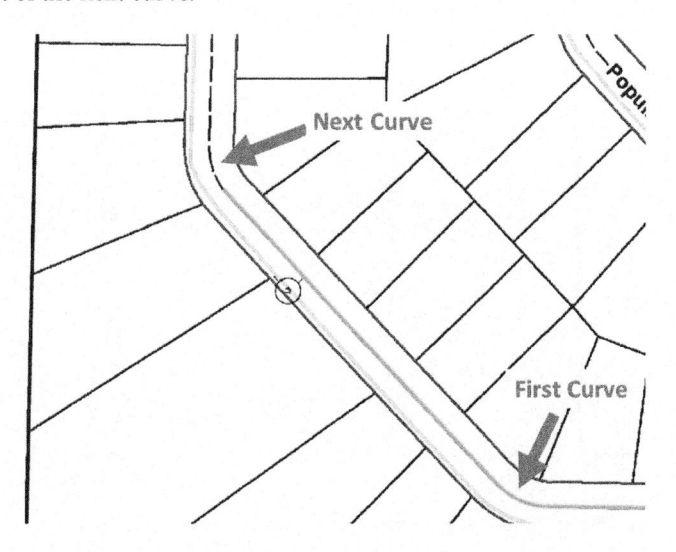

Figure 5.21 – The location of the next curve and the end of the new straight segment

10. After drawing the straight segment between the two curves, select the **Arc Segment** tool again. Using the same method you used for the previous curve, trace the arc of the road centerline as it turns toward the north. The following diagram illustrates the approximate locations of the start, midpoint, and end of the curve:

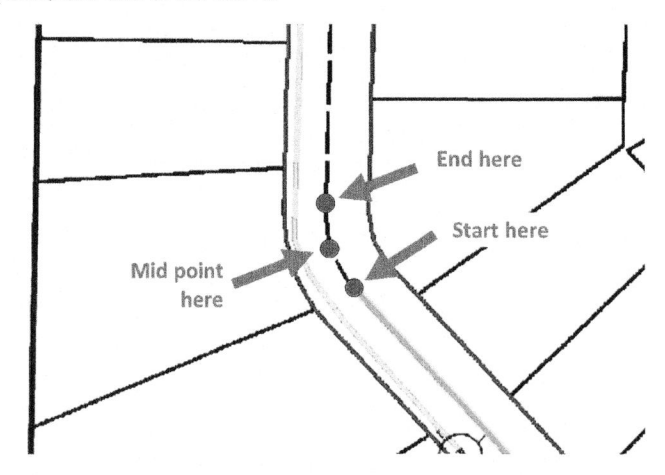

Figure 5.22 – Drawing the second curve with the Arc Segment tool

11. Continue drawing the centerline using this same process until you have completely traced the centerline of Oak Place until it intersects again with GA HWY 50 on the north side of the subdivision. When you've done this, your centerline should look similar to the following:

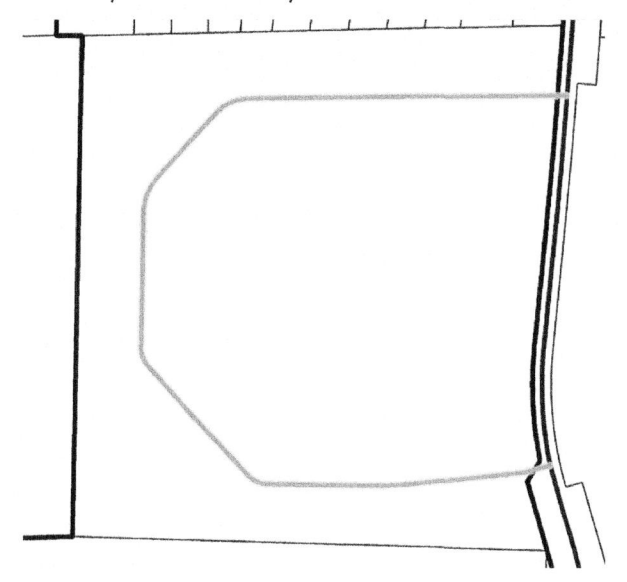

Figure 5.23 – Completed street centerline for Oak Place

12. Click the **Save** button on the **Edit** tab in the ribbon to save the new `Street_Centerlines` segments you have created.
13. Click the **Clear** button on the **Edit** tab to deselect any selected features.

You have just created a complex polyline feature. It includes straight segments and arcs so that it represents real-world features accurately. You created this feature by tracing the shape of the line from the scanned subdivision plat.

Creating new features based on specific measurements

Now, you will use measurements to create some of the other centerlines shown in the plat:

1. Click on the drop-down arrow located below the **Snapping** button on the **Edit** tab and set the snapping locations to **Vertex**, **Edge**, and **Intersection**.
2. Activate the **Map** tab in the ribbon and select the **Explore** tool.
3. Hold down the *Shift* key and draw a box around the short cul-de-sac located on the northwestern side of the subdivision. Your map should zoom into the area of the box you drew and should look similar to the following:

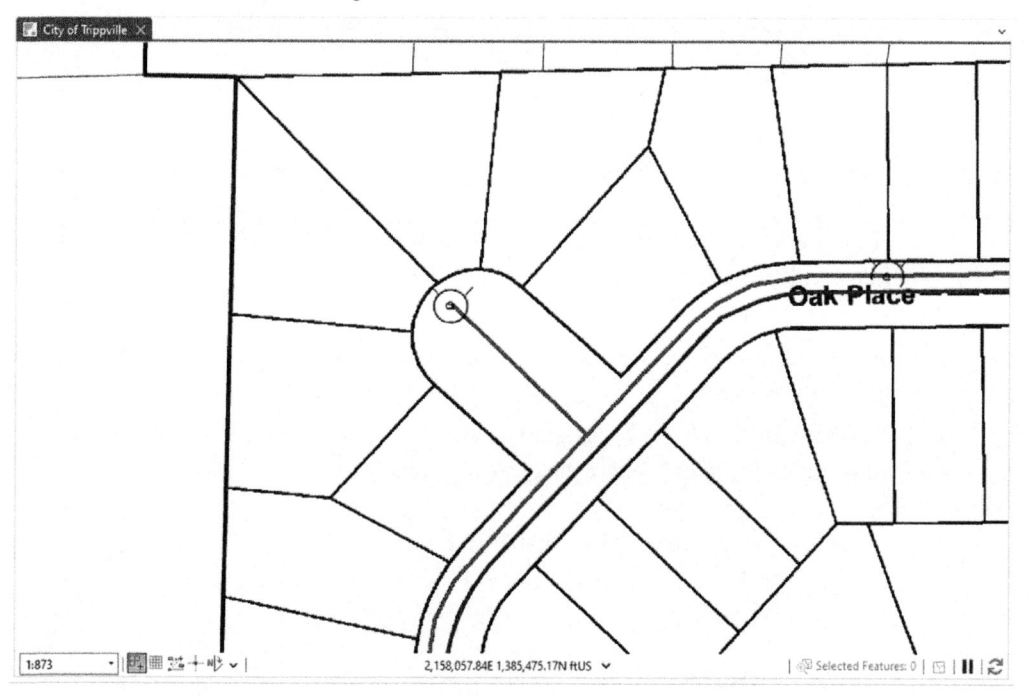

Figure 5.24 – Zooming into the area of the box to create the next new Street_Centerlines segment

4. Activate the **Edit** tab in the ribbon and select the Street_Centerlines feature template in the **Create Features** pane.

You are about to draw the centerline for the cul-de-sac. This centerline is not shown on the plat, so you will be provided with its measurements:

1. Click on the approximate intersection between the centerline for Oak Place and the cul-de-sac, as shown here:

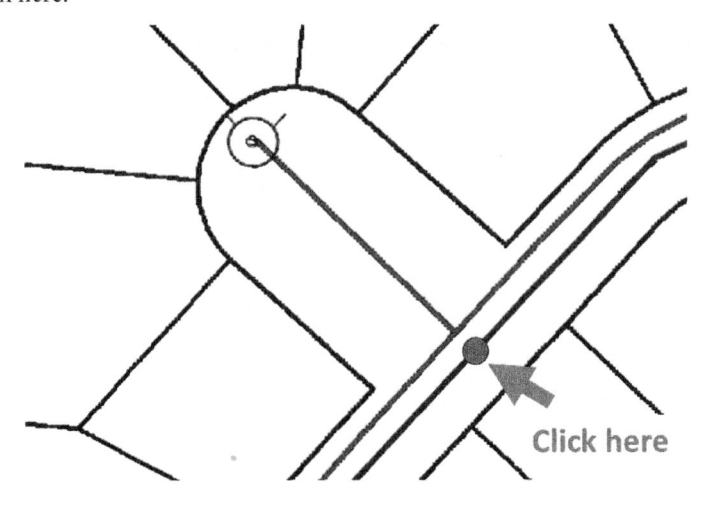

Figure 5.25 – Location to start drawing the new centerline for the cul-de-sac

2. Move your mouse to the northwest in the general direction of the approximate centerline of the cul-de-sac. Right-click once you are near the end of the waterline; this is shown as a blue line on the scanned plat.
3. From the menu that appears, select the **Direction/Distance** tool. The **Direction and Distance** tool window should open.
4. Set **Direction** to N47-06-50W QB and **Distance** to 125 ftUS, as shown in the following screenshot, and press *Enter* to apply the measurements:

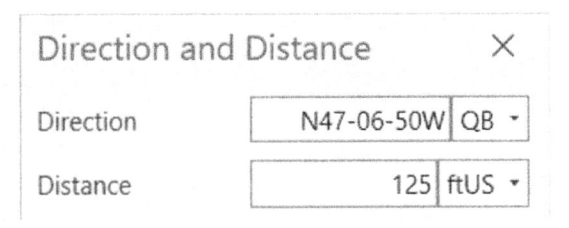

Figure 5.26 – The Direction and Distance parameters filled in

5. Press the *F2* key to finish creating the new feature.

> **Information**
>
> **QB** next to **Direction** stands for **quadrant bearing**. This is a type of measurement commonly used by surveyors and engineers to identify direction. Quadrant bearings divide the compass into four quadrants – north-east, south-east, south-west, and north-west. Then, the direction within that quadrant is measured based on the angle from either due north or due south to a maximum of 90 degrees. So, a feature heading directly due east could be identified as either north 90 east or south 90 east. Typically, the measurements are shown using degrees, minutes, and seconds as opposed to decimal degrees. In the example used in this recipe, 47 is the degrees, 6 is the minutes, and 50 is the seconds within the northwest quadrant.

6. Click on the **Save** button on the **Edit** tab to save the newly created `Street_Centerlines` feature you created. When you're asked if you wish to save all edits, click **Yes**.
7. Using the **Explore** tool on the **Map** tab, zoom back so you can see the entire subdivision area once more. Then, zoom in so you can see Pine Drive, which is shown on the Forrest Park Subdivision plat. This will be the next centerline you'll draw.
8. Once again, click on the `Street_Centerlines` feature template in the **Create Features** pane.
9. Click on the northern intersection of Pine Drive and Oak Place, as indicated on the plat. It should be very close to the location of the manhole.
10. Move your mouse along the centerline of Pine Drive, as shown on the plat, until you reach the intersection with Popular Circle, as illustrated in the following diagram. Then, right-click and select **Distance** from the **Context** menu:

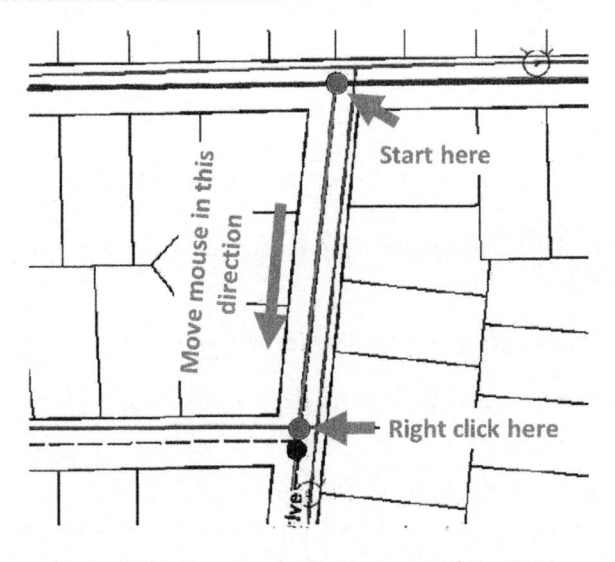

Figure 5.27 – Drawing the first segment of Pine Drive

11. Enter a distance of 350 ftUS and press *Enter*. This will lock the length of the line segment to 350 feet but still allow you to control the direction.
12. Move your mouse until the new line segment is in line with the centerline shown on the plat. You may need to zoom in using the scroll wheel on your mouse to see the location. Once you are happy with the alignment, click with your mouse to set the direction of the new segment.
13. Use the same process mentioned previously to draw the next segment, which goes to the southern intersection of Pine Drive and Popular Circle. Set **Distance** to 342 ftUS.
14. Lastly, click on the intersection of Pine Drive and Oak Place. You should see a snapping tip appear, indicating that you are snapping the end of the new segment to the intersection with Oak Place. Your new centerline should look like this:

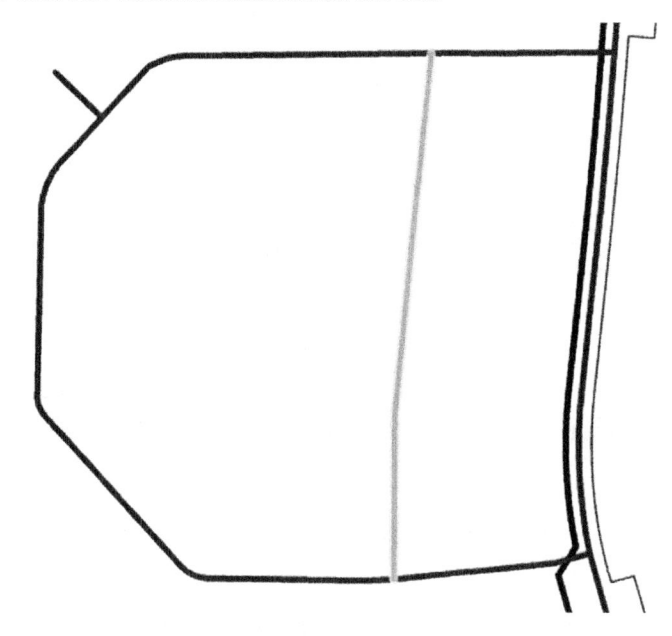

Figure 5.28 – Completed Pine Drive

With that, you've created the centerline that represents Pine Drive based on specific distance and direction measurements.

Creating line features using the methods you've learned

You have one last centerline to add: Popular Circle. It is located to the west of Pine Drive. Follow these steps:

1. Use the **Explore** tool on the **Map** tab to zoom into Popular Circle, which is shown on the Forrest Park Subdivision.jpg plat. This is the road that loops into Pine Drive that you just created.

2. Using the tools and methods you have learned to create the centerline segments for Oak Place and Pine Drive, create new centerline segments for Popular Circle. Save your edits once you have completed the new centerline representing Popular Circle.

Once you are done creating new street centerlines that represent all the new roads in the new Forrest Park Subdivision, your map should look similar to the following. The `Forrest Park Subdivision.jpg` layer has been turned off so that you can easily see the new line segments:

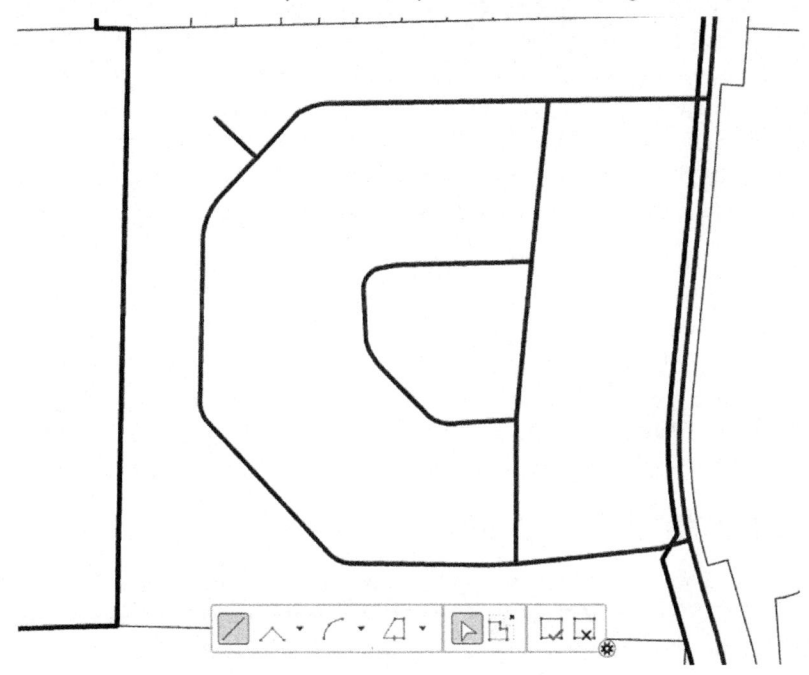

Figure 5.29 – New street centerlines completed

1. Click the **Save** button in the **Manage Edits** group on the **Edit** tab. The icon looks like a floppy disk with a pencil. When you're asked whether you wish to save your edits, click **Yes**.
2. Close the **Create Features** pane and save your project using the **Save Project** button located on the **Quick Access** toolbar.
3. Close ArcGIS Pro.

With that, you have successfully created new line features representing the centerlines of streets located in a new subdivision.

How it works...

In this recipe, you created several new lines that represented the centerlines of streets that are part of a new subdivision that is being built in the City of Trippville. Each of these lines included multiple segments; some were straight, while others were curved. So, you used various editing tools and methods to create these.

To start, you added the `Street_Centerlines` feature class from the `Trippville_GIS.gdb` geodatabase to your map as a layer. Then, you adjusted the snapping locations to ensure you snapped the new centerlines to the existing ones to make sure they were connected. Once you did this, you opened the **Create Features** pane by clicking on the **Create** button in the **Edit** tab so that you could access the feature template for the `Street_Centerlines` layer. The feature template defines various properties that are used to create new features, such as what layer is it created in, what symbology is assigned, what default attribute values are populated, and more.

From there, you started drawing the new centerline segments nu tracing them from the `Forrest Park Subdivision.jpg` scanned plat, initially using the **Line** tool for the straight sections and the **Arc Segment** tool for the curves. Once you completed the primary street for the new subdivision, you started to create the interior and cul-de-sac roads using those same methods, but you enhanced these methods by using specific measurements for the length of various segments to increase the overall accuracy of the data you created.

Creating polygon features

So far, you've learned how to create point and line features using ArcGIS Pro. It is now time to learn how to create polygon features. Polygons are very similar to polylines in that they consist of multiple vertices. The big difference is that a polygon must form a closed figure and a polyline does not.

A polygon is constructed with a minimum of four vertices. This may seem confusing since a triangle is a polygon but only has three sides. So, why are four vertices required? Well, that is because the polygon must be closed. So, the first and last vertices are in the same location. When creating new polygons in ArcGIS Pro, the software automatically does this. So, you never have to worry about whether your polygon is closed or not.

In this recipe, you will create polygons that represent building footprints. You will use several different methods to do this. Many will be like the methods you used to create lines.

Getting ready

Before starting this recipe, you will need to have completed the *Configuring editing options* recipe of *Chapter 4*. It is also recommended that you complete the *Creating new line features* recipe in this chapter, though it is not required. However, completing that recipe will help teach you some of the tools and skills you will need to create new polygons.

Creating polygon features

This recipe can be completed with all licensing levels of ArcGIS Pro. You will also need to ensure you have internet access to make use of the Esri-provided basemaps.

How to do it...

Follow these steps to create new polygon features that will represent the area covered by buildings you see in aerial photography you have access to via ArcGIS Online as a basemap:

1. Launch ArcGIS Pro and open the `CreateNewFeatures.aprx` project. This project is located in `C:\Student\ArcGISPro3Cookbook\Chapter5\CreateNew` if you installed the data in the default location. If you completed the *Creating new point features* recipe in this chapter, this project should be in your list of **Recent Projects**.

2. Activate the **Map** tab in the ribbon. Then, click on the **Basemap** button and select the `Imagery` basemap from the presented options, as illustrated in the following screenshot:

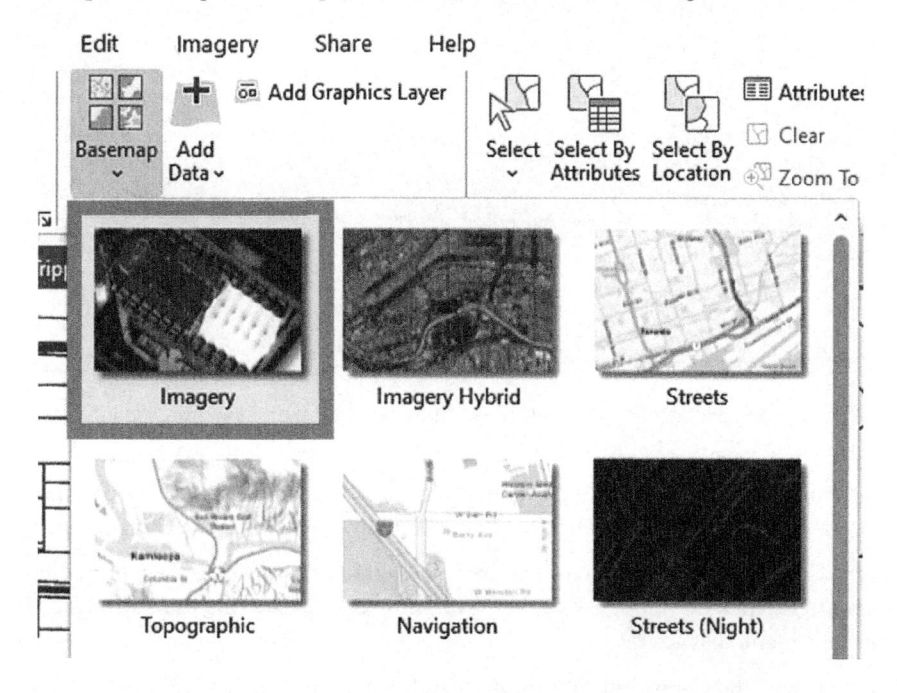

Figure 5.30 – Changing the basemap in your map to Imagery

3. In the **Contents** pane, select the **List by Drawing Order** button. Then, turn off the `Parcels`, `Sewer Lines`, `Manholes`, and `Forrest Park Subdivision.jpg` layers so that they aren't visible on the map.

4. Active the **Map** tab in the ribbon and click on the **Bookmarks** button. Select the `Building 1` bookmark from the options presented. Your map will zoom into the location of the first building footprint you will create, as illustrated in the following screenshot:

Figure 5.31 – Building 1 bookmark results

5. In the **Catalog** pane, expand the **Databases** folder so that you can see its contents. Then, expand the `Trippville_GIS.gdb` geodatabase and the `Base` feature dataset.

6. Right click on the `Buildings` polygon feature class and select **Add to Current Map** from the menu that appears. The `Buildings` layer should now appear in the **Contents** pane.

7. In the **Contents** pane, click on the **List by Selection** button (the blue and white polygon icon). Then, right-click on the `Buildings` layer and select **Make this the only selectable layer** from the menu that appears.

8. Next, click on the **List by Editing** button in the **Contents** pane. Right-click on the `Buildings` layer once again and select **Make this the only editable layer** from the menu.

9. Active the **Edit** tab in the ribbon, then click on the **Snapping** button to disable snapping. The button should no longer be highlighted in blue when it is disabled.

Creating polygon features 187

So far, you have prepared ArcGIS Pro and your map for editing by adding the appropriate layer, setting that layer as the only editable layer, and disabling snapping.

Creating new building polygons

You are now ready to begin creating new polygons that will represent building footprints based on their outline that you can see in aerial photography:

1. Click on the **Create** button to open the **Create Features** pane.
2. In the **Create Features** pane, select the Buildings feature template. Then, select the **Polygon** tool located beneath the template, as shown here:

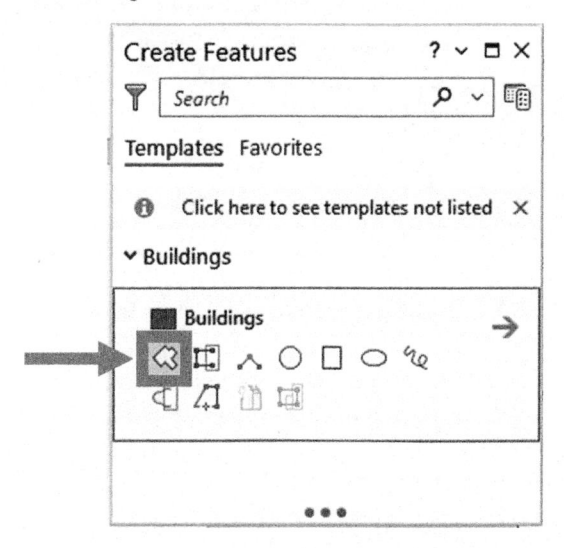

Figure 5.32 – Selecting the Polygon tool in the Create Features pane

3. In the **Edit** toolbar, select the **Right Angle Line** tool, as indicated in the following screenshot:

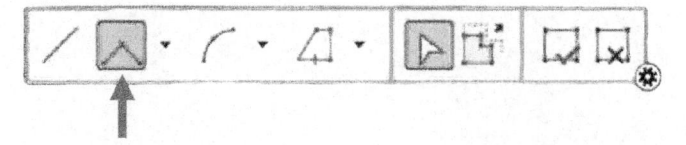

Figure 5.33 – The Right Angle tool in the Edit toolbar

188   Creating New Spatial Data

4. Now, click on the southwestern corner of the building, move your mouse along the southern edge of the building, and click on the southeastern corner closest to the road that runs north to south, as shown here:

Figure 5.34 – Drawing the southern side of the building polygon

5. Move your mouse along the eastern side of the building shown in the aerial image and click at the northeast corner. Then, continue following the outline of the building by clicking at the various corners, as shown in the following figure. If you forget to double-click on the last vertex location, you can press the *F2* key to finish the polygon:

Figure 5.35 – Digitizing a building footprint polygon

Creating polygon features | 189

You have just created your first polygon using a very simple method. Using the **Right Angle Line** tool, you constructed one side by tracing it from the aerial photo. Then, you continued to trace the building using the **Right Angle Line** tool to ensure the polygon had right-angled corners. This generated an accurate representation of the building. Now, it's time for you to work on the next building you need to digitize.

1. Activate the **Map** tab in the ribbon and click on the **Bookmarks** button. Select the `Building 2` bookmark from the list that appears. This will zoom your map into the correct location.
2. In the **Create Features** pane, select the **Buildings** feature template and click on the **Rectangle** tool, as shown in the following screenshot:

> **Tip**
> If you closed the **Create Features** pane, activate the **Edit** tab in the ribbon. Then, click on the **Create** button to reopen the **Create Features** pane.

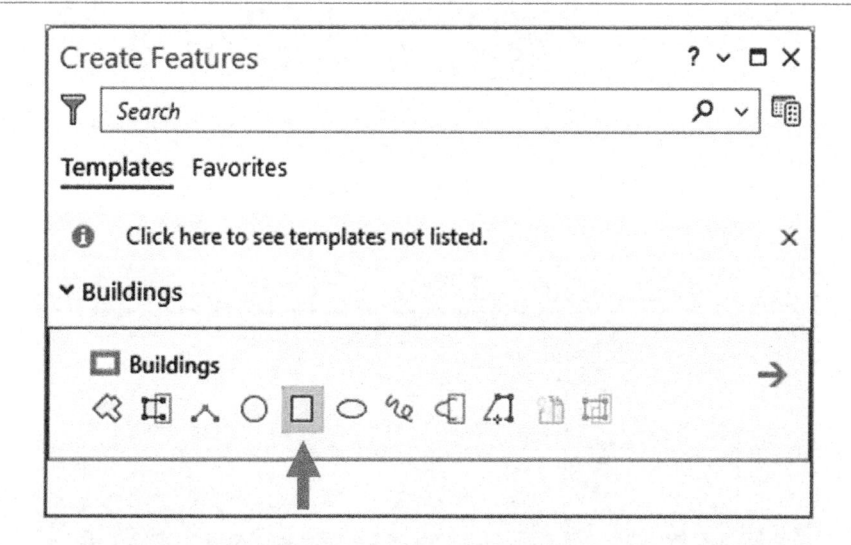

Figure 5.36 – Accessing the Rectangle tool from the Create Features pane

3. Click on the southwestern corner of the building. Then, move your mouse along the southern edge of the building and click at the southeastern corner, as shown in the following figure:

Figure 5.37 – Starting the second building polygon using the Rectangle tool

4. Drag your mouse toward the northeast corner of the building shown in the aerial photo. Click on the northeast corner to create your new polygon, as shown in the following figure:

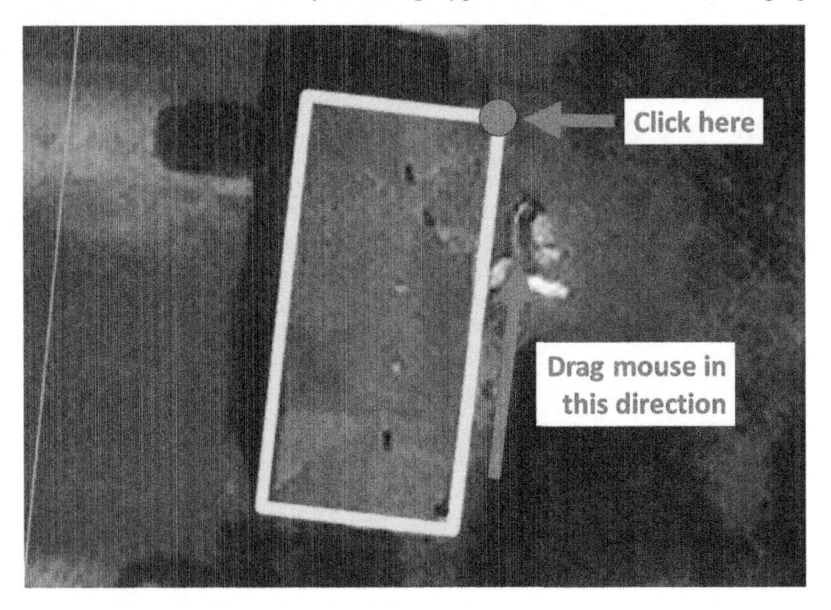

Figure 5.38 – Finishing the second building polygon using the Rectangle tool

With that, you have created a second polygon representing a building footprint using a different method.

1. Activate the **Edit** tab in the ribbon and click on the **Save** button to save the polygon you just created back to the `Trippville_GIS.gdb` geodatabase. If you're asked whether you want to save all edits, click the **Yes** button.
2. Close the **Create Features** pane.
3. Save your project using the **Save Project** button on the **Quick Access** toolbar, which is located in the top-left corner of the ArcGIS Pro interface.
4. Close ArcGIS Pro.

In this recipe, you created two new polygons representing building outlines or footprints based on their outline seen from an aerial photograph that was provided by Esri via ArcGIS Online.

How it works...

In this recipe, you created two new polygons that represented buildings found on aerial photography provided by Esri through ArcGIS Online basemaps. You did this using tools found in the **Create Features** pane.

First, you changed the basemap layer being displayed on the map to the `World Imagery` basemap. To do this, you activated the **Map** tab and clicked on the **Basemap** button. Then, you selected the new basemap from the list of available options provided.

Next, you configured which layers were visible on the map to reduce clutter. You turned off all but the `RW`, `Street_Centerlines`, `City Limits`, and `Railroad` layers by using the **List by Contents** option in the **Contents** pane. Then, you added the `Buildings` feature class to the map as a new layer and zoomed the map into the location of the first building you needed to create using a bookmark. Once there, you opened the **Create Features** pane so that you could access the `Building` feature template. This feature template defines properties for creating new features, as well as provides access to editing tools. From there, you used the **Right Angle Line** tool to create a new polygon by tracing the outline of the building, as shown in the `World Imagery` basemap.

To create the second building polygon, you used a bookmark to zoom into its location. You then selected the **Rectangle** tool from the **Create Feature** pane to create a simple rectangle that represented the building shown in the **World Imagery** basemap. This allows you to get measurements between manholes, as well as associate other information with those specific segments.

Both methods created accurate and complete representations of the area covered by both buildings. Which tools and methods are best depends on the source data available and the overall shape of the polygon you are creating. As you will see in the next recipe, there are other ways to create new polygons.

Creating a new polygon feature using the Autocomplete Polygon tool

You now know how to create new polygon features that are not connected or adjacent to other polygons. But how do you create new polygon features that are adjacent to other existing polygon features? You would want to construct these in a way that wouldn't create gaps or overlaps between the new polygon and existing ones. Using the **Autocomplete Polygon** tool is one way to successfully do this.

In this recipe, you will create a new parcel polygon located just outside the city limits of Trippville. The city council is considering annexing this parcel and wants to see how it relates to the existing city limits. You will create this new parcel using the **Autocomplete Polygon** tool.

Getting ready

Before starting this recipe, you will need to have completed the *Configuring editing options* recipe of *Chapter 4*. It is recommended that you have completed the *Creating polygon features* recipe in this chapter as well to ensure you have some foundational experience with the concepts contained within this recipe.

This recipe can be completed with all licensing levels of ArcGIS Pro. You will also need to ensure you have an active internet connection to use the Esri-provided basemaps.

How to do it…

You will now create a new parcel polygon that adjoins the city limits of Trippville using the **Autocomplete Polygon** tool:

1. Launch ArcGIS Pro and open the `CreatingNewFeatures.aprx` project. This should be in your **Recent Projects** list. The project should open with the `City of Trippville` map that you used in previous recipes in this chapter.

2. Active the **Map** tab in the ribbon and click on the **Bookmarks** button. Then, select the `Autocomplete Polygon` bookmark from the list that appears. This will zoom your map into the correct location.

3. In the **Contents** pane, click on the **List by Draw Order** button. Then, turn on the `Parcels` and `City Limits` layers so that they are visible on the map.

4. Active the **Edit** tab in the ribbon and enable snapping by clicking on the **Snapping** button so that it is highlighted in light blue.

5. Click on the drop-down arrow located below the **Snapping** button. Verify that the snapping locations are set to **End Point**, **Vertex**, **Edge**, and **Intersection**, as shown here:

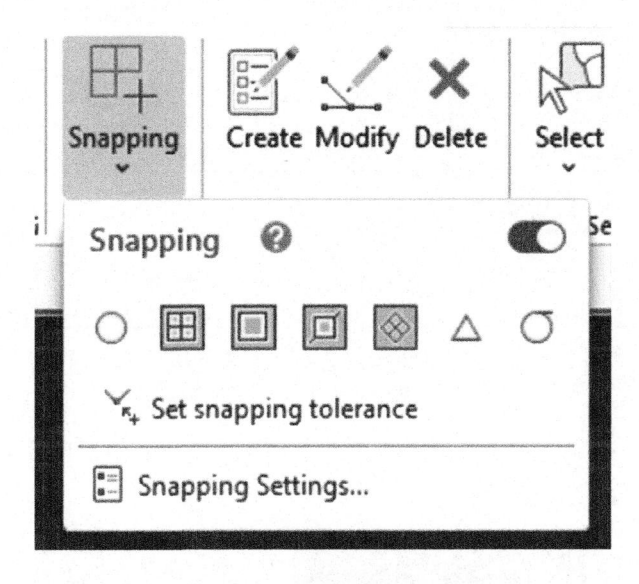

Figure 5.39 – Snapping locations set

6. In the **Contents** pane, click on the **List by Snapping** button and set the `RW`, `Parcels`, and `City Limits` layers as the only three snappable layers.
7. Again, in the **Contents** pane, click on the **List by Editing** button and set the `Parcels` layer as the only editable layer.

The `Parcels` layer is now ready for editing.

Creating a new parcel using Autocomplete Polygon

Now that you have the map and layer prepared for updating, it is time to create the new parcel polygon using the **Autocomplete Polygon** tool to ensure the new parcel is created without any gaps or overlaps with existing parcel polygons:

1. Activate the **Edit** tab in the ribbon and click on the **Create** button to open the **Create Features** pane.
2. In the **Create Features** pane that just opened, select the `Parcels` feature template. Then, select the **Autocomplete Polygon** tool, as shown in the following screenshot:

194   Creating New Spatial Data

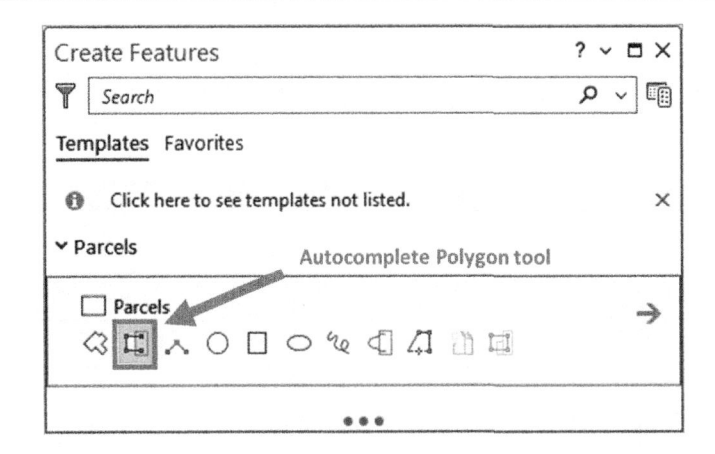

Figure 5.40 – Selecting the Autocomplete Polygon tool

3. On the map, click on the intersection of the city limits and RW lines, as illustrated in the following figure:

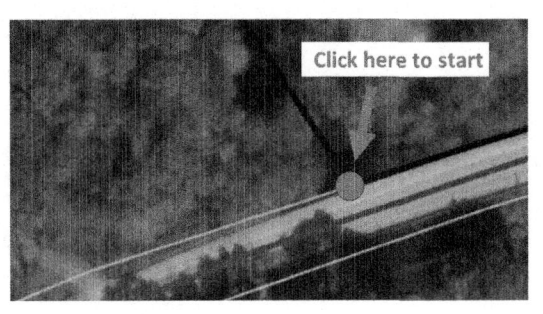

Figure 5.41 – Starting the Autocomplete Polygon tool process

4. Then, select the **Trace** tool from the **Edit** toolbar. The **Trace** tool is the fourth button from the left on the **Edit** toolbar:

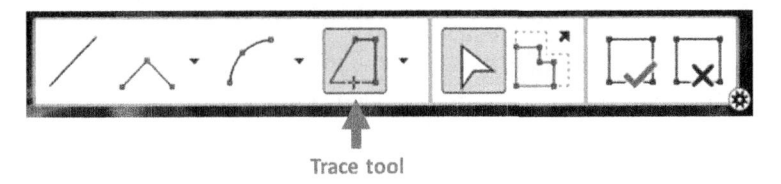

Figure 5.42 – Selecting the Trace tool on the Edit toolbar

Creating a new polygon feature using the Autocomplete Polygon tool 195

5. With the **Trace** tool selected, press the *O* key to open the **Trace Options** window.

6. In the **Trace Options** window, enable the **Limit trace length** option and set it to `125 ftUS`, as illustrated in the following screenshot. Click **OK** to apply the new settings:

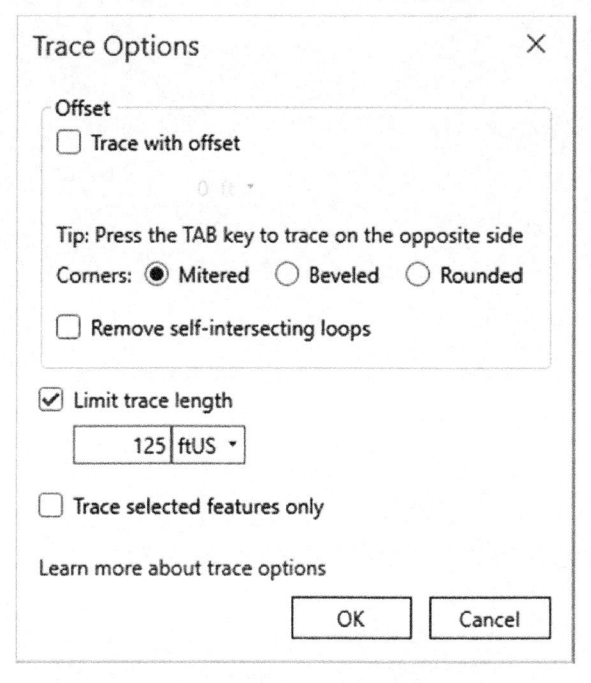

Figure 5.43 – Enabling the Limit trace length option for the Trace tool

7. Next, click on the RW line, as shown in the following figure, to start using the **Trace** tool to trace the existing RW line so that the southern boundary of the new polygon follows the existing RW line exactly:

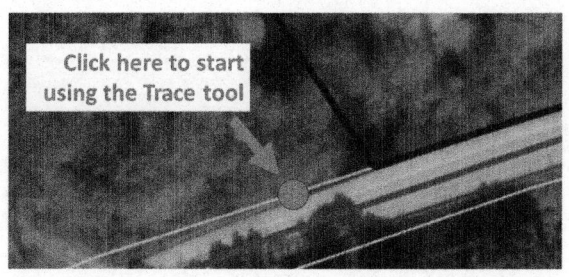

Figure 5.44 – Starting to use the Trace tool

196 Creating New Spatial Data

8. Move your mouse pointer along the RW line until the **Trace** tool stops drawing the new polygon boundary. This means you have reached the `125 ftUS` value you set under **Trace Options**:

Figure 5.45 – The Trace tool's stop location

9. Select the **Line** tool from the **Edit** toolbar. This is the same toolbar where you selected the **Trace** tool. The **Line** tool is the first tool from the left of the **Edit** toolbar.

10. Move your mouse pointer in the map to the north/northwest and right-click. Select **Direction/Distance** from the menu that appears.

11. Set **Direction** to `N33-34-00W QB` and **Distance** to `290.00 ft`, as shown in the following screenshot. You may need to press *Enter* to apply the values:

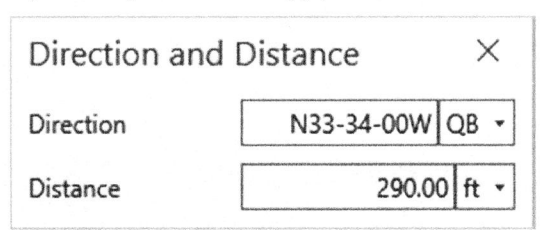

Figure 5.46 – Direction and Distance settings

12. Move your mouse pointer to the east toward the existing parcel located inside the `City Limits` boundary and right-click. Select **Direction** from the menu that appears.

13. Set **Direction** to N60-03-45E QB. You might need to press *Enter* to apply the new value. The new boundary should now be locked into the direction you just entered:

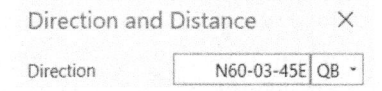

Figure 5.47 – Setting Direction

14. Move your mouse pointer in the map so that it is located inside the parcel to the east of the City Limits boundary, as shown in the following figure, and double-click to create the new parcel polygon:

Figure 5.48 – Finishing the Autocomplete Polygon tool process

The new parcel polygon should now be complete. By using the **Autocomplete Polygon** tool, it now shares a common boundary with the adjoining City Limits boundary on its eastern side.

1. Click on the **Save** button in the **Manage Edits** group on the **Edit** tab to save the edit you just made. Click **Yes** to confirm that you with to save your edits when you're asked.
2. Close the **Create Features** pane.
3. You can save your project by clicking on the **Project** tab in the ribbon and then clicking on the **Save** option that is shown.
4. Close ArcGIS Pro.

With that, you have successfully created a new polygon that shares a common boundary with both the existing city limits and RW features.

How it works...

When creating polygons that represent features such as parcels, political boundaries, emergency response boundaries, and others, it is important to create them in such a way as to ensure they do not have overlaps and gaps. The **Autocomplete Polygon** tool is a great way to ensure the new polygons you create follow existing boundaries exactly. This means they do not have gaps or overlaps.

In this recipe, you created a new parcel polygon by defining three of four sides using traditional methods you learned in other recipes in this chapter. The fourth side was automatically created because the **Autocomplete Polygon** tool traces the boundary of the adjoining existing parcel automatically.

When using the **Autocomplete Polygon** tool, you need to ensure the sketch you create starts and ends by touching or crossing the boundary of the existing adjacent polygon. If your sketch does not touch or cross the existing adjacent boundary, the tool will fail to create a new polygon.

6
Editing Tabular Data

In *Chapters 4* and *5*, you learned how to edit and create spatial features including points, lines, and polygons. Spatial data is only one half of the GIS puzzle. Each layer you see in GIS has two components, **spatial** and **tabular**. The tabular component contains additional information about features you see in the map or scene. This might include information such as the feature's size, when it was created, coordinate values, and more. Of course, tabular data used in a GIS is not limited to just information directly related to features you see. It can also include standalone tables as well.

As with creating or editing spatial data, ArcGIS Pro contains a host of tools you can use to update data stored in a table. These tools allow you to edit individual values one at a time or multiple features at the same time. ArcGIS Pro also provides multiple locations in which you can edit tabular data so that you can use the tools and locations that are most effective in your environment.

In this chapter, you will learn how to edit attributes and tabular data using various tools and methods. You will see how to edit single values as well as ways you can perform mass edits using a single value or an expression. The recipes included in this chapter are the following:

- Editing individual attributes using the **Attributes** pane
- Editing multiple attributes with a single edit using the **Attributes** pane
- Editing individual attributes in the table view
- Using the **Calculate Field** tool to populate multiple features
- Using the **Calculate Geometry** tool to populate values for multiple features

Editing individual attributes using the Attributes pane

Now that you have the basic skills to create and edit spatial features, it is time to look at the other half of GIS data: **attributes**. Attributes are stored in a tabular format and contain additional information about features you see on the map. The information included depends on several factors, such as what the feature is, what fields are included in the table, and any integrations or links with other data that might exist. For example, if you were looking at parcel features, these would typically be a polygon feature type, and you might find attribute fields that would store the area and perimeter, the owner's name, address, total value, and the parcel identification number of each parcel. If you were looking at a sanitary sewer line or manhole, these would be different feature types and have a different list of attributes. Sewer lines might have attributes to identify the pipe material and size, install date, depth underground, slope, and manholes it connects to. Manholes might have attributes that store the rim elevation, depth, and construction material. These are just a few examples so that you can see that the attributes you need to update will vary. ArcGIS Pro provides many tools and methods you can use to update attributes associated with features.

In this recipe, you will learn how to use the **Attributes** pane to update values associated with features. You will update attributes for the road centerlines you created in *Chapter 5*'s *Creating new line features* recipe.

Getting ready

Before starting this recipe, you will need to have completed *Chapter 4*'s *Configuring editing options* and *Splitting line features* recipes and *Chapter 5*'s *Creating new line features* recipe. This recipe can be completed with all licensing levels of ArcGIS Pro. You will also need to ensure you have internet access to make use of the Esri-provided basemaps coming from ArcGIS Online.

How to do it...

Now, you will update attributes for the street centerlines you created in *Chapter 5*'s *Creating new line features* recipe using the **Attributes** pane. This will include updating values for the `ST_NAME`, `RD_CLASS`, and `Condition` fields found in the attribute table for the `Street_Centerlines` layer:

1. Launch ArcGIS Pro and open the `EditingAttribute.aprx` project file located in `C:\Student\ArcGISPro3Cookbook\Chapter6\` by clicking on the **Open another project** button in the ArcGIS Pro start window and navigating to the indicated folder.

   The project will open with a single map named `Forrest Park Subdivision`. The map contains four layers (`Street_Centerlines`, `City Limits`, `Parcels`, and `Forrest Park Subdivision.jpg`) plus the `World Topographic` map. You should be zoomed in on a familiar area, the new Forrest Park subdivision.

2. In the **Contents** pane, click on the **List by Editing** button and set the `Street_Centerlines` layer to be the only editable layer. Do this by right-clicking on the layer and selecting **Make this layer the only editable layer** from the menu that appears.

3. Click on the **List by Selection** button in the **Contents** pane. Set the `Street_Centerlines` layer as the only selectable layer.

4. Activate the **Edit** tab in the ribbon and click on the **Select** tool.

5. Click on the centerline segment for Oak Place, as illustrated in the following screenshot. If the centerline for Oak Place consists of multiple lines, you can use your *Shift* key to select multiple segments:

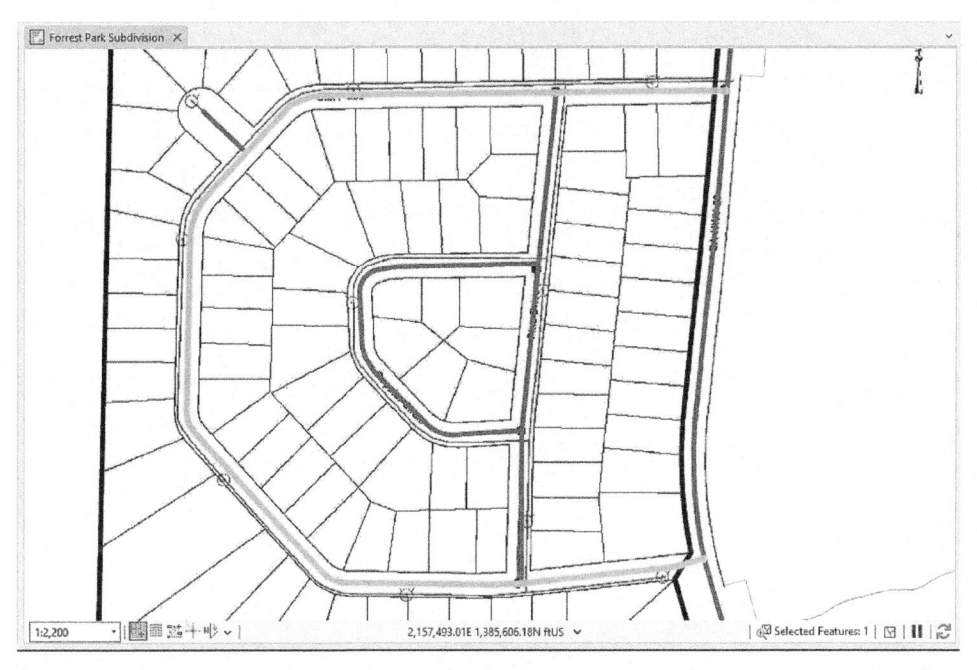

Figure 6.1 – Selecting the Oak Place centerline

6. Click on the **Attributes** button located on the **Edit** tab in the ribbon. This will open the **Attributes** pane on the right-hand side of the interface by default.

The **Attributes** pane is made up of two panels, as shown in the following screenshot:

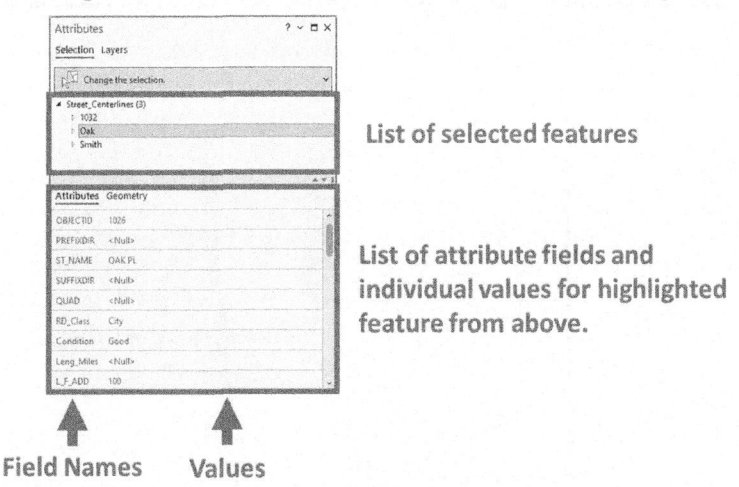

Figure 6.2 – Attributes pane layout

As you can see, the top panel displays a list of selected features. In your case, you should see only one. The bottom panel displays a list of fields and the associated value for each field.

7. Click in the cell located next to the ST_NAME field and type OAK PL. Press the *Enter* key when done.

8. Using the same process, fill in the values for the following fields in the **Attributes** pane. Then, click the **Apply** button:

   - RD_Class – City
   - Condition – Good
   - L_F_Add – 100
   - L_T_Add – 302
   - R_F_Add – 101
   - R_T_Add – 299
   - Name – Oak
   - TYPE – PL
   - Toll – N
   - Speed – 25
   - One_Way – N

9. Save your attribute edits by clicking on the **Save** button on the **Edit** tab in the ribbon.

10. Next, use the **Select** tool on the **Edit** tab to select the centerline for Pine Drive, which runs north to south in connection to Oak Place, as shown in the following screenshot:

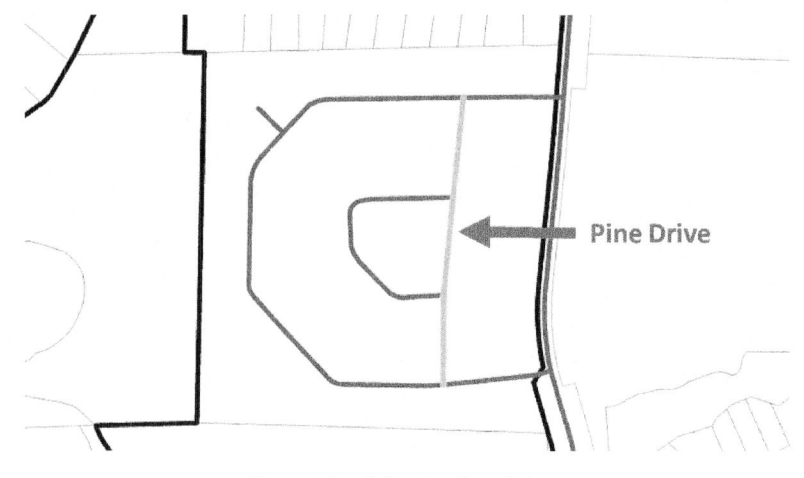

Figure 6.3 – Selecting Pine Drive

11. In the **Attributes** pane, update the following values for Pine Drive. Remember to click the **Apply** button when you have updated all values:

    - ST_NAME – PINE DRIVE
    - RD_Class – City
    - Condition – Good
    - L_F_Add – 100
    - L_T_Add – 152
    - R_F_Add – 101
    - R_T_Add – 150
    - Name – Pine
    - TYPE – Dr
    - Toll – N
    - Speed – 25
    - One_Way – N

12. Save your attribute edits by clicking on the **Save** button on the **Edit** tab in the ribbon.

13. Select **Popular Circle** and update its attributes using the same process. Use the following values:

    - ST_NAME – POPULAR CIR
    - RD_Class – City
    - Condition – Good
    - L_F_Add – 100
    - L_T_Add – 122
    - R_F_Add – 101
    - R_T_Add – 119
    - Name – Popular
    - TYPE – Cir
    - Toll – N
    - Speed – 25
    - One_Way – N

14. On the **Edit** tab in the ribbon, click on the **Clear** button to deselect all current features.
15. Close the **Attributes** pane.
16. Save your project and close ArcGIS Pro.

You have just updated attribute values for street centerline segments for the new Forrest Park subdivision using the **Attributes** pane. This is not the only way to update attributes, as you will see in recipes later in this chapter.

How it works...

In this recipe, you updated attribute values for street centerline features. Attributes are the second component of GIS data and provide descriptive information about features you see on the map. These can be used to query data and perform analysis. So, it is important to ensure they are updated.

To accomplish this, you first ensured the `Street_Centerlines` layer was both selectable and editable in the **Contents** pane. Next, you activated the **Edit** tab in the ribbon and used the **Select** tool to select the centerline representing Oak Place. Then, you clicked on the **Attributes** button to open the **Attributes** pane. In the **Attributes** pane, you updated values for several attribute fields by clicking in the cell located to the right of the field name and typing in the new value. Once you updated all the values, you clicked the **Apply** button and then saved your edits.

It should be mentioned that it is a best practice to update attribute values as you create new features and not wait till later. That ensures you do not forget to make those updates.

Editing multiple attributes with a single edit using the Attributes pane

Now that you know how to edit attributes for single features, it is time to look into how you can edit attributes for multiple features at a single time using the **Attributes** pane. It is not uncommon to find yourself in a situation where you apply the same value to multiple features. Being able to do this with a single edit instead of having to do it one feature at a time can greatly increase the efficiency of your editing workflow.

In this recipe, you will update attributes for the multiple sewer lines that you created in *Chapter 4*'s *Configuring editing options* and *Splitting a line feature* recipes with a single edit using the **Attributes** pane.

Getting ready

To complete this recipe, you need to have first completed the previous *Configuring editing options* and *Splitting a line feature* recipes found in *Chapter 4*. This recipe can be completed with any license level of ArcGIS Pro (**Basic**, **Standard**, or **Advanced**).

How to do it...

You will now use the **Attributes** pane to update attributes for several sewer lines with a single edit:

1. Launch ArcGIS Pro and open the `EditingAttribute.aprx` project. This project is located in `C:\Student\ArcGISPro3Cookbook\Chapter6\` if you installed the data in the default location. If you completed the previous recipe from this chapter, this project should be in your **Recent Projects** list.
2. In the **Contents** pane, select the **List by Drawing Order** button at the top of the pane. The icon for this button resembles a folder directory tree.
3. Turn off the `Street_Centerlines` layers in the **Contents** pane so that they are not displayed in the **City of Trippville** map.
4. If needed, turn on the `Forrest Park Subdivision.jpg` layer in the **Contents** pane so that it is visible on the map.
5. Activate the **Map** tab in the ribbon. Then, click on the **Add Data** button in the **Layer** group.
6. Select the **Databases** folder under **Project** in the panel on the left-hand side of the **Add Data** window.
7. Double-click on the `Trippville_GIS.gdb` geodatabase in the panel on the right-hand side of the **Add Data** window.
8. Double-click on the `Sewer` feature dataset. Feature datasets are symbolized by an icon that looks like three overlapping squares.
9. Select the `sewer_lines` feature class using a single click, then click **OK** to add `sewer_lines` as a layer to your map.

You have now added the layer you need to update to your map. However, you are not quite ready to begin editing attributes just yet.

Setting selectable and editable layers

Now that you have added the `sewer_lines` layer to your map, you will need to set the layer to be selectable and editable:

1. Select the newly added `sewer_lines` layer in the **Contents** pane. Then, click on the **List by Editing** button. It is the button that resembles a pencil.
2. Right-click on the `sewer_lines` layer and select **Make this the only editable layer** from the menu that appears.

Your **Contents** pane should only show the `sewer_lines` layer with a checkmark next to it, as shown in the following screenshot. This means it is the only layer in the map that you will be able to edit. This prevents you from accidentally editing another layer by accident:

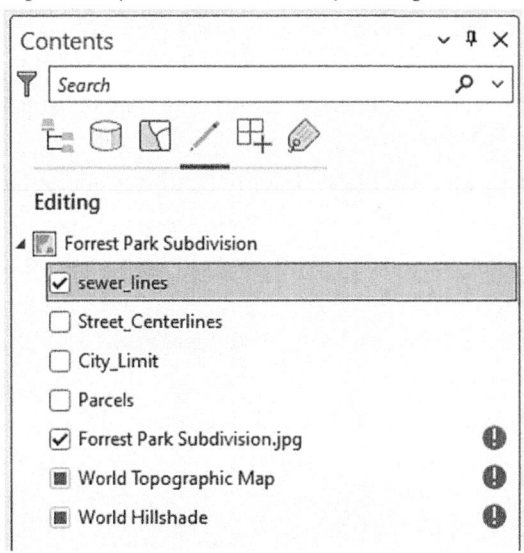

Figure 6.4 – Setting the sewer_lines layer as the only editable layer

3. Click on the **List by Selection** button in the **Contents** pane. This button looks like three polygons, with one being light blue and two being white.

4. Right-click on the `sewer_lines` layer and select **Make this the only selectable layer** from the menu that appears.

Similar to when you set this layer to be the only editable layer, this should be the only layer with a checkmark to the left of it, as shown in the following screenshot:

Editing multiple attributes with a single edit using the Attributes pane   207

Figure 6.5 – Setting the sewer_lines layer as the only selectable layer

5. Save your project using the **Save Project** button on the **Quick Access** toolbar located in the upper left-hand corner of the ArcGIS Pro interface.

Editing attributes for multiple features from the Attributes pane

You are now ready to begin editing attributes for the sewer lines. Since these are new sewer lines, they have many of the same values. You will use the **Attributes** pane to update attribute values for several sewer-line segments at a single time:

1. Activate the **Edit** tab in the ribbon and click on the **Select** tool.
2. In the map, click at a location to the southeast of all the sewer lines and drag your mouse toward the northwest, as shown in the following screenshot:

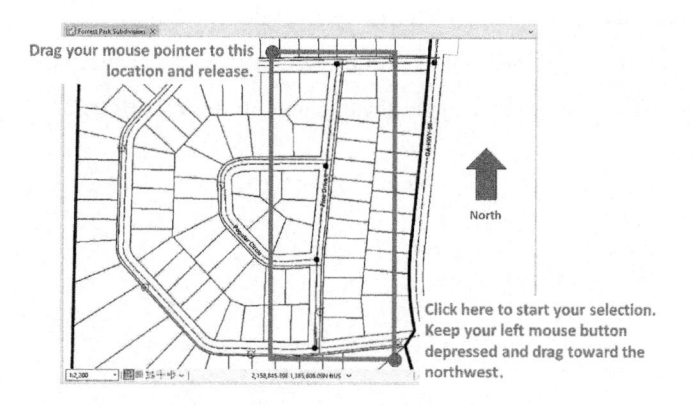

Figure 6.6 – Selecting sewer-line segments in the map

3. On the **Edit** tab, click on the **Attributes** button to open the **Attributes** pane.

   In the **Attributes** pane, you should see that four sewer line features have been selected. You will now update several attributes associated with these selected features. You could update these individually using the method described in the previous recipe, *Editing individual attributes using the Attributes pane*. However, since these sewer lines are all new and have the same values for several attributes, there is a quicker and easier way.

4. In the **Attributes** pane, select where it says `sewer_lines (4)`, as illustrated in the following screenshot:

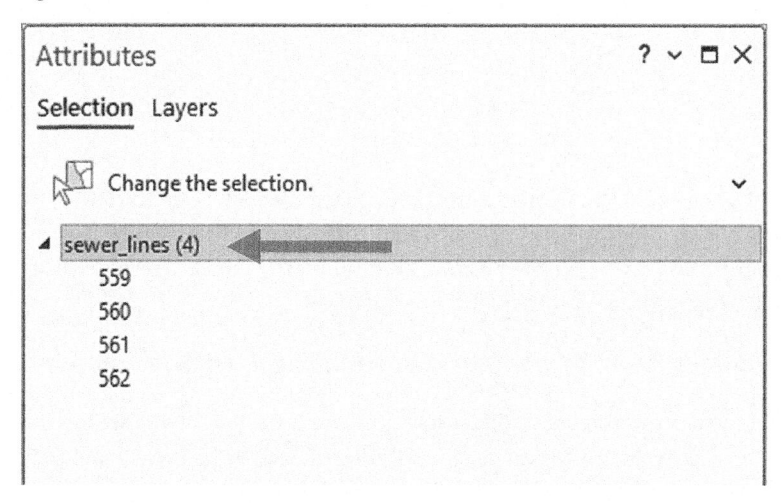

Figure 6.7 – Selecting sewer_lines (4) in the Attributes pane

5. In the **Attributes** pane, click on the cell located to the right of **PIPE SIZE**. Type `10` and press the *Enter* key.
6. Click on the cell to the right of **MATERIAL** and select `PVC` from the list that appears.
7. Next, click in the cell to the right of **Condition** and select `Good` from the list that appears.
8. Verify your **Attributes** pane matches the following and click **Apply**:

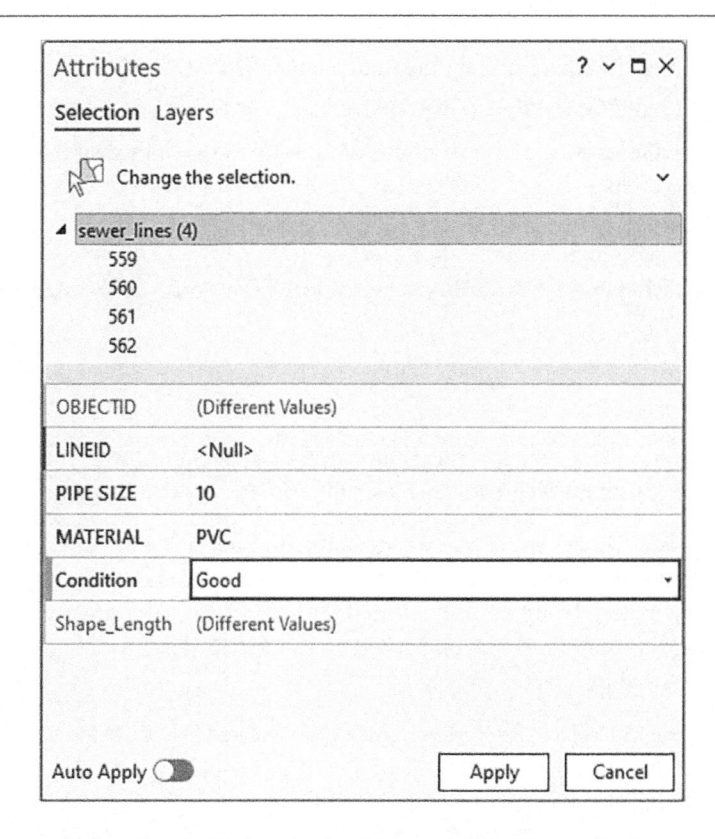

Figure 6.8 – Attributes pane with sewer-line attributes filled in

9. On the **Edit** tab in the ribbon, click on the **Save** button to save the edits you just applied.

Verifying the results of the edits

You have just updated the values for the **PIPE SIZE**, **MATERIAL**, and **Condition** attributes for all four selected sewer lines by typing the desired values in once for each field. Let's verify that this worked and that all four pipes are updated:

1. In the **Attributes** pane, select the first value located below `sewer_lines (4)` so that you can see the values associated with that specific sewer line.

2. Look at the attribute values for the **PIPE SIZE**, **MATERIAL**, and **Condition** fields. Verify they match the values you entered previously. **PIPE SIZE** should be `10`, **MATERIAL** should be `PVC`, and **Condition** should be `Good`.

3. Select the other three records in the **Attributes** pane and compare their values to what you had entered to see whether they match.

If you performed the edits correctly, the values in the **PIPE SIZE**, **MATERIAL**, and **Condition** fields should be the same for all four features.

4. Close the **Attributes** pane and save your project using the **Save Project** button located on the **Quick Access** toolbar.
5. Close ArcGIS Pro.

You have just successfully updated attributes for four different features in a single layer using the **Attributes** pane.

How it works...

In this recipe, you were able to edit attribute values for four different features with a single edit using the **Attributes** pane. This method greatly increases the efficiency of editing data.

In order to do this, you first set layers that were editable and selectable in the **Contents** pane to the layer you wished to update. For this recipe, this was the sewer_lines layer. Once you completed this, you next enabled the **Editing** tab and selected the four sewer-line segments that you added in *Chapter 4* using the **Select** tool. Next, you clicked on the **Attributes** button on the **Edit** tab in the ribbon to open the **Attributes** pane.

In the **Attributes** pane, you selected sewer_lines (4) located above the four individual segments you had selected. Then, you updated the values for the **PIPE SIZE**, **MATERIAL**, and **Condition** fields in the cells located in the bottom panel of the **Attributes** pane. Lastly, you clicked the **Apply** button to commit those updates back to the features. Of course, you clicked the **Save** button on the **Edit** tab to save the updates back to the Trippville_GIS.gdb geodatabase.

You could have updated attribute values for each segment one at a time. That would have been much more time-consuming and repetitive. This method was much more efficient.

Editing individual attributes in the table view

If you have completed the previous recipes in this chapter, you have learned how to update attribute values using the **Attributes** pane. You experienced editing attributes for individual features and multiple features. This is just one way you can edit attributes.

ArcGIS Pro also allows you to edit attributes directly in a table when you view the attribute table of an entire layer. As with the **Attributes** pane, you can edit individually or for multiple features at one time.

In this recipe, you will learn how to edit attributes for individual features or records in the table view. You will update attributes for the manholes you created in *Chapter 4*'s *Creating new point features* recipe. As you will see, editing attributes for individual features in the table view is very similar to editing in the **Attributes** pane.

Getting ready

Before starting this recipe, you will need to have completed *Chapter 4*'s *Configuring editing options* and *Creating new point features* recipes. This recipe can be completed with all licensing levels of ArcGIS Pro. You will also need to ensure you have internet access to make use of the Esri-provided basemaps.

How to do it...

Now, it is time for you to update the attributes for the sewer manholes you created in *Chapter 4*. You will do that in the table view, which displays the attribute table for layers:

1. Launch ArcGIS Pro and open the `EditingAttributes.aprx` project. This project is in `C:\Student\ArcGISPro3Cookbook\Chapter6\` if you installed the data in the default location. If you completed previous recipes in this chapter, this project should be in your **Recent Projects** list.

2. You should still be zoomed in on the new Forrest Park Subdivision area that contains the Oak Place and Pine Drive streets. If not, activate the **Map** tab in the ribbon. Then, click on the **Bookmarks** button and select the `Creating Manholes` bookmark from the presented list. If you are zoomed in on the correct location, continue to the next step.

3. On the **Map** tab in the ribbon, click on the **Add Data** button to start the process of adding the `Manhole` feature class as a layer in the map.

4. In the left panel of the **Add Data** window, click on **Databases** so that you can see a list of connected databases in the right panel.

5. In the right panel, double-click on the `Trippville_GIS.gdb` geodatabase to display its contents.

6. Double-click on the `Sewer` feature dataset in the right panel. Then, select the `Manhole` feature class and click the **OK** button to add it as a new layer to your map.

> **Tip**
>
> Do not select the `Manhole_Z` feature class if you see it in the feature dataset as well. That feature class will be used in later recipes in this book when you explore 3D data and how to visualize, create, and maintain it.

Setting the symbology for the manhole layer

Now that you have added the `Manhole` feature class from the `Trippville_GIS.gdb` geodatabase to your map as a new layer, you will update the symbology so that the manholes are easier to see. You will do this by importing the symbology from a layer file (`.lyrx`) that already exists:

1. In the **Contents** pane, select the `manhole` layer you just added. When you select the layer, you should see three new tabs appear in the ribbon: **Feature Layer**, **Labeling**, and **Data**.

2. Activate the **Feature Layer** tab in the ribbon. Then, click on the **Import** button to open the **Import Symbology** window.

3. The **Input Layer** parameter should already be set to the manhole layer because that is the layer you selected on the **Contents** pane. To set the **Symbology Layer** parameter, click on the **Browse** button located on the right. It resembles a manila file folder.

4. In the left panel of the **Symbology Layer** window, select **Folders**. Then, double-click on Chapter6 in the right panel.

5. Select the manhole.lyrx layer file and click **OK**.

6. Verify your **Import Symbology** tool matches the following and click **OK**:

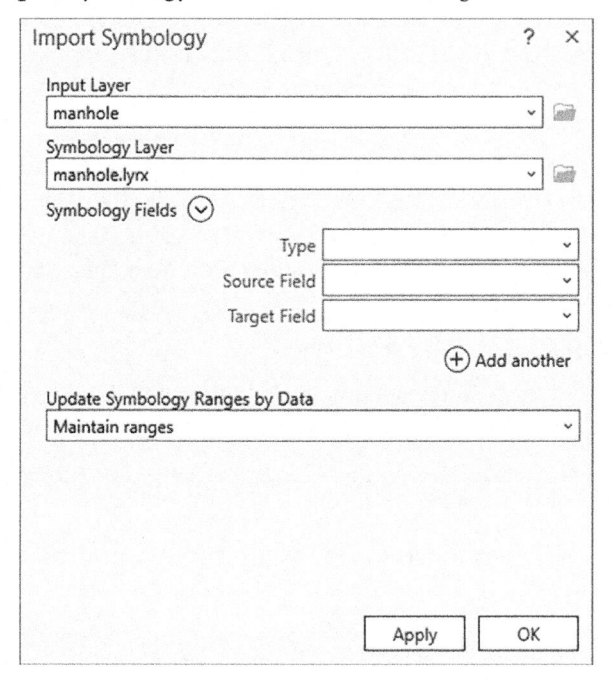

Figure 6.9 – Importing symbology from a layer file

7. Save your project by clicking on the **Save Project** button located on the **Quick Access** toolbar. at the upper left of the ArcGIS Pro interface.

> **Information**
> Layer files are very useful for standardizing various layer properties. In addition to symbology, which you have just seen, layer files can be used to apply standard settings to any layer property, such as visible scale range, definition queries, label settings, elevation, and others.

Configuring layers for editing

Now that you have added the `manhole` layer and updated its symbology, you next need to configure this layer as being selectable and editable. It is best to make this the only selectable and editable layer to ensure you do not accidentally select or edit other layers:

1. In the **Contents** pane, click on the **List by Editing** button. It resembles a pencil.
2. Right-click on the `manhole` layer and select **Make this the only editable layer** from the menu that appears.
3. In the **Contents** pane, click on the **List by Selection** button (blue and white polygon icon). Then, right-click on the `manhole` layer and select **Make this the only selectable layer** from the menu that appears.
4. Next, click on the **List by Editing** button in the **Contents** pane. Right-click on the `Buildings` layer once again and select **Make this the only editable** layer from the menu.
5. In the **Contents** pane, right-click on the `manhole` layer and select **Attribute Table** from the menu that appears, as illustrated in the following screenshot. This will open the attribute table for the `manhole` layer:

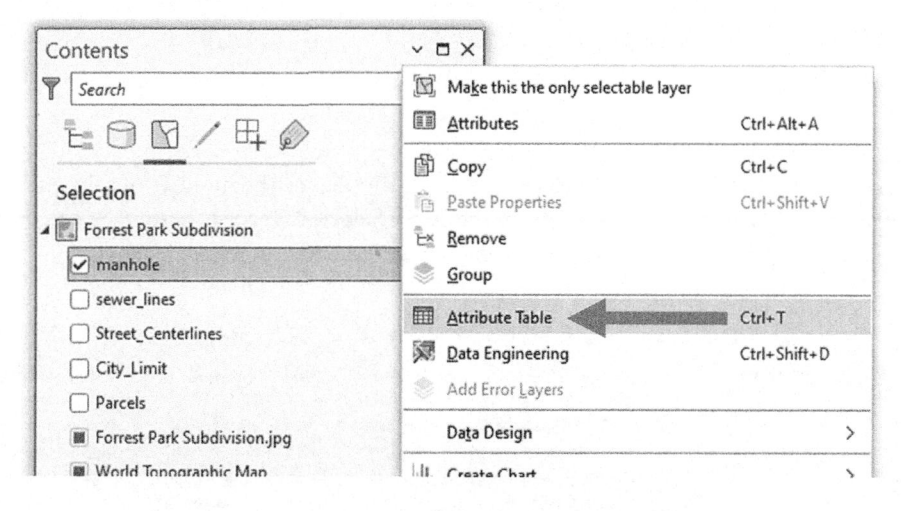

Figure 6.10 – Opening the attribute table for the manhole layer

> **Tip**
> The table will open at the bottom of the ArcGIS Pro interface. You can undock it and move it to another location including a different monitor if desired. This can make seeing the table and the map at the same time easier. To undock the table, simply click on the tab containing the name of the open table and drag it to the new location.

Selecting manholes and editing attributes

Now, you are ready to select manholes and update their attributes in the attribute table:

1. Activate the **Edit** tab in the ribbon and click on the **Select** tool.
2. Click on the manhole located at the northern intersection of Oak Place and Pine Drive, as shown in the following screenshot:

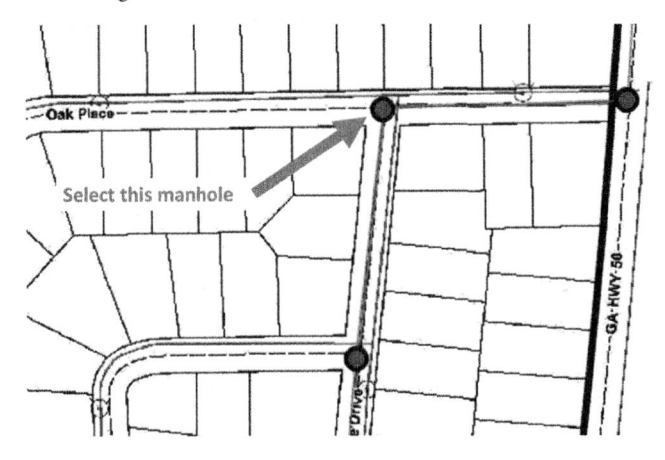

Figure 6.11 – Selecting a manhole

3. Look at the bottom of the **Table** window. You should see the number of selected features displayed, showing 1 of 545. Click on the **Show Selected Records** button located to the left of the text saying **1 of 545 selected**, as illustrated in the following screenshot:

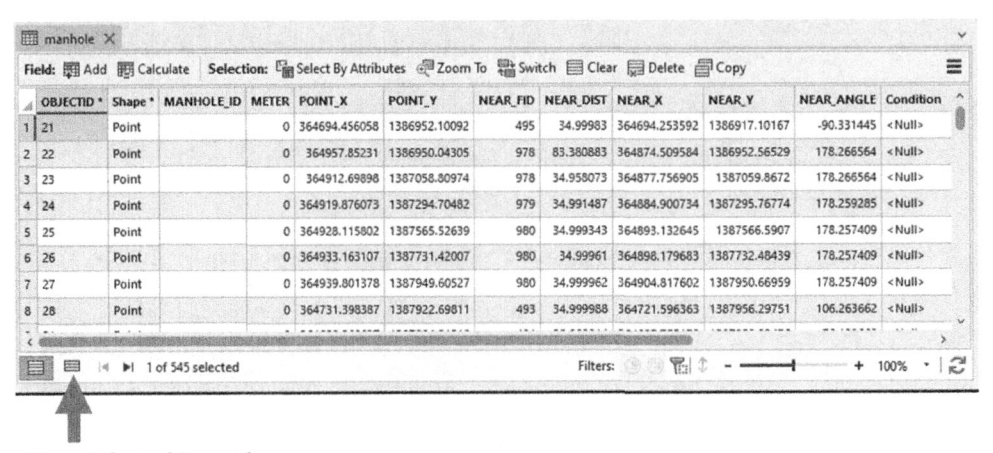

OBJECTID *	Shape *	MANHOLE_ID	METER	POINT_X	POINT_Y	NEAR_FID	NEAR_DIST	NEAR_X	NEAR_Y	NEAR_ANGLE	Condition
1	21	Point		0	364694.456058	1386952.10092	495	34.99983	364694.253592	1386917.10167	-90.331445	<Null>
2	22	Point		0	364957.85231	1386950.04305	978	83.380883	364874.509584	1386952.56529	178.266564	<Null>
3	23	Point		0	364912.69898	1387058.80974	978	34.958073	364877.756905	1387059.8672	178.266564	<Null>
4	24	Point		0	364919.876073	1387294.70482	979	34.991487	364884.900734	1387295.76774	178.259285	<Null>
5	25	Point		0	364928.115802	1387565.52639	980	34.999343	364893.132645	1387566.5907	178.257409	<Null>
6	26	Point		0	364933.163107	1387731.42007	980	34.99961	364898.179683	1387732.48439	178.257409	<Null>
7	27	Point		0	364939.801378	1387949.60627	980	34.999962	364904.817602	1387950.66959	178.257409	<Null>
8	28	Point		0	364731.398387	1387922.69811	493	34.999988	364721.596363	1387956.29751	106.263662	<Null>

1 of 545 selected

Show Selected Records

Figure 6.12 – Show Selected Records button location

4. In the `manhole` attribute table, scroll until you see the **Depth** field. Click in the cell located below **Depth**, type 4.0, and press the *Enter* key.
5. Next, click in the cell located below **Basin** and type Sweetwater. Press the *Enter* key when you are done.
6. Using the same process, select each manhole located along Pine Drive and update the values for the **Depth** and **Basin** fields. The values for the **Depth** field should decrease by .5 feet as you move south along Pine Drive. The **Basin** value should be Sweetwater for each manhole.
7. Save your edits by clicking on the **Save** button on the **Edit** tab in the ribbon.
8. On the **Edit** tab, click on the **Clear** button to deselect any selected features you might have.
9. Save your project and close ArcGIS Pro.

You have just updated the attributes for the manholes from the attribute table. As you have seen, this is not much different from using the **Attributes** pane.

How it works...

In this recipe, we updated the **Depth** and **Basin** values for up to four manholes directly within the `manhole` layer's attribute table. To do this, we followed several steps.

First, you had to add the `Manhole` feature class from the `Trippville_GIS.gdb` geodatabase and the `Sewer` feature dataset to the map as a new layer. You then updated the symbology of the `manhole` layer so that you could see it better by importing a layer file (.lyrx). Next, you set the newly added `manhole` layer as being the only selectable and editable layer in the map from the **Contents** pane.

With the prep work done, you then proceeded to edit the data by selecting one manhole and opening the attribute table. You then set the attribute table so that it only showed selected records, clicked in the cells for the depth of the selected manhole, and typed the depth of the manhole, which was 4 feet. Next, you updated the value for the **Basin** field to Sweetwater by typing that into the cell located below **Basin**.

You then selected the other manholes located along Pine Drive and updated the depth and basin values for each manhole individually. When you saved all your edits, they were committed back to the `Trippville_GIS.gdb` geodatabase.

You now know two different methods to update attribute values for individual features. You can use the **Attributes** pane or go directly to the attribute table for the layer. Both methods accomplish the same purpose and take very similar effort. So, you can use whichever one works best for you and your situation.

Using the Calculate Field tool to populate multiple features

You have now seen how to edit individual attributes for single features from within the attribute table. But what if you need to update several features at the same time with the same value or use a calculation in the attribute **Table** window? ArcGIS Pro includes at least a couple of tools that allow you to accomplish this, including the **Calculate Field** and **Calculate Geometry** tools.

In this recipe, we will focus on the **Calculate Field** tool. You will use this tool to populate the pipe material for several selected sewer lines. Next, you will populate a field that combines values found in other fields within the table that will be used for labeling.

Getting ready

Before starting this recipe, you will need to have completed *Chapter 4*'s *Configuring editing options* and *Splitting line features* recipes. This recipe can be completed with all licensing levels of ArcGIS Pro. You will also need to ensure you have internet access to make use of the Esri-provided basemaps.

How to do it...

You will use the **Calculate Field** tool to update the pipe material for sewer lines. This is a simple example showing what you can do with this tool to begin getting comfortable using it:

1. Launch ArcGIS Pro and open the `EditingAttributes.aprx` project. This should be in your **Recent Projects** list. The project should open with the `Forrest Park Subdivision` map that you have used in previous recipes in this chapter.

2. In the **Contents** pane, make sure the `sewer_lines` layer is visible.

3. Click on the **List by Editing** button in the **Contents** pane and right-click on the `sewer_lines` layer. Select **Make this the only editable layer** from the menu.

4. In the **Contents** pane again, click on the **List by Selection** button. Then, right-click on the `sewer_lines` layer and select **Make this the only selectable layer**.

5. Right-click on the `Forrest Park Subdivision.jpg` layer and select **Zoom to Layer** from the menu that appears. Your map should look similar to the following, depending on your monitor size and resolution:

Using the Calculate Field tool to populate multiple features   217

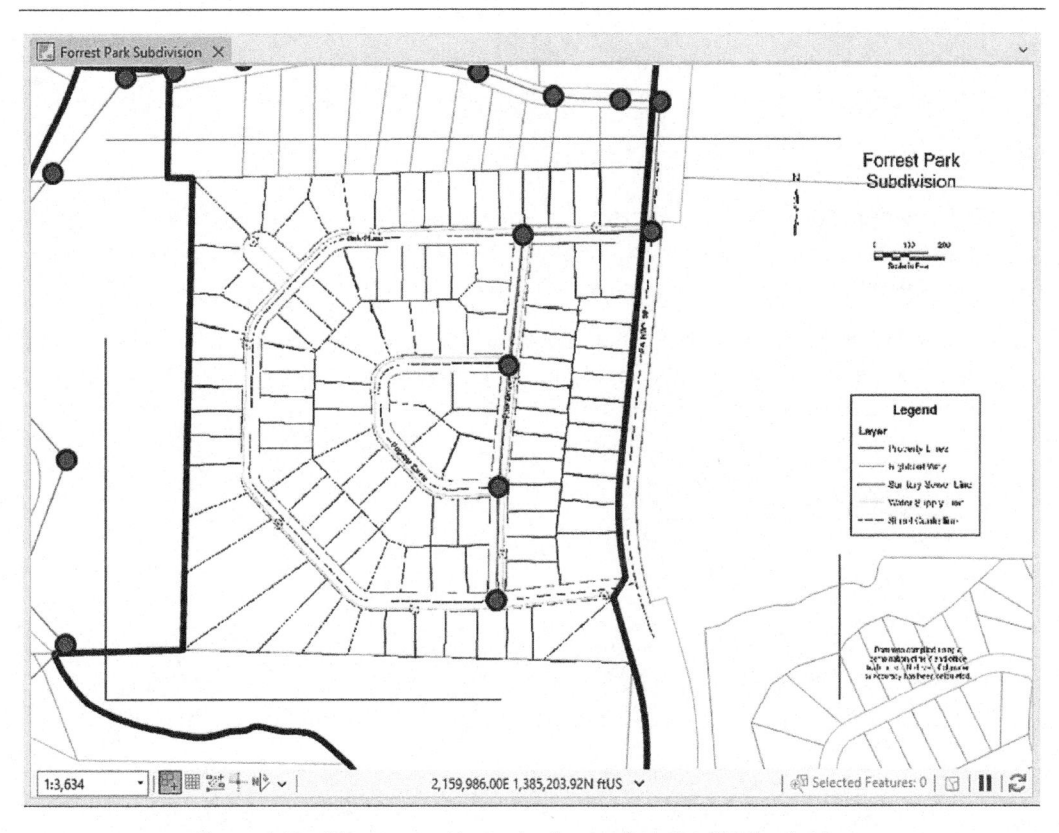

Figure 6.13 – Map zoomed in on the Forrest Park Subdivision.jpg layer

6. Activate the **Edit** tab in the ribbon and click on the **Select** button.
7. In the map, draw a selection rectangle that matches the following representation as closely as possible:

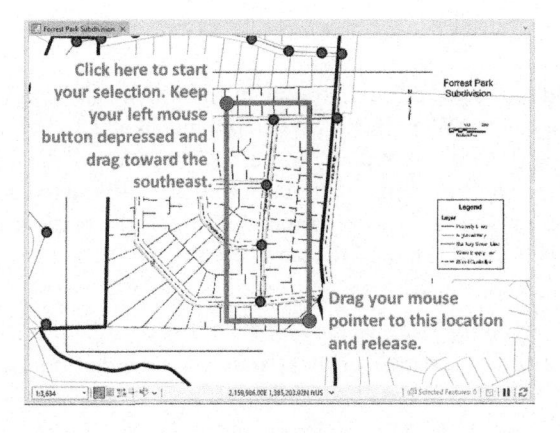

Figure 6.14 – Selecting sewer-line features in the map

You should have four features selected in the `sewer_lines` layer. You can verify this by looking at the lower right of the map where it says **Selected Features**.

Opening the attribute table and using the Calculate Field tool

Now that you have the sewer-line features that you will need to edit selected, you will open the attribute table for the `sewer_lines` layer and use the **Calculate Field** tool to update the material for the selected pipes:

1. At the bottom of the attribute table for the `sewer_lines` layer, click on the **Show Selected Records** button so that you only see the four sewer-line features you have selected. The attribute table should look like the following now:

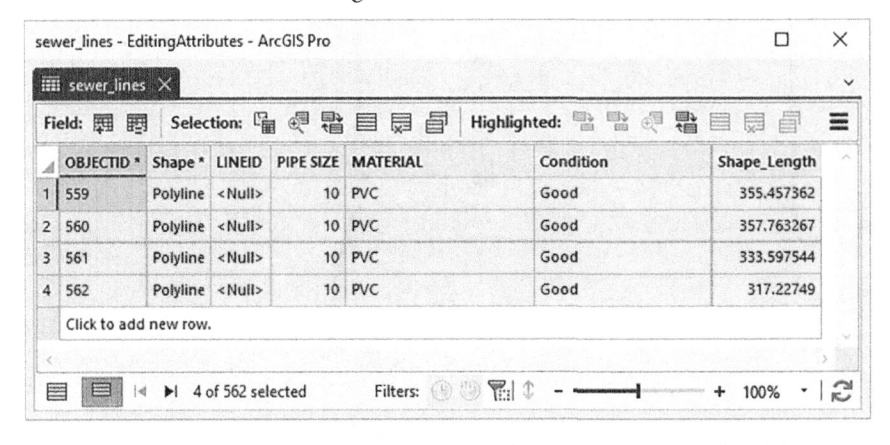

OBJECTID *	Shape *	LINEID	PIPE SIZE	MATERIAL	Condition	Shape_Length
559	Polyline	<Null>	10	PVC	Good	355.457362
560	Polyline	<Null>	10	PVC	Good	357.763267
561	Polyline	<Null>	10	PVC	Good	333.597544
562	Polyline	<Null>	10	PVC	Good	317.22749

4 of 562 selected

Figure 6.15 – Showing selected records in the sewer_lines attribute table

2. Right-click on the **MATERIAL** field and select **Calculate Field** from the menu that appears. The **Calculate Field** tool window should open.

> **Tip**
>
> If the **Calculate Field** tool is grayed out in the menu and will not allow you to select it, there are a few reasons you might encounter this issue. First, you might not have write or edit permissions to the data because of where it is stored, such as on a server. Second, the field might be marked as read-only, not allowing edits. Third, you may have selected **Enabled and disabled editing** from the **Edit** tab option on the **Project** tab and **Options**. This will require you to click on the **Edit** button, which will appear on the **Edit** tab in the ribbon.

The **Input Table** parameter should be automatically populated with the `sewer_lines` layer. The **Field Name (Existing or New)** parameter should also be automatically populated with the **MATERIAL** field. You will now build a simple expression to update the values in the **MATERIAL** field from `PVC` to `DI`, which stands for **Ductile Iron**.

3. In the cell located below MATERIAL =, type "DI", as shown in the following screenshot. Then, click the **Validate** button, which looks like a green checkmark, located just above the **Apply** button:

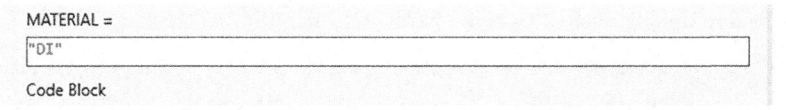

MATERIAL =
"DI"
Code Block

Figure 6.16 – Entering a new value to be populated in the table

4. When ArcGIS Pro finishes validating your expression, it should say Expression is valid. When you see it says that, click the **OK** button to populate the table with the new value.

> **Warning**
>
> If you run the **Calculate Field** tool with no features selected, it will update values for all features in the input data or layer. If you do not have **Enable Undo** turned on, then you will not be able to undo the results of the tools if you make a mistake.

The values in the **MATERIAL** field should now all show DI in place of the former PVC value. You have just updated the values for the four selected features using the **Field Calculate** tool.

Creating a new field and populating it with values from other existing fields

Next, you will add a new field to the sewer_lines attribute table and use the **Calculate Field** tool to populate it with values from other existing fields in the table.

1. In the **Contents** pane, ensure the sewer_lines layer is still selected. Then, click the **Data** tab in the ribbon.
2. Select the **Fields** button in the **Data** tab. This will open a **Fields: sewer_lines** tab in the **Table** window, as illustrated in this screenshot:

Figure 6.17 – The Fields: sewer_lines tab in the Table window

3. Click where it says **Click here to add a new field.**. In the **Field Name** cell, which should be automatically selected, type Map_Label and press the *Enter* key.

4. Click in the cell located under **Data Type** and select Text using the drop-down arrow. You may need to double-click in the cell to access the drop-down arrow.

5. Scroll over and select the cell located under **Length**. Type 100 and press the *Enter* key.

6. Verify that your **Fields: sewer_lines** window matches the following and click the **Save** button on the **Data** tab in the ribbon to create your new field:

Visible	Read Only	Field Name	Alias	Data Type	Allow NULL	Highlight	Number Format	Domain	Default	Length
✓	✓	OBJECTID	OBJECTID	Object ID			Numeric			
✓		Shape	Shape	Geometry	✓					
✓		LINEID	LINEID	Double	✓		Numeric			
✓		SIZE_	PIPE SIZE	Float	✓		Numeric			
✓		MATERIAL	MATERIAL	Text	✓			pipe_mat		16
✓		Condition	Condition	Text	✓			Condition		50
✓	✓	Shape_Length	Shape_Length	Double	✓		Numeric			
✓		Map_Label		Text	✓					100

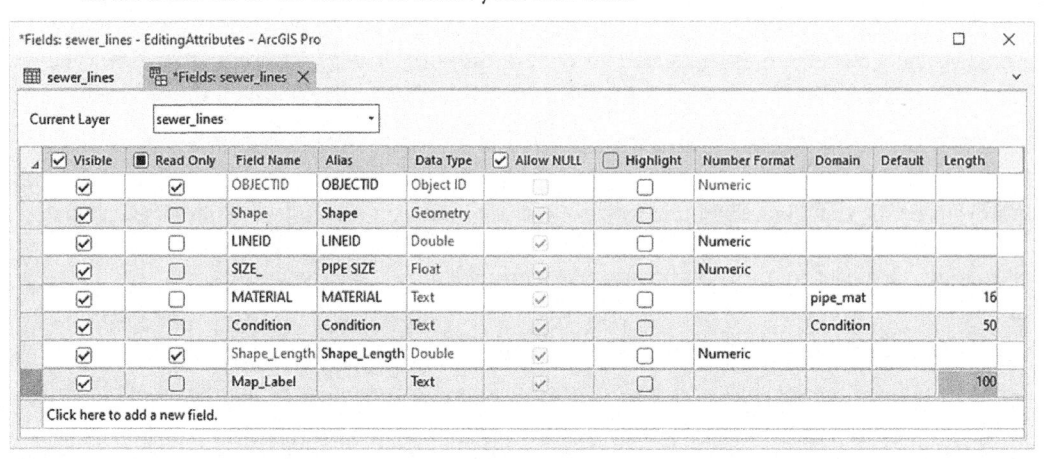

Figure 6.18 – New Map_Label field added to sewer_lines layer attribute table

7. In the **Table** window, click on the tab for the sewer_lines table to check the attribute table is visible.

8. Activate the **Edit** tab in the ribbon and click on the **Clear** button to deselect all features. The **Table** window should appear blank because it is set to show selected records and you just cleared your selection, so nothing is selected.

9. Click on the **Show all records** button located in the lower-left corner of the **Table** window, as shown in the following screenshot. Once you click on the button, you should see records appear in the table. They were always there. They were just hidden because you had none selected:

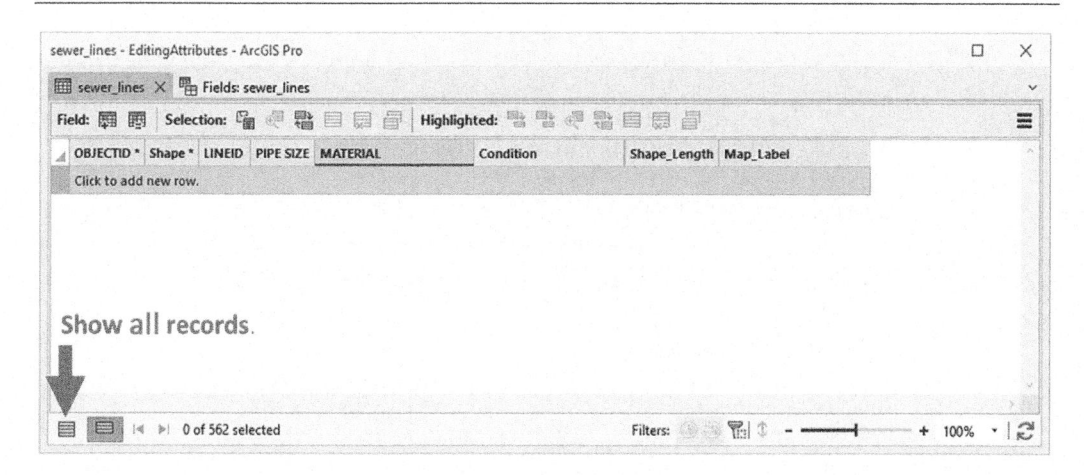

Figure 6.19 – Show all records button

10. Right-click on the `Map_Label` field you just created. Select the **Calculate Field** tool from the menu that appears.

    Once again, the **Calculate Field** tool will open with the parameters for **Input Table** and **Field Name (Existing or New)** automatically populated.

11. Ensure the **Expression Type** parameter is set to `Python`. If it is not, use the drop-down arrow to do so.

12. In the cell located below `Map_Label =`, type `"Pipe Size and Material = "+str(!SIZE_!)+"in "+!MATERIAL!.upper()`, as illustrated in the following screenshot. There is a space after `=` and after `in`:

    ```
    Map_Label =
    "Pipe Size and Material = "+str(!SIZE_!)+"in "+!MATERIAL!.upper()
    Code Block
    ```

    Figure 6.20 – Map_Label expression in Calculate Field tool

13. Click the **Validate** button located below the **Code Block** area. Remember that it looks like a green checkmark.

14. When the **Validate** process completes and says **Expression is valid**, click the **OK** button to run the tool.

    When the **Calculate Field** tool finishes, you should now see values appear in the `Map_Label` field, which had been empty before. Values contained in the field should show something such as `Pipe Size and Material = 8in DI`.

Let's quickly break down the expression so that you understand exactly what it did:

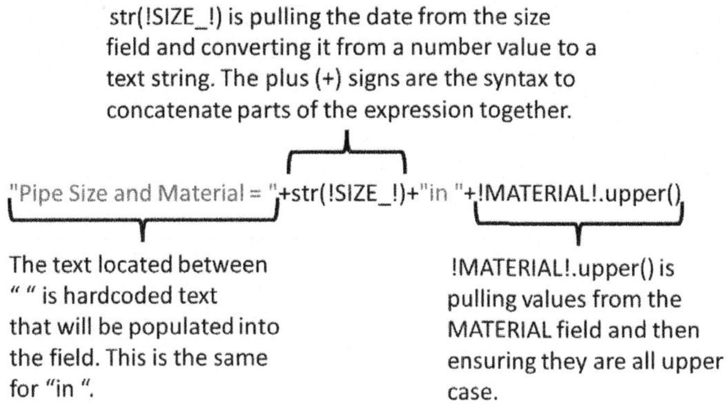

str(!SIZE_!) is pulling the date from the size field and converting it from a number value to a text string. The plus (+) signs are the syntax to concatenate parts of the expression together.

"Pipe Size and Material = "+str(!SIZE_!)+"in "+!MATERIAL!.upper()

The text located between " " is hardcoded text that will be populated into the field. This is the same for "in ".

!MATERIAL!.upper() is pulling values from the MATERIAL field and then ensuring they are all upper case.

Figure 6.21 – The Calculate Field expression explained

15. Close the **Table** window and save your project using the **Save Project** button on the **Quick Access** toolbar in the upper-left corner of the ArcGIS Pro interface.
16. Close ArcGIS Pro.

You have now successfully updated attribute values using the **Calculate Field** tool.

How it works...

You have now learned how you can use the **Calculate Field** tool to update attributes for multiple features from within the attribute table itself. If you have features selected in the indicated input table, this tool will only update those you have selected. If no features are selected, it will update all features or records within the input table. You also learned that the **Calculate Field** tool can use a simple or complex expression to update values within a table.

To use the **Calculate Field** tool, you first had to ensure the desired layer was set to be both selectable and editable in the **Contents** pane. Next, you opened the attribute table for the layer you wished to update; in this recipe, that was the sewer_lines layer. Next, in the table, you right-clicked on the field you were updating and selected **Calculate Field** from the menu that appeared. With the **Calculate Field** tool window open, you updated the **MATERIAL** field by creating a simple expression that replaced the existing value in the table, PVC, with a new value of DI. This expression is simply "DI".

Next, you created a new field called Map_Label in the sewer_lines attribute table from the **Fields** view that lists all attribute fields and their properties for the layer. You then used the **Calculate Field** tool to populate values into the new field you created. The new values included hardcoded text and values found in other fields within the table. This required a much more complex expression that leveraged the capabilities found in the Python scripting language.

It is important to remember that the **Calculate Field** tool does not permanently embed or update values in the table, so the field is automatically updated going forward.

Using the Calculate Geometry tool to populate values for multiple features

You have just seen in the last recipe how to use the **Calculate Field** tool to update attribute values directly in the table itself for multiple features. There is another tool that can do that as well but is focused on spatial values or unit conversions: the **Calculate Geometry** tool. You can use this tool to convert standard measures to metric, convert units to other units, populate different coordinate values into the table, and more. As you will see, this is a very powerful tool.

In this recipe, we will use the **Calculate Geometry** tool to populate x and y coordinates for sewer manholes and to calculate the area of parcels in square meters.

Getting ready

Before starting this recipe, you will need to have completed *Chapter 4's Configuring editing options* and *Creating point features* recipes. This recipe can be completed with all licensing levels of ArcGIS Pro. You will also need to ensure you have internet access to make use of the Esri-provided basemaps.

How to do it...

You will use the **Calculate Geometry** tool to populate x and y coordinate values for sewer manholes and calculate the area of parcels in square meters. These two tasks will demonstrate the power of the **Calculate Geometry** tool:

1. Launch ArcGIS Pro and open the `EditingAttributes.aprx` project. This should be in your **Recent Projects** list. The project should open with the `Forrest Park Subdivision` map you have used in previous recipes in this chapter.
2. In the **Contents** pane, make sure the `manhole` and `Parcels` layers are visible.
3. Click on the **List by Editing** button on the **Contents** pane and right-click on the `manhole` layer. Select **Make this the only editable layer** from the menu.
4. Right-click on the `manhole` layer and select **Attribute Table** from the menu that appears.

The attribute table for the `manhole` layer contains several fields. This includes `POINT_X` and `POINT_Y`, as shown in the following screenshot:

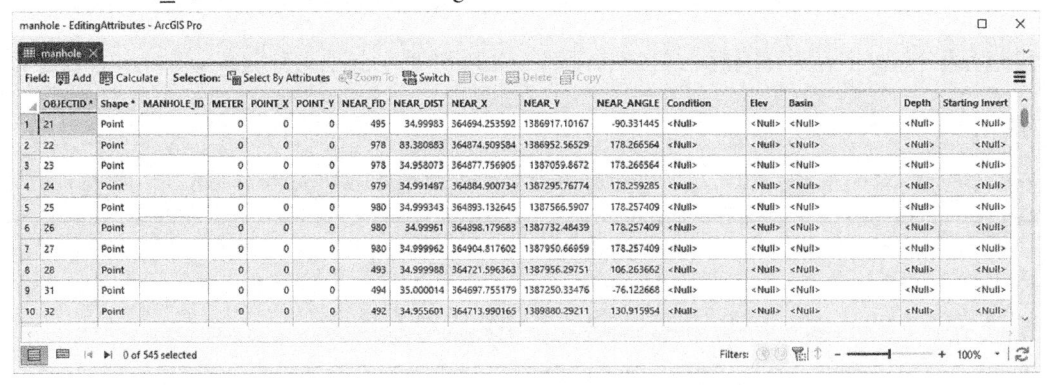

OBJECTID	Shape	MANHOLE_ID	METER	POINT_X	POINT_Y	NEAR_FID	NEAR_DIST	NEAR_X	NEAR_Y	NEAR_ANGLE	Condition	Elev	Basin	Depth	Starting Invert
21	Point		0	0	0	495	34.99983	364694.253592	1386917.10167	-90.331445	<Null>	<Null>	<Null>	<Null>	<Null>
22	Point		0	0	0	978	83.380883	364874.509584	1386952.56529	178.266564	<Null>	<Null>	<Null>	<Null>	<Null>
23	Point		0	0	0	978	34.958073	364877.756905	1387059.8672	178.266564	<Null>	<Null>	<Null>	<Null>	<Null>
24	Point		0	0	0	979	34.991487	364884.900734	1387295.76774	178.259285	<Null>	<Null>	<Null>	<Null>	<Null>
25	Point		0	0	0	980	34.999343	364893.132645	1387566.5907	178.257409	<Null>	<Null>	<Null>	<Null>	<Null>
26	Point		0	0	0	980	34.99961	364898.179683	1387732.48439	178.257409	<Null>	<Null>	<Null>	<Null>	<Null>
27	Point		0	0	0	980	34.999962	364904.817602	1387950.66959	178.257409	<Null>	<Null>	<Null>	<Null>	<Null>
28	Point		0	0	0	493	34.999988	364721.596363	1387956.29751	106.263662	<Null>	<Null>	<Null>	<Null>	<Null>
31	Point		0	0	0	494	35.000014	364697.755179	1387250.33476	-76.122668	<Null>	<Null>	<Null>	<Null>	<Null>
32	Point		0	0	0	492	34.955601	364713.990165	1389880.29211	130.915954	<Null>	<Null>	<Null>	<Null>	<Null>

Figure 6.22 – Attribute table for the manhole layer

These two fields are there to store the *x* and *y* coordinate values that represent the location of manholes. This allows that information to be included if the data is exported into a standalone table or spreadsheet.

5. Right-click on the `Point_X` field and select **Calculate Geometry** from the menu that appears. This will open the **Calculate Geometry** tool.

6. The **Input Features** parameter should automatically be populated with `manhole`. If not, use the drop-down arrow to do so.

7. Under **Geometry Attributes** and the **Field (Existing or New)** column, you should see the first cell on the first row automatically set to `POINT_X` because that is the field you right-clicked on. In the cell under the **Property** column, use the drop-down arrow to set the parameter to `Point x-coordinate`, as illustrated in the following screenshot:

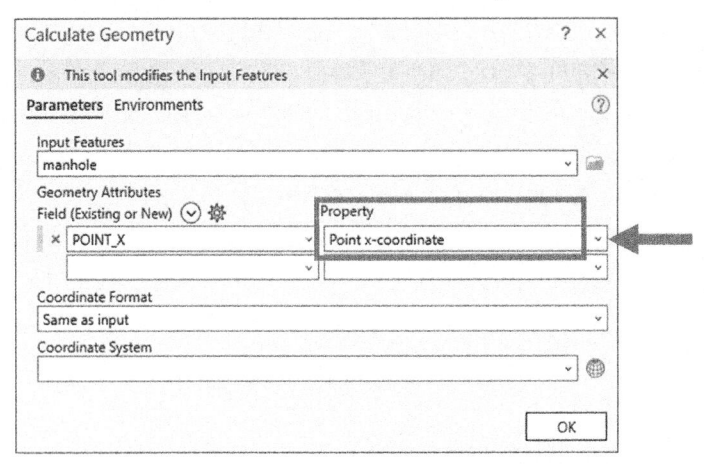

Figure 6.23 – Setting the property to update in the Calculate Geometry tool

8. In the cell located below POINT_X in the **Field (Existing or New)** column, click the drop-down arrow and select POINT_Y.

9. In the cell located to the right of where you just set the POINT_Y value under the **Property** column, select Point y-coordinate using the drop-down arrow.

10. Leave the **Coordinate Format** value set to the Same as input default value.

11. Use the drop-down arrow under **Coordinate System** to select manhole from the list that appears. This will set the coordinate system to NAD_1983_StatePlane_Georgia_West_FIPS_1002_Feet / VCS:NAVD_1988, which is the coordinate system the manhole layer is set to in the geodatabase.

12. Verify your **Calculate Geometry** tool matches the following and click **OK** to run the tool:

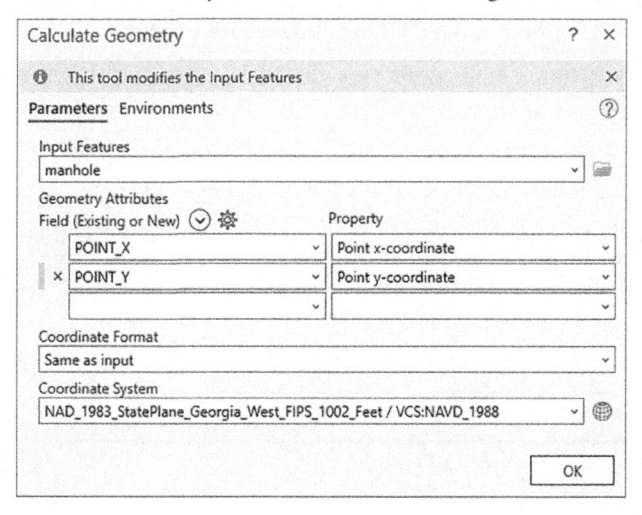

Figure 6.24 – Calculate Geometry tool with parameters filled in

13. When the tool finishes, review the values stored in the POINT_X and POINT_Y fields. They should now have values that are not zero.

14. Close the manhole attribute table and save your project.

You have just used the **Calculate Geometry** tool to populate *x* and *y* coordinates for features in the manhole layer. In this case, these coordinates are in the local state-plane coordinate system.

Calculating the area in square meters from data in feet

The **Calculate Geometry** tool can be used to convert units as well when calculating values. You will see that in action next:

1. In the **Contents** pane, click on the **List by Editing** button and set the Parcels layer as the only editable layer.

2. Right-click on the `Parcels` layer in the **Contents** pane and select **Attribute Table** to open the attribute table for that layer.

3. Locate the **Square Meters** field in the attribute table. This field should contain `<Null>` values. You are going to populate this field with the area of each parcel feature using the **Calculate Geometry** tool.

4. In the **Contents** pane, right-click the `Parcels` layer and select **Properties** from the menu.

5. In the **Layer Properties: Parcels** window, select **Source** from the left panel. Then, scroll down and expand **Spatial Reference**.

6. Look for the **Linear Unit** property and verify it indicates `US Survey Feet`. This is the native unit this data is stored in.

7. Click the **Cancel** button to close the **Layer Properties** window.

8. In the **Parcels** attribute table, right-click on the **Square Meters** field and select **Calculate Geometry** from the menu.

9. Once again, the **Input Features** parameter should automatically be set to `Parcels`. Under the **Property** column, select `Area` using the drop-down arrow.

10. Set the **Area Unit** parameter to `Square Meters` using the drop-down arrow.

11. For the **Coordinate System** parameter, select `Current Map (Forrest Park Subdivision)` using the drop-down arrow.

12. Verify your **Calculate Geometry** tool matches the following and click **OK** to run the tool:

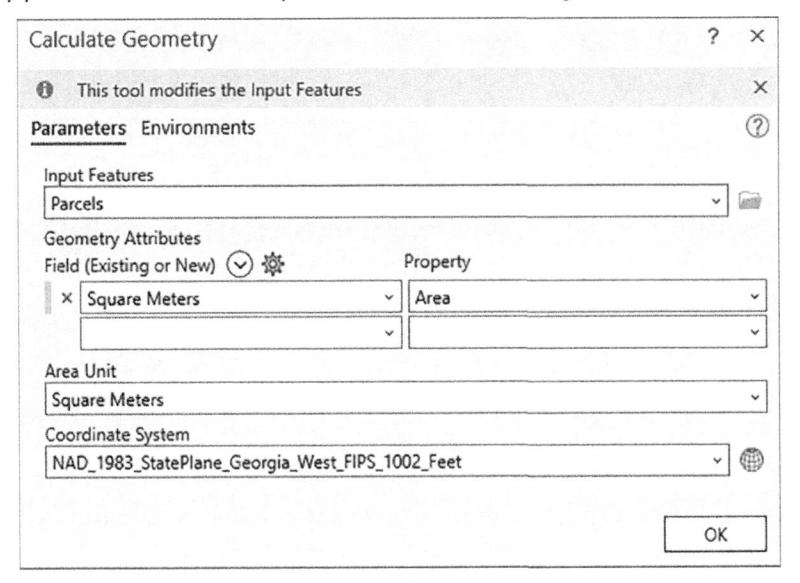

Figure 6.25 – Calculate Geometry for Parcels area in square meters complete

13. When the **Calculate Geometry** tool is completed, review the values that now appear in the **Square Meters** field. They should no longer be <NULL>.
14. Save your project and close ArcGIS Pro.

You have successfully updated the **Square Meters** attribute for all parcels.

How it works...

You have just used the **Calculate Geometry** tool to update attribute values for all features in the manhole and Parcels layers. First, you populated the POINT_X and POINT_Y fields for the manhole using the **Calculate Geometry** tool to extract those values from the spatial location of each manhole. Second, you used the **Calculate Geometry** tool to calculate the area of each parcel and convert that value to square meters, which was then written to the table for each feature.

To start, you had to make the manhole layer editable so that you would be allowed to update values in the attribute table. Next, you opened the attribute table for the manhole layer so that you could see the field and data it contained. Then you right-clicked on the POINT_X field, which was one of the two you needed to update, and selected **Calculate Geometry** from the menu that appeared. In the **Calculate Geometry** tool, you set the property you wanted to populate the POINT_X field with to Point x-coordinate and also added the POINT_Y field with the Point y-coordinate property to the tool. Lastly, you set the **Coordinate System** parameter to the same one used by the manhole layer and clicked **OK** to run the tool.

Next, you switched to the Parcels layer. You followed the same general sequence but set the field to be updated to **Square Meters** and the property to the area. Then, you set the units to square meters and coordinated the system to the current map. When the **Calculate Geometry** tool ran, it populated new values into the **Parcels** attribute table and the **Square Meters** table that represents the total area of each parcel polygon.

You learned several methods that can be used to update attributes for features. These methods allow you to update attributes for single features or multiple features at a time. Which method works best will depend on what you are doing at the time. If you are creating new features or making spatial changes at the same time, it might be easier to use the **Attributes** pane. If you are just focused on updating attribute values and are not making spatial changes, then making updates in the table directly combined with the **Calculate Field** tool or the **Calculate Geometry** tool might be best.

7
Projection and Coordinate System Basics

One of the things that makes GIS such a powerful visualization and analytical tool is its ability to overlay multiple layers of information on a map. Putting our GIS data into real-world coordinate systems and projections is how we do this. By placing our data in a real-world system, we tie it to the Earth. This allows us to locate features anywhere on the Earth's surface and then bring them onto a map so that we can see how those features are related spatially.

There are two basic types of coordinate systems we can use in ArcGIS Pro: geographic and projected. A geographic coordinate system is based on a 3D model of the Earth, called the ellipsoid or spheroid. The ellipsoid is then tied back to the physical Earth by the datum. Geographic coordinate systems use degrees as their primary unit of measurement.

Each degree can then be broken down into different sub-units, such as decimal degrees, minutes, decimal minutes, or minutes and seconds. You will often hear geographic coordinate systems referred to as latitude and longitude:

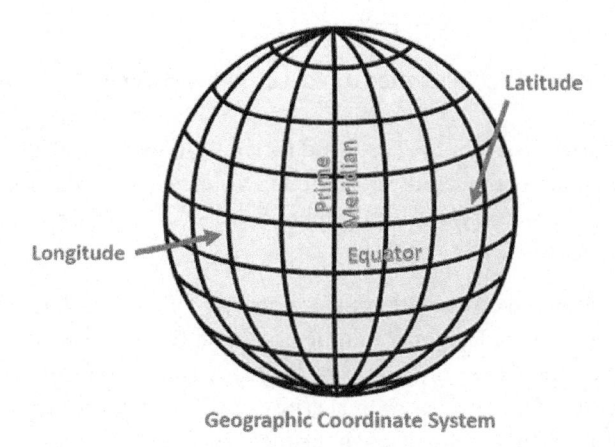

Figure 7.1 – Geographic coordinate system illustration

The primary lines of reference used in a geographic coordinate system are the equator and the prime meridian. These are the starting points for determining the latitude and longitude values required to locate a feature's position. Geographic coordinate systems can be used to identify the location of a feature on the Earth very accurately. However, as the diagram illustrates, the distance between the lines of latitude and longitude is not consistent. It varies depending on where on the Earth you are located. This makes it difficult to measure distances and areas. This is where projected coordinate systems come in.

Projected coordinate systems locate and display data on a 2D plane using a Cartesian coordinate system. This allows us to apply standard units of linear measurement, such as feet and meters. The origin of the projected coordinate system is tied to a geographic coordinate system. This is what ties the projected coordinate system to the Earth. From the origin, an x axis and a y axis are defined, along with the units, as illustrated in the following figure. As projected coordinate systems use uniform units, we can easily measure distances and areas:

Projected Coordinate System

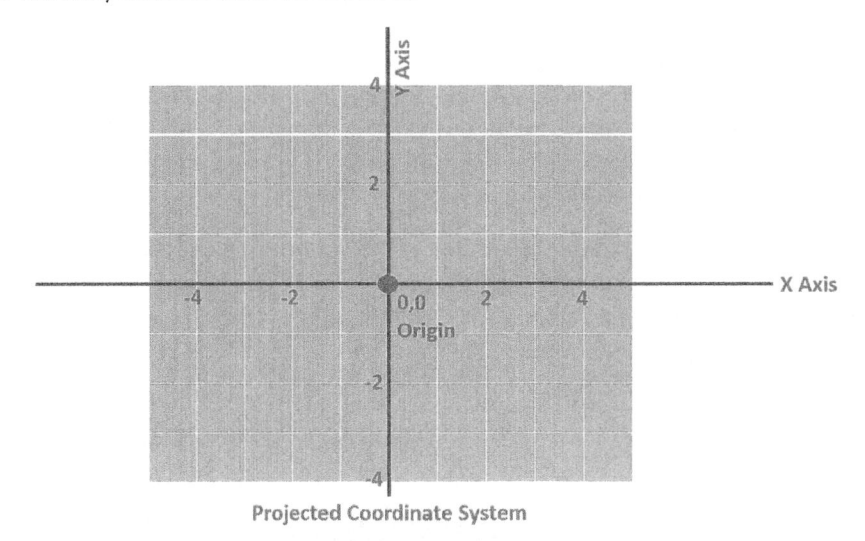

Figure 7.2 – Projected coordinate system illustrated

Of course, taking a 3D surface such as the Earth and flattening it does introduce distortions into our data. Therefore, we have different methods of projecting the data from a 3D model to a 2D plane to reduce these distortions. There are three basic types of projections: planar, conic, and cylindrical. From there, they get more advanced.

If all your data is in a real-world coordinate system, regardless of type, ArcGIS Pro can display it on a single map or scene, even if the layers are in different coordinate systems. It will project the data on the fly on that map or scene to a common coordinate system. Therefore, it is important that all your spatial data is assigned the correct coordinate system. Otherwise, your data might be displayed incorrectly, and the results of your analysis might be incorrect.

In this chapter, you will learn how to work with coordinate systems within ArcGIS Pro. You will do this by completing the following recipes:

- Determining the coordinate system for an existing map
- Setting the coordinate system for a new map
- Changing the coordinate system of a map
- Defining a coordinate system for data
- Projecting data to a different coordinate system

Determining the coordinate system for an existing map

Every map and scene you create in ArcGIS Pro makes use of a coordinate system in order to display data in a location. However, it not only impacts how and where features are displayed but also what you can do with the map. The coordinate system used by a map can impact whether you can share it as a web map via ArcGIS Online, how well layers overlay one another, and more. Therefore, it is important to know what coordinate system your map is using.

The first layer you add to the map will define the coordinate system for that map, but what if you are working with an existing map, perhaps even one that was created by someone else? How do you determine the coordinate system for the existing map you are working with? In this recipe, you will learn how to determine which coordinate system has been assigned to an existing map.

Getting ready

This recipe does require the sample data to be installed on the computer. You are not required to have completed any previous recipes. However, it is recommended that you complete the recipes in *Chapter 1, ArcGIS Pro Capabilities and Terminology*, or have some experience using ArcGIS Pro before you start this recipe. You can complete this recipe with any ArcGIS Pro licensing level.

How to do it...

Now that you have a general understanding of coordinate systems, you will learn the steps for determining which coordinate system has been assigned to an existing map in ArcGIS Pro:

1. Launch ArcGIS Pro, open the **In the ArcGIS Pro** start window, and click on **Open another project**.
2. Navigate to `C:\Student\ArcGISPro3Cookbook\Chapter7\CoordinateSystems` and open the `CoordinateSytems.aprx` project.

The project will open with the basic map of Trippville. This simple map contains four layers, including railroads, county line, city limits, and topographic.

3. Look at the status bar below the map. This will display several values and tools, including the map scale, the coordinate location of your mouse pointer, and the buttons to enable functionality, as shown in the following screenshot:

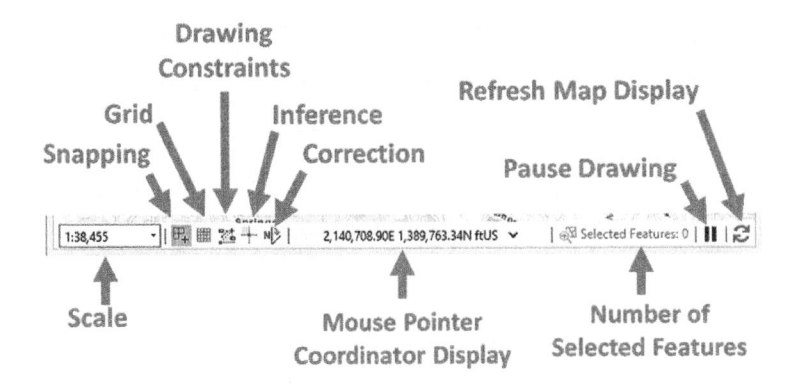

Figure 7.3 – Status bar located below the map view area

4. Move your mouse around in the map while watching the mouse pointer coordinate display. Then answer the following question:

Question: What units are being displayed on the map?

Answer:

As you moved your mouse within the map while watching the coordinate display, you should have seen the values change. The first value displayed is the X value for the location of your mouse pointer within the map. The second value is the Y value for the location of your pointer. The display is in US feet, as indicated by the **ftUS** text located to the right of the coordinates.

Looking at the values in the coordinate display can help you determine what type of coordinate system your map may be in: geographic or projected. X values greater than 180 and Y values greater than 90 might indicate that your map is using a projected coordinate system.

> **Tip**
> You can control the units used in the mouse pointer coordinate display within the **Map Properties** window. In the **General** option, you can set the display units to be any value you wish. This can differ from the primary map units based on the coordinate system assigned to the map.

Now, it is time to determine exactly what coordinate system is being used in the basic map of Trippville.

5. In the **Contents** panel, right-click on the basic map of Trippville and select **Properties** from the menu that appears, as illustrated in the following screenshot:

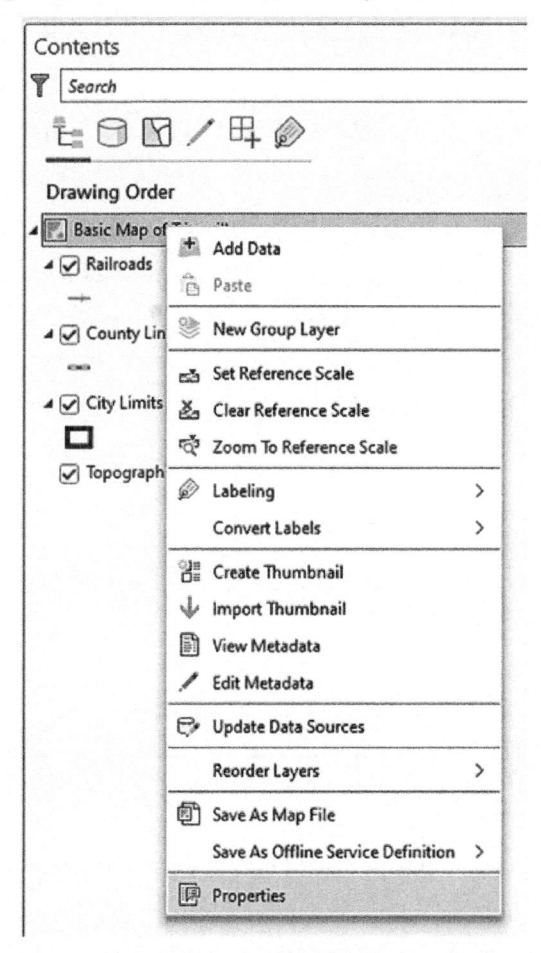

Figure 7.4 – Selecting the map properties option from the menu

6. In the **General** option in the **Map Properties** window, review the settings and answer the following questions:

    Question: What are the map units set to?

    Answer:

    Question: What are the display units set to?

    Answer:

7. Click on **Coordinate Systems** in the left panel of the **Map Properties** window.

8. Look in the right panel of the **Map Properties** window under **Current XY** and answer the following question:

   Question: What is the current XY coordinate system for the map?

   Answer:

9. Click on **Details**, located to the right of **Current XY**, as shown in the following screenshot:

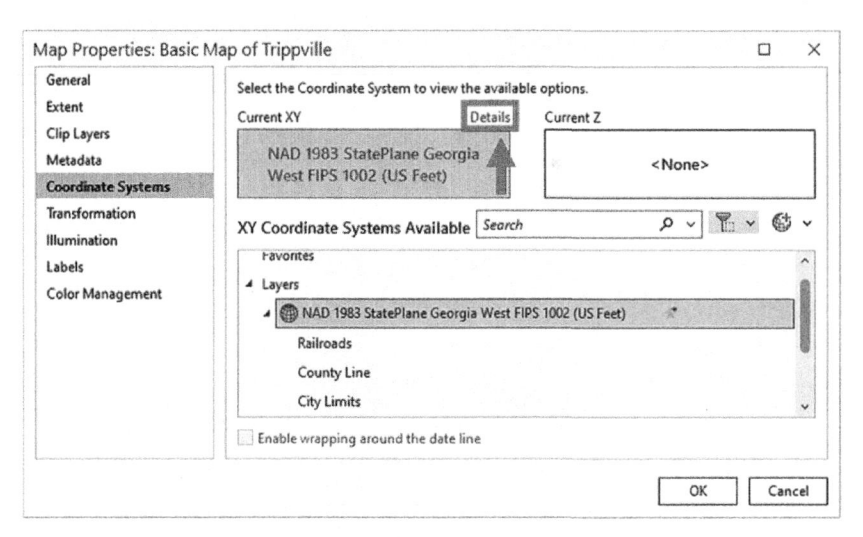

Figure 7.5 – Clicking on Details

10. Examine the details associated with the coordinate system assigned to the map and answer the following questions:

    Question: What type of coordinate system is assigned to the map?

    Answer:

    Question: What datum is being used with the assigned coordinate system?

    Answer:

    Question: What spheroid or ellipsoid is referenced for the assigned coordinate system?

    Answer:

    As you can now see, the map is using the **Georgia State Plane West** zone, which references the **North American Datum of 1983 (NAD 83)**. This is a projected coordinate system that uses the GRS 80 spheroid.

11. Close the **Coordinate System** details and **Map Properties** windows.
12. Save your project and close ArcGIS Pro.

You now know how to determine the coordinate system assigned to a map in ArcGIS Pro. Coordinate systems are at the heart of GIS and allow different data to be successfully overlaid on one another.

How it works...

In this recipe, you learned how to determine what units are being displayed on your map as you move your mouse pointer around the map. This can help you determine what type of coordinate system might be assigned to your map. If it is geographic, these values will often be less than 180. If the values are over that, then there is a good chance it has been assigned a projected coordinate system.

You also learned how to determine the exact coordinate system that is assigned to an existing map. It is important to know what coordinate system you are working with because it forms the foundation for how layers interact with each other spatially. If a map is not assigned a coordinate system, then you will not be able to bring layers together into a single map and expect them to overlay one another properly.

Setting the coordinate system for a new map

You now know how to determine the coordinate system assigned to existing maps, but how do you assign one to a new map? That is easy. As was mentioned earlier, the first layer you add to a new map will set the coordinate system for the map.

In this recipe, you will create a new map in the project you used in the previous recipe. You will then see what coordinate system has been assigned to the new map. Next, you will add a new layer to the map and see what that does to the coordinate system for your map. Lastly, you will add another layer to the map that is in a different coordinate system and check to see whether that changes the coordinate system for your map.

Getting ready

This recipe requires that the sample data be installed on the computer. It is recommended that you complete the previous recipe in this chapter along with the recipes in *Chapter 1, ArcGIS Pro Capabilities and Terminology*, before you begin this recipe. This will ensure that you have the skills and understanding required to complete this recipe successfully. You can complete this recipe with any ArcGIS Pro licensing level.

How to do it

You will set the coordinate system for a new map you add to an ArcGIS Pro project in this recipe:

1. Start ArcGIS Pro and open the `CoordinateSystems.aprx` project located in `C:\Student\ArcGISPro3Cookbook\Chapter7\CoordinateSystems` using the skills you have learned in previous recipes.

Projection and Coordinate System Basics

> **Tip**
> This is the same project you used in the previous recipe. If you completed that recipe, you should see this project included in your recently opened project list. If you do see it in that list, you can just double-click on it to open the project.

2. Click on the **Insert** tab on the ribbon.

3. Click on the **New Map** button located on the **Insert** tab. This will create a new blank 2D map in your project.

4. In the **Contents** panel, right-click on **Map** and select **Properties** from the menu that appears, as illustrated in the following screenshot:

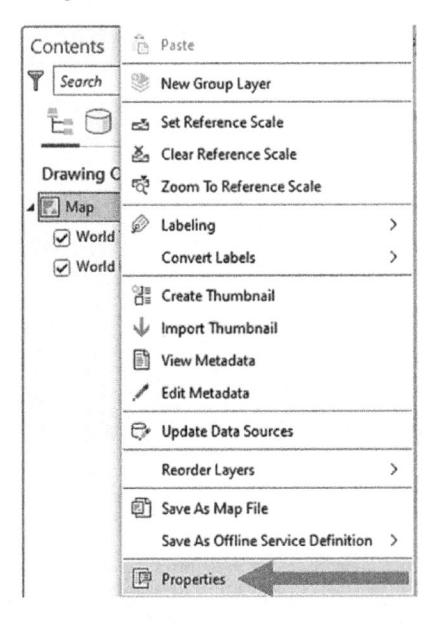

Figure 7.6 – Opening a map's properties

5. In the left panel of the **Map Properties** window, click on **Coordinate Systems** and answer the following question:

   Question: What coordinate system is currently assigned to the new map you created?

   Answer:

   As you can see, the new map you added has been assigned a coordinate system even though you have not added a layer yet. The assigned coordinate system is based on the basemap automatically added by default. If the basemap is one of the Esri-provided ones from ArcGIS Online, then the coordinate system assigned to your map will be the WGS 1984 Web Mercator (auxiliary sphere).

> **Note**
> WGS 1984 Web Mercator (auxiliary sphere) is the standard coordinate system used by most commercial web mapping systems, including Google Maps, Bing Maps, and Esri ArcGIS Online content.

If you are using a custom basemap, the coordinate system will be whatever is assigned to the custom basemap. If you have configured ArcGIS Pro to not automatically use a basemap, the coordinate system will be undefined.

Adding a new layer to the map

Now, let's see what happens when you add a layer to your new map. Will the coordinate system assigned to the map change or remain the same?

1. Close the **Map Properties** window by clicking on **Cancel**.
2. In the **Catalog** panel, expand the `Databases` folder so you can see its contents.
3. Expand the `Trippville_GIS.gdb` geodatabase.
4. Right-click on the `Zoning` feature class in the `Trippville_GIS.gdb` geodatabase and select **Add to Current Map**. Your map should now look like the following screenshot. Keep in mind that your colors may be different:

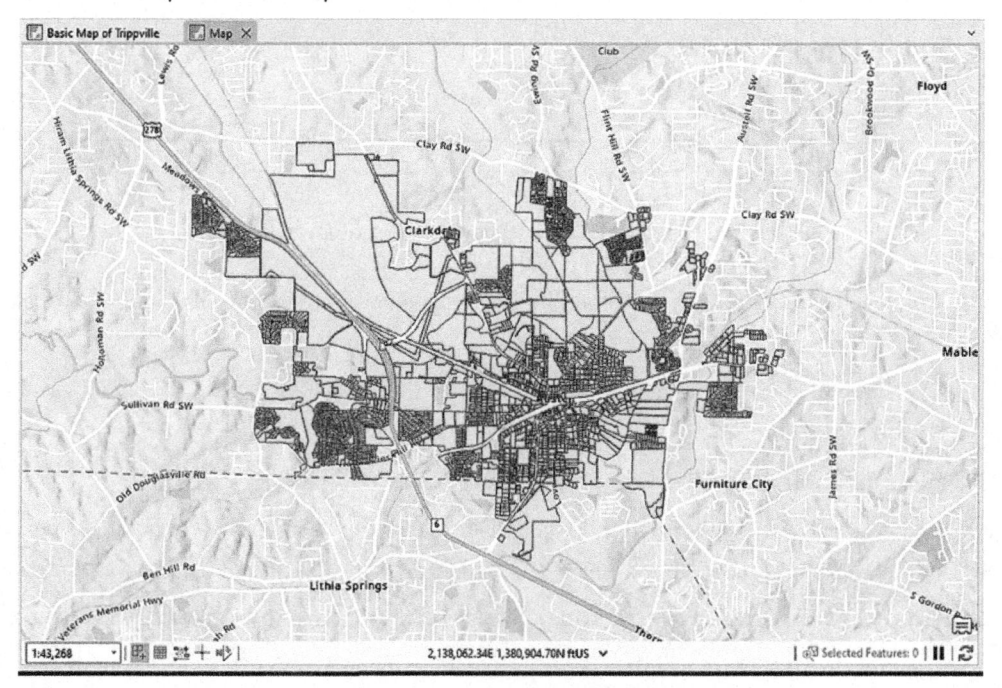

Figure 7.7 – Map with Zoning layer added

5. Save your project.
6. In the **Contents** panel, right-click on the **Zoning** layer you just added and select **Properties** from the menu that appears.
7. In the **Layer Properties** window, select **Source** from the left panel.
8. In the right panel of the **Layer Properties** window, scroll down until you see **Spatial Reference** and expand that option, as illustrated in the following screenshot:

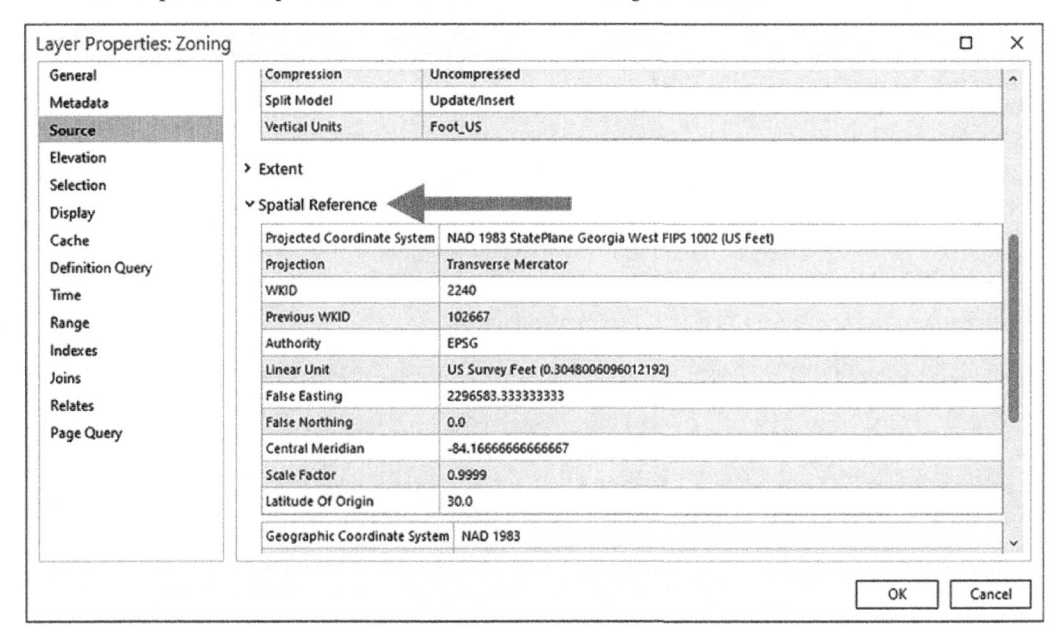

Figure 7.8 – Viewing layer properties and spatial reference

9. Examine the information listed about the spatial reference for the zoning layer and then answer the following question:

   Question: What is the projected coordinate system for the zoning layer?

   Answer:

10. Close the **Layer Properties** window by clicking on the **OK** button.
11. Right-click on **Map** in the **Contents** panel, then select **Properties** from the menu that appears.
12. In the **Map Properties** window, select **Coordinate Systems** on the left panel.
13. Look in the box below **Current XY** in the right panel of the window. Answer the following questions:

    Question: What is the coordinate system assigned to the map now?

    Answer:

Question: Has the coordinate system for the map changed since you added the new layer?

Answer:

As you can see, the coordinate system for your map has changed. It is now using the NAD 1983 State Plane Georgia West FIPS 1002 Feet coordinate system.

Adding an additional layer to the map

You have seen that adding a new layer to a new map will change the assigned coordinate system. Now, let's see whether it changes again if you add another layer to the map:

1. Close the **Map Properties** window by clicking on the **OK** button.
2. In the **Catalog** pane, expand `Folders` so that you can see its contents.
3. Expand the `CoordinateSystems` folder to reveal its contents.
4. Select the `Elevation Contours.shp` shapefile. Then drag and drop it onto the map to add it as a new layer:

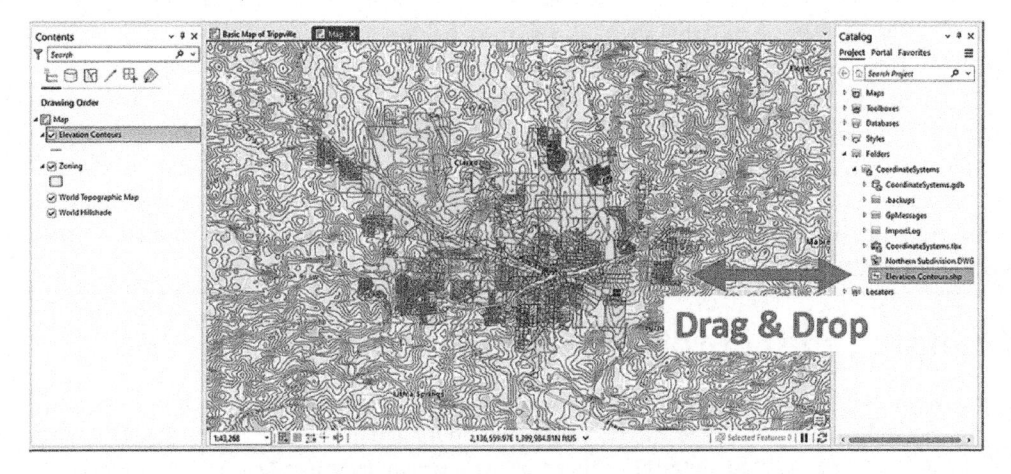

Figure 7.9 – Drag and drop a shapefile to add a new layer

5. Using the same method you did for the zoning layer, examine the spatial reference for the elevation contours layer you just added to the map. Then answer the following question:

   Question: What coordinate system is assigned to the elevation contours layer?

   Answer:

6. Close the **Layer Properties** window by clicking on the **OK** or **Cancel** button.

7. Open the **Properties** for the map again using the same method you used previously. Verify the coordinate system assigned to the map and answer the following question:

    Question: Did the coordinate system assigned to the map change when you added a new layer that used a different spatial reference?

    Answer:

8. Close the **Map Properties** window by clicking on the **OK** or **Cancel** button.
9. Save your project and close ArcGIS Pro.

Even though the elevation contours layer is in the GCS WGS 1984 geographic coordinate system, the coordinate system for your map has not changed. Only the first layer added to a map will change the coordinate system. All other layers are projected on the fly to match the coordinate system of the map.

How it works...

In this recipe, you learned how to assign a real-world coordinate system to a map by adding the first layer. You started by inserting a new map into an ArcGIS Pro project. Then you verified the coordinate system for the new map by reviewing the map's properties. You saw that this new map used the default WGS 84 Web Mercator (auxiliary sphere) coordinate system that is also used by other online maps, such as Google, Bing, and ArcGIS Online. Next, you added the zoning layer to your new map. When comparing the spatial reference assigned to the layer and the coordinate system assigned to the map after you added the layer, you saw that the coordinate system for the map was changed to match the spatial reference of the zoning layer because it was the first layer added to the map. Other layers added afterward were projected to match the coordinate system assigned to the map.

Changing the coordinate system of a map

You now know how to determine the coordinate system for an existing map and how to set one for a new map, but how do you change the coordinate system for a map or scene? You might be wondering why you would want to change the coordinate system for a map. There are several reasons why you might want to do so. As mentioned in the introduction, some coordinate systems are better suited to specific operations. If you need to measure areas or distances for analysis or editing, a projected coordinate system is best, especially for small areas such as a county, city, or district. If you are trying to show locations of features across large areas, such as a country or the world, then a geographic coordinate system is often best. If you are publishing as a web map, then it might require the map to be in the WGS 1984 Web Mercator (auxiliary sphere) coordinate system to conform with other datasets. Therefore, if you need to do one of these and your map is in a different coordinate system, you might need to change it. Luckily, it is not hard to change the coordinate system for a map, as you are about to see.

Changing the coordinate system for a map does not change the coordinate system for the underlying data referenced in the map. This means that if you have a shapefile in your map as a layer and that shapefile is in NAD 83 UTM Zone 16N Meters, and if you change the map coordinate system to the

WGS 1984 Web Mercator (auxiliary sphere), this will not change the coordinate system the shapefile is stored in. In other words, changing the coordinate system of the map does not project the data it references to a new coordinate system. You will learn how to project data to a different coordinate system later in this chapter.

In this recipe, you will learn how to change the coordinate system used within a map.

Getting ready

It is recommended that you complete all previous recipes in this chapter before starting this one. This will ensure that you have the basic knowledge and understanding required to successfully complete this recipe. This recipe can be completed with all license levels of ArcGIS Pro. You do need to have the sample data installed as well.

How to do it

You will now work through the steps needed to change the coordinate system that has been assigned to a map in ArcGIS Pro:

1. Launch ArcGIS Pro and open the `CoordinateSystems.aprx` project you have been using in the previous recipes in this chapter. You should see it on your recently opened projects list in the ArcGIS Pro start window.

2. Make sure that the **Basic Map of Trippville** is the active map in the map view by clicking on the tab for the **Basic Map of Trippville** at the top of the map view area, as illustrated in the following screenshot:

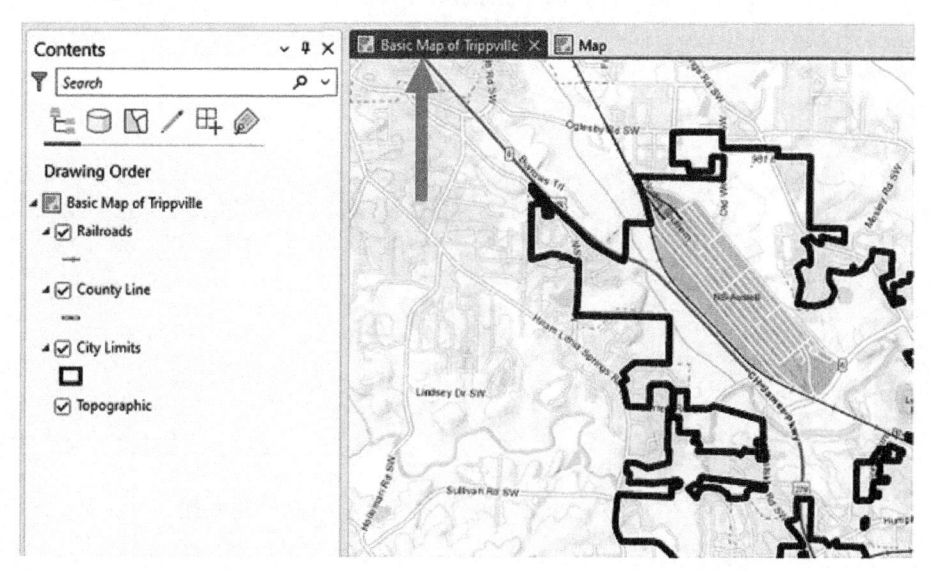

Figure 7.10 – Ensuring the basic map of Trippville is active

> **Tip**
> If you have closed the basic map of Trippville, you need to reopen it. To reopen the map, expand the maps folder in the catalog panel, then right-click on the basic map of Trippville in the maps folder, and select **Open** from the menu that appears.

3. Move your mouse pointer to the approximate center of the map and answer the following question. Try to remember the location you use. You will need to know it for a later task in this recipe:

   Question: What are the X and Y coordinates for the location of your mouse pointer? Remember that you can see these values in the status bar below the map.

   Answer:

   As you can see, the X coordinate will be somewhere around 2,150,000, and the Y coordinate somewhere around 1,390,000. Of course, your exact values will be different, but they should be close to those values.

4. In the **Contents** panel, double-click on the **Basic Map of Trippville** to open the **Map Properties** window.

5. Select **Coordinate Systems** in the left panel of the **Map Properties** window. It should now look like the following screenshot:

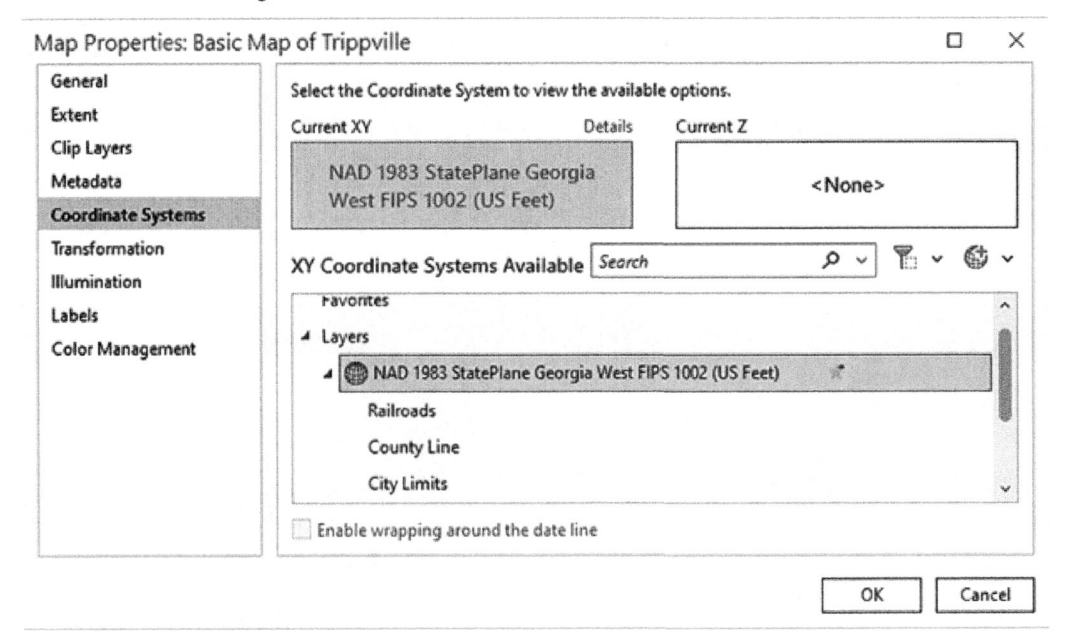

Figure 7.11 – Map properties for the basic map of Trippville

Changing the coordinate system of a map   243

You are now going to change the map's coordinate system from NAD 1983 StatePlane Georgia West to WGS 1984 Web Mercator (auxiliary sphere), so it will be ready to publish to a web map later.

6. In the box located below **XY Coordinate Systems Available**, scroll down so that you can see **Projected Coordinate System**. Then expand the projected coordinate system options so that you can see the available choices as illustrated in the following screenshot:

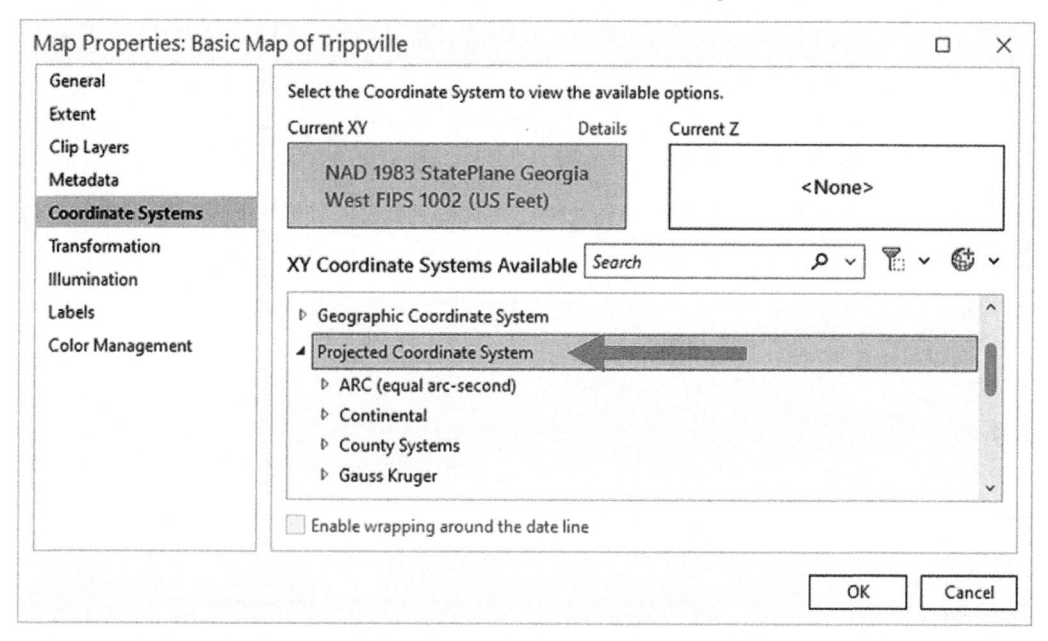

Figure 7.12 – Projected coordinate systems list displayed

7. Scroll down to **World** and expand that option. Then scroll down and select **WGS 1984 Web Mercator (auxiliary sphere)**. It will be located near the bottom of the list.

   The value in the **Current XY** box should change to match the new coordinate system you just selected.

8. Click on **OK** to apply the newly selected coordinate system to the basic map of Trippville.
9. Save your project.
10. Move your mouse pointer to the same location you did in *Step 3* as the approximate center of the map. Then answer the following question:

    Question: Are the coordinates for the approximate center of the map different or the same now that you have changed the coordinate system of the map?

Answer:

As you can see, the coordinates for the approximate center of the map have indeed changed. They are very different. Has this also changed the coordinate system for the layers and their source data? It was said earlier that they should not have changed, but let's verify just to make sure:

11. In the **Contents** panel, right-click on the **City Limits** layer and select **Properties** from the menu that appears.
12. In the **Layer Properties** window, click on the **Source** option in the left panel.
13. Scroll down in the right panel of the window until you see **Spatial Reference**. Then expand the **Spatial Reference** section. Your **Layer Properties** window should now resemble the following screenshot:

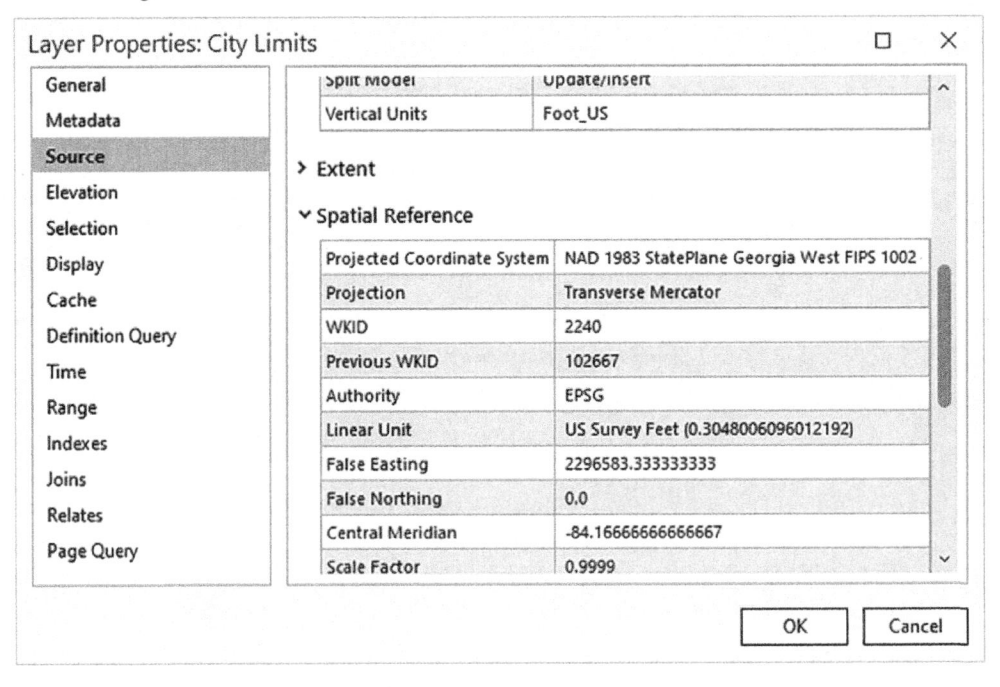

Layer Properties: City Limits		
General	Split Model	Update/Insert
Metadata	Vertical Units	Foot_US
Source	> Extent	
Elevation		
Selection	∨ Spatial Reference	
Display	Projected Coordinate System	NAD 1983 StatePlane Georgia West FIPS 1002
Cache	Projection	Transverse Mercator
Definition Query	WKID	2240
Time	Previous WKID	102667
Range	Authority	EPSG
Indexes	Linear Unit	US Survey Feet (0.3048006096012192)
Joins	False Easting	2296583.333333333
Relates	False Northing	0.0
Page Query	Central Meridian	-84.16666666666667
	Scale Factor	0.9999

Figure 7.13 – Layer properties source being displayed

14. Look at the value for the **Projected Coordinate System** and answer the following question:

    Question: Is the coordinate system assigned to the layer the same or different from the one assigned to the map?

    Answer:

15. Close the **Layer Properties** window by clicking on the **Cancel** button.
16. Save your project and close ArcGIS Pro.

As you have just verified, the coordinate system for the layer source has not changed. It is still being stored in the same coordinate system it has always been in. Just changing the coordinate system for the map does not change the source data's coordinate system. It only changes how it is displayed on that specific map. ArcGIS Pro projects the data on the fly from the coordinate system and it is stored into the map that is displayed. This is very useful when working with data from multiple organizations that may not use the same coordinate system as yours. One thing to remember though is that projecting on the fly does use more computer resources and will slow down the rendering of the data every time you pan or zoom on the map.

How it works...

In this recipe, you learned how to change the coordinate system used by a map. This can be useful if you have data in a map that is in different coordinate systems, and you want it all to be displayed on a single system that you designate. It can also be helpful if you need to perform a specific task that is not easily done using the native coordinate system for the data, for example, if you want to measure the area or size of a feature but the data is in a geographic coordinate system that uses degrees as its native unit. Setting the map to a projected coordinate system such as **Universal Transmercator (UTM)** would make that task much easier.

You changed the coordinate system for the map by first opening its properties. Then you went to the coordinate system property settings. From there, you looked under projected coordinate systems and found the WGS 1984 Web Mercator (auxiliary sphere) coordinate system and selected it. This is the coordinate system used by most online mapping solutions, including ArcGIS Online, Google Maps, and Bing Maps. You then clicked on the OK button to apply your changes to the map.

You then verified that changing the coordinate system used by the map did not change the native coordinate system assigned to the layers contained in the map. It only changed the coordinate system the layers were displayed in. In the next recipe, you will learn how to assign the coordinate system for data.

Defining a coordinate system for data

You now know how to assign and change the coordinate system used in a map. Maps are not the only things that are assigned a coordinate system. Our data is also assigned a coordinate system, or at least it should be. If your data is not assigned a coordinate system, then ArcGIS Pro may not be able to display the correct location with all your other data. This is also true if your data is assigned the wrong coordinate system.

Where is the coordinate system information stored within your data? Well, this depends on what type of data you are working with. If it is a feature class stored in a geodatabase, then the coordinate system information is stored within the geodatabase itself. If you are working with shapefiles, then the assigned coordinate system is stored in a file with a .prj file extension.

If you are working with CAD data, meaning `.dwg`, `.dxf`, or `.dgn` files, then you may also need a `.prj` file for ArcGIS Pro to know what coordinate system it is in. While some CAD files can be assigned a coordinate system within the drawing itself, not all are. You may need to create an external `.prj` file so that ArcGIS Pro knows what coordinate system the file is in and how to properly display it.

As you saw in the *Creating a new geodatabase* recipe in *Chapter 2, Creating and Storing Data*, when you create a new feature class in a geodatabase, you are asked to assign the coordinate system as part of the create feature class tool. In most cases, when you create new feature classes or shapefiles, you will assign the coordinate system for that data as you create it. However, you will find that many existing or legacy datasets do not have a coordinate system defined. This is true mostly for shapefiles and CAD files.

Just because a dataset does not have an assigned coordinate system, this does not mean it was not created in a real-world coordinate system. The majority of shapefiles are created in a real-world coordinate system. They are just missing the `.prj` file required by ArcGIS Pro and other Esri products to know what the coordinate system is. The same is true of many CAD files as well, though you will encounter a larger number of CAD files that are not in a real-world coordinate system. You will learn more about that in a later chapter. If the data that is missing from the coordinate system was created in the same coordinate system as the rest of your data, you can still use it for maps, performing analyses, and more without defining the coordinate system for the data. It is when you try to mix data in different coordinate systems that you will start to encounter problems if you try to publish the data to a web service or convert it to another format. It is generally considered best practice to always define a coordinate system for all your data.

In this recipe, you will learn how to define a coordinate system for data that does not have one. In this case, it will be a CAD file that was created in a real-world coordinate system, which is different from all your other data.

Getting ready

It is not required that you complete any previous recipes to do this one. This recipe can be completed with all license levels of ArcGIS Pro. You will need to ensure that the sample data is installed before starting.

How to do it

You will now work through the steps required to assign a coordinate system to a CAD file that does not have one assigned. It was created in a coordinate system but that coordinate system was not assigned to the file in a way ArcGIS Pro can understand:

1. Launch ArcGIS Pro and open the `CoordinateSystems.aprx` project located in `C:\Student\ArcGISPro3Cookbook\Chapter7\CoordinateSystems` using the skills you have learned in past recipes.

2. Make sure that **Basic Map of Trippville** is the active map in the map view. If you are not sure how to do this, refer to *Step 2* in the previous recipe.

3. In the **Catalog** panel, expand `Folders` so that you can see its contents.

4. Expand the `CoordinateSystems` folder.

5. Expand the `Northern Subdivision.DWG` file. This is an AutoCAD drawing file that contains information about a new subdivision being built near the City of Trippville.

> **Note**
> DWG files, also called drawings or drawing files, are the native file format for AutoCAD. AutoCAD is an application created by Autodesk and widely used by land surveyors, engineers, architects, and other design professionals. It is not unusual for GIS users to pull data from drawing files into a GIS database.

6. Right-click on the **Polyline** feature class within the `Northern Subdivision.DWG` file and select **Add to Current Map**.

   The new layer should appear in the **Contents** panel; however, you will not see it drawn on your map. You will now begin to investigate why. You should also get a warning message that says a source layer is missing a spatial reference. That is a hint as to why the new layer is not visible in your map area.

7. Right-click on the **Northern Subdivision-Polyline** layer and select **Zoom To Layer**, as illustrated in the following screenshot:

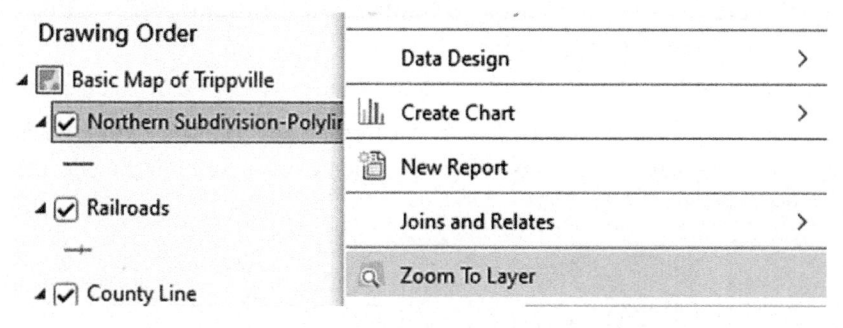

Figure 7.14 – Zoom to northern subdivision – polyline layer

8. Move your mouse pointer to the map view and watch the map display the coordinate readout in the status bar at the bottom of the map view.

   As you can see, the layer was drawn on the map. It is just not appearing in the right location. Let's continue to investigate a little more to see why it is not drawing the layer in the correct location in relation to all your other data.

9. Double-click on the **Northern Subdivision-Polyline** layer to open the **Layer Properties** window.
10. Select the **Source** option from the panel on the left side of the window and scroll down in the panel on the right until you see **Spatial Reference**.
11. Expand **Spatial Reference** and answer the following question:

    Question: What is the coordinate system assigned to the Northern Subdivision-Polyline layer?

    Answer:

    As you can see, ArcGIS Pro indicates that this new layer has an unknown coordinate system. This means that as far as ArcGIS Pro is concerned, the data is not in a real-world coordinate system. However, after talking with the person who created the drawing file, you have found out it is in the UTM 16N coordinate system. ArcGIS Pro just does not know this because there is no projection file to tell it that the data is in that coordinate system. You need to create the projection file for the drawing file.

Assigning a coordinate system to data

You will now assign the coordinate system and projection to the drawing file using the define projection tool in ArcGIS Pro. This will create a PRJ file that tells ArcGIS Pro what coordinate system the drawing file is in. The steps are as follows:

1. Close the **Layer Properties** window by clicking on the **Cancel** button.
2. Click on the **Analysis** tab in the ribbon. Then click on the **Tools** button to open the **Geoprocessing** pane.
3. In the **Geoprocessing** pane, click on the **Toolboxes** tab. This will display a list of the system toolboxes included with ArcGIS Pro.
4. In the **Geoprocessing** pane, scroll down and expand the contents of the **Data Management** toolbox.
5. Scroll down and expand the contents of the **Projections and Transformations** toolset. You will see several tools that deal with assigning a coordinate system or projecting data to a different coordinate system.

6. Select the **Define Projection** tool as illustrated in the following screenshot:

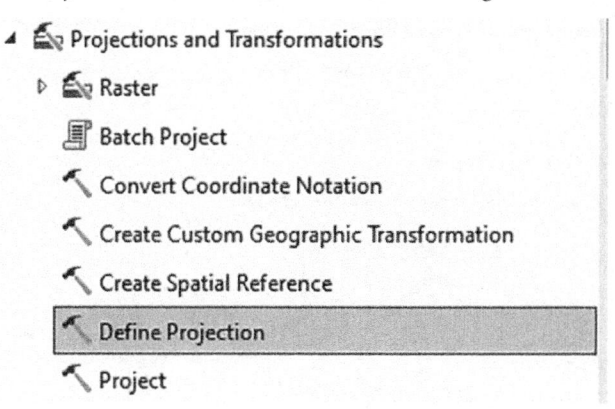

Figure 7.15 – Selecting the Define Projection tool

7. Set the **Input Dataset or Feature Class** parameter to **Northern Subdivision-Polyline** using the drop-down arrow.
8. For **Coordinate System**, click on the **Select Coordinate System** icon, which looks like a small wireframe globe. This will open the **Coordinate System** window.
9. In the area available under **XY Coordinate Systems Available**, expand **Projected Coordinate System**, as illustrated in the following screenshot:

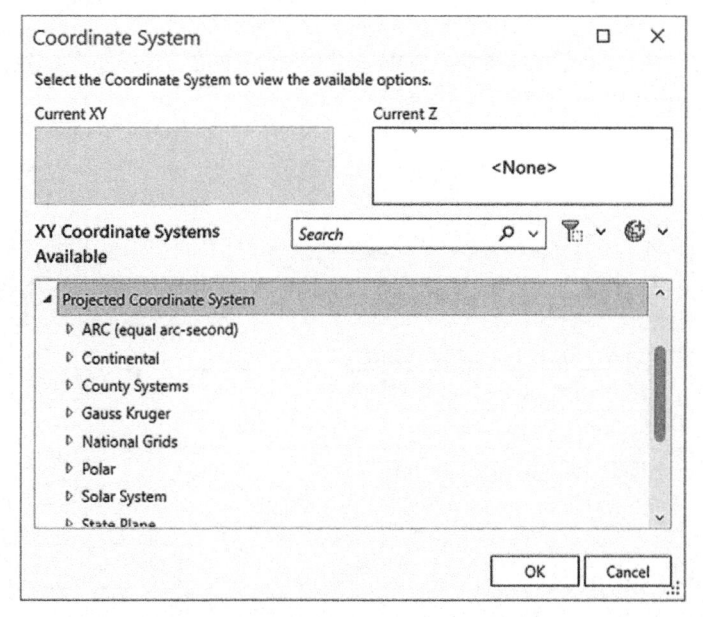

Figure 7.16 – Projected Coordinate System list expanded

250 Projection and Coordinate System Basics

10. Scroll down until you locate **UTM**. Then expand it so you can see its contents.
11. Scroll down and expand **NAD 1983**.
12. Scroll down the list and select **NAD 1983 UTM Zone 16N** as shown in the following screenshot. Then click **OK**:

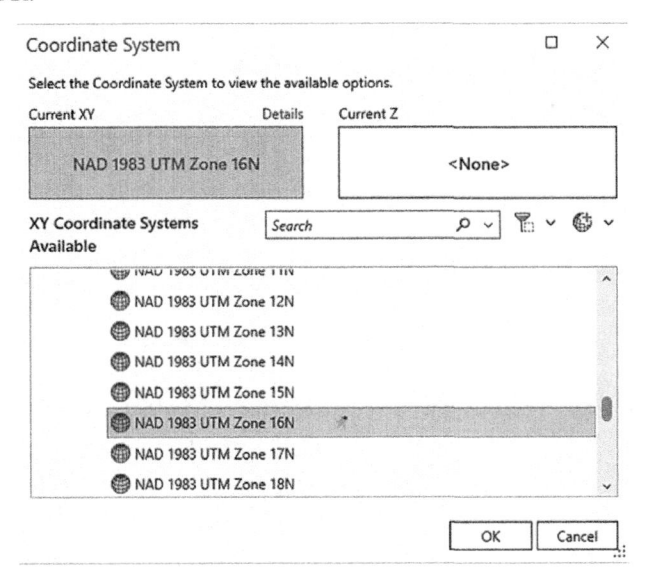

Figure 7.17 – Selecting NAD 1983 UTM Zone 16N from the list

13. Verify your **Define Projection** tool looks like the following screenshot and click **Run**:

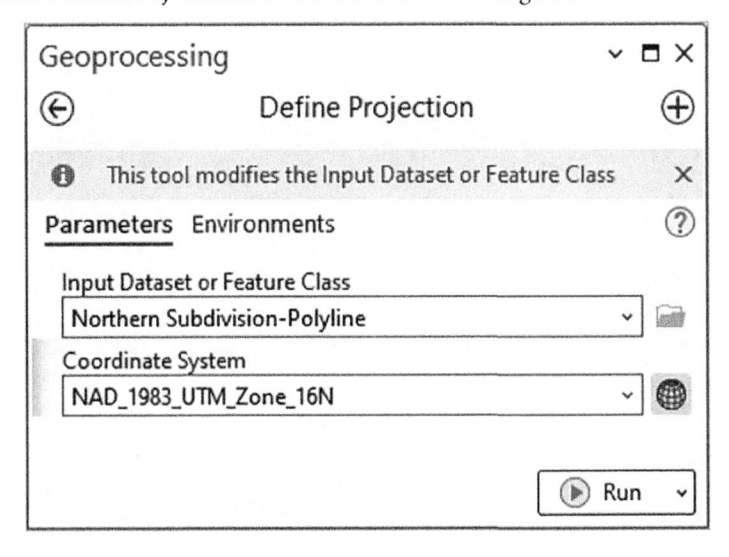

Figure 7.18 – The Define Projection tool is ready to run

Defining a coordinate system for data   251

When the **Define Projection** tool finishes, the **Northern Subdivision-Polyline** layer should disappear from the map view. This is because ArcGIS Pro is now able to project the layer on the fly to match the coordinate system of the map since it now knows what coordinate system the layer is stored in. But let's verify that.

14. Close the **Geoprocessing** panel to free up screen space.
15. Right-click on the **Northern Subdivision–Polyline** layer in the **Contents** pane and select **Zoom To Layer**.
16. Zoom out on the map using the scroll wheel on your mouse so that you can see the layer in relation to the City of Trippville area, as shown in the following screenshot:

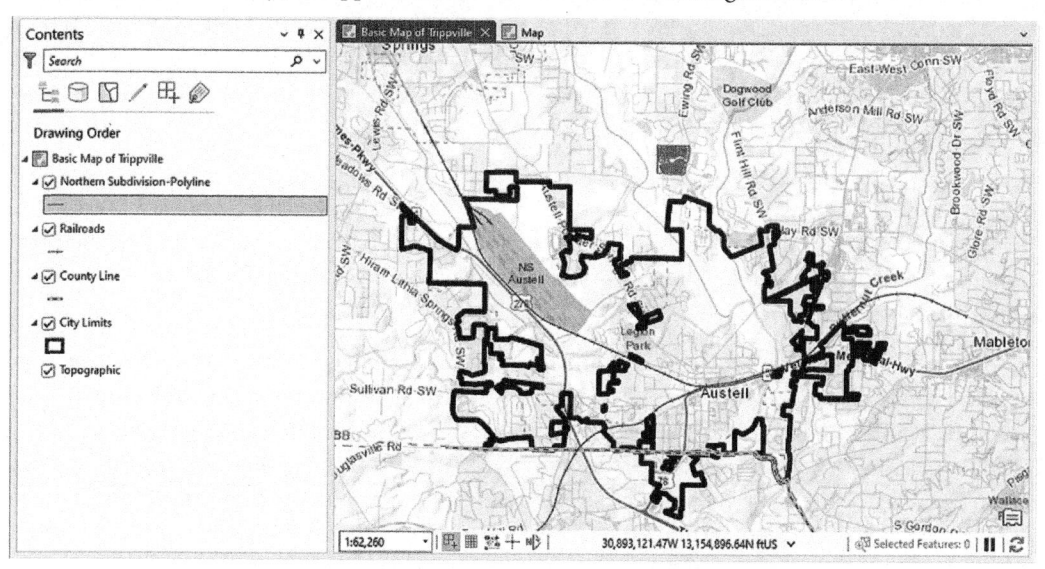

Figure 7.19 – Map zoomed out so you can see the relationship

17. Save your project and close ArcGIS Pro.

This recipe illustrates the importance of having a coordinate system assigned to your GIS data. As you saw, without it, the data will not display in the correct location. This makes it impossible to use as part of a map or to perform an analysis. It is also not unusual to get data that has been assigned the wrong coordinate system. The behavior and process described in this recipe can also be used to assign the correct coordinate system to data that has been wrongly assigned.

The define projection tool you used in this recipe will also assign a coordinate system and generate a `.prj` file for shapefiles, DXF, and DGN files. You do need to ensure that the data is in a real-world coordinate system and know that the coordinate system is before you use the tool. The define projection tool will not georeference data that was not created in a real-world coordinate system, nor will it project data from one coordinate system to another.

How it works...

In this recipe, you assigned a coordinate system to data that was missing the assignment. You did this using the define projection tool located in the data management toolbox and the projections and transformations toolset. This tool generated a PRJ file that ArcGIS Pro was able to reference to know what coordinate system the DWG file you were trying to use was in. Once you generate that PRJ file using the tool, the data was displayed in its proper location, allowing you to see the location of this new subdivision in relation to the City of Trippville.

For the define projection tool to work, you needed to verify the data had actually been created in a real-world coordinate system and what that system was. In this recipe, that coordinate system happened to be the UTM NAD 1983 Zone 16N. If the data was not created in a real-world coordinate system such as UTM, then the define projection tool would not work. You would need to georeference the data, which is a much more complicated process.

Projecting data to a different coordinate system

You now know how to change the coordinate system for a map and how to assign one to data that is missing a coordinate system, but how do you change the coordinate system for data? This process is called projecting data.

When you need to move or project data from one coordinate system to another, it requires a lot more than just redefining the assigned coordinate system as you did in the previous recipe. It requires ArcGIS Pro to recalculate all the coordinate values for the features within that dataset. This means all those features will have new coordinates that represent values in the coordinate system it has moved to. To calculate the new coordinate values, ArcGIS Pro must perform many calculations for each feature to account for possible changes in datums, units, reference locations, Earth models, and more. Therefore, it can take a long time for larger datasets.

The tool you use to move data from one coordinate system to another is the project tool. This tool is located in the data management tools toolbox and the projections and transformations toolset. In this recipe, you will use the project tool to move data from one coordinate system to another. You will project the Trippville city limits from its current coordinate system to the WGS 1984 Web Mercator (auxiliary sphere) coordinate system for publishing to ArcGIS Online.

Getting ready

It is recommended that you have completed the other recipes in this chapter before you continue with this one. This ensures that you have a fundamental understanding of coordinate systems and how they are handled in ArcGIS Pro. This understanding will allow you to better understand the process described in this recipe.

Projecting data to a different coordinate system

This recipe can be completed with all license levels of ArcGIS Pro. To complete the last part of this recipe, you will also need a connection to ArcGIS Online. Your user account will need to be the Creator type and have publisher rights.

How to do it

You will work through the process required to transform data from one coordinate system to another:

1. Start ArcGIS Pro and open the `CoordinateSystems.aprx` project located in `C:\Student\ArcGISProCookbook\Chapter7\CoordinateSystems` using the skills you have learned in past recipes.
2. Make sure that **Basic Map of Trippville** is the active map in the map view.
3. In the **Contents** panel, double-click on the **City Limits** layer to open **Layer Properties**.
4. Click on the **Source** option in the left panel of the window.
5. In the right panel of the **Layer Properties** window, scroll through the options until you get to **Spatial Reference**.
6. Expand the **Spatial Reference** option and answer the following question:

    Question: What is the projected coordinate system for the city limits layer?

    Answer:

    You can see that the city limits layer is being stored in the NAD 1983 State Plane Georgia West coordinate system. You want to project this layer to the WGS 1984 Web Mercator (auxiliary sphere) coordinate system, so it will be easier to integrate with other online content. The WGS 1984 Web Mercator (auxiliary sphere) coordinate system is the one used by most public mapping websites, such as Google Earth, Bing Maps, and Esri ArcGIS Online.

Projecting data to a new coordinate system

You will now use the project tool to project the city limits layer from its current coordinate system to the WGS 1984 Web Mercator (auxiliary sphere) coordinate system:

1. Close the **Layer Properties** window by clicking on **Cancel**.
2. Click on the **Analysis** tab in the ribbon.
3. Click on the **Tools** button in the **Geoprocessing** group on the **Analysis** tab to open the **Geoprocessing** panel.

4. Click on the **Toolboxes** tab at the top of the **Geoprocessing** panel, as shown in the following screenshot:

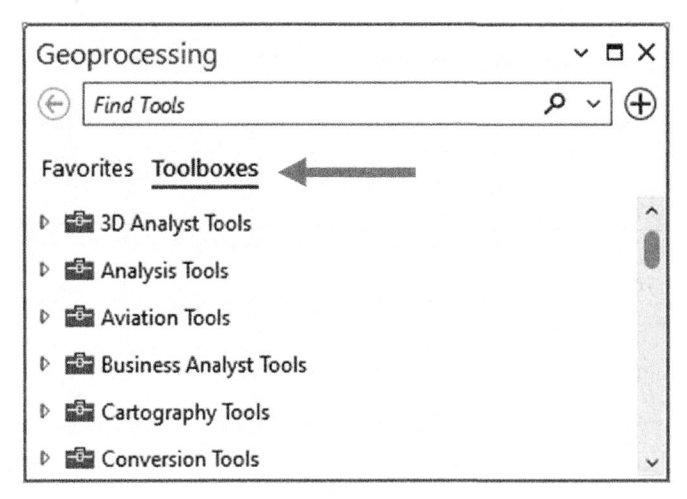

Figure 7.20 – Toolboxes tab in Geoprocessing panel

5. Expand the **Data Management Tools** toolbox by clicking on the small arrowhead located next to it.
6. Scroll down and expand the **Projections and Transformations** toolset by clicking on the small arrowhead next to it.
7. Click on the **Project** tool to open it.

> **Note**
> The batch project tool does the same thing as the project tool but allows you to input multiple feature classes or datasets.

8. Click on the drop-down arrow for **Input Dataset or Feature Class** and select `City Limits` from the list.

Projecting data to a different coordinate system   255

9. Click on the **Browse** button located to the right of the **Output Dataset or Feature Class** cell, as indicated:

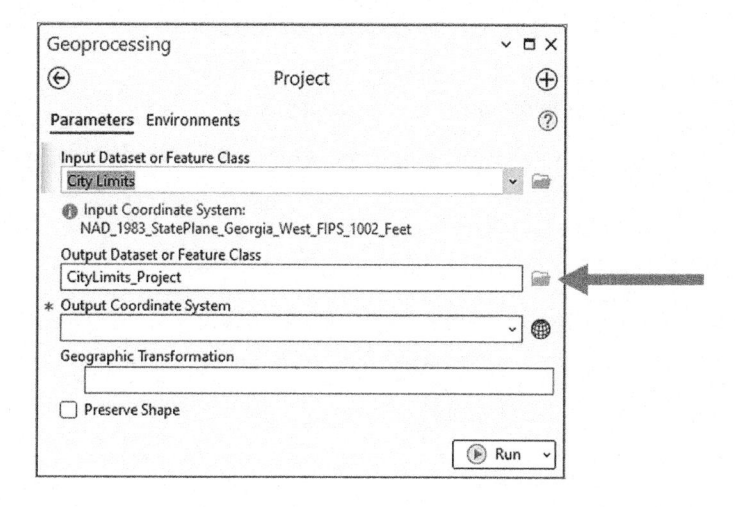

Figure 7.21 – Browse button location indicated

10. In the **Output Dataset or Feature Class** window, select **Databases** from the panel on the left. You may need to expand the **Project** option first.
11. Double-click on the `Trippville_GIS.gdb` geodatabase in the right panel.
12. In the **Name** cell located at the bottom, type `City_Limits_Web`.
13. Verify that your **Output Dataset or Feature Class** window looks like the following screenshot and click **Save**:

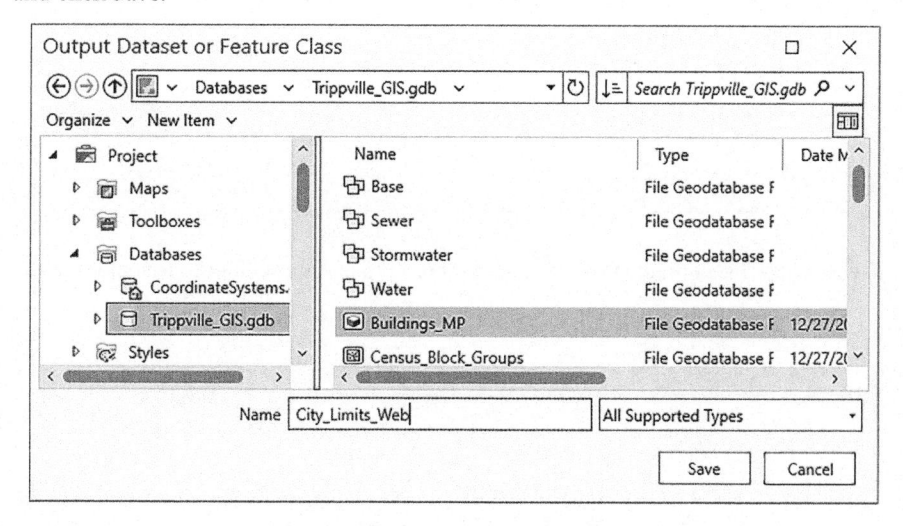

Figure 7.22 – Output filename

14. Click on the small wireframe globe, located to the right of **Output Coordinate System,** to open the **Coordinate System** window.
15. In the bottom panel, expand **Projected Coordinate Systems**.
16. Scroll through the list of different types of projected coordinate systems until you see **World**. Expand the **World** option, so you can see its contents.
17. Scroll down until you see **WGS 1984 Web Mercator (auxiliary sphere)** and select it.
18. Verify that your **Coordinate System** window looks like the screenshot and click on **OK**:

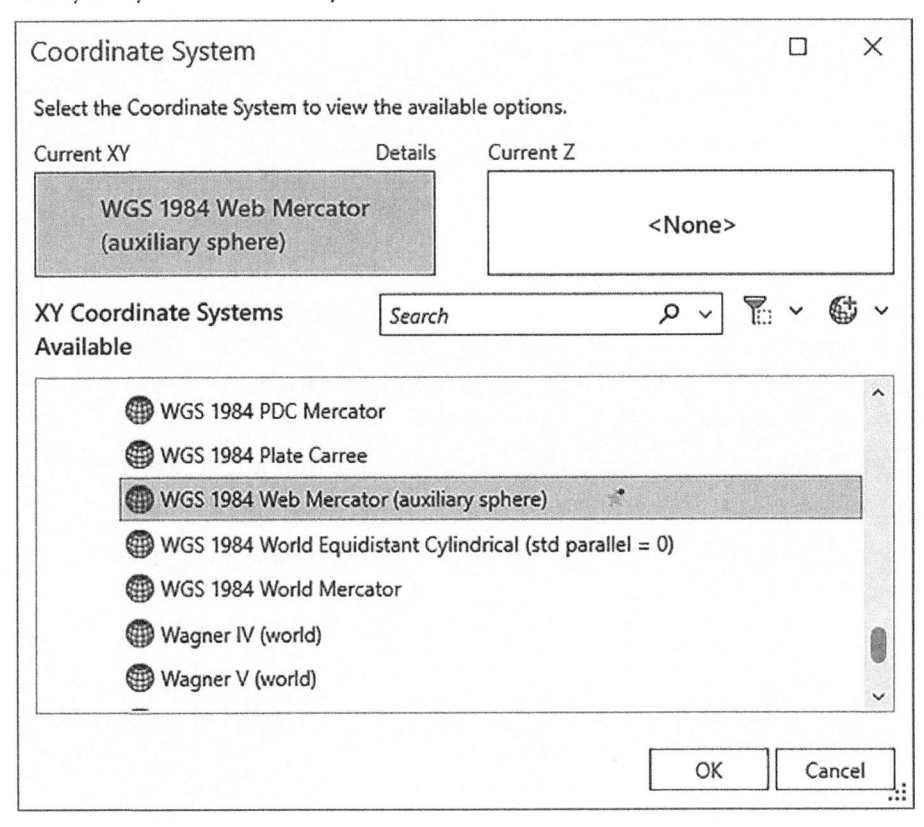

Figure 7.23 – Selecting the WGS 1984 Web Mercator (auxiliary sphere) coordinate system

19. Verify that your **Project** tool looks like the following screenshot, then click on the **Run** button at the bottom:

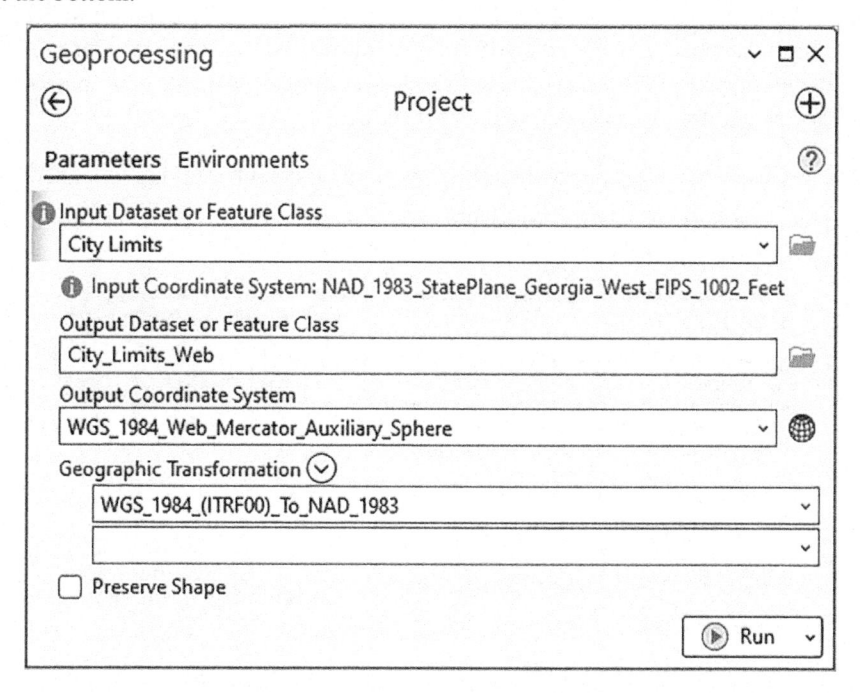

Figure 7.24 – Project tool completely filled out

The new **City_Limits_Web** layer will be added to your map when it is complete. Notice that a geographic transformation is automatically applied. Transformations apply mathematical adjustments when projecting data to account for differences in datums. The original coordinate system for the city limits layer referenced NAD 1983. The WGS 1984 Web Mercator (auxiliary sphere) coordinate system references the WGS 84 datum. While these are both Earth-centered datums, they are slightly different. That difference must be accounted for when the data is projected. That is what the transformation process does.

Note that the project tool creates a new feature class, leaving the original one intact. This provides a safety net in case you were to select the wrong output coordinate system.

Sharing a new layer to ArcGIS Online

Now that you have projected the city limits layer to the new coordinate system, you will create a new map containing it and then share it via ArcGIS Online. This will require you to have an ArcGIS Online user account and to have publishing rights with your user account to create new content. If you do not have that you can stop and move on to the next recipe:

1. Close the **Geoprocessing** pane when the project tool is complete and save your project.
2. Right-click on the **City_Limits_Web** layer in the **Contents** panel and select **Copy**, as illustrated in the following screenshot:

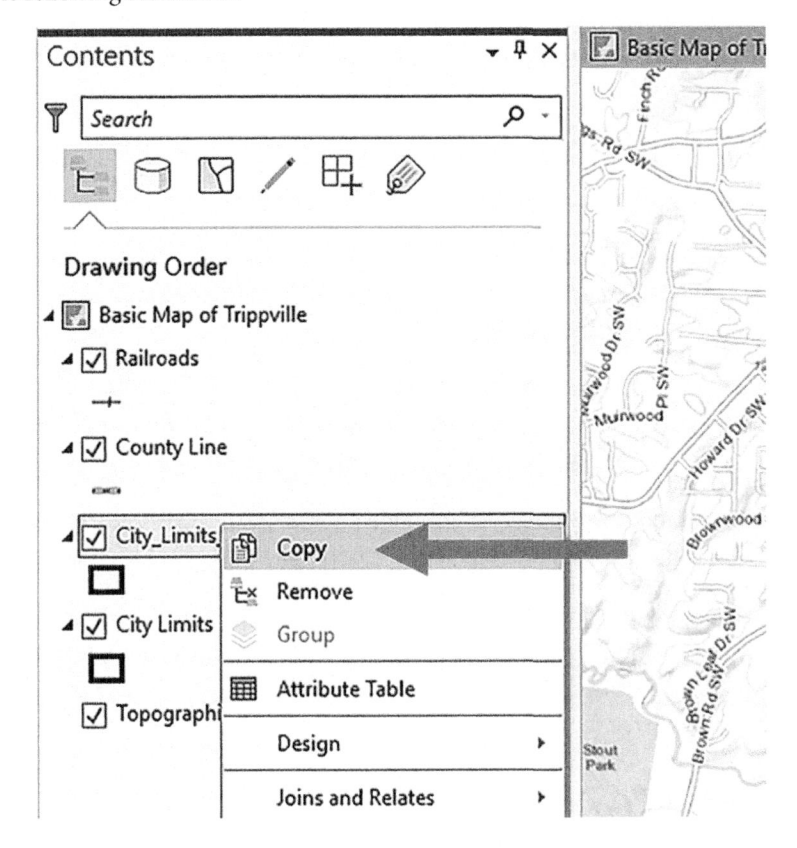

Figure 7.25 – Copying the layer from the menu

3. Click on the **Insert** tab on the ribbon and select the **New map** button.

4. Right-click on **Map** in the **Contents** panel and select **Paste**, as shown in the following screenshot:

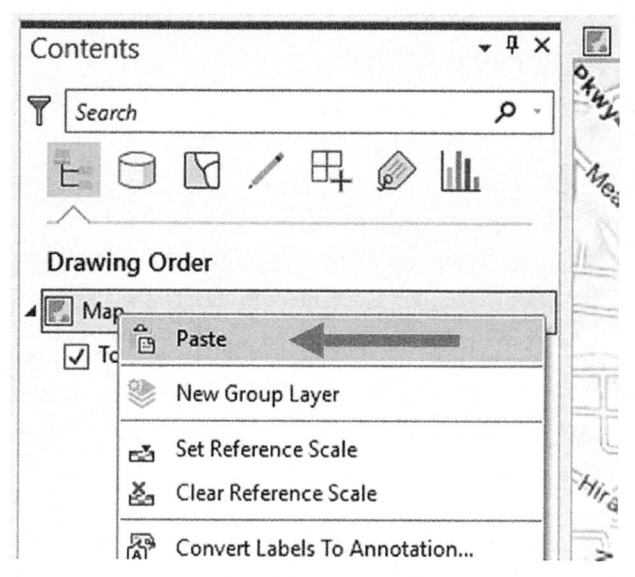

Figure 7.26 – Paste a layer onto a map

> **Tip**
> If the **Paste** option is grayed out, you may need to go back to the basic map of Trippville and copy the layer again.

5. Click on the **Share** tab on the ribbon.
6. Click on the **Web Layer** button located in the **Share As** group on the **Share** tab. This will open the share as web layer tool.

> **Tip**
> If this button is grayed out or not accessible, it may be because you are not logged in to ArcGIS Online. Look in the upper-right corner of the ArcGIS Pro interface. If it says **Sign in** you are not connected to ArcGIS Online and need to do so. Click on **Sign in** to log in to your ArcGIS Online account.

7. In the **Name** cell, type `City_of_Trippville_Limits`. Spaces and special characters are not allowed in the name of a web layer, except for underscores.
8. Type the following for **Summary**: `This layer represents the current boundary for the City of Trippville.`

260   Projection and Coordinate System Basics

9. Type the following for **Tags**, and make sure that you press the *Enter* key between each one: `Trippville, City, Boundary, Limits`.

10. Ensure that **Layer Type** is set to `Feature`.

11. Set **Location** to `Username (root)`. This is where the new web layer will be created within your content.

12. Leave the **Share with** options unselected. Since this is just a learning exercise, you do not need to share the web layer you are creating.

13. Verify that your **Share As Web Layer** tool looks like the following screenshot, with the exception of the location, which will have your username, and click on the **Configuration** tab near the top:

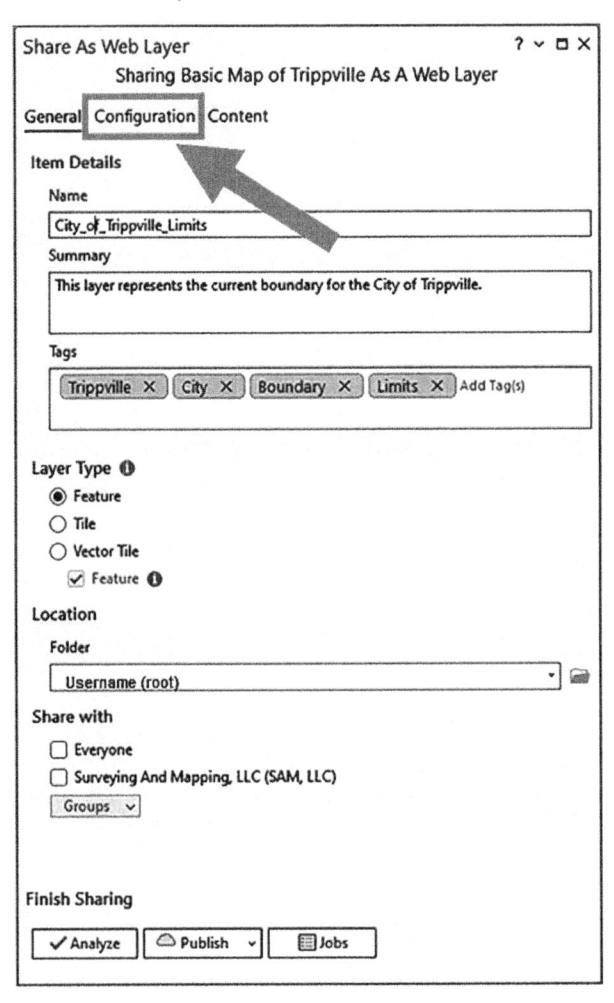

Figure 7.27 – Share as web layer parameters filled out

14. Review the options under **Configuration**; you are not going to change these. When you are done reviewing the options, click on the **Content** tab near the top of the panel.
15. Verify that your web layer will include `City_of_Trippville_Limits` with `City_Limits_Web`.
16. Click on the **Analyze** button at the bottom of the panel to see whether there are any issues that might prevent the web layer from being published successfully.

    When you click **Analyze**, it should return one error and one or two warnings depending on what basemap your organization defaults to:

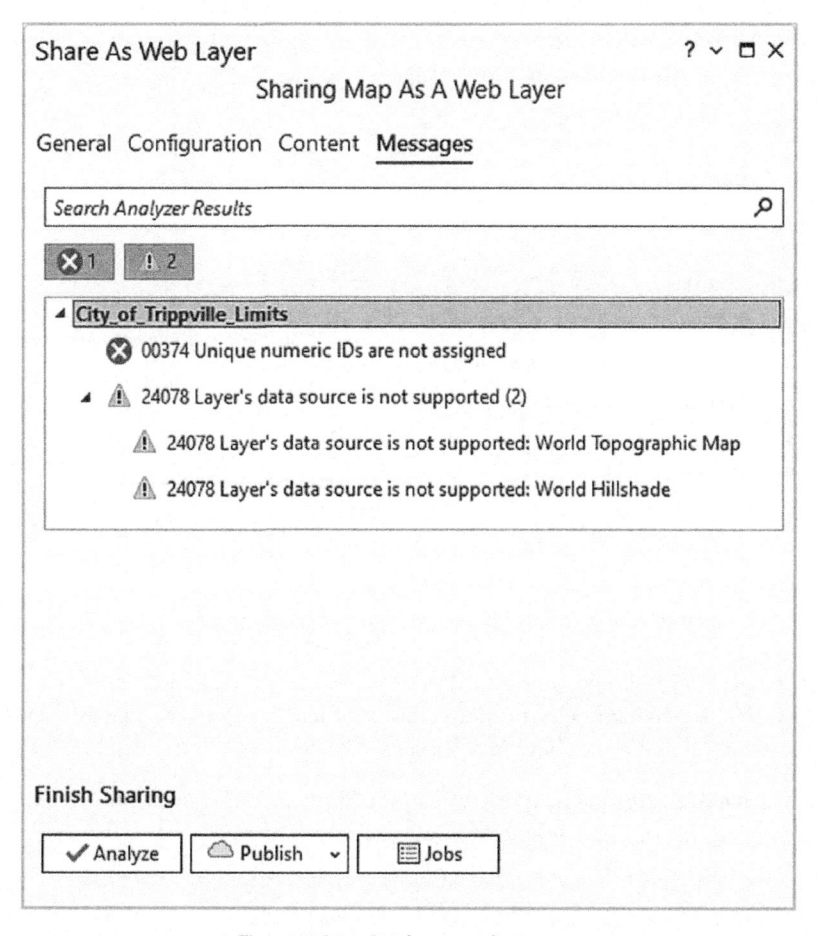

Figure 7.28 – Analyze results report

The warnings indicate that the basemap being used is not supported. You can ignore this warning as ArcGIS Pro will just skip over the basemap when publishing the web layer as if it does not exist. The error you must fix. The error says that unique numeric IDs are not assigned to the city limits web layer. You will correct that error next.

1. Move your mouse pointer over the error indicated by a red circle containing a white X and saying **00374 Unique numeric IDs are not assigned**. Three dots should appear to the right of the error. Click on the three dots to open a menu.
2. In the menu that appears, select **Auto-Assign IDs Sequentially** as illustrated in the following screenshot:

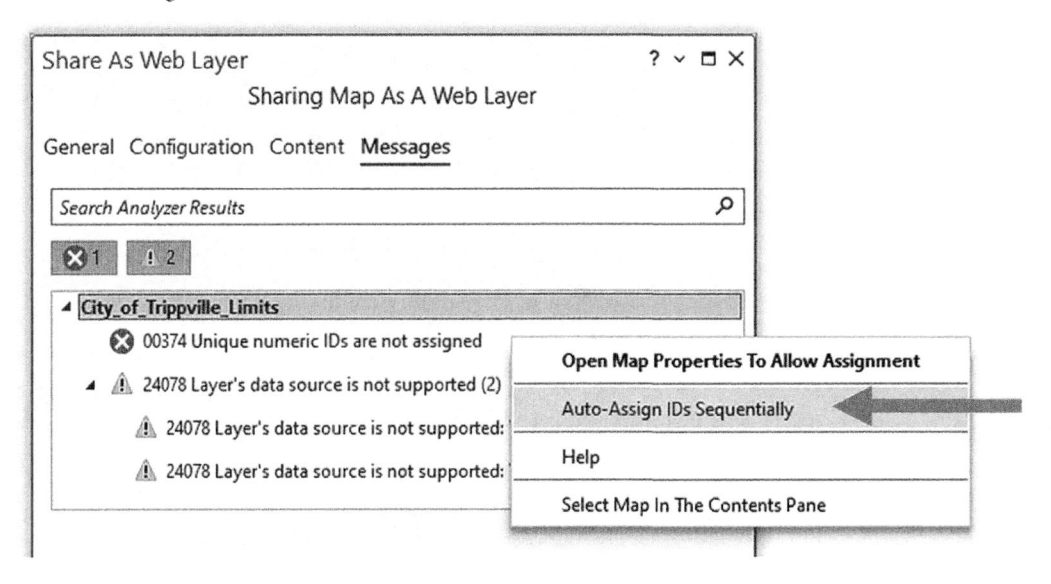

Figure 7.29 – Correcting an error while publishing the web layer

A green checkmark should appear next to the error indicating that it was corrected. However, it is possible that fixing one error may create another, so you need to analyze one more time to make sure no new errors have been created.

3. Click the **Analyze** button one more time in the **Share as Web Layer** tool. This time no errors should appear. You should only see the warnings.

> **Note**
> Warnings are signified by exclamation points in a yellow triangle. Warnings indicate something that violates the recommended settings or practices but that will not stop the web layer from publishing. Errors, on the other hand, will cause the publishing of the web layer to fail. All errors must be resolved before you can publish the web layer.

Projecting data to a different coordinate system   263

4. Click the **Publish** button at the bottom of the **Share As Web Layer** panel. This will publish this layer to ArcGIS Online. Depending on your web connection, this operation might take some time to complete.

5. Once the tool has successfully published the web layer, click on **Manage the web layer**, located at the bottom of the tool, as illustrated in the following screenshot. This will launch your web browser and take you to your ArcGIS Online login screen:

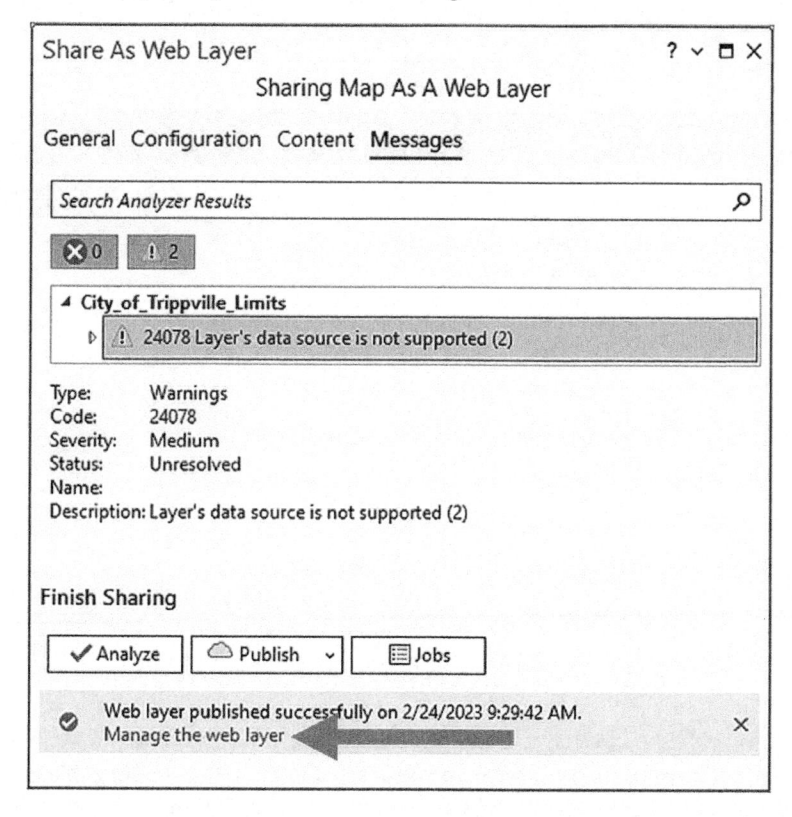

Figure 7.30 – Clicking on Manage the web layer after publishing

6. Log in to your ArcGIS Online account with your username and password in the web browser that opened.

Once logged in, you will be taken to the web layer you just published to ArcGIS Online. This means you were successful. The web layer is a copy of your data now stored in ArcGIS Online and served out as a web service. You will now examine the new web layer in ArcGIS Online using the Map Viewer web application included with ArcGIS Online.

Using ArcGIS Online Map Viewer to verify results

With the web layer successfully published to ArcGIS Online, you will now verify that it is working by looking at it in the Map Viewer application that is built into ArcGIS Online. ArcGIS Online is Esri's cloud solution for ArcGIS Platform. Perform the following actions:

1. Click on the **Open in Map Viewer** button located on the right side of your browser, as shown in the following screenshot:

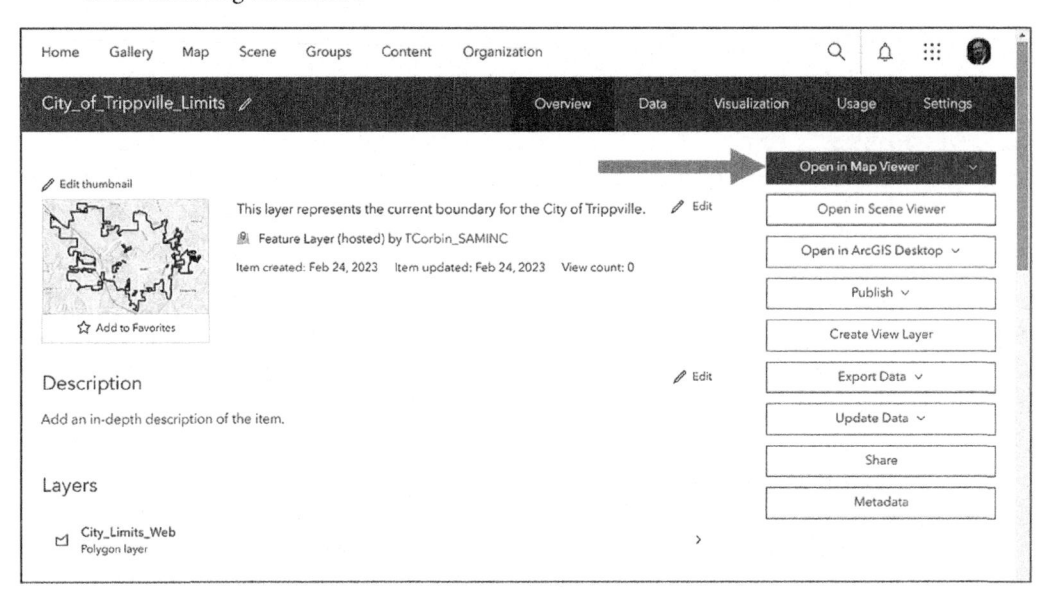

Figure 7.31 – Opening the web layer in the Map Viewer application

Your newly published web layer should open in the ArcGIS Online Map Viewer web application. It should look very similar to the map you created in ArcGIS Pro. The difference is that you see the city limits for Trippville as being stored in ArcGIS Online and not on your computer. It is now all in the ArcGIS Online cloud.

2. Take some time to explore the Map Viewer application and your web layer. Once you are done, close your web browser and return to ArcGIS Pro.
3. Close the **Share As Web Layer** tool.
4. Save your project and close ArcGIS Pro.

You have now successfully published a web layer to ArcGIS Online. As a web layer, that data can be accessed from web applications, mobile devices, and desktop GIS applications such as ArcGIS Pro.

How it works...

You have just projected a layer to a new coordinate system using the project tool. You projected the city limits layer to the coordinate system used by most web mapping applications, so it could be more easily integrated with other web-based GIS data in the future, and you published that layer to ArcGIS Online. You then viewed the new layer that you published in a web map via ArcGIS Online Map Viewer.

8
Creating a Geodatabase

Throughout the other chapters in this book, you have worked with existing data from the `Trippville_GIS.gdb` geodatabase. This has made things easy because you did not need to build the data or database from scratch. While it is often the case that much of the data you will use in ArcGIS Pro already exists, you may find the need at some point to create your own geodatabase. How do you do that?

Before we answer that question, we need to understand what a geodatabase is. A **geodatabase** is a spatially enabled database used to store GIS information and is the primary storage format for Esri's ArcGIS platform. Geodatabases allow you to store your GIS data in a single repository, as it supports storage of both vector and raster data. In addition to storing data, geodatabases also support the inclusion of data validation objects, such as topologies, domains, and subtypes that can be used to help ensure your data is correct. Geodatabases also support the storage of advanced data models, such as utility networks and parcel fabrics. These advanced data models include a combination of data rules and tools for data editing and analysis. To learn more about geodatabases, you can go to `https://pro.arcgis.com/en/pro-app/latest/help/data/geodatabases/overview/what-is-a-geodatabase-.htm`.

In this chapter, you will learn how to create a new geodatabase to support an organization's general operations. In addition to creating a new geodatabase, you will learn how to create basic items that can be stored inside a geodatabase. Recipes included in this chapter are as follows:

- Creating a new geodatabase and feature classes
- Creating a feature dataset
- Creating a coded values domain
- Creating subtypes

Creating a new geodatabase and feature classes

As mentioned, a geodatabase is the primary storage format for the ArcGIS platform, which includes ArcGIS Pro. This format has many advantages over other data formats, such as shapefiles, coverages, or DWG AutoCAD files.

Geodatabases can store a range of data used in a GIS including tabular, raster, and vector data in a single database. This makes it easier to manage and access all your GIS data because it is in one place. Then, they support data validation using topologies, advanced data models, domains, and subtypes. This allows you to find and correct errors in your data. Geodatabases are scalable, so they can grow as your organization does. All of this makes geodatabases extremely powerful and flexible in supporting an organization's GIS needs.

In this recipe, you will learn the basic skills needed to create a new geodatabase and add new feature classes stored in it.

Getting ready

It is recommended that you complete at least all the recipes in *Chapter 1, Getting to Know ArcGIS Pro Capabilities and Terminology*, and the user interface before beginning this chapter. This will ensure you have the foundational knowledge required to successfully complete this chapter. You will need to ensure you have ArcGIS Pro 3.x installed and a license assigned. This recipe can be completed with any license level of ArcGIS Pro, Basic, Standard, or Advanced.

How to do it...

You will now walk through the tools in ArcGIS Pro that will allow you to create a new file geodatabase and add feature classes to it:

1. Launch ArcGIS Pro and open the `Creating a Geodatabase.aprx` project file located in `C:\Student\ArcGISPro3Cookbook\Chapter8\`, by clicking on the **Open another project** button in the ArcGIS Pro **Start** window and navigating to the indicated folder.

    The project will open in the **Catalog** view. This project does not contain any maps. It only contains connections to databases, toolboxes, styles, and folders. You are about to create a new file geodatabase for use in this project.

2. Right-click the **Databases** folder in either the **Contents** or **Catalog** pane, as illustrated in the following screenshots. Select **New File Geodatabase** from the menu that appears.

Creating a new geodatabase and feature classes   269

In the Contents pane   **In the Catalog pane**

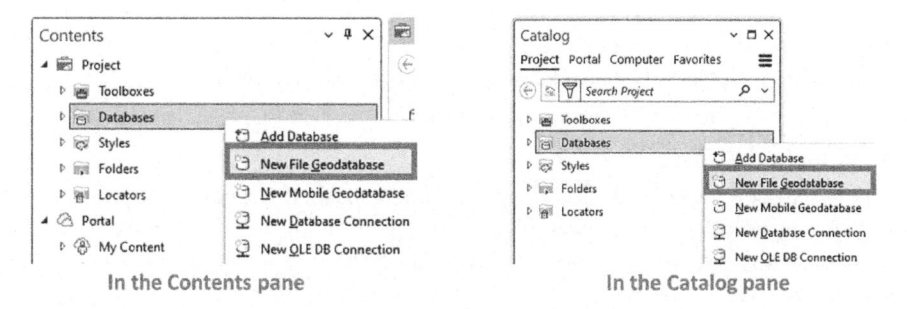

Figure 8.1 – Selecting New File Geodatabase in either the Contents or Catalog pane

The **New File Geodatabase** window should now be open on your screen.

3. In the left panel of the **New File Geodatabase** window, expand the **This PC** option.

4. Scroll down and click on **Local Disk (C:)** so that you can see its contents in the right panel of the window. The local disk might also be identified as **OS** or **Local Drive**.

5. In the right panel of the **New File Geodatabase** window, scroll down and double-click in the Student folder.

6. Continue navigating to the MyProjects folder located in the ArcGIS Pro3Cookbook folder by double-clicking on each folder. Name your new geodatabase MyGeodatabase.

7. Verify that your **New File Geodatabase** window matches the following, and click the **Save** button to create the new geodatabase.

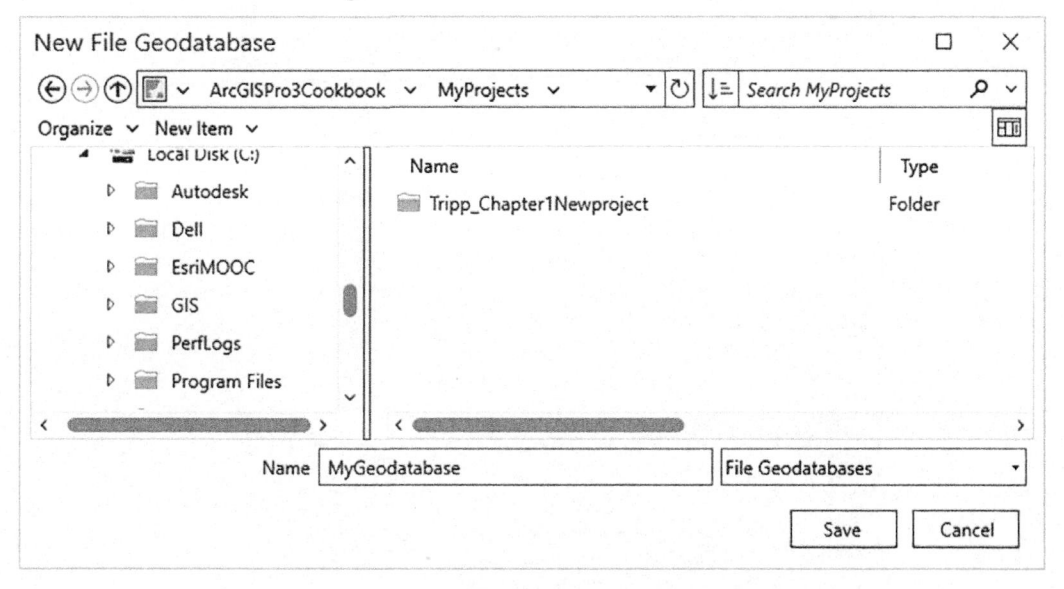

Figure 8.2 – The New File Geodatabase window

The new `MyGeodatabase.gdb` file geodatabase should now appear in the **Catalog** view. This means you have just successfully created a new file geodatabase in ArcGIS Pro.

> **Information**
>
> A file geodatabase is just one of several types of geodatabases that can be used in ArcGIS Pro. A file geodatabase is designed to be used by small organizations with one or two primary GIS power users. Other types of geodatabases include personal (not supported in ArcGIS Pro), mobile, and enterprise. To learn more about the types of geodatabase supported in ArcGIS Pro, go to `https://pro.arcgis.com/en/pro-app/latest/help/data/geodatabases/overview/types-of-geodatabases.htm`.

Creating a new feature class

Now that you have created a new geodatabase, you need to create an object that will store data. You will start with a simple object called a **feature class**. A feature class will store either points, lines, or polygons that share a common attribute table and spatial reference. A geodatabase can contain multiple feature classes:

1. Right-click on the new `MyGeodatabase.gdb` file geodatabase you just created in either the **Catalog** view or pane. Then, select **New | Feature Class** from the menu that appears, as shown in the following screenshots.

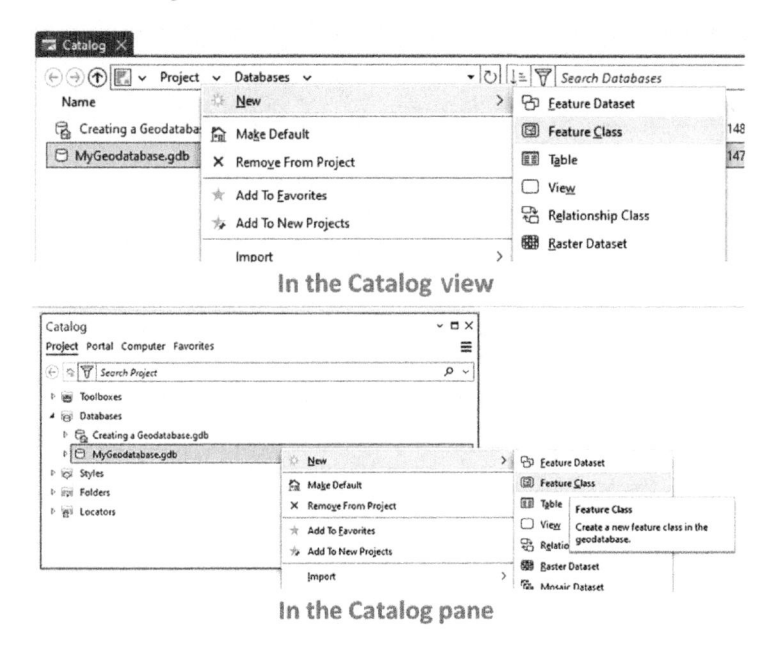

Figure 8.3 – Creating a new feature class in either the Catalog view or pane

2. The **Create Feature Class** tool should now be open. Name the new feature class you are creating `Side_Walks`. For **Alias**, type `Side Walks`.

> **Information**
>
> Feature class names in a geodatabase have some limitations you should be aware of. They must start with a letter. Although the name can contain a number, it cannot start with one. Also, the name cannot contain spaces or special characters. The only exception is an underscore. The length of the name is limited by the operating system of the computer where you store the geodatabase file. The alias does not have these limitations, so it can be more descriptive of the data the feature class contains. The alias is what will display when the feature class is added to a map as a layer.

3. Set **Feature Class Type** to **Line** using the drop-down arrows.
4. Under **Geometric Properties**, ensure that **M Values** is not selected, and **Z Values** is enabled, as shown in the following screenshot, and click **Next**.

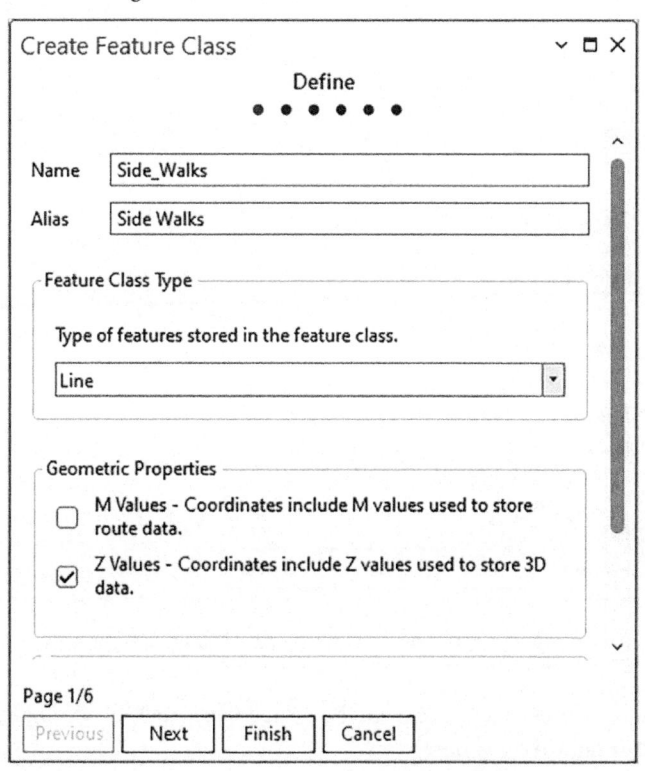

Figure 8.4 – The Create Feature Class tool, defining initial properties

5. In the **Fields** screen of the **Create Feature Class** tool, click where it says **Click here to add a new field**.
6. In the **Field Name** column, name the new field `Street`. Then, set **Data Type** to **Text** by clicking in the cell and using the drop-down arrow.
7. In the lower panel of the **Fields** screen, set **Alias** for the new field to `Street Sidewalk Follows`.
8. Set the **Length** to `50` in place of the default **255**.
9. Verify that your **Create Feature Class** tool fields screen matches the following, and then click **Next**.

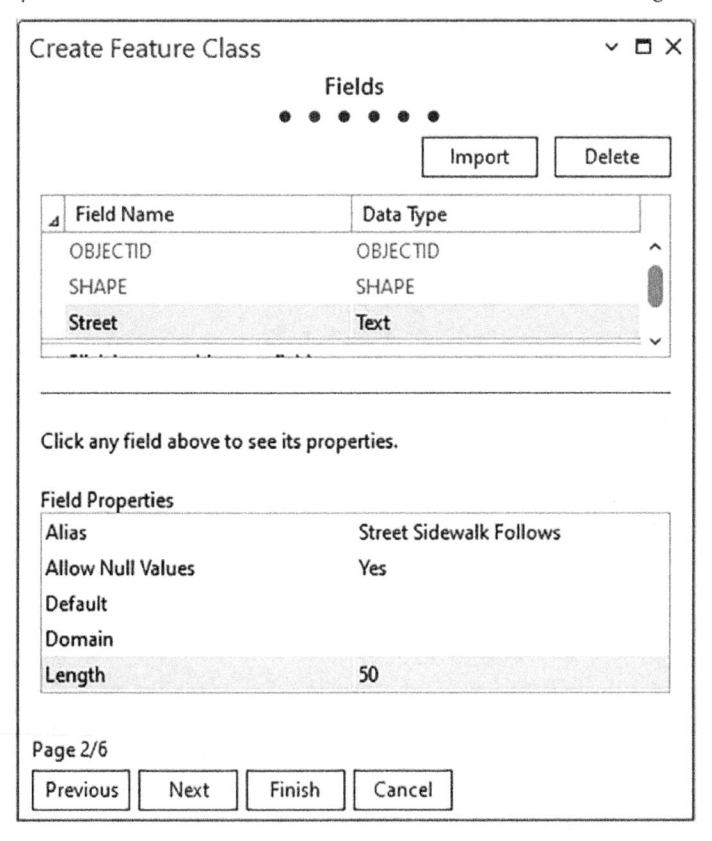

Figure 8.5 – The Create Feature Class tool fields screen

10. On the **Spatial Reference** screen, accept the default **Current XY** coordinate system, **WGS 1984 Web Mercator (auxiliary sphere)**.

> **Information**
>
> The WGS 1984 Web Mercator (auxiliary sphere) coordinate system is the *x* and *y* (horizontal) coordinate system used by most web-based mapping applications and tools, including ArcGIS Online, Google Maps, and Bing Maps. This makes it a great choice for any data that will be accessed, edited, or analyzed via web or mobile applications.

11. Then, click in the box that shows <**None**> under **Current Z** to assign the vertical coordinate system, since this new feature class has *Z* values enabled.
12. In the lower panel of the **Spatial Reference** screen, scroll down and expand the **Vertical Coordinate System**. Next, expand **Gravity – Related** so that you can see the values underneath.
13. Scroll down and expand **North America**. Continue scrolling down, and select **NAVD88 height (ftUS)**.
14. Verify that your **Spatial Reference** screen matches the following, and then click **Next**.

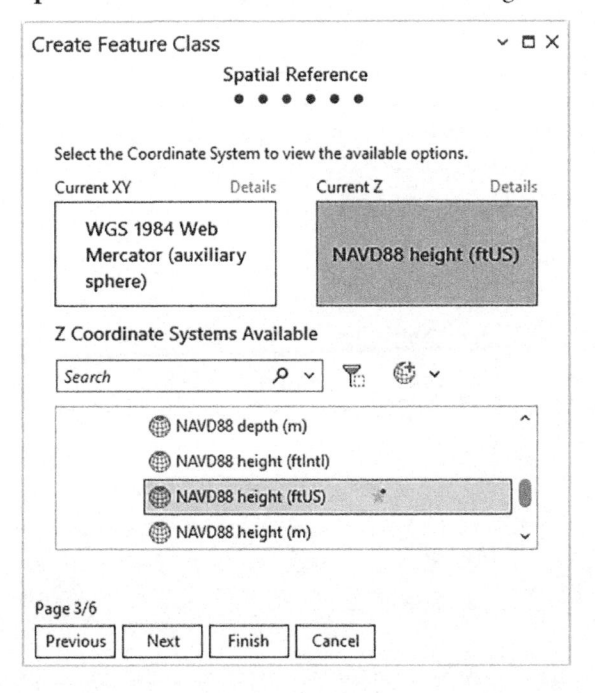

Figure 8.6 – The Create Feature Class Spatial Reference screen

15. Accept the default values on the **Tolerance** screen by clicking the **Next** button. It is generally considered a best practice by Esri to allow the software to calculate these values and not to try to adjust them manually.
16. Do the same thing for **Resolution**.

17. On the **Storage Configuration** screen, click the **Finish** button to create the new feature class in `MyGeodatabase.gdb`.
18. In the **Catalog** pane, expand `MyGeodatabase.gdb` to see its contents.

You should now see the new `Side_Walks` feature class you just created located inside the `MyGeodatabase.gdb` file geodatabase you created earlier.

Adding new fields to the attribute table

After a feature class is created, you often need to add new fields to the attribute table so that you can capture additional information. You will add two new fields to the feature class you just created that will allow you to capture the data on the sidewalk that was installed and its pavement type:

1. In that **Catalog** pane, right-click on the `Side_Walks` feature class, click on **Data Design**, and then **Fields**, as shown in the following.

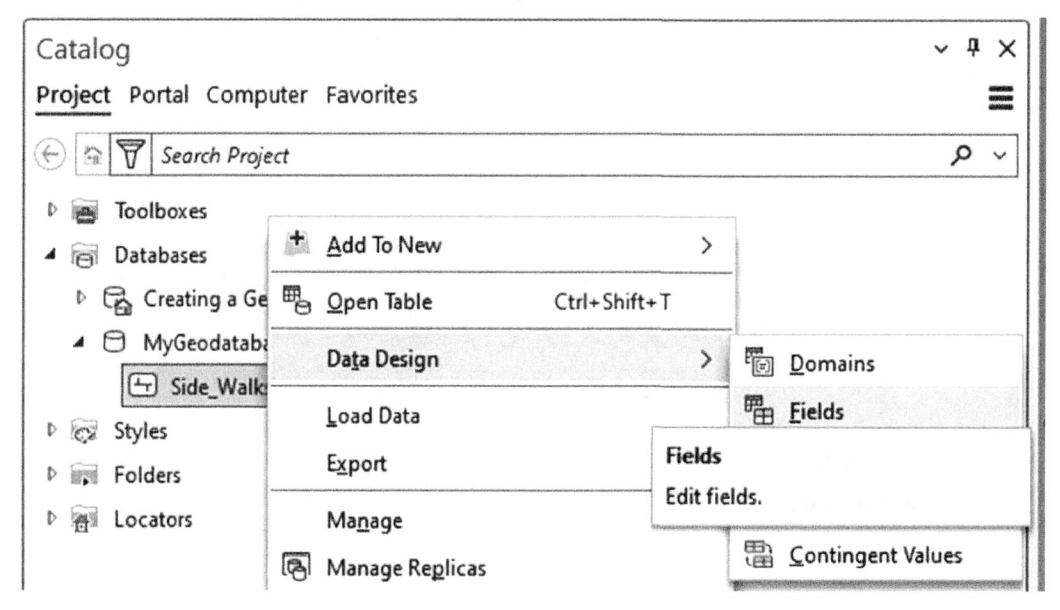

Figure 8.7 – Accessing Data Design | Fields in the Catalog pane

The **Fields: Side Walks** tab should open in the middle of the ArcGIS Pro interface. You should see a list of fields that are included in the attribute table for this new feature class you just created. This will include three system fields, `ObjectID`, `Shape`, `Shape_Length`, and one user-defined field, `Street`. System fields are fields that are automatically created and maintained by ArcGIS Pro. User-defined fields are fields users create and are responsible for maintaining.

You are now going to create two new user-defined fields. This will require you to provide a name and data type at a minimum.

2. Click on the row where it says **Click here to add a new field**.
3. In **Field Name**, type `Install_Date`. Then, give it an alias of **Date Sidewalk was installed**. You can use your **Tab** key to cycle through the different field properties you are defining.

> **Information**
>
> Database fields have several limitations. The first is the name; it has the same limitations as the name of a feature class. Also, you must designate a data type. Field data types typically include text or string, long integer, short integer, single or float, double, date, blob, guid, and raster. To learn more about these different field types, go to `https://pro.arcgis.com/en/pro-app/latest/help/data/geodatabases/overview/arcgis-field-data-types.htm`.

4. Set **Data Type** to **Date Only** using the drop-down arrow. You will leave the rest of the field properties set as default.
5. Click on **Click here to add a new field** again to add another field to the table.
6. Name the new field `Pave_Type` with an alias of `Pavement Type`. Next, set **Data Type** to Text with a length of 15.
7. Verify that your **Fields: Side Walks** tab matches the following. Then, click the **Save** button on the **Fields** tab in the ribbon to save these two new fields to the attribute table for the `Side_Walks` feature class.

Field Name	Alias	Data Type	Allow NULL	Domain	Default	Length
OBJECTID		Object ID				
Shape	SHAPE	Geometry	✓			
Shape_Length		Double	✓			
Street	Street Sidewalk Follows	Text	✓			50
Install_Date	Date Sidewalk was installed	Date Only	✓			
Pave_Type	Pavement Type	Text	✓			15
Click here to add a new field.						

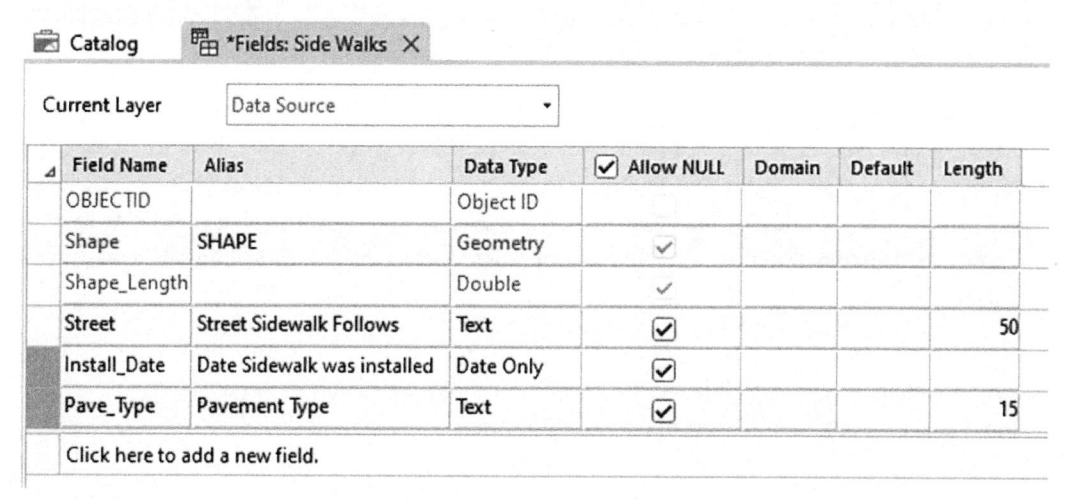

Figure 8.8 – Creating the Install_Date and Pave_Type fields

8. Close the **Fields: Side Walks** tab, and then save your project.
9. Close ArcGIS Pro.

You have now created a new file geodatabase, added a new feature class to it, and then added two new user-defined fields to the attribute table for the new feature class.

How it works...

In this recipe, you created a new empty file geodatabase called `MyGeodatabase.gdb`. You then started to add new items to this geodatabase, starting with a new feature class that will store data about sidewalks. Once you created the new feature class, you then added two new fields to the attribute table for the `sidewalks` feature class, allowing you to capture data about when the sidewalk was installed and the type of material it is made of.

To do this, you right-clicked on the `Databases` folder in either the **Contents** pane or the **Catalog** pane and selected **New File Geodatabase** from the menu that appeared. You then navigated to the folder on your local computer that you wished to save the new file geodatabase in, giving it the name of `MyGeodatabase.gdb`.

With the new geodatabase created, it was time to create a new feature class. You did this by right-clicking on the new geodatabase and going to **New | Feature Class** in the menus that appear. From there, you worked through the **New Feature Class** tool, defining various properties in the multiple screens. This included naming the new feature class, providing an alias, setting the spatial data type, creating a single user-defined attribute field, assigning the spatial reference, and accepting the defaults for tolerance and resolution per Esri's recommended best practice.

Lastly, you added two new user-defined fields, `Install_Date` and `Pave_Type`, after you created the new feature class. To do this, you right-clicked on the `Side_Walks` feature class in the **Catalog** pane and selected **Data Design** and **Fields** from the menus that appeared. This opened the **Fields: Side Walks** tab in the middle of the interface. You then clicked on the row that says **Click here** to add a new field. That allowed you to input the properties for the new fields, including the name, alias, and data type.

To apply these new fields to the attribute table, you clicked on the **Save** button located in the **Fields** tab in the ribbon. Once you did that, those two new fields were available in the attribute table for you to input data associated with individual sidewalk segments.

Creating a feature dataset

You now know how to create a new file geodatabase and add feature classes to it. However, as you might imagine, it is possible for a single geodatabase to include the many feature classes required to support the operations of an organization. That might make it hard to locate and manage the feature classes that a single department, group, or individual might need to use. So, it would be helpful if you could group related feature classes together in a geodatabase, similar to how we organize computer files in the folder. You can do that using **feature datasets**.

Feature datasets allow you to organize related feature classes together, making them easier to locate and manage. You can basically think of them as folders within your geodatabase. All feature classes stored in a feature dataset will have common properties, such as a coordinate system, resolution, and tolerance. To learn more about feature datasets, you can visit https://pro.arcgis.com/en/pro-app/latest/help/data/feature-datasets/feature-datasets-in-arcgis-pro.htm.

In this recipe, you create a new feature dataset in the `MyGeodatabase.gdb` file geodatabase that you created in the previous recipe. Then, you will create a new feature class within that feature dataset.

Getting ready

To complete this recipe, you will need to have finished the previous *Creating a geodatabase* recipe in this chapter. It is also recommended that you complete all the recipes in *Chapter 1* of this book to ensure you have a foundational understanding of the terminology and interface for ArcGIS Pro. This recipe can be completed with all the licensing levels of ArcGIS Pro, Basic, Standard, or Advanced.

How to do it...

You will create a new feature dataset in the `MyGeodatabase.gdb` file geodatabase you created in the first recipe in this chapter:

1. Launch ArcGIS Pro, and open the `Creating a Geodatabase.aprx` project. This project is located in `C:\Student\ArcGISPro3Cookbook\Chapter8\Creating a Geodatabase\` if you installed the data in the default location. If you completed the previous recipe from this chapter, this project should be on your **Recent Projects** list.

2. In the **Catalog** pane, expand the `Databases` folder and right-click on the `MyGeodatabases.gdb` file geodatabase. Select **New** | **Feature Dataset** from the menu that appears, as shown in the following screenshot.

Creating a Geodatabase

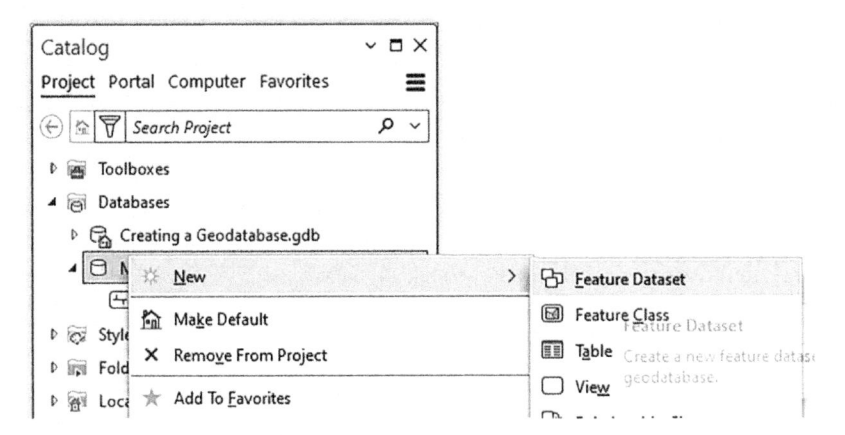

Figure 8.9 – Creating a new feature dataset in the Catalog pane

The **Geoprocessing** pane should now be open with the **Create Feature Dataset** tool. You will now input the parameters required to create the new feature dataset.

3. In the **Create Feature Dataset** tool, the **Output Geodatabase** parameter should already be defined as the `MyGeodatabase.gdb` file geodatabase because you right-clicked on it to open the tool. In **Feature Dataset Name**, type `Property`.

4. Accept the default for **Coordinate System**. It should be set to `PROJCS["WGS_84_Web_Mercator_Auxilary_Sphere", GEOGCS`.

5. Verify that your **Create Feature Dataset** tool matches the following screenshot, and then click **Run** to create the new feature dataset.

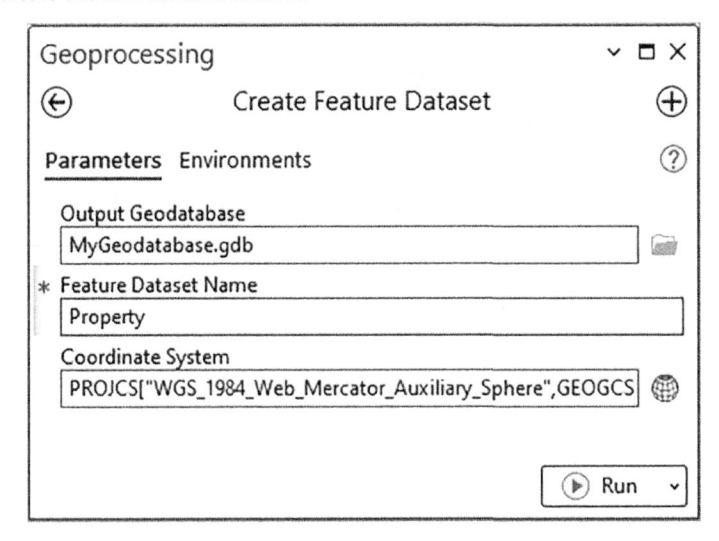

Figure 8.10 – The Create Feature Dataset tool with completed parameters

> **Tip**
> If you get an error indicating that a parameter or value is missing, make sure to press *Enter* or click on another parameter after typing the feature dataset name into the tool.

6. Close the **Geoprocessing** pane when complete.

You should now see the new `Property` feature dataset appear in the `MyGeodatabase.gdb` file geodatabase. It will have an icon of three overlapping squares to the left of the name. If you do not see it, right-click on the `MyGeodatabase.gdb` file and click **Refresh** on the menu that appears.

Creating a new feature class in a feature dataset

You will now create a new feature class in the `Property` feature dataset you just created. You should notice that this process is a bit easier than what you experienced in the previous recipe:

1. In the **Catalog** pane, right click on the `Property` feature dataset you just created. Select **New** and **Feature Class** from the menu that appears. This will open the **Create Feature Class** tool that you used in the previous recipe.

2. Name the new feature class `Property_Lines`, and set **Feature Class Type** to **Line** using the drop-down arrow. You can leave the **Alias** parameter blank. It will automatically inherit the **Name** value if **Alias** is left blank.

3. Under **Geometric Properties**, disable both the *M* and *Z* values, and then click **Next**.

4. In the fields screen, add the following two fields with the associated properties:

   - Field one:
     - **Field Name**: `Prop_Line_Type`
     - **Alias**: `Property Line Type`
     - **Data Type**: `Text`
     - **Length**: `10`
   - Field two:
     - **Field Name**: `Length_FT`
     - **Alias**: `Length in Feet`
     - **Data Type**: `Float`

5. Verify that your **Fields** screen matches the following screenshot, and then click **Next**.

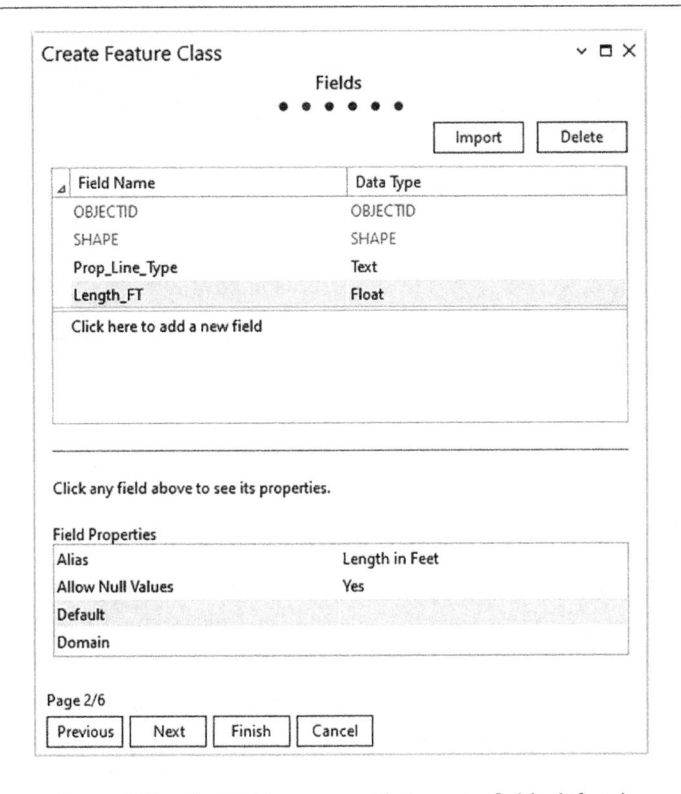

Figure 8.11 – The Fields screen with two new fields defined

6. Now, click the **Finish** button, as all other properties, including coordinate system, tolerance and resolution, are automatically inherited from the Property feature dataset, as you can see from the blue warning text at the top of the **Coordinate System** screen.

7. Save your project and close ArcGIS Pro.

You have successfully created a new feature dataset in a geodatabase and then added a new feature class to the feature dataset.

How it works...

In this recipe, you created a new feature dataset in the MyGeodatabase.gdb geodatabase that you created in the first recipe. Feature datasets are useful for many purposes. First, they can help you organize data within your geodatabase so related feature classes are grouped together making it easier to locate data. Second, feature datasets enable the use of more advanced data modeling capabilities such as topologies and utility networks. Also, when working in an enterprise environment, feature datasets can make it easier to control access to data for users. These are just a few of the advantages of using feature datasets.

Creating a new feature dataset is fairly simple and easy. Before you get started, you need to know a couple of things:

- First, what is the name and purpose of the feature dataset?
- Second, what coordinate system will you use when creating the feature database? All the feature classes you create inside of it will inherit that coordinate system.

With those two things determined, you can begin the steps to create the new feature dataset inside a geodatabase.

In this recipe to create the new feature dataset, you right-clicked on the `MyGeodatabase.gdb` file geodatabase and selected **New | Feature Dataset** from the menus that appeared. This opened the **Create Feature Dataset** geoprocessing tool in the **Geoprocessing** pane. In the tool, you verified that the output database was the one you right-clicked on, `MyGeodatabase.gdb`. Then, you named it `Property`. Lastly, you assigned the desired coordinate system. In this case, you accepted the default coordinate system, which was WGS 1984 Web Mercator (auxiliary sphere). Then, you pressed the **Run** button. When the tool finished, the new feature dataset appeared inside the geodatabase.

Creating a coded values domain

The adage *garbage in equals garbage out* applies to GIS. This means if the data in your GIS is not correct, then the results of any analysis will be wrong. To help ensure GIS data is as error-free as possible, ArcGIS includes several tools and methods to validate data and reduce the ability of users to enter erroneous information. Domains are one of them.

Domains constrain the values that allowed for entry into a field within a database table. There are two types, **Code Values** and **Range**. Range domains are applied to numeric fields and restrict the number values that can be entered between a minimum and maximum number. The Code Values domain creates a pick list of values that the user can select from when populating the value in the field. Domains limit the ability of users to enter any values that are outside those that are acceptable, which helps reduce errors.

In this recipe, you will create a new coded values domain that will contain a list of pavement types. You will then assign the domain to the `Pave_Type` field in the attribute table of the `Side_Walk` feature class you created earlier in this chapter.

Getting ready

Before starting this recipe, you will need to have completed all the previous recipes in this chapter. While not required, it is recommended that you complete the recipes in *Chapter 1* to help ensure you have a foundational understanding of the terminology and user interface for ArcGIS Pro. This recipe can be completed with all licensing levels of ArcGIS Pro.

How to do it...

Now, it is time for you to update the attributes for the sewer manholes you created in *Chapter 4*. You will do that in the table view that displays the attribute table for layers:

1. Launch ArcGIS Pro and open the `Creating a Geodatabase.aprx` project. This project is in `C:\Student\ArcGISPro3Cookbook\Chapter8\Creating a Geodatbase\` if you installed the data in the default location. If you completed previous recipes in this chapter, this project should be on your **Recent Projects** list.

2. In the **Contents** pane, select the `Databases` folder so that you can see its contents in the **Catalog** view located in the middle of the interface.

3. In the **Catalog** view, right-click on the `MyGeodatabase.gdb` file geodatabase and select **Domains** from the menu that appears, as illustrated in the following screenshot.

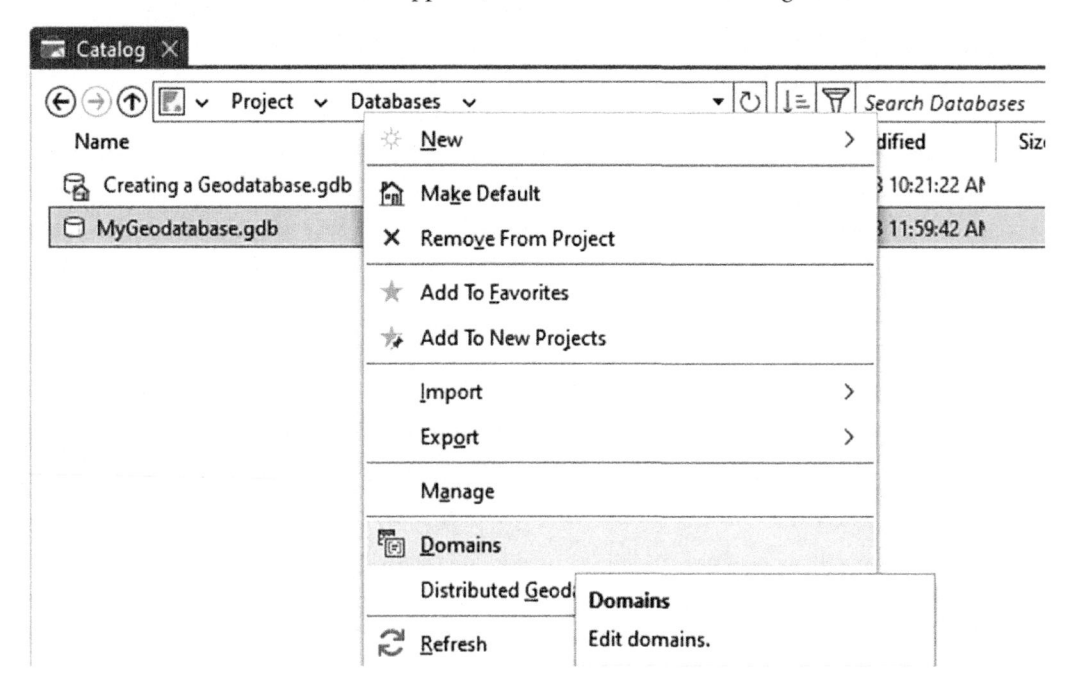

Figure 8.12 – Accessing Domains in the Catalog view

The **Domains: MyGeodatabase** tab should now be open in the main view area of the ArcGIS Pro interface. This is where you will create your new domain.

> **Information**
>
> Domains are considered a property of the entire geodatabase so that you can assign it to any appropriate field in any table within the geodatabase. This reduces duplication and makes domains easier to manage.
>
> You are about to create a domain that will contain a list of accepted pavement types for sidewalks. Think about how many other things are paved, such as streets, parking lots, parking decks, and running tracks. You could use the domain you are about to create with any or all of those as well. To learn more about domains, visit https://pro.arcgis.com/en/pro-app/latest/help/data/geodatabases/overview/an-overview-of-attribute-domains.htm.

4. Set **Domain Name** to Pavement_Typ and **Description** to List of different possible pavement types.

   When you enter the new domain name, you should notice a new smaller section appears with two columns, **Code** and **Description**. This is where you will create the list of accepted values for the domain later in the recipe.

5. Ensure that **Field Type** is set to Text. **Field Type** must match the field type for the database fields you wish to assign it to.

6. Leave **Split Policy** and **Merge Policy** set to Default. Then, click the **Save** button on the **Domains** tab in the ribbon.

   Your **Domains: MyGeodatabase** tab should not look like this.

Domain Name	Description	Field Type	Domain Type	Split Policy	Merge Policy	Code	Description
Pavement_Typ	List of different possible pavement types	Text	Coded Value Domain	Default	Default		

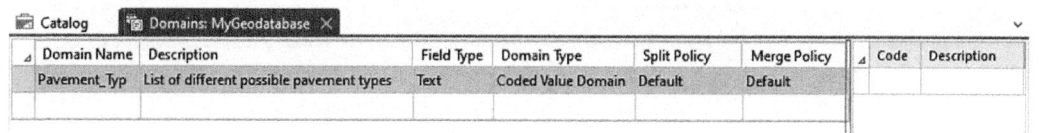

Figure 8.13 – A new domain created

You have just created the domain in your geodatabase. The domain is just an empty shell right now.

Adding coded values to the domain

You will now add the values to the domain that will be acceptable as possible pavement types for the sidewalks. As mentioned, these might also be applicable to other things that are paved, such as streets or parking lots:

1. In the right section of the **Domains: MyGeodatabase** tab under **Code**, type Conc, and under **Description**, type Concrete, as illustrated in the following screenshot. You will need to press the *Enter* key for each value.

Code	Description
Conc	Concrete

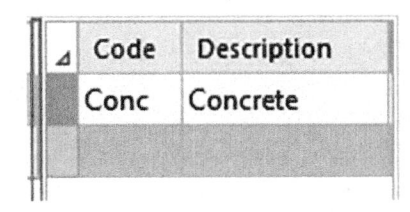

Figure 8.14 – The first coded value entered into the domain

2. In the row that appears below the **Conc** and **Concrete** values, type `Asph` under **Code** and `Asphalt` under **Description**.

3. Continue entering the following codes and descriptions:

   - `Grav: Gravel`
   - `Dirt: Dirt`
   - `Unk: Unknown`

4. Verify that you have successfully entered the list of coded values into the domain, as shown in the following screenshot. Then, click the **Save** button on the **Domains** tab in the ribbon.

Code	Description
Conc	Concrete
Asph	Asphalt
Grav	Gravel
Dirt	Dirt
Unk	Unknown

Figure 8.15 – All acceptable pavement types entered into the domain

You have just added all the types of pavements you expect to encounter for the sidewalks.

Assigning the domain to an attribute field

With the domain created and values added, it is time to assign the domain to an attribute field. That will then limit the possible values a user can enter into that field to those included in the domain:

1. Close the **Domains: MyGeodatabase** tab in the main view area of the ArcGIS Pro interface.

2. In the **Contents** pane, expand the `Databases` folder, and then select the `MyGeodatabase.gdb` file geodatabase so that you can see its contents in the **Catalog** view.

Creating a coded values domain   285

3. In the **Catalog** view, right-click on the `Side_Walks` feature class. Then, select **Data Design** and **Fields** from the menus that appear, as shown in the following screenshot.

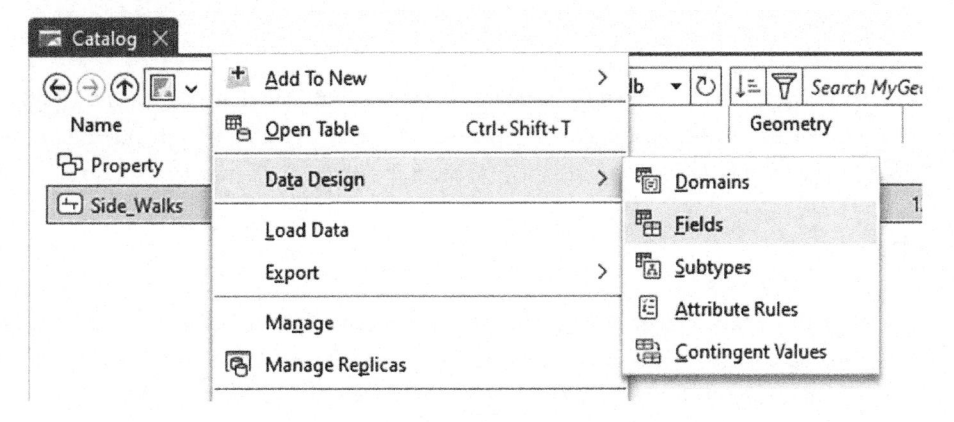

Figure 8.16 – Accessing fields for the Side_Walks feature class in the Catalog view

4. In the **Fields: Side Walks** tab, locate the `Pave_Type` field from the list of attribute fields.
5. Click on the cell in the **Domain** column, as shown in the following screenshot. Then, select the `Pavement_Typ` domain from the list that appears.

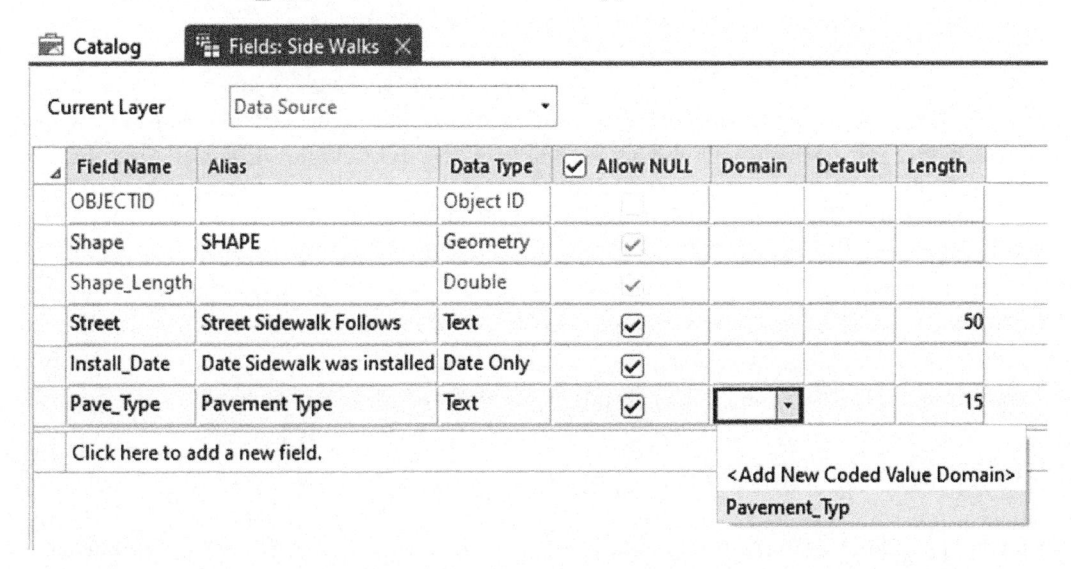

Figure 8.17 – Assigning the Pavement_Typ domain to the Pave_Type field in the Side_Walks attribute table

6. Click on the **Save** button located on the **Fields** tab in the ribbon.
7. Close the **Fields: Side Walks** tab in the main view area and save your project. Then, close ArcGIS Pro.

You have now assigned the domain you created to the `Pave_Typ` field in the `Side_Walk` attribute table. This means when someone edits data in that feature class, as they are updating the pavement type, they will only be able to select values that are included in the `Pavement_Typ` domain.

How it works...

You have just created a new domain and assigned it to an attribute field, limiting the values that can be entered into that field when editing the data. This helps reduce errors and increases data consistency. Because this domain is part of the entire database, it can be used multiple times as needed with other fields in the database. This makes domains very flexible and powerful.

To create the new domain, you selected the `MyGeodatabase.gdb` file geodatabase and selected the **Domain** option from the menu that appeared. Then, in the main view area of ArcGIS Pro, you provided a name for the new domain, along with a description to let users know its intended purpose. Lastly, you selected a data type that matches the fields you want to assign the domain to and accepted the default split and merge policies.

With the domain created, you then needed to create a list of acceptable or approved values it will contain. This was done in the panel that appears to the right of the domain tab in the main view area of ArcGIS Pro. You entered a code and description for each approved value included in the domain.

The last step was to then assign the domain to a field in a table. In this case, it was the `Pave_Type` field in the `Side_Walk` feature class attribute table. To do this, you expanded the `Databases` folder in the **Contents** pane and selected the `MyGeodatabase.gdb` file. Then, you right-clicked on the `Side_Walks` feature class in the **Catalog** view, selecting **Data Design and Fields** from the menus. You then clicked on the **Domain** column for the `Pave_Type` field and selected the `Pavement_Typ` domain from the list that appears. To apply this new setting, you clicked the **Save** button in the ribbon.

Creating subtypes

Subtypes provide a way to create categories of data within a single feature class or standalone table. You can then assign different behavior to the data based on its subtype. This can include things such as specific default field values, domains, and topology rules per subtype. This can, in turn, increase the overall accuracy of your data and improve analysis and editing.

Subtypes are based on a field in the table that identifies the specific subtype or category that the features or records in the table belong to. The field must be either a long or short integer data type. Once a subtype has been defined in a feature class, when you add that feature class to a map as a layer, the symbology for that layer will be based on the subtypes. To learn more details about subtypes and their capabilities, visit https://pro.arcgis.com/en/pro-app/3.0/help/data/geodatabases/overview/an-overview-of-subtypes.htm.

In this recipe, you will add a new parcel polygon feature class to the MyGeodatabase.gdb file geodatabase and load data into it, using a model that was created in ArcGIS ModelBuilder. You will then create a field in the attribute table that will be used to identify the subtype of each parcel.

Getting ready

Before starting this recipe, you will need to have completed the recipes in *Chapter 1* to ensure you understand ArcGIS Pro terminology and how to navigate the interface. You also need to have completed *Chapter 4*'s *Configuring editing options* recipe and the *Creating a feature dataset* recipe in this chapter.

This recipe can be completed with all licensing levels of ArcGIS Pro. You will also need to ensure you have internet access to make use of the Esri-provided basemaps.

How to do it…

You are now ready to begin the process of creating subtypes in a geodatabase feature class. You will start by running a model that will create a new parcel polygon feature class. Then, you will create the subtypes that will identify different types of parcels:

1. Launch ArcGIS Pro and open the CreatingGeodatabase.aprx project. This should be in your **Recent Projects** list.
2. Click on the **Map** tab at the top of the main view area. It should be located to the right of the **Catalog** tab. If the **Map** tab is not available, in the **Catalog** pane, expand the Maps folder. Then, right-click on **Map** and select **Open** from the menu that appears.

    You should now see a map in the main view area, as shown in the following screenshot. Your scale and exact location may vary, depending on your monitor and resolution. Later, you will add a new layer, showing parcels that you will identify by their type.

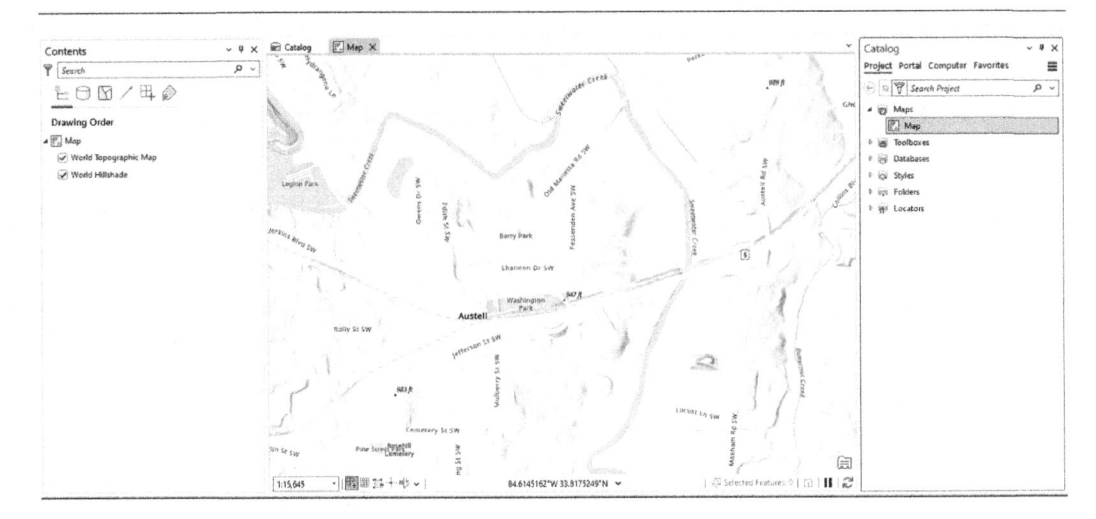

Figure 8.18 – A map open in the ArcGIS Pro project

3. In the **Catalog** pane, expand the `Databases` folder. Then, expand the `MyGeodatabase.gdb` file geodatabase.

4. Expand the `Property` feature dataset in the `MyGeodatabase.gdb` file geodatabase so that you can see its contents. It should only contain a single feature class, `Property_line`, which you made in the *Creating a feature dataset* recipe earlier in this chapter. If you do not see either the `Property` feature dataset or the `Property_line` feature class, you need to go back and complete the *Creating a feature dataset* recipe in this chapter.

Creating a new feature class using a model

Now, you will use a model to create a new `Parcel_Polygon` feature class in the `MyGeodatabase.gdb` file geodatabase. This new feature class will be the one you add subtypes to:

1. In the **Catalog** pane, expand the `Toolboxes` folder. Then, expand the `Creating a Geodatabase.tbx` toolbox.

2. Double-click on the **Add Parcel Polygon Feature Class** model you see in the **Catalog** pane to open the model.

Creating subtypes 289

> **Information**
>
> Models are custom tools that you can create using ArcGIS ModelBuilder, which is part of ArcGIS Pro. ModelBuilder uses a graphical representation of the custom tool you create, allowing you to build tools without needing to be a programmer or understand how to write programming code. Models are identified by an icon that is a series of different color boxes linked together as if they are part of a flow chart. Models can be very simple or complex, depending on their operation. Models can be used to automate a regular process or workflow to ensure that everyone does it the same way each time. To learn more about ModelBuilder and models, visit `https://pro.arcgis.com/en/pro-app/latest/help/analysis/geoprocessing/modelbuilder/what-is-modelbuilder-.htm`.

The **Geoprocessing** pane should now be open, with the **Add Parcel Polygon Feature Class** model open to run. When you run this model, it will do several things, all of which you could do manually. The model, however, does it all automatically, making it easier and faster. Here is a look at the steps and tools included in the model to help you better understand what this tool does.

Creates new Parcel_Polygon Feature Class.

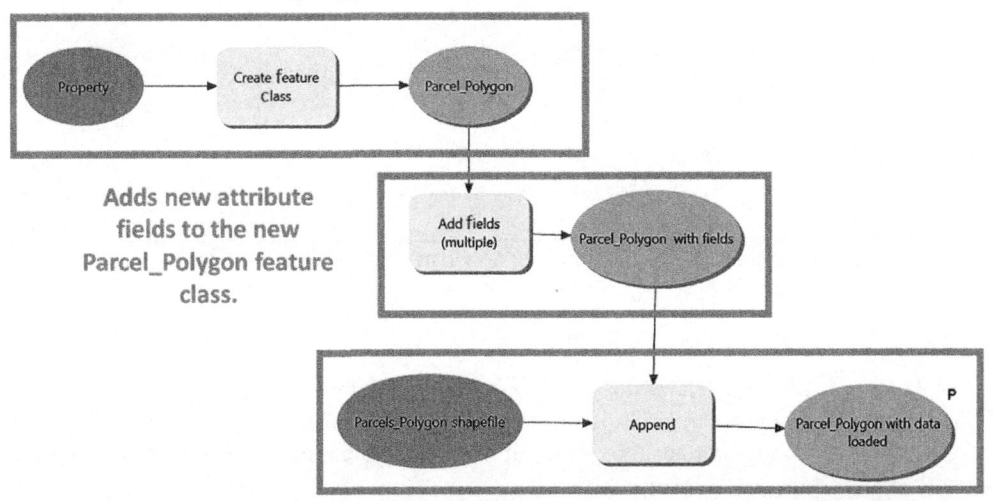

Adds new attribute fields to the new Parcel_Polygon feature class.

Loads data into new feature class from an existing shapefile.

Figure 8.19 – Looking at the Add Parcel Polygon Feature Class model in ModelBuilder

As illustrated in the preceding diagram, this model first creates the new feature class. Then, it uses the **Add Fields (multiple)** geoprocessing tool to add new attribute fields to the `Parcel_Polygon` feature class that was created in the line above. Lastly, it uses the **Append** geoprocessing tool to load data in from an existing shapefile that contains parcels. As you are about to see, when you run this model, it does all of this with a single click on the **Run** button.

Creating a Geodatabase

3. Click the **Run** button located at the bottom of the **Add Parcel Polygon Feature Class** model in the **Geoprocessing** pane.

4. In the **Catalog** pane, look in the `Property` feature dataset in the `MyGeodatabase.gdb` file geodatabase. You should now see that a new feature class named `Parcel_Polygon` has been added. You should also see that the new feature class has been added as a new layer in your map, as illustrated in the following screenshot.

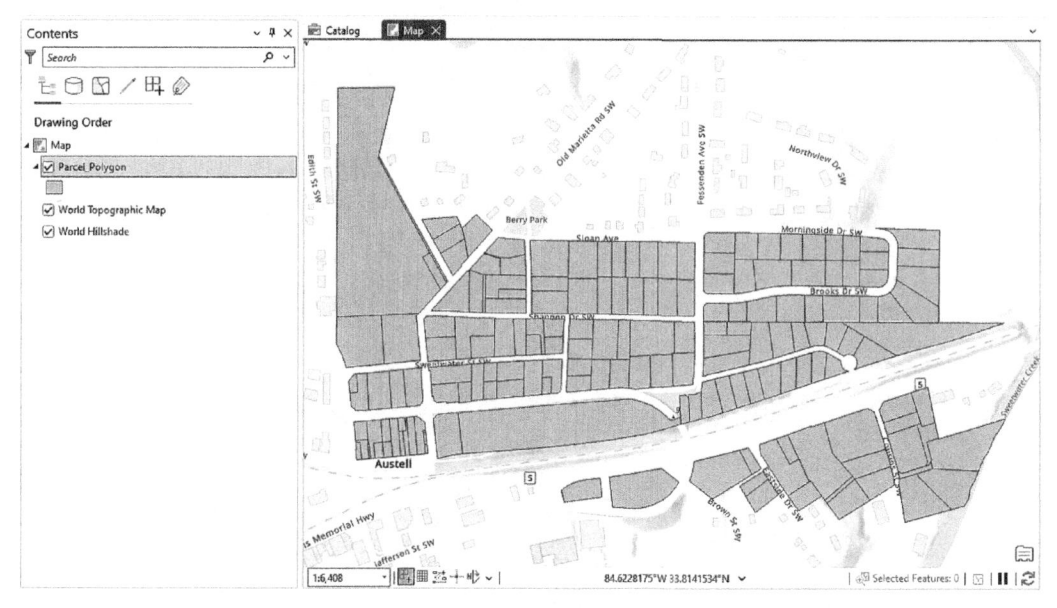

Figure 8.20 – Parcel_Polygons added as a new layer in the map

> **Tip**
> If the `Parcel_Polygon` layer does not appear automatically in the map, that is because you do not have the Geoprocessing results added to Map option enabled. You will need to add the new feature class to your map. You can do this by simply dragging it from the **Catalog** pane and dropping it into the map.

5. In the **Contents** pane, select the newly added `Parcel_Polygon` layer. Then, click on the **Data** tab in the ribbon.

6. Click on the **Fields** button in the ribbon to open the **Fields** view. This displays a list of all the attribute fields for this layer and their properties, as shown in the following screenshot.

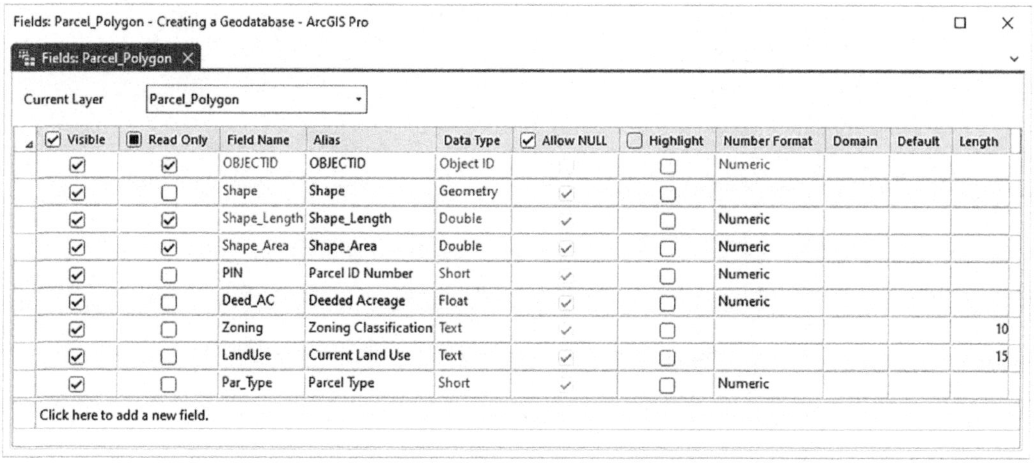

Visible	Read Only	Field Name	Alias	Data Type	Allow NULL	Highlight	Number Format	Domain	Default	Length
✓	✓	OBJECTID	OBJECTID	Object ID		☐	Numeric			
✓	☐	Shape	Shape	Geometry	✓	☐				
✓	✓	Shape_Length	Shape_Length	Double	✓	☐	Numeric			
✓	✓	Shape_Area	Shape_Area	Double	✓	☐	Numeric			
✓	☐	PIN	Parcel ID Number	Short	✓	☐	Numeric			
✓	☐	Deed_AC	Deeded Acreage	Float	✓	☐	Numeric			
✓	☐	Zoning	Zoning Classification	Text	✓	☐				10
✓	☐	LandUse	Current Land Use	Text	✓	☐				15
✓	☐	Par_Type	Parcel Type	Short	✓	☐	Numeric			

Figure 8.21 – The Parcel_Polygon layer attribute fields

7. Review the list of attribute fields. Locate one that has a data type of either a long or short integer.

You should only see two fields that have a data type of a long or short integer. They are the `PIN` and `Par_Type` fields. The `PIN` field is used to store the parcel identification number. `PIN` is a unique number assigned to the parcel to identify it for taxation purposes. The `Par_Type` field is used to identify what type of organization owns the parcel. This will be the field you will use to set up and identify the different subtypes for the parcel polygons. Subtypes must be based on an integer type field. They will not work with any other type of field.

Setting up subtypes

Now, you will set up subtypes for the `Parcel_Polygon` layer, using the `Par_Type` field to identify the subtype for each polygon in that layer. In this recipe, we want to identify what type of organization owns a particular parcel. This is because each type is taxed differently:

1. In the ribbon, click on the **Subtypes** button within the **Data Design** group on the **Fields** tab. This should open a new tab in the window that contains the **Fields** view you were just reviewing. This new tab should be named `Subtypes: Parcel_Polygon`.

Creating a Geodatabase

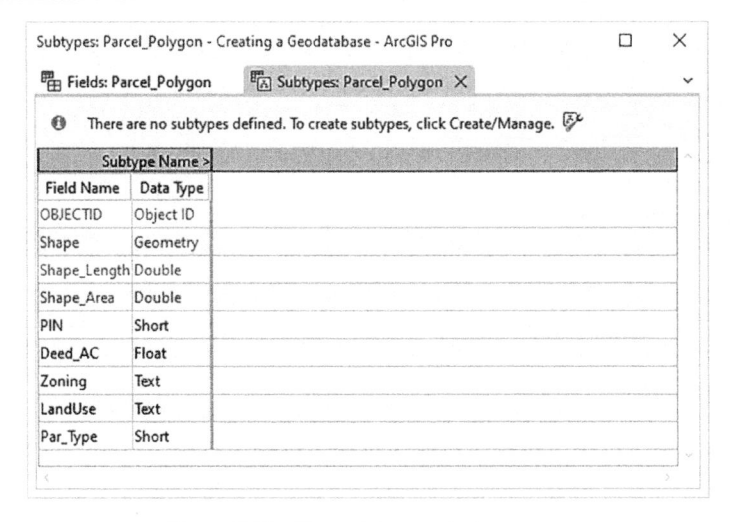

Figure 8.22 – The subtypes window open

2. Click the **Create/Manage** button on the **Subtypes** tab that appears in the ribbon. This will open the **Manage Subtypes** window.
3. Using the drop-down arrow, set **Subtype Field** to `Par_Type`.
4. Under the **Subtypes** section of the **Manage Subtypes** window below the **Code** column, double-click on the cell and type 0 (number zero). Then, double-click in the cell below the **Description** column, and type `Unknown`.
5. Continue adding the additional codes and descriptions, as shown in the following screenshot.

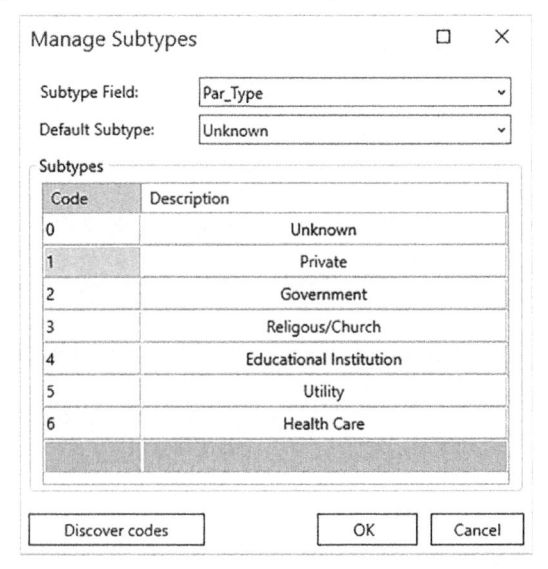

Figure 8.23 – The Manage Subtypes window with all codes added

6. Using the drop-down arrow, set **Default Subtype** to `Private`. This is the default subtype that will be assigned to any new parcel polygon features you might add later. This was chosen as the default because most new parcels will be privately owned by either an individual or business. Then, click the **OK** button to create the new subtypes.

The **Subtypes** tab should change to show the new subtypes you just created for the `Par_Type` field, as illustrated in the following screenshot.

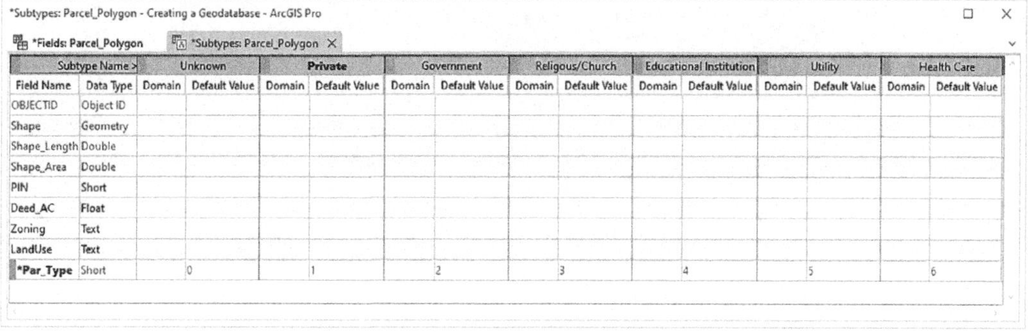

Subtype Name >		Unknown		Private		Government		Religous/Church		Educational Institution		Utility		Health Care	
Field Name	Data Type	Domain	Default Value	Domain	Default Value	Domain	Default Value	Domain	Default Value	Domain	Default Value	Domain	Default Value	Domain	Default Value
OBJECTID	Object ID														
Shape	Geometry														
Shape_Length	Double														
Shape_Area	Double														
PIN	Short														
Deed_AC	Float														
Zoning	Text														
LandUse	Text														
*Par_Type	Short		0		1		2		3		4		5		6

Figure 8.24 – New subtypes added to the Par_Type field

7. In the ribbon on the **Subtype** tab, click the **Save** button to push the changes you just made to the `MyGeodatabase.gdb` geodatabase and the `Parcel_Polygon` feature class.

8. Close the window that contains the **Fields: Parcel_Polygon** and **Subtypes: Parcel_Polygon** tabs.

The subtypes have now been created for use with the `Parcel_Polygon` layer, but the existing parcels have not been assigned to any of the new subtypes. You will do that next.

Populating the subtype values

There are several ways you can populate the subtype values for the existing features in the attribute table for the `Parcel_Polygon` layer. You could edit the data manually using the skills learned in *Chapter 6, Editing Tabular Data*. In this scenario, there is a database table that contains the parcel identification number and the parcel ownership type, making this much easier. This has been further simplified with a model that automates populating the subtype values, as follows:

1. In the **Catalog** pane, if needed, expand the `Toolboxes` folder and the `Creating a Geodatabase.tbx` toolbox.

2. Double-click on the **Populate Subtype Values from Stand Alone Table** model.

   This model does several things, as indicated in the following diagram. First, it joins the `Parcel_Owner_Type.dbf` table to the `Parcel_Polygon` layer, using the common values in the `Parcel Identification` fields located in both. This makes the data in the `Parcel_Owner_Type.dbf` table available in the `Parcel_Polygon` attribute table.

Then, the model uses a **Calculate Field** tool to copy the owner type values from the `Parcel_Owner_Type.dbf` table to the `Parcel_Polygon` layer. Lastly, it removes the join because it is no longer needed after the values have been copied over.

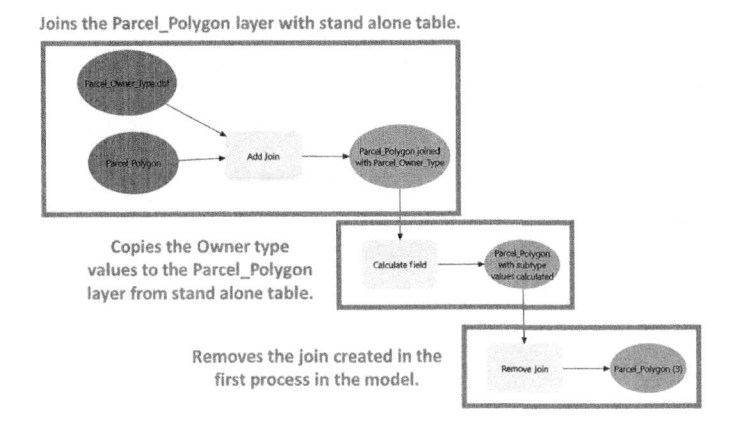

Figure 8.25 – The model to populate subtype values in the Parcel_Polygon layer

> **Information**
>
> If you want to see the processes included in a model, you can open it in the **ModelBuilder** window. This allows you to see all the included tools, parameters, comments, connections, and other parts of the model. It also allows you to edit the model if needed. To open an existing model in the **ModelBuilder** window, right-click on the model in the **Catalog** pane and select **Edit** from the menu that appears. This will display the model like the graphics you have seen throughout this chapter. If you do this with the models used in this chapter, be careful not to edit them. Editing them might break them.

To verify that the model worked, you will remove the `Parcel_Polygon` layer and add it back to the map.

3. Right-click on the `Parcel_Polygon` layer in the **Contents** pane, and select **Remove** from the menu that appears. This will remove the layer from the map but does not delete it from the geodatabase.
4. Activate the **Map** tab in the ribbon. Then, click on the **Add Data** button.
5. In the left panel of the **Add Data** window, under **Project**, expand the `Databases` folder.
6. In the right panel of the **Add Data** window, double-click on the `MyGeodatabase.gdb` file geodatabase. Then, double-click on the `Property` feature dataset.
7. Select the `Parcel_Polygon` feature class, and click the **OK** button to add this as a layer in the map.

8. Note how the newly added layer is symbolized. It should have a symbol for each subtype value, as illustrated in the following screenshot.

Figure 8.26 – A map showing the subtypes for the Parcel_Polygon layer

9. Save your project and close ArcGIS Pro.

You have just populated each parcel polygon feature with the appropriate subtype, and they are now displayed accordingly in the map.

How it works...

You have created subtypes for the `Parcel_Polygon` feature class and populated values that now show each polygon feature in its appropriate subtype. This shows you how subtypes can be used to create subcategories within a feature class. These subcategories can then be used for many purposes, such as to define symbology, assign different domains to fields, and apply different topology rules. The ability to assign different behaviors based on subtypes makes them very powerful and can help ensure data quality.

To create subtypes for the `Parcel_Polygon` feature class, you first needed to create the new feature class and needed attribute fields. This was done by running a model called the **Add Parcel Polygon Feature** class. This model contained several geoprocessing tools that were put together to form an automated workflow. This model created the new `Parcel_Polygon` feature class in the `MyGeoDatabase.gdb` file geodatabase and the `Property` feature dataset. It then added multiple fields to the new feature class. Lastly, it loaded data into the new feature class from an existing shapefile using the **Append** tool.

With the new feature class created, it was time to create the subtypes it would use. This first required you to have an attribute field that was either a short or long integer data type. By opening the fields list associated with the `Parcel_Polygon` feature class, you were able to identify two fields that were integer fields, `PIN` and `Par_Type`. The `Par_Type` field was identified as the one you would use to identify the subtype for each parcel. Then, you clicked on the **Subtype** button, and then the **Create/Manage** tool to open the **Manage Subtype** window.

In the **Manage Subtype** window, you entered the numeric code and the text description for each subtype you needed to generate the appropriate subcategories within the `Parcel_Polygon` feature class. In this scenario, you created subtypes that identified different types of landowners, including unknown, private, government, educational institutions, utilities, and healthcare.

With the subtypes defined, the next step was to populate the values using another database table that contained the appropriate subtype values, along with the parcel identification number. You used another model that joined the `Parcel_Polygon` feature class with the standalone table. The model then copied the values from the standalone table into the subtype field in the `Parcel_Polygon` attribute table. Lastly, it removed the join because it was no longer needed.

To verify the subtype values had been copied correctly, you removed the `Parcel_Polygon` layer from the map and added it, using the **Add Data** button on the **Map** tab in the ribbon. When the `Parcel_Polygon` feature class was added back to the map, it was automatically symbolized, based on the subtype assigned to each parcel polygon.

You learned how to create a new geodatabase and create new contents that can be stored in a geodatabase. These contents included creating a new feature class and feature dataset. You also learned how to create domains that form pick lists, which are used to edit attribute data and limit the values that can be entered. As a result, domains can help improve data quality by reducing the chance of human error and different value entries. Lastly, you learned how to create subtypes that can be used to subcategorize data within a feature class or table. Once created, you can use subtypes to control different types of behavior, including symbology and domain assignments.

9
Enabling Advanced Functionality in a Geodatabase

When you completed the recipes in *Chapter 8*, you learned how to create your own geodatabase as well as how to add content such as feature classes and feature datasets. You also learned how to add simple behaviors including domains and subtypes. These simple behaviors improve data quality at a minimum. Subtypes can be further used to apply more advanced behaviors to subgroups of data within the geodatabase.

Geodatabases have the capability to include more advanced behaviors that can allow you to track who creates and edits data, capture GPS attributes beyond just location, apply rules to attribute fields, and more. These capabilities are one of the reasons the geodatabase format has become the standard storage format for the ArcGIS Platform. Other formats such as shapefiles, Google Earth KML files, or AutoCAD DWG files do not support this level of functionality within the data itself. If you would like to learn more about some of these advanced capabilities of the geodatabase, you can go to `https://pro.arcgis.com/en/pro-app/latest/help/data/geodatabases/overview/view-and-edit-fields-domains-and-subtypes.htm`.

In this chapter, you will learn how to enable and configure some of these advanced capabilities in a geodatabase. Recipes included in this chapter are as follows:

- Enabling editor tracking
- Adding GPS metadata capture
- Creating attribute rules
- Creating contingent values

Enabling editor tracking

It is not unusual for more than one person in an organization to edit or create data in a geodatabase. You might have field staff collect or inspect features, and then have the office staff review and approve updates. So, how do you identify who created the feature or was the last person to update the information associated with a feature? That is where editor tracking comes in.

Editor tracking automatically updates fields in the table or attribute table that are related to creating and editing data. This includes fields that identify who created the feature or record, the date and time it was created, who was the last person to update it, and the date and time it was last updated. This can then be used to track production, identify where errors might be coming from, and more.

Editor tracking is not enabled by default. You must enable it on a feature class or table. Once enabled, it will automatically add fields to the feature class or table. These fields will be `created_user`, `created_date`, `last_edited_user`, and `last_edited_date`. Once editor tracking is enabled, these fields will automatically be populated by ArcGIS.

In this recipe, you will enable editor tracking for the Light `Poles` layer. This will allow you and others to see when new light poles are added, who added them, and the last person to edit them.

Getting ready

You must have completed the recipes in *Chapter 8* before you can do this recipe. It is recommended that you complete at least all the recipes in *Chapter 1* as well to ensure you have the foundational knowledge required to successfully complete this chapter. You will need to ensure you have ArcGIS Pro 3.x installed and a license assigned. This recipe can be completed with any license level of ArcGIS Pro, Basic, Standard, or Advanced.

How to do it...

You will now work through the process needed to enable editor tracking on a feature class in ArcGIS Pro:

1. Launch ArcGIS Pro and open the `AdvancedGeodatabaseFunctionality.aprx` project file located in `C:\Student\ArcGISPro3Cookbook\Chapter9\` by clicking on the **Open another project** button in the ArcGIS Pro Start window and navigating to the indicated folder.

    The project will open with the *Trippville Parcels* map. This map will contain several layers including `City Limits`, `Light Poles`, and `Parcels`. The city of Trippville is currently working to collect and update the location of light poles throughout the city. To track the productivity of this project, it wants to know who creates each light pole and when, as well as who is updating the data. Before we enable editor tracking, let's investigate what attribute fields already exist.

2. In the **Contents** pane, right-click the `Light Poles` layer and select **Attribute Table** from the menu that appears.

3. Review the fields that are included in the attribute table for the `Light Poles` layer. As you should see, it only contains two fields: `OBJECTID` and `SHAPE`.

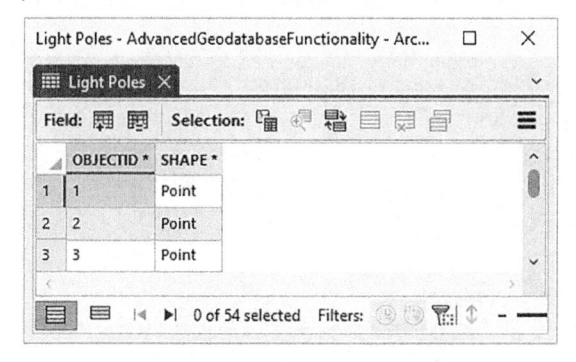

Figure 9.1 – Light Poles attribute table

4. Close the `Light Poles` attribute table.
5. In the **Catalog** pane, expand the `Databases` folder so you can see the databases attached to this project.
6. Expand the `Trippville_GIS.gdb` file geodatabase.
7. Scroll down and locate the `light_poles` feature class. This is the feature class referenced by the `Light Poles` layer.
8. Right-click on the `light_poles` feature class and select **Properties** from the menu that appears. This will open the **Feature Class Properties** window, as illustrated in the following figure:

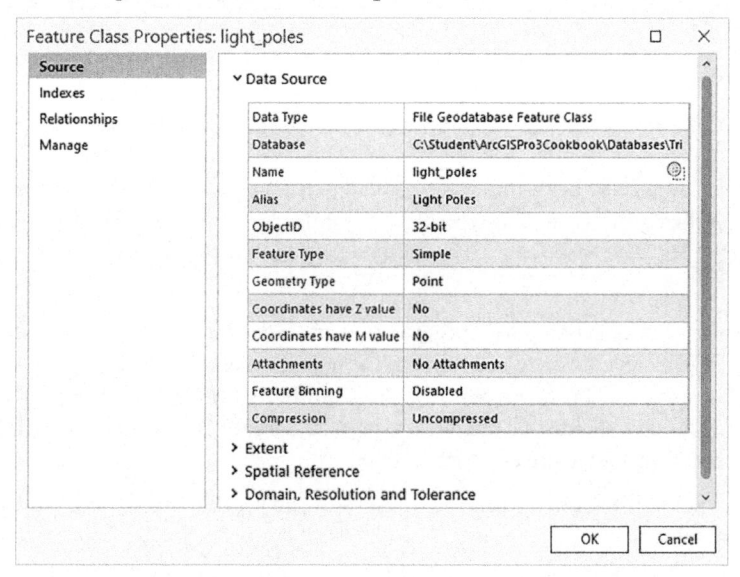

Figure 9.2 – The light_poles feature class properties

9. In the left panel of the **Feature Class Properties** window, select **Manage**.
10. Click in the box located to the left of **Editor tracking** to enable the editor tracking functionality for the `light_poles` feature class.

When you enable editor tracking, you should see additional information appear in the **Feature Class Properties** window:

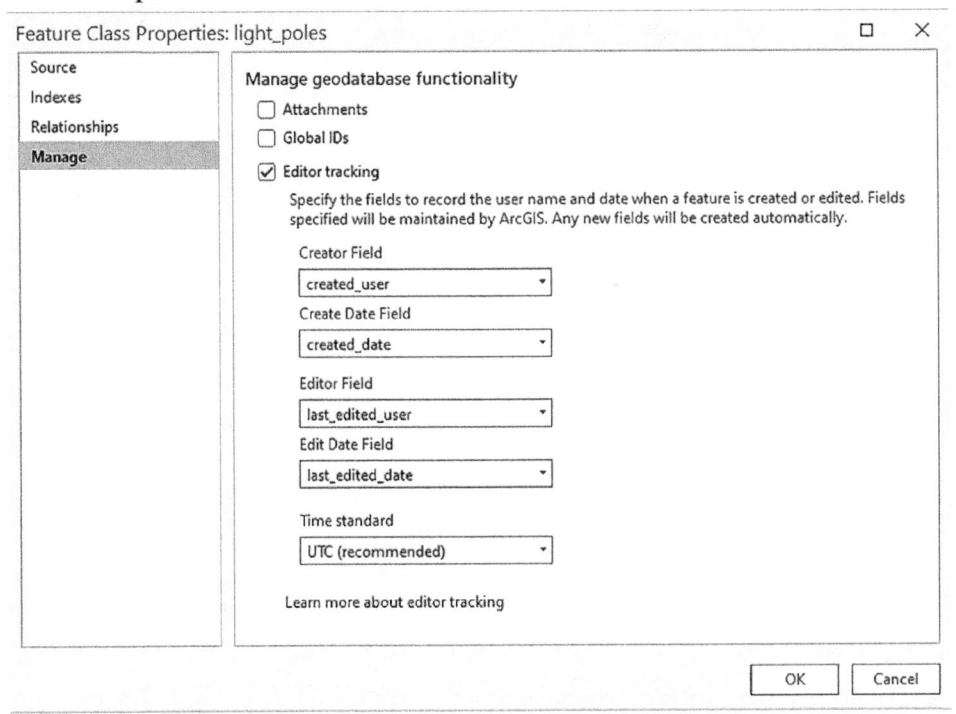

Figure 9.3 – Editor tracking enabled for the light_poles feature class

This shows the specific fields that will be added to the attribute table and used to store the editor tracking information, as well as which time standard will be used.

> **Tip**
> If the table already has existing fields that you want to use for tracking the creator or editor data and time, you can choose to use those existing fields instead of having ArcGIS Pro create new ones. Just use the drop-down arrows to select different fields. **Creator Field** and **Editor Field** can only be set to fields that are a text data type. **Create Date Field** and **Edit Date Field** can only be set to fields that are date data types.
>
> If you don't want to track one or more of these, you can set them to <**None**> again using the drop-down arrow.

11. You will accept all the default values for the fields. Click **OK** to apply the settings to the `light_poles` feature class.

Editor tracking should now be enabled on the `light_poles` feature class. This means that any time a new light pole is created or an existing one is updated, those four fields should automatically be updated.

Verifying that editor tracking is enabled

Now, you will verify that editor tracking was successfully enabled on the `light_poles` feature class. You will first check to see whether the editor tracking fields were added to the attribute table:

1. In the **Contents** pane, right-click on the `Light Poles` layer and select **Attribute Table**.
2. Review the attribute table for the `Light Poles` layer. As shown in the following screenshot, it now contains the editor tracking fields:

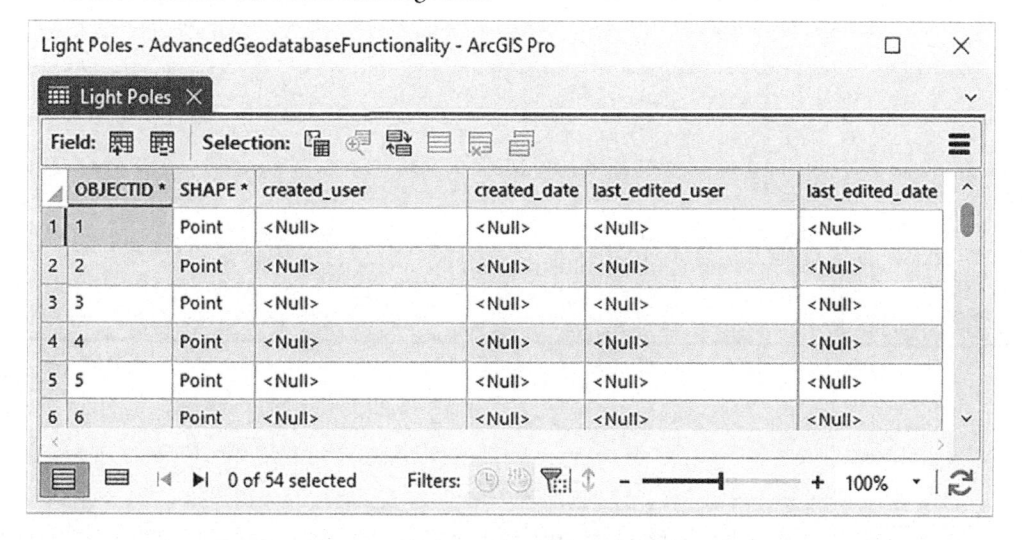

OBJECTID *	SHAPE *	created_user	created_date	last_edited_user	last_edited_date
1	Point	<Null>	<Null>	<Null>	<Null>
2	Point	<Null>	<Null>	<Null>	<Null>
3	Point	<Null>	<Null>	<Null>	<Null>
4	Point	<Null>	<Null>	<Null>	<Null>
5	Point	<Null>	<Null>	<Null>	<Null>
6	Point	<Null>	<Null>	<Null>	<Null>

Figure 9.4 – Light Poles attribute table with editor tracking fields

As you can see, four fields were added to the table. These were created when you enabled editor tracking. Now, you will create a new light pole to test the functionality of editor tracking.

3. Activate the **Map** tab in the ribbon. Then, click on the **Bookmarks** button and select the `Add new light pole` bookmark to zoom the map to the location where you will add the new light pole.

302   Enabling Advanced Functionality in a Geodatabase

4. Active the **Edit** tab in the ribbon and click on the **Create** button to open the **Create Features** pane.

5. In the **Create Features** pane, select the `Light Poles` feature template, as shown in the following figure:

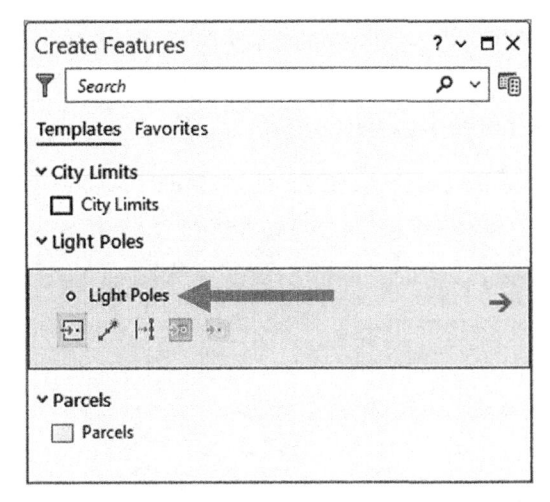

Figure 9.5 – The Create Features pane

6. Move your mouse pointer to the southeast intersection of *Spring Street SW* and *Highway 5* and click to place the new light pole, as illustrated in the following screenshot:

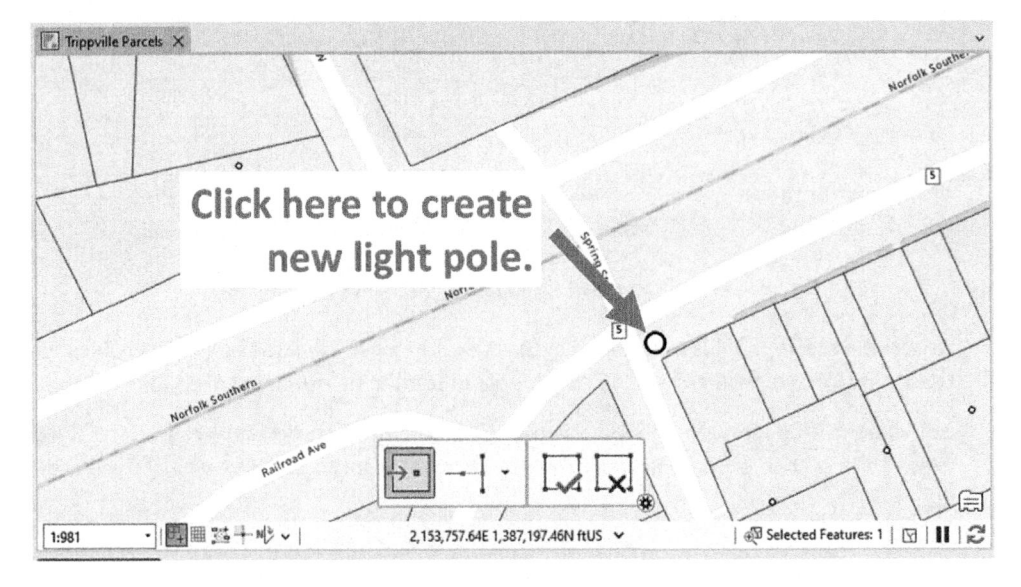

Figure 9.6 – Location of the new light pole to create

With the new light pole created, you will next verify that the editor tracking fields were populated.

7. In the `Light Poles` attribute table that you opened in *step 2* of this section of the recipe, scroll down to the bottom until you see the record for the light pole you just created. If you accidentally closed the attribute table, refer to *step 2* to reopen it.

8. Look to see whether there are values in the editor tracking fields. As illustrated in the following figure, you should see values have been automatically populated into those fields:

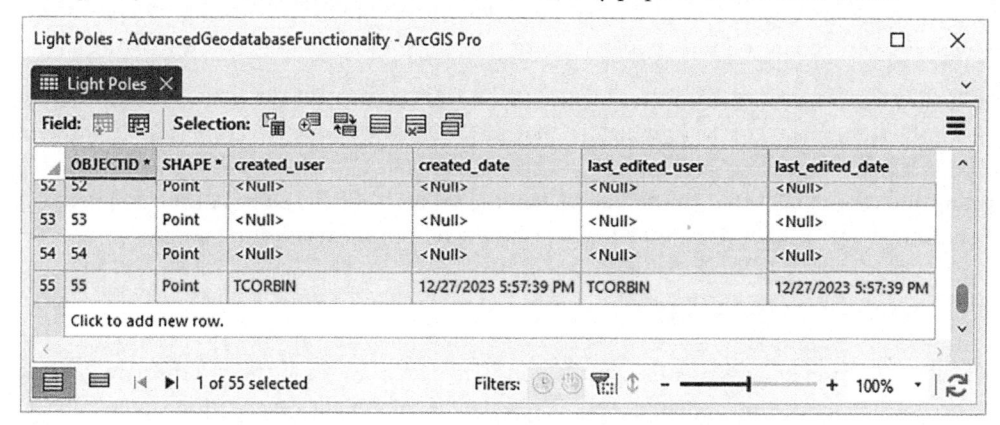

OBJECTID *	SHAPE *	created_user	created_date	last_edited_user	last_edited_date
52	Point	<Null>	<Null>	<Null>	<Null>
53	Point	<Null>	<Null>	<Null>	<Null>
54	Point	<Null>	<Null>	<Null>	<Null>
55	Point	TCORBIN	12/27/2023 5:57:39 PM	TCORBIN	12/27/2023 5:57:39 PM

Figure 9.7 – Light Poles attribute table with new light pole added

If you successfully enabled editor tracking and added the new light pole correctly, you should see your username populated into both the `created_user` and `last_edited_user` fields. The current date should also be populated into both the `created_date` and the `last_edited_date` fields. This means that editor tracking is working as expected.

9. On the **Edit** tab in the ribbon, click the **Save** button. If asked to save all edits, click **Yes**.
10. Close the `Light Poles` attribute table and the **Create Features** pane.
11. On the **Edit** tab in the ribbon, click on the **Clear** button to deselect the light pole you just created.
12. Activate the **Map** tab and click the **Zoom Full Extents** button located in the **Navigate** group.
13. Save your project and close ArcGIS Pro.

You have now successfully enabled editor tracking for a feature class in a geodatabase.

How it works...

In this recipe, you enabled editor tracking on the `light_poles` feature class in the `Trippville_GIS.gdb` geodatabase. This allows you to see who creates new light pole features and when. It also allows you to track who edits existing data in the feature class and when they do it.

To enable editor tracking, you had to open the properties of the feature class. You did this by expanding the `Databases` folder that showed which databases were connected to the project in the **Catalog** pane. Next, you expanded the `Trippville_GIS.gdb` geodatabase and located the `light_poles` feature class. You then right-clicked on the `light_poles` feature class and selected properties from the menu that appeared to open the **Feature Class Properties** window.

In the **Feature Class Properties** window, you clicked on the **Manage** option in the left panel of the window. From there, you clicked in the box located to the left of **Editor tracking**, accepted the field defaults, and clicked **OK** to finish enabling editor tracking for the `light_poles` feature class.

Once editor tracking was enabled, you verified that it was successful. First, you opened the attribute table for the `Light Poles` layer, which referenced the `light_poles` feature class, to ensure that the fields for editor tracking were added to the attribute table. Lastly, you added a new light pole in the map and checked its attributes to ensure that the values had been automatically populated into the editor tracking fields. This verified that you successfully enabled editor tracking.

Adding GPS metadata capture

Many of the features and assets we capture in our GIS databases are located in the field using GPS or other satellite navigation systems. GPS is the United States satellite navigation system. There are others, including Glonass, Galileo, and BeiDou. Collectively, these are known as **Global Navigation Satellite Systems (GNSSs)**. For the sake of simplicity, we will focus on GPS for this recipe.

Not all GPS data is created equal. GPS comes in different grades, which can impact the accuracy of the data collected. For example, a normal smartphone has GPS capability but is only accurate on average to within 15 to 20 feet horizontally. A dedicated survey-grade unit can be as accurate as 1 cm horizontally. There are also sources of error such as overall locations of the satellites, atmospheric conditions, and others. So, how can we know the overall quality of the data in our GIS database that was collected with GPS?

Geodatabases allow you to automatically capture the metadata associated with any feature collected with GPS. This can include information about the position source, receiver name, latitude, longitude, altitude, horizontal accuracy, vertical accuracy, and much more. These values will allow you to determine the overall confidence in the accuracy of the data that was collected. You can even determine whether the data was collected with GPS or not.

In this recipe, you will learn how to add the GPS metadata fields to the `light_poles` feature class you worked with in the previous recipe to enable editor tracking.

Getting ready

To complete this recipe, you will need to have finished the previous *Enabling editor tracking* recipe in this chapter. It is also recommended that you complete all the recipes in *Chapter 1* and *Chapter 8* of this book to ensure you have the foundational understanding of the terminology and interface for

ArcGIS Pro. This recipe can be completed with all licensing levels of ArcGIS Pro, Basic, Standard, or Advanced.

How to do it...

You will now add the GPS metadata fields to the `light_poles` feature class. Esri has included a geoprocessing tool to make this much easier:

1. Launch ArcGIS Pro and open the `AdvancedGeodatabaseFunctionality.aprx` project. This project is located in `C:\Student\ArcGISPro3Cookbook\Chapter9` if you installed the data in the default location. If you completed the previous recipe from this chapter, this project should be on your list of recent projects.

   This project should open with a map that contains layers representing the city limits of Trippville, light poles, and parcels (`City Limits`, `Light Poles`, and `Parcel`, respectively). As mentioned in the previous recipe, the city has undertaken a project to locate the light poles within the city limits. It wants the location of each pole captured using GPS. It has several different GPS units; some are capable of high accuracy and others are less capable.

   The City Manager wants to make sure everyone who uses the light pole data has some idea of the accuracy of each light pole feature. This will require the capture of the GPS metadata at the time each location is collected.

2. If it has been a while since you completed the previous recipe, right-click on the `Light Poles` layer and select **Attribute Table** from the menu that appears.

3. Review the names of the fields you see in the table. It should contain six fields, as shown in the next screenshot:

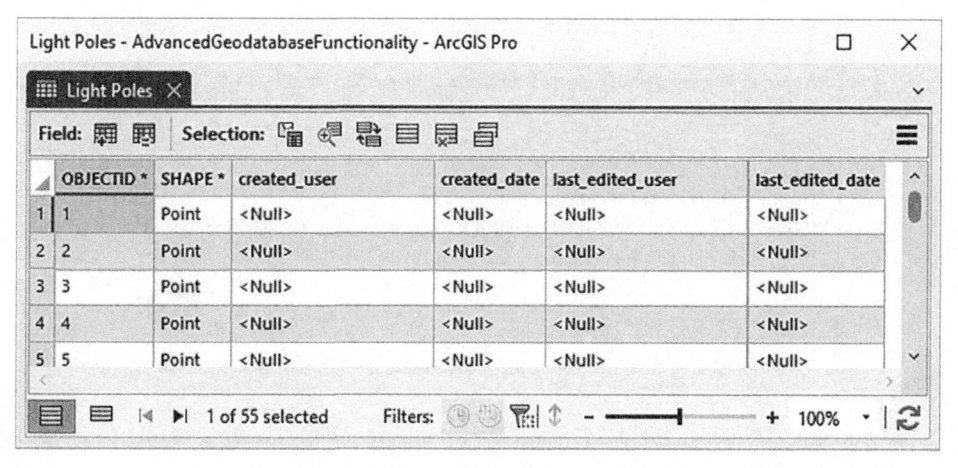

Figure 9.8 – Light Poles attribute table

Now, you will add the GPS metadata fields to this table.

4. Click on the **Analysis** tab in the ribbon. Then, click the **Tools** button to open the **Geoprocessing** pane.
5. In the **Geoprocessing** pane, click on the **Toolboxes** tab.
6. In the **Find Tools** cell located at the top of the **Geoprocessing** pane, type `GPS Metadata Fields`.
7. In the search results that are returned, double-click on the **Add GPS Metadata Fields** tool, as indicated in the following figure:

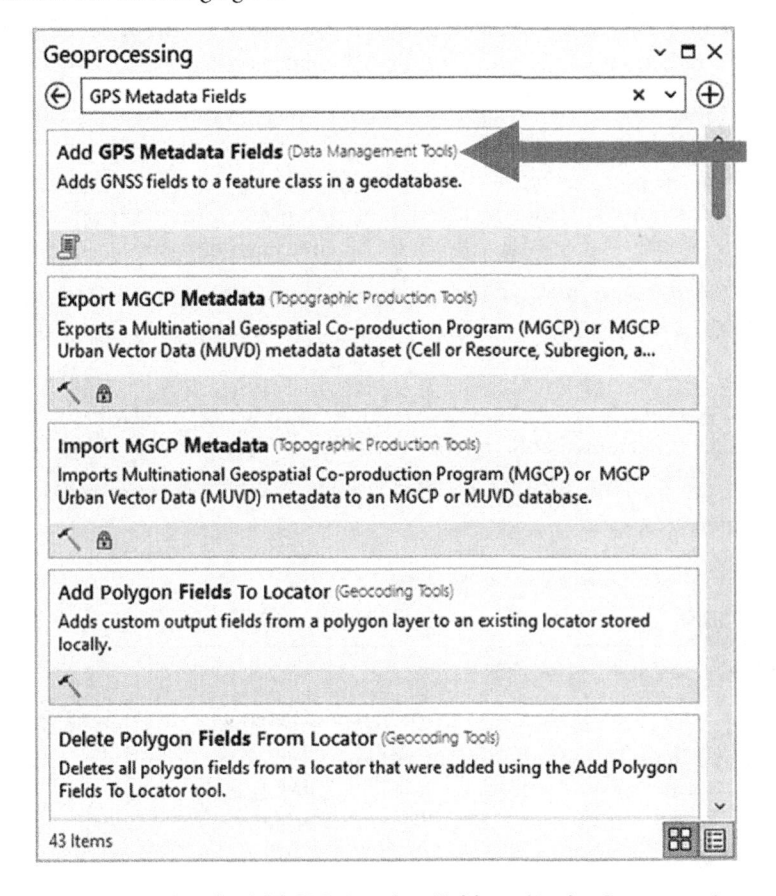

Figure 9.9 – Accessing the Add GPS Metadata Fields tool in the Geoprocessing pane

8. The **Add GPS Metadata Fields** tool should now be open in the **Geoprocessing** pane. Set **Input Features** to **Light Poles** using the drop-down arrow.

9. Verify that your **Add GPS Metadata Fields** tool matches the following and click **Run**:

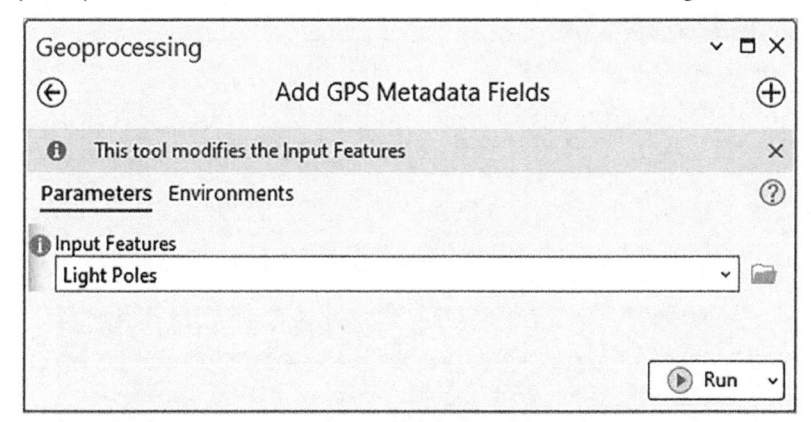

Figure 9.10 – The Add GPS Metadata Fields tool

When the **Add GPS Metadata Fields** tool completes, it will add numerous fields to the attribute table that represent various pieces of information associated with GPS data collection.

Verifying the result of the Add GPS Metadata fields tool

Let's verify that these fields were indeed added. To do this, you will open the **Fields** list to see what fields were added:

1. Close the **Geoprocessing** pane when the tool has been completed.
2. In the **Content** pane, ensure the `Light Poles` layer is selected. Click on the **Data** tab in the ribbon.
3. Click on the **Fields** button to open the **Fields: Light Poles** list in a tab next to the attribute table for the layer.
4. Scroll through the list of fields now contained in the `Light Poles` attribute table. The **Add GPS Metadata Fields** tool should have added over 20 new fields to the table, as shown in the following figure:

Enabling Advanced Functionality in a Geodatabase

Visible	Read Only	Field Name	Alias
✓	✓	OBJECTID	OBJECTID
✓	☐	SHAPE	SHAPE
✓	✓	created_user	created_user
✓	✓	created_date	created_date
✓	✓	last_edited_user	last_edited_user
✓	✓	last_edited_date	last_edited_date
✓	☐	ESRIGNSS_POSITIONSOURCETYPE	Position source type
✓	☐	ESRIGNSS_RECEIVER	Receiver Name
✓	☐	ESRIGNSS_LATITUDE	Latitude
✓	☐	ESRIGNSS_LONGITUDE	Longitude
✓	☐	ESRIGNSS_ALTITUDE	Altitude
✓	☐	ESRIGNSS_H_RMS	Horizontal Accuracy (m)
✓	☐	ESRIGNSS_V_RMS	Vertical Accuracy (m)
✓	☐	ESRIGNSS_FIXDATETIME	Fix Time
✓	☐	ESRIGNSS_FIXTYPE	Fix Type
✓	☐	ESRIGNSS_CORRECTIONAGE	Correction Age
✓	☐	ESRIGNSS_STATIONID	Station ID
✓	☐	ESRIGNSS_NUMSATS	Number of Satellites
✓	☐	ESRIGNSS_PDOP	PDOP
✓	☐	ESRIGNSS_HDOP	HDOP
✓	☐	ESRIGNSS_VDOP	VDOP
✓	☐	ESRIGNSS_DIRECTION	Direction of travel (°)
✓	☐	ESRIGNSS_SPEED	Speed (km/h)
✓	☐	ESRISNSR_AZIMUTH	Compass reading (°)
✓	☐	ESRIGNSS_AVG_H_RMS	Average Horizontal Accuracy (m)
✓	☐	ESRIGNSS_AVG_V_RMS	Average Vertical Accuracy (m)
✓	☐	ESRIGNSS_AVG_POSITIONS	Averaged Positions
✓	☐	ESRIGNSS_H_STDDEV	Standard Deviation (m)

Figure 9.11 – Fields added by the Add GPS Metadata Fields tool

> **Information**
> You can also add these fields to the attribute table manually as you would any other field. To ensure they work to automatically capture the appropriate metadata values, you must name the fields exactly as the tool does and match the data type for each field. The **Add GPS Metadata Fields** tool is just an easier way to add them. Also, if you do not need to capture all the metadata fields, you can delete the ones you do not need to reduce the overall number of fields included in the table.

5. Close the **Fields: Light Poles** tab after you have reviewed all the newly added fields.
6. Save your project and close ArcGIS Pro.

You have now successfully added the GPS metadata fields to the `Light Poles` attribute table.

Challenge (optional)

If you happen to have an external Bluetooth or USB GPS antenna and are using a laptop to run ArcGIS Pro, you can connect the antenna to your laptop and configure ArcGIS Pro to use the GPS to collect data in the field.

To learn how to connect a GPS receiver to ArcGIS Pro, go to `https://pro.arcgis.com/en/pro-app/latest/help/projects/connect-to-a-gnss-device.htm`. Once you have the receiver connected, try going outside and collecting locations. Then, look in the attribute table to see what values have been populated into the GPS metadata fields. Which fields get populated will be dependent on what receiver you use. Here is an example of what you might see populated when using a smart device such as an iPhone:

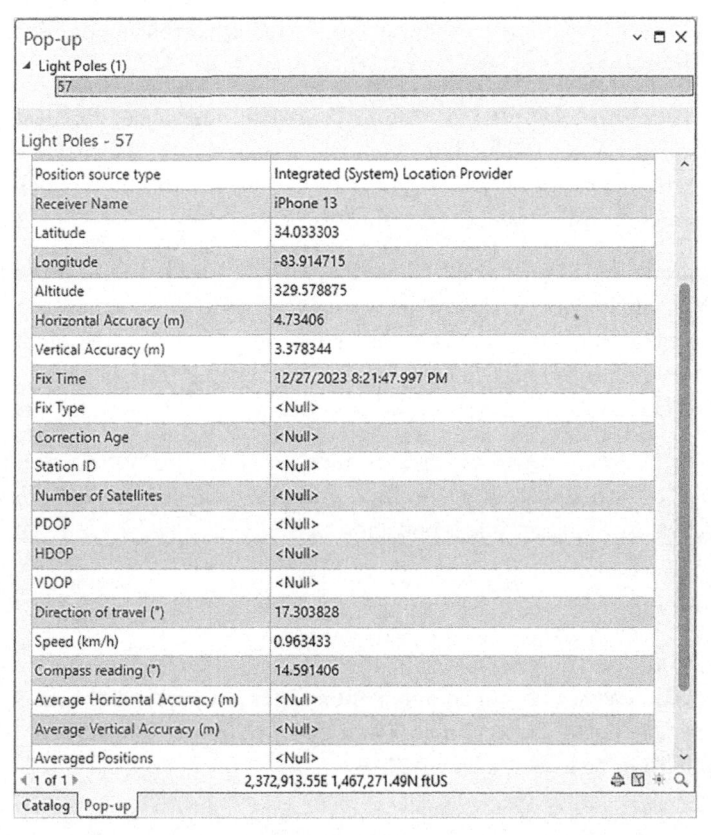

Light Poles - 57	
Position source type	Integrated (System) Location Provider
Receiver Name	iPhone 13
Latitude	34.033303
Longitude	-83.914715
Altitude	329.578875
Horizontal Accuracy (m)	4.73406
Vertical Accuracy (m)	3.378344
Fix Time	12/27/2023 8:21:47.997 PM
Fix Type	<Null>
Correction Age	<Null>
Station ID	<Null>
Number of Satellites	<Null>
PDOP	<Null>
HDOP	<Null>
VDOP	<Null>
Direction of travel (°)	17.303828
Speed (km/h)	0.963433
Compass reading (°)	14.591406
Average Horizontal Accuracy (m)	<Null>
Average Vertical Accuracy (m)	<Null>
Averaged Positions	<Null>

Figure 9.12 – Light pole location collected with GPS

As you can see, several GPS metadata fields were not populated. This is due to limitations with the integrated GPS receiver on the iPhone 13 that was used.

How it works...

In this recipe, you added fields to the attribute table for the Light Poles layer. These fields allow you to capture data about the GPS unit at the time data is collected. This information can be used to assess the overall quality of the data that was collected to ensure it meets data accuracy requirements.

To add these fields, you use the **Add GPS Metadata Fields** geoprocessing tool. This tool is in the **Data Management** toolbox and the **Fields** toolset. When we run this tool, it adds over 20 different fields that capture pertinent data about the GPS unit used and conditions at the time of collection.

Creating attribute rules

Attribute rules are another advanced behavior you can enable in a geodatabase using ArcGIS Pro. Attribute rules can be used to automatically calculate field values, perform data quality checks, or restrict invalid edits when updating data. They can be used to further enhance the capabilities associated with domains and subtypes, which were covered in *Chapter 8*.

There are three types of attribute rules: calculation, constraint, and validation. A calculation rule will automatically calculate values for a field in a table increasing editing efficiency. For example, if you have an acreage field, you can apply an attribute rule to that field that automatically calculates the acreage for a feature each time to add a new feature or edit an existing one.

A constraint rule limits values that can be entered into a field. This reduces the chance of invalid or bad data entry. Validation rules check attribute or spatial data to see whether they violate specified requirements. These rule types can be used together or individually to improve your overall data quality. To learn more about attribute rules, go to https://pro.arcgis.com/en/pro-app/latest/help/data/geodatabases/overview/an-overview-of-attribute-rules.htm.

In this recipe, you will create a couple of attribute rules for the Parcel_Polygon feature class you created in *Chapter 8*. The first will automatically calculate the mapped acreage for each parcel when a new one is created, or an existing one is updated.

Getting ready

Before starting this recipe, you will need to have completed all the recipes in *Chapter 8*. While not required, it is recommended that you complete the recipes in *Chapter 1* to help ensure you have a foundational understanding of the terminology and user interface for ArcGIS Pro. This recipe can be completed with all licensing levels of ArcGIS Pro.

How to do it...

You will now create an attribute rule that will calculate the mapped acreage for newly created or edited parcel polygon features:

1. Launch ArcGIS Pro and open the `AdvancedGeodatabaseFunctionality.aprx` project. This project is in `C:\Student\ArcGISPro3Cookbook\Chapter9\` if you installed the data in the default location. If you completed previous recipes in this chapter, this project should be on your list of recent projects.

> **Tip**
> If you see red exclamation marks to the left of the layer names in the **Catalog** pane, this indicates that the layers cannot find their data source. If this is the case, make sure you completed all the recipes in *Chapter 8* of this book. You cannot complete this recipe without the database and data created in those recipes. If you did complete those recipes and you still see the red exclamation marks, you can repair the data sources for the layers by doing the following.

2. Right-click on the layer in the **Contents** pane and select **Properties** from the menu. If you see a red exclamation point to the left of any of the layers, then follow *steps 3-5*; otherwise, skip to *step 6*.
3. In the **Properties** window, select **Source** from the left panel. Then, click on the **Set Data Source** button in the right pane.
4. In the **Change data source** window, expand the `Databases` folder.
5. Expand the `MyGeodatabase.gdb` geodatabase you created in *Chapter 8* and select the correct feature class that matches the layer. Then, click the **Open** button.

> **Information**
> You have now repaired the data source for the layer and the red exclamation mark should disappear. Continue to use this process for any other layers that have broken source connections. To learn more about repairing or updating data sources, you can go to `https://pro.arcgis.com/en/pro-app/latest/help/projects/update-data-sources.htm`.

6. In the **Catalog** pane, expand the `Maps` folder so you can see its contents. Right-click on the **Attribute Rules** map and select **Open**. This will open a map that contains the `Parcels` layer that references back to the `Parcel_Polygon` feature class in the `MyGeodatabase.gdb` geodatabase and the `Property` feature dataset you created in *Chapter 8*.
7. In the **Contents** pane, select the `Parcels` layer. Then, select the **Data** tab in the ribbon.
8. You need to review the fields that are included in the attribute table for this layer. So, click the **Fields** button so you can see the list of fields included in the table, as shown in the following figure:

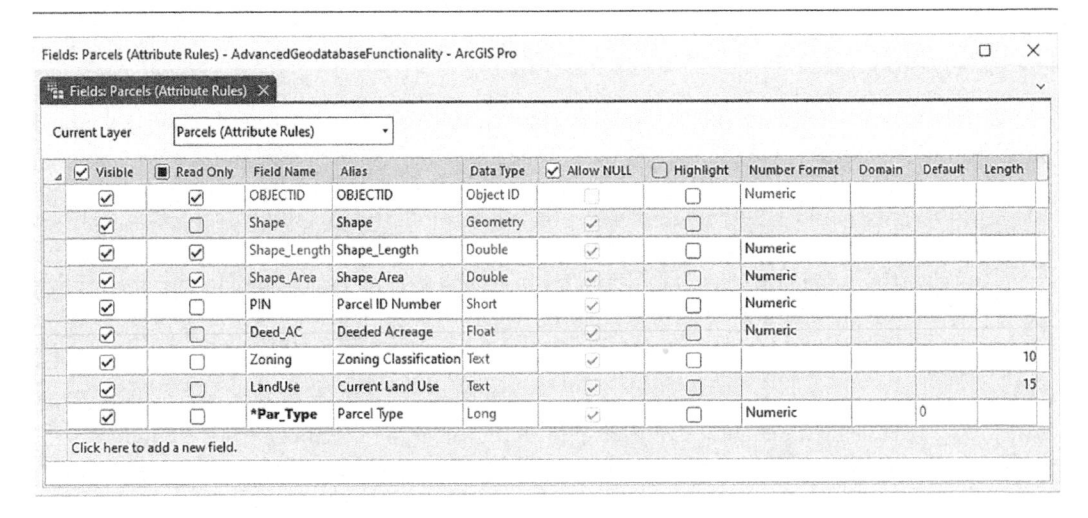

Figure 9.13 – Parcel layer attribute field list

9. Review the list of fields included in the attribute table for the `Parcels` layer. Check to see whether there is a field that will store the mapped acreage of each parcel.

As you review the list of available fields, you will see a field to store the deeded acreage but not one for the mapped acreage. The deeded acreage is the number of acres that is listed on the legal documents associated with a piece of property. That number and the actual mapped acreage should be the same, or close, but that is not always the case. There are several valid reasons for these two numbers to be different, such as the legal documents being vague using language such as plus or minus, the description being incorrect, and more. So, it is good to have both values in order to evaluate the data.

Adding a new field

You will now create a new field to store the mapped acreage values:

1. In the **Fields: Parcels (Attribute Rules)** window, click where it displays **Click here to add a new field**.
2. Create the new field with the following parameters.

   - **Field Name**: `Mapped_AC`
   - **Alias**: `Mapped Acreage`
   - **Data Type**: `Float`

3. Verify that your **Fields: Parcels (Attribute Rules)** values match the following. Then, click the **Save** button on the **Fields** tab in the ribbon.

Creating attribute rules 313

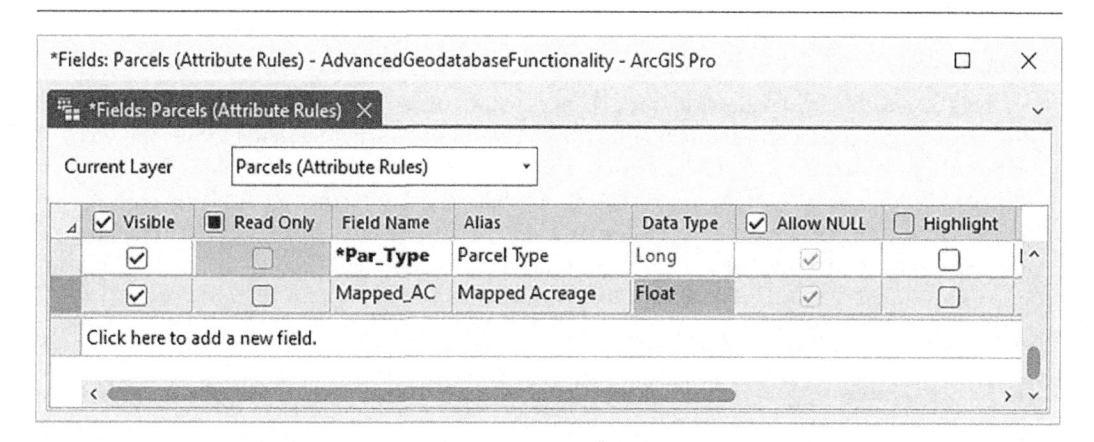

Visible	Read Only	Field Name	Alias	Data Type	Allow NULL	Highlight
✓	☐	*Par_Type	Parcel Type	Long	✓	☐
✓	☐	Mapped_AC	Mapped Acreage	Float	✓	☐

Figure 9.14 – Mapped_AC field added to the Parcels layer

4. Close the **Fields: Parcels (Attribute Rules)** window.
5. Save your project.

   You have just created the field in the `Parcels` layer attribute table, which will be used to store the acreage of each feature based on the dimensions of how it is drawn in the map.

   There is one more preparatory step before you can create the attribute rule to automatically calculate the mapped acreage values. You must enable **Global IDs** for the feature class the layer references.

6. In the **Catalog** pane, expand the `Databases` folder to reveal its contents.
7. Expand the `MyGeodatabase.gdb` geodatabase and the `Property` feature dataset.
8. Right-click on the `Parcel_Polygon` feature class and select **Manage** from the menu that appears. This will open the **Feature Class Properties** window to the **Manage** option.
9. Click in the box to the left of **Global IDs**, as shown in the following figure:

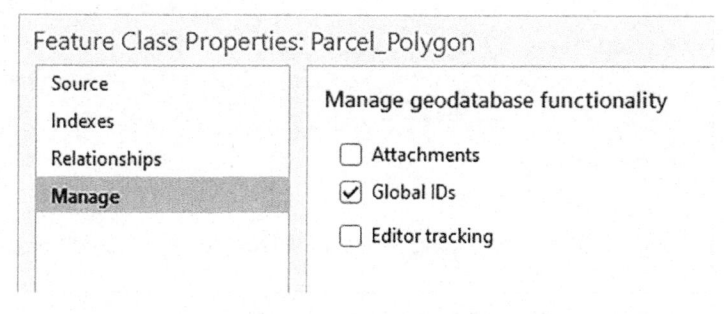

Figure 9.15 – Enabling Global IDs

Information

Global IDs are used by databases to track changes within the entire database. They are required for many advanced functions such as attribute rules and editing data via mobile and web applications. When enabled, a `GlobalID` or `GUID` field will appear in the table. The values stored in these fields are unique across the entire database, meaning the values are not repeated anywhere within the database. This allows the database to easily identify what has changed throughout the entire database.

10. Click **OK** to apply this setting.

11. In the **Contents** pane, right-click on the `Parcels` layer and select **Attribute Table** from the menu.

12. Review the columns/fields included in the table to ensure you see the `Mapped Acreage` and `GlobalID` fields, as illustrated in the following figure. You may need to scroll to the right to see them.

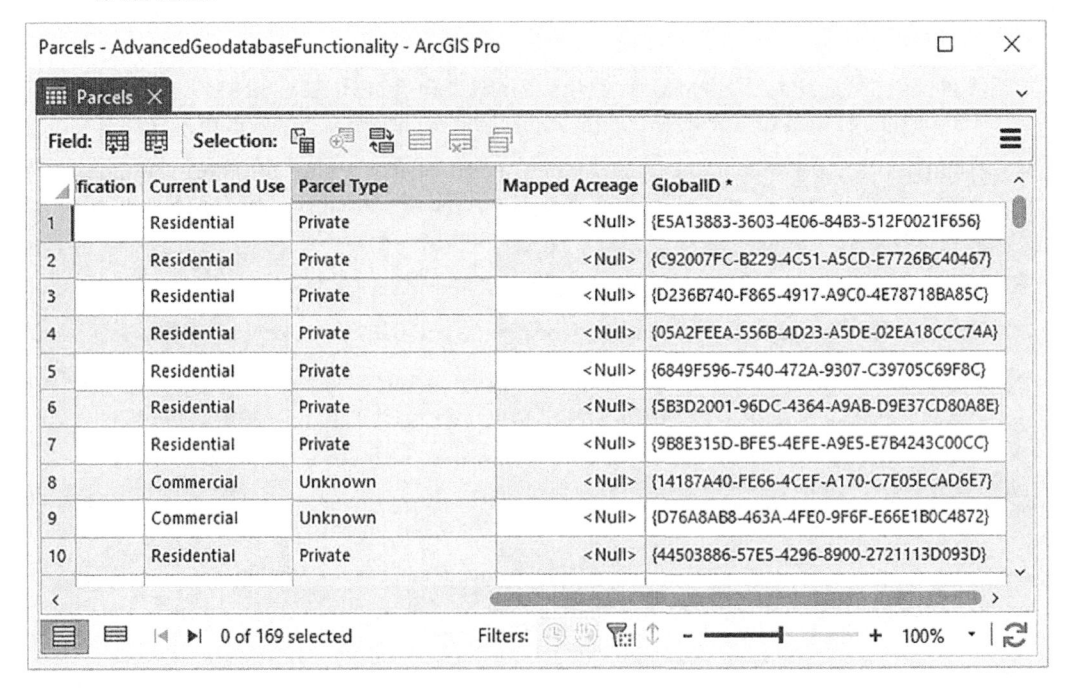

	fication	Current Land Use	Parcel Type	Mapped Acreage	GlobalID *
1		Residential	Private	<Null>	{E5A13883-3603-4E06-84B3-512F0021F656}
2		Residential	Private	<Null>	{C92007FC-B229-4C51-A5CD-E7726BC40467}
3		Residential	Private	<Null>	{D236B740-F865-4917-A9C0-4E78718BA85C}
4		Residential	Private	<Null>	{05A2FEEA-556B-4D23-A5DE-02EA18CCC74A}
5		Residential	Private	<Null>	{6849F596-7540-472A-9307-C39705C69F8C}
6		Residential	Private	<Null>	{5B3D2001-96DC-4364-A9AB-D9E37CD80A8E}
7		Residential	Private	<Null>	{9B8E315D-BFE5-4EFE-A9E5-E7B4243C00CC}
8		Commercial	Unknown	<Null>	{14187A40-FE66-4CEF-A170-C7E05ECAD6E7}
9		Commercial	Unknown	<Null>	{D76A8AB8-463A-4FE0-9F6F-E66E1B0C4872}
10		Residential	Private	<Null>	{44503886-57E5-4296-8900-2721113D093D}

Figure 9.16 – Parcels layer attribute table with Mapped Acreage and GlobalID fields added

13. Close the `Parcels` attribute table and save your project.

Creating the attribute rule to calculate mapped acreage

Now that the `Parcels` layer is prepared, you will create the attribute rule that will automatically calculate the mapped acreage values for any parcel feature that is added or updated:

1. In the **Contents** pane, verify the `Parcels` layer is still selected. Then, click on the **Data** tab in the ribbon.
2. Click on the **Attribute Rules** button on the **Data** tab in the ribbon. This will open the **Attribute Rules: Parcel_Polygon** tab in the main view area, as shown in the following figure:

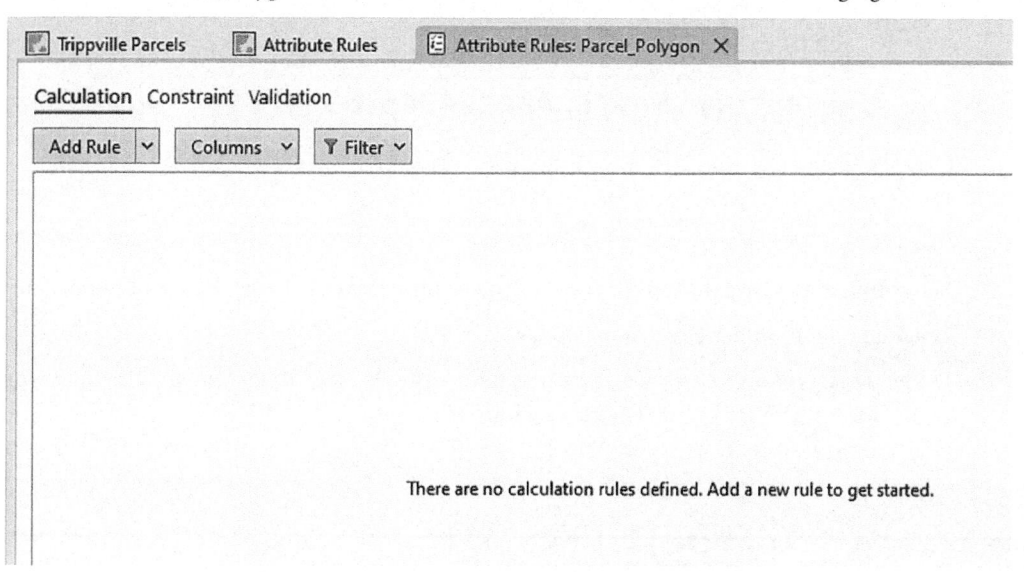

Figure 9.17 – The Attribute Rules: Parcel_Polygon tab open

3. In the **Attribute Rules: Parcel_Polygon** tab, click on the **Calculation** option at the top of the tab.
4. Click on the **Add Rule** button to open the **New Rule** panel where you configure the rule.
5. For **Rule Name,** enter a value of `Calculate Mapped Acres`.
6. Give it a **Description** value of `Calculate the acreage of each polygon based on the area drawn in the map`.
7. Set **Subtype** to <All> using the drop-down arrow.

> **Information**
> You can set different rules for each subtype. This provides more granular control over how your data behaves. It also illustrates how these different types of behaviors can work together to decrease the chance of errors.

8. Set **Field** to Mapped_AC using the drop-down arrow.
9. Click the **Expression** button to open the **Expression Builder** window.
10. On the **Fields** panel, locate and double-click on the Shape_Area field so it appears below.
11. In the **Expression** area, click to the right of the Shape_Area field and type /4046.8564224. This will divide the Shape_Area value, which is in square meters, by 43,560. That is the conversion factor when going from square meters to acres.
12. Verify that your expression matches the following and click **OK** in the **Expression Builder** window:

Expression

$feature.Shape_Area/4046.8564224

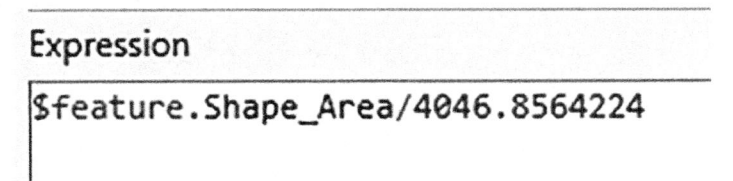

Figure 9.18 – Attribute rule expression

13. Select **Insert** and **Update** under **Triggers**. This controls when the rule is applied to the data.
14. Leave all options under **Execution** deselected.
15. Under **Tags**, type meters to acres and conversion, as illustrated in the following figure. Press the *Enter* key after typing each value.

Tags

meters to acres
conversion

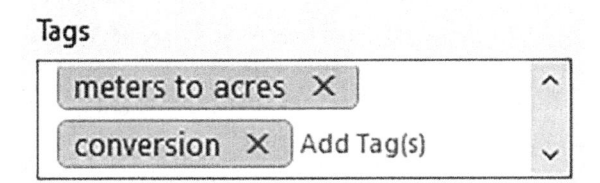

Figure 9.19 – Attribute rules tags

16. Verify that your **Attribute Rules: Parcel_Polygon** tab matches the following and click the **Save** button located on the **Attribute Rules** tab in the ribbon to save the rule you just created.

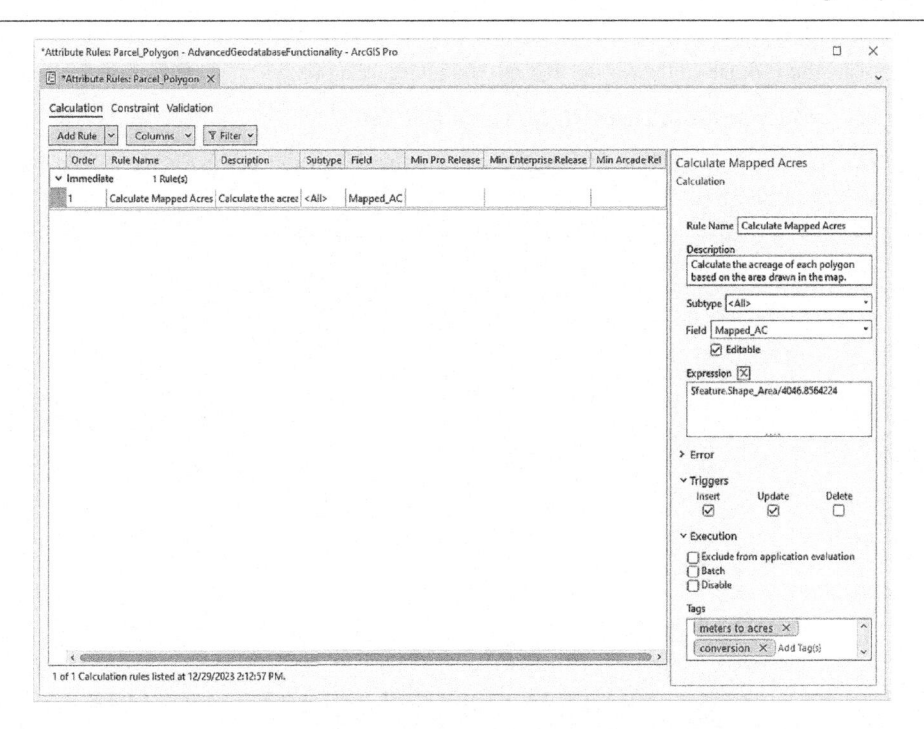

Figure 9.20 – Attribute rule created with complete parameters

17. Close the **Attribute Rules: Parcel_Polygon** tab and save your project.

You have just added an attribute rule to the `Parcels` layer that will update the value stored in the `Mapped_AC` field if you create a new parcel or edit one. Next, you will test the rule to see whether it is working.

> **Warning**
>
> It should be pointed out that attribute rules are not compatible with older Esri applications such as ArcMap, Collector, or ArcGIS Pro 2.0 and earlier. Once you add attribute rules to a geodatabase, it may be unusable on these older applications. They might also cause issues with third-party integrations or custom applications. Make sure to verify that they are compatible before you use attribute rules on a production database.

Testing the attribute rule

To test the rule you just created, you will create a new parcel polygon feature. If the rule is working, the `Mapped_AC` field should automatically be populated:

1. Activate the **Map** tab in the ribbon. Then, click on the **Bookmarks** button and select the **New parcel location** option from the list. This will zoom you to the east side where you will create a new parcel polygon.

318   Enabling Advanced Functionality in a Geodatabase

2. Activate the **Edit** tab in the ribbon and select the **Create** button to open the **Create Features** pane.
3. Select the **Private** feature template located under **Parcels**.
4. In the **Private** feature template, select the **Autocomplete Polygon** tool, as shown in the following screenshot:

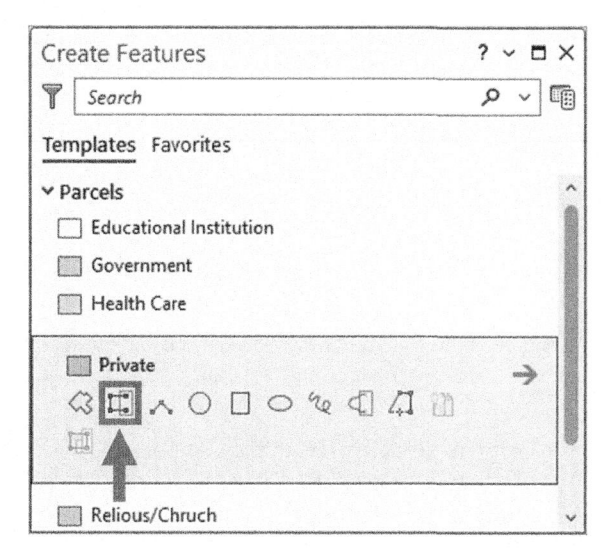

Figure 9.21 – Selecting the Autocomplete Polygon tool

5. On the map, click on the northeast corner of the parcel located to the west of the empty area, as shown in the following figure:

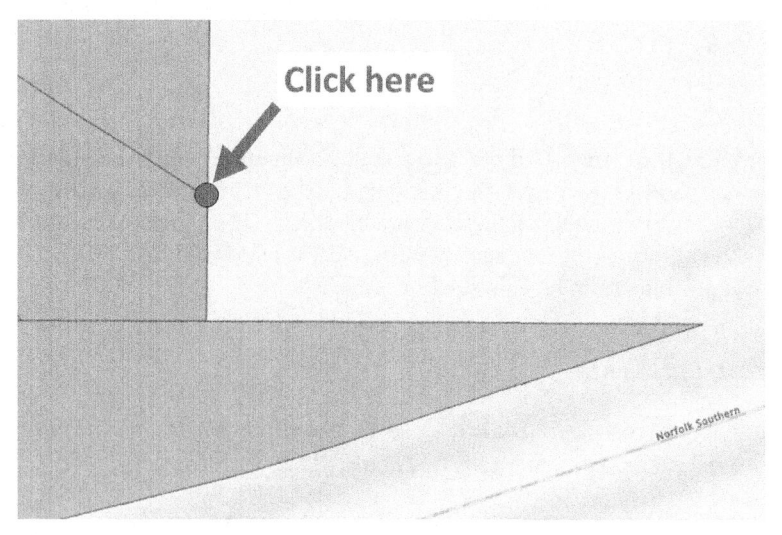

Figure 9.22 – Starting point for the new parcel polygon

6. Move your mouse pointer parallel to the northern border of the southern parcels and click when you are even with the easternmost point, as illustrated in the following figure:

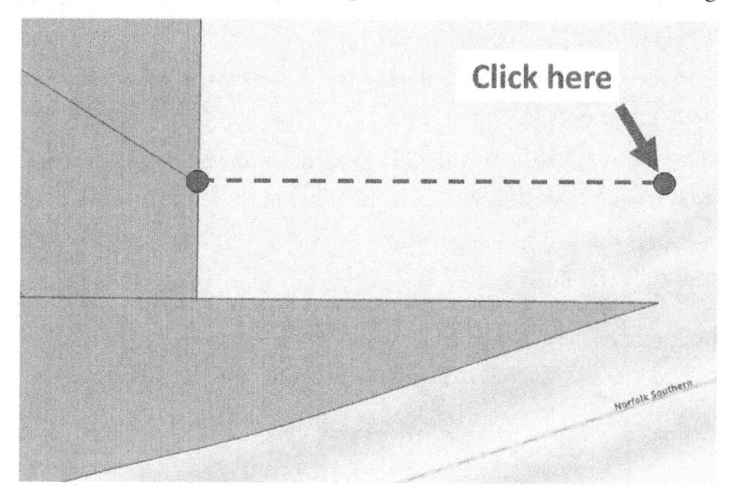

Figure 9.23 – Second point for the new parcel polygon

7. Move your mouse pointer to the easternmost corner of the southern parcel and double-click to create the new parcel polygon, as shown in the following figure:

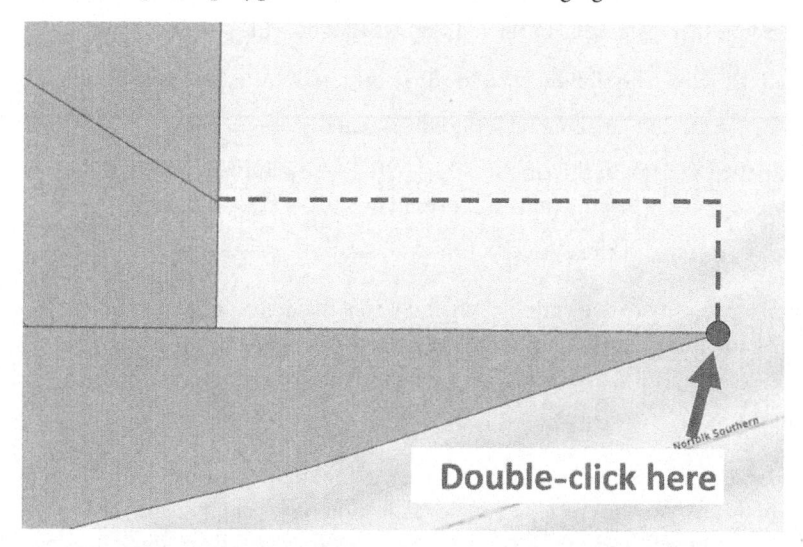

Figure 9.24 – Last point location to create the new parcel polygon

The new parcel polygon should now be created on the map. Next, you will look at the attribute table to see whether the value for the Mapped Acres field has been populated.

8. In the **Contents** pane, right-click on the `Parcels` layer and select **Attribute Table**.
9. At the bottom of the table select the **Show Selected Records** button so only the newly created parcel record is displayed.
10. Scroll over to the right until you see the `Mapped Acreage` field. See whether the field has been populated with a value.

    If the attribute rule you created worked as expected, you should see the `Mapped Acreage` field contains a value close to 1.066.

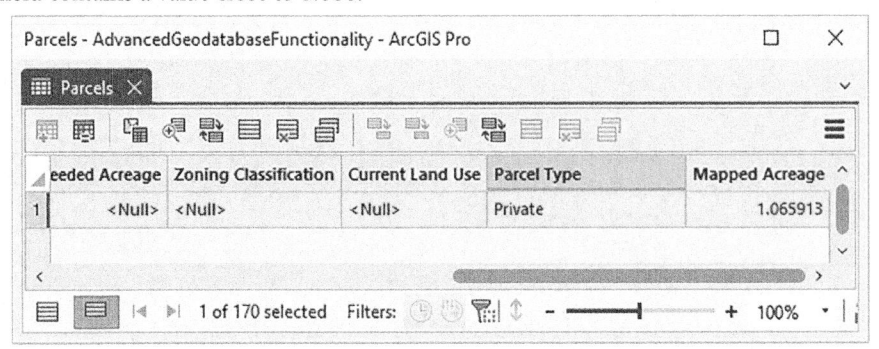

    Figure 9.25 – Results of the attribute rule in action

    The actual number will vary somewhat depending on exactly where you clicked your points. The fact that there is a value in the field shows that the rule is working.
11. On the **Edit** tab in the ribbon, click the **Save** button. If asked to save all edits, click **Yes**.
12. Close the `Parcels` attribute table and the **Create Features** pane.
13. Activate the **Map** tab in the ribbon. Then, click on the **Full Extent** button to zoom the map back out so you can see the full area covered by the `Parcels` layer.
14. Save your project and close ArcGIS Pro.

Congratulations, you just created your first attribute rule that auto-populates values into the `Mapped Acreage` field any time a new parcel is created or an existing one is updated.

How it works...

As you have just seen, attribute rules can be a powerful tool to help ensure your data remains updated and accurate. In this recipe, you created a simple calculation rule that populates the `Mapped Acreage` field by taking the value found in the `Shape_Area` field, which is in square meters, and dividing it by 4,046.8564224 to convert it to acres, an area measurement used frequently in the United States for land parcels.

Creating this attribute rule required several steps. First, you needed to create a new attribute field called `Mapped_AC` with an alias of `Mapped Acreage`. This is the field that is used to store the total area of a parcel in the `Parcel_Polygon` feature class. You did this by selecting the `Parcels` layer in the **Contents** pane. Next, you clicked on the **Data** tab in the ribbon and selected the **Field** button to open the list of existing attribute fields for the `Parcels` layer. At the bottom of the list of existing fields, you clicked on **Click here to create a new field**. From there, you created the new field named `Mapped_AC` with an alias of `Mapped Acreage` and a data type of `Float`.

With the field created, next, you needed to enable **Global IDs** for the `Parcel_Polygon` feature class that the `Parcels` layer referenced. You did this by first locating the `Parcel_Polygon` feature class in the `MyGeodatabase.gdb` geodatabase in the **Catalog** pane. You then opened the properties for `MyGeodatabase.gdb` and went to the **Manage** option. There, you clicked in the box to the left of the **Global IDs** parameter to enable them. This created a new field in the `Parcels` attribute table called `GlobalID`. Global IDs allow the database to easily track changes that occur and are required for certain advanced functionality such as attribute rules.

Lastly, you created the attribute rule. You verified the `Parcels` layer was selected in the **Contents** pane and clicked on the **Data** tab in the ribbon again. Then, you clicked on the **Attribute Rules** button to open the **Attribute Rules** tab in the view area. Next, you ensured that the **Calculation** tab was selected and clicked on the **New Rule** button. This opened a new panel to the right of the **Attribute Rules** tab in the main view area. You then filled in the parameters for the new rule including name, description, subtype, field, triggers, and expression. With those parameters complete, you saved the new rule, so it was then active within the `Parcels` layer.

You then tested the rule to verify that it worked by creating a new parcel feature in the `Parcels` layer. Upon drawing the new parcel, you verified its attribute values and saw the value for the `Mapped Acreage` field was automatically populated with a value based on the expression you entered when creating the rule. This showed that the rule was working as expected.

This was a very simple but powerful attribute rule and makes updating parcels much easier. Remember that attribute rules can be used to do more than just calculate values. They can also be used to constrain values and perform data validation checks on data as well.

Creating contingent values

Contingent values are another advanced geodatabase function you can implement to help improve data accuracy and editing efficiency. Contingent values limit the data values that can be entered into a database field based on a value that has been selected in another field. For example, if you select concrete as the material for a new power pole, this limits the height of the pole to a specific set of values because a wood power pole comes in specific lengths. Because these contingent values limit options, this helps ensure that only valid data values are populated into your data for improved integrity.

Contingent values require domains and/or subtypes to function. Domains and subtypes provide the initial limiting values that can be entered into the designated fields. Those fields, with their domains or subtypes, are then added to a field group. A field group is a selection of fields contained in a table or feature class that will have contingent values behavior applied to them.

Like attribute rules, contingent values are not compatible with older versions of Esri applications such as ArcMap. So, if you add contingent values to a geodatabase, those older applications will not be able to work with the data any longer. To learn more about contingent values, you can go to `https://pro.arcgis.com/en/pro-app/latest/help/data/geodatabases/overview/contingent-values.htm`.

In this recipe, you will add contingent values to the `Side_Walks` feature class you created in the *Creating a new geodatabase and feature class* recipe in *Chapter 8*. You will limit the pavement type based on who owns the sidewalk. You will need to first create a new field that identifies who owns the sidewalk. Then, you will need to create a domain containing a list of possible owners. Lastly, you will create and configure the contingent values that limit the pavement type based on who owns the sidewalk.

Getting ready

Before starting this recipe, you will need to have completed the recipes in *Chapter 8*. It is also recommended you complete the recipes in *Chapter 1* to ensure you understand the ArcGIS Pro terminology and how to navigate the interface.

This recipe can be completed with all licensing levels of ArcGIS Pro. You will also need to ensure you have internet access to make use of the Esri-provided basemaps.

How to do it...

You are now ready to start the workflow required to enable contingent values on the `Side_Walks` feature class. As you will see, this takes a few steps to do:

1. Launch ArcGIS Pro and open the `AdvancedGeodatabaseFunctionality.aprx` project. This should be in your **Recent Projects** list.

2. Ensure the **Attribute Rules** map is open and active in the main view. This is the same map you were using in the previous recipe in this chapter. If it is not open, go to the **Catalog** pane and expand **Maps**. Then, right-click on the **Attribute Rules** map and select **Open** from the menu that appears.

   The **Attribute Rules** map should contain three layers, as shown in the next screenshot:

Creating contingent values   323

Figure 9.26 – Attribute Rules map

You will notice it does not include the `Side_Walks` layer. You will add that next.

3. Activate the **Map** tab in the ribbon. Then, click on the **Add Data** button.
4. In the **Add Data** window, select `Databases` under **Project** in the left panel.
5. Double-click on the `MyGeodatabase.gdb` geodatabase file.
6. Select the `Side_Walks` feature class and click the **OK** button to add this to the **Attribute Rules** map as a layer.

The `Side Walks` layer should now appear in the **Contents** pane. This layer references the `Side_Walks` feature class you selected in the **Add Data** window.

Figure 9.27 – Attribute Rules map with the Side Walks layer added

Adding a new field and domain

Now that the `Side Walks` layer has been added to the map, you will need to add a new field to identify who owns the sidewalk and a domain containing a list of possible owners:

1. In the **Contents** pane, right-click on the `Side Walks` layer and select **Data Design** and **Fields** from the menus that appear, as illustrated in the following figure. This will open the list of existing fields contained in the attribute table for the `Side Walks` layer.

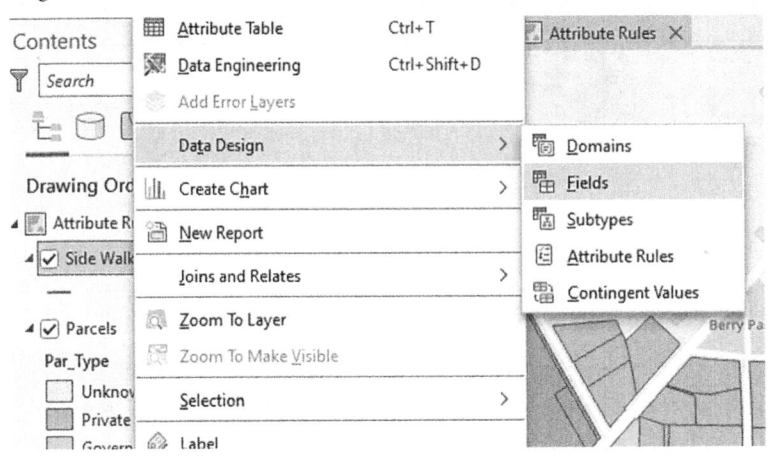

Figure 9.28 – Data Design and Fields from the menus

> **Tip**
> You could have also gone to the **Data** tab and clicked the **Fields** button as you have done in past recipes. Either method would have opened the list of fields in the attribute table for the `Side Walks` layer. As you may have determined, ArcGIS Pro often has multiple methods or workflows to accomplish the same task.

2. Click where it says **Click here to add a new field to create a new field** in the table.
3. Create the new field with the following parameters:

    - **Field Name**: Owner
    - **Alias**: Sidewalk Owner
    - **Data Type**: Text
    - **Length**: 50
    - Accept all other default values

4. Double-click on the **Domain** cell and select **<Add new Coded Value domain>** from the list that appears. This will open the **Domains: MyGeodatabase** tab in the main view area.

5. Give the new domain a **Domain Name** value of `SidewalkOwners` and a **Description** value of `List of owners that are responsible for building and maintaining sidewalks`.

6. Ensure **Field Type** is set to **Text** and accept all other defaults.

7. In the panel located to the right of the domain list, you should see two columns: **Code** and **Description**. Add the following values to the list. The first value is the code. The second value is the description. You can use the *Tab* key to move between columns and to the next row. Press the *Enter* key when you have keyed in the last value in the list:

   - `City:` `City owned`
   - `County:` `County owned`
   - `Private:` `Privately owned`
   - `Unk:` `Unknown owner`

8. Verify that the new `SidewalkOwner` domain matches the following screenshot. Then, on the **Domains** tab in the ribbon, click the **Save** button to save your newly created domain.

Domain Name	Description	Field Type	Domain Type	Split Policy	Merge Policy	Code	Description
Pavement_Typ	List of different possible pavement types	Text	Coded Value Domain	Default	Default	City	City owned
SidewalkOwners	Description of List of owners that are responsible for building and maintaining sidewalks	Text	Coded Value Domain	Default	Default	County	County owned
						Private	Privately owned
						Unk	Unknown owner

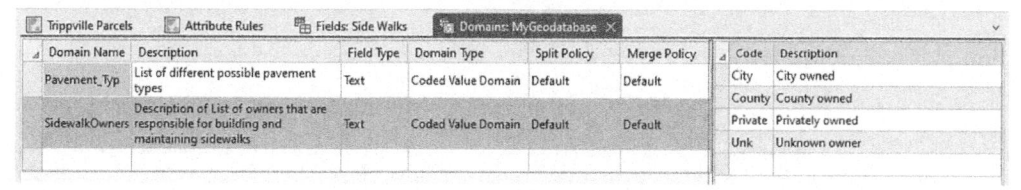

Figure 9.29 – New domain added

9. Close the **Domains: MyGeodatbase** tab and activate the **Fields: Side Walks** tab in the main view area.

10. Click the **Save** button on the **Fields** tab in the ribbon to save the new field you have created. Close the **Fields: Side Walks** tab once the save is complete.

11. Save your project.

Creating contingent values

You are now ready to create the contingent values, starting with the field group:

1. In the **Contents** pane, ensure the `Side Walks` layer is selected. Then, click on the **Data** tab in the ribbon.

2. On the **Data** tab, click the **Contingent Values** button. This will open the **Contingent Values** tab in the ribbon and the **Contingent Values: Side Walks** tab in the main view area.

3. In the main view area, click on the **Click here to add a new field group** button, as illustrated in the following figure. This will open the **Field Groups** window.

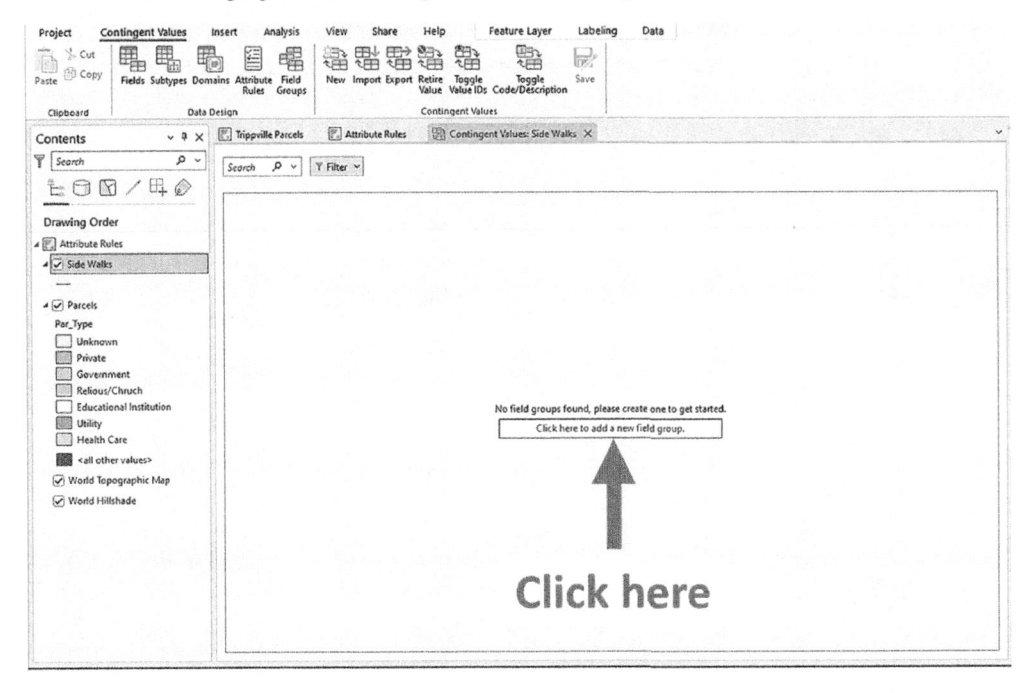

Figure 9.30 – Creating a new field group

4. Below **Name** in the **Field Groups: Side Walks** window, click where it says **Click here to add a new field group**.

5. Then, type `SidewalkOwners and pavementtype` and press *Enter*. You should see the **Add Field** button appear in the lower part of the window.

6. Click on the **Add Field** button so a drop-down list appears. Select the `Pave_Type` and `Owner` fields, as illustrated in the following figure. Then, click the **Add** button.

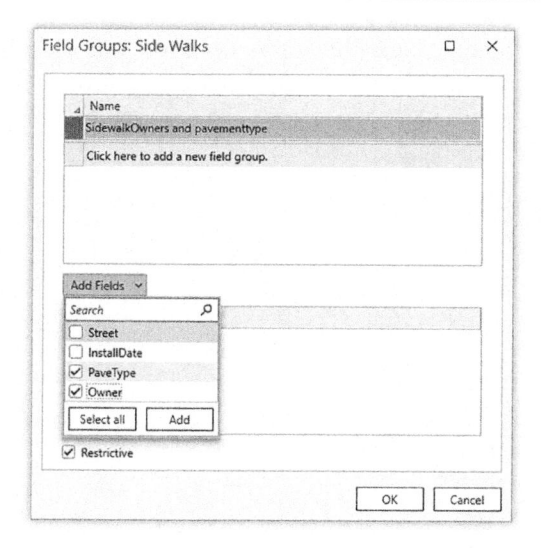

Figure 9.31 – Adding fields to the field group

7. With the `Pave_Type` and `Owner` fields added to the group, click the **OK** button to create the new field group.

   The **Contingent Values** tab in the main view area should now show two columns: `Pave_Type` and `Owner`, as shown in the next screenshot. Under that, it should say **Click here to add a new contingent value**.

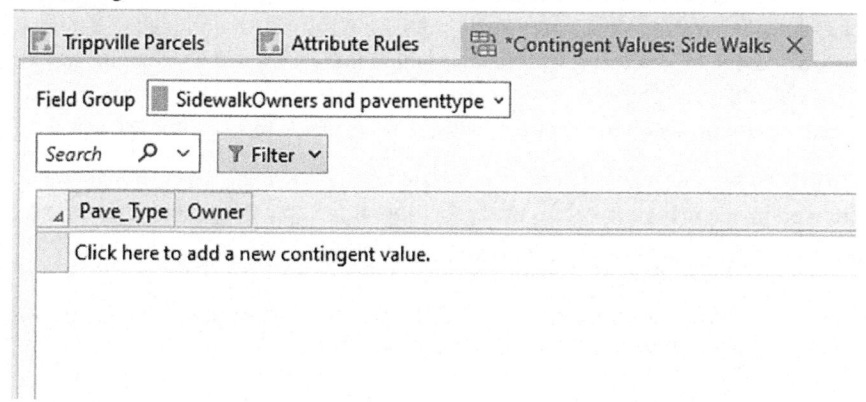

Figure 9.32 – Ready to start adding contingent values

This is where you start creating the list of restricted values based on what is input into one of those two fields.

8. Click where it says **Click here to add a new contingent value**.

9. Under the `Owner` column, click the drop-down arrow and select **City** from the list. Then, under the `Pave_Type` column, select **Concrete** using the drop-down arrow. This is your first contingent value. When the owner is set to **City**, the pavement type will be limited to **Concrete** as the only possible value to select.

10. Continue to add the values as shown in the following figure:

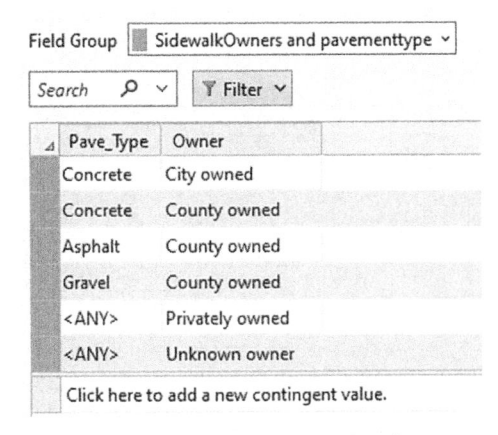

Pave_Type	Owner
Concrete	City owned
Concrete	County owned
Asphalt	County owned
Gravel	County owned
<ANY>	Privately owned
<ANY>	Unknown owner

Figure 9.33 – Contingent values list

This list of contingent values means that if the owner is set to **County owned**, then for the pavement type, the only available values that can be selected are **Concrete**, **Asphalt**, and **Gravel**. If the owner is set to **Privately Owned** or **Unknown owner**, any pavement type can be entered that is included in the `Pavement_Type` domain. As already mentioned, if the owner is set to **City owned**, then **Concrete** is the only available value for `Pave_Type`. This works the other way as well. If `Pave_Type` is set to **Asphalt**, then the owner must be **County owned**, **Privately owned**, or **Unknown owner**. The city will not be available to select for `Owner`.

11. Verify that your list of contingent values matches the previous screenshot and click the **Save** button on the **Contingent Values** tab in the ribbon to apply them to the `Side Walks` layer.

12. Close the **Contingent Values: Side Walks** tab and save your project.

You have now created the contingent values for the `Side Walks` layer. Next, you will test the contingent values to see whether they work as expected.

Testing the contingent values

To test the contingent values you just created, you will create a new sidewalk feature in the `Side Walks` layer and then try to update values for the `Owner` and `Pave_Type` fields. When updating the field values, you will see whether they are constrained as expected:

1. In the ribbon, activate the **Map** tab and click on the **Bookmarks** button. Select the **New parcel location** option from the displayed list. The map should zoom to the same location where you created the new parcel in the previous recipe.

2. Activate the **Edit** tab in the ribbon and select the **Create** button to open the **Create Features** pane.

3. In the **Create Features** pane, notice the `Side Walks` feature template has an error indicator (red circle with white exclamation point) next to it. This indicates that the layer is not editable because a condition exists that is blocking it. Hover your mouse pointer over the error indicator to see what the issue is.

   As you can see, the error indicates that the `Side Walks` feature template violated the contingent values you just finished setting up.

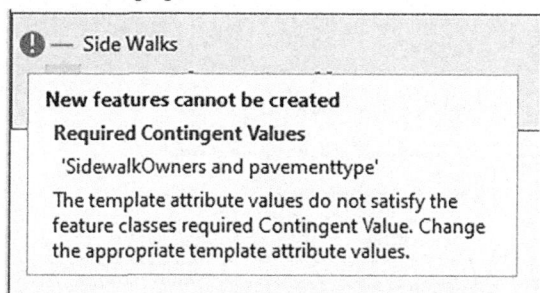

   Figure 9.34 – The Side Walks feature template edit error

   Before you can create a new sidewalk feature, you will need to update the feature template, so its default attribute values don't conflict with the contingent values you created.

4. In the **Create Features** pane, right-click on the `Side Walks` feature template and select **Properties**, as illustrated in the following figure:

   Figure 9.35 – Opening feature template properties

5. In the **Template Properties** window, select **Attributes** in the left panel.

Notice that the **Pavement Type** and **Sidewalk Owner** fields are highlighted in red. That is because both are set to **<Null>** as the default value. This is the cause of the error. You will now update these default values.

6. Click to the right of **Pavement Type** where it shows **<Null>**. Select `Unknown` from the list that appears.

7. Click to the right of **Sidewalk Owner** where it shows `<Null>` and select **Unknown owner** from the list that appears. The red highlight should disappear once the new default values have been chosen.

8. Verify that your **Template Properties** window matches the following, then click the **OK** button to apply the changes:

Template Properties: Side Walks		
General	☑ Show Non-Visible Fields	
Tools	Street Sidewalk Follows	<Null>
Attributes	Date Sidewalk was installed	<Null>
	Pavement Type	Unknown
	Sidewalk Owner	Unknown owner

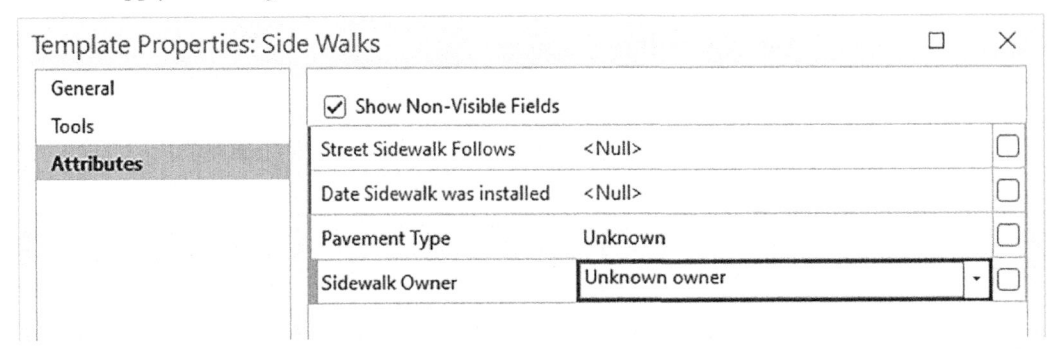

Figure 9.36 – Feature template with updated default values

9. Save your project.

10. In the **Create Features** pane, ensure the `Side Walks` feature template is still selected. Then, click on the map, as shown in the following figure, to draw a new sidewalk feature:

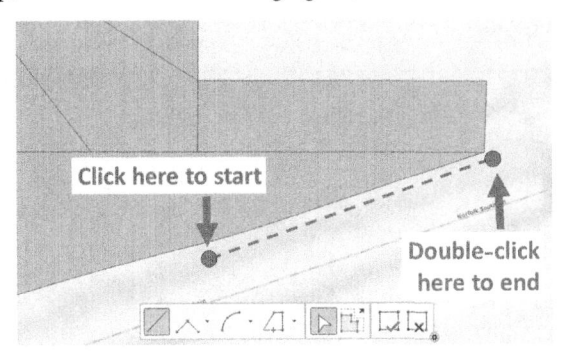

Figure 9.37 – Drawing a new sidewalk feature

11. On the **Edit** tab in the ribbon, select the **Attributes** button to open the **Attributes** window.

Creating contingent values | 331

12. In the **Attributes** window, click on the cell to the right of **Sidewalk Owner** and select **<Show All>** from the list that appears. Then, select **City owned** from the list.
13. When you select **City owned**, **Pavement Type** and **Sidewalk Owner** should be highlighted in red. Also, an error should appear at the top of the **Attributes** window.

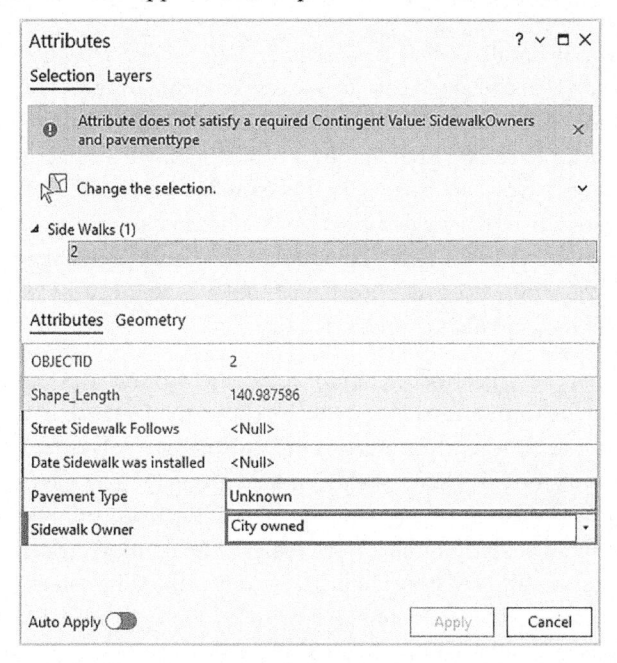

Figure 9.38 – Attribute window with an error indicated

The reason the error is indicated is that if **Sidewalk Owner** is set to **City owned**, you cannot have an **Unknown** pavement type. This violates your contingent values. Next, you will fix this error.

14. Click in the cell to the right of **Pavement Type** so you can see the list of allowed values. Notice that **Concrete** is the only value presented. Select **Concrete** from the list.
15. The red highlights and error should disappear. If they have, click the **Apply** button.
16. On the **Edit** tab in the ribbon, click the **Save** button to save your new feature and attribute changes to the geodatabase.
17. Save your project and close ArcGIS Pro.

You have successfully tested your contingent values. When you changed **Sidewalk Owner** to **City owned**, an error was indicated showing the **Pavement Type** value violated the contingent values you specified. When you changed **Pavement Type** to **Concrete**, the error disappeared because there was no longer a violation.

How it works...

You have just created contingent values for the `Side Walks` layer that validated the combination of values found in the **Sidewalk Owner** and **Pavement Type** fields. This helps ensure that when a new sidewalk is added or an existing one is updated, then the erroneous values cannot be entered.

You first had to add the **Sidewalk Owner** field to the `Side Walks` attribute table, as well as create the coded values domain that contained a list of owners. To create the new field, you right-clicked on the `Side Walks` layer and then selected **Data Design** and **Fields** from the menus that appeared. This opened a list of fields currently found in the attribute table for the `Side Walks` layer. Next, you clicked in the bottom row where it said **Click here to add a new field**. Then, you populated the parameters required to create the new field, such as name, alias, and data type. To create the new domain, you clicked in the domain cell for the new field and selected **Create new coded values domain**. This allowed you to create the new `SidewalkOwners` domain containing the list of those owners who are responsible for building and maintaining sidewalks.

The next step in creating contingent values was to create a field group. This was done by selecting the `Side Walks` layer in the **Contents** pane and then selecting the **Data** tab in the ribbon and clicking on the **Contingent Values** button, which opened the **Contingent Values** tab in the main view area. From there, you clicked on the **Click here to add new field group** button. This opened the **Field Group** window. Here, you named the new field group and selected the fields from the `Side Walks` layer attribute table that would take part in the group. With the field group created, next, you added the acceptable combination of values that would be allowed based on the values found in the domains assigned to the **Sidewalk Owner** and **Pavement Type** fields. For example, if **Sidewalk Owner** was set to **City owned**, then the only allowed **Pavement Type** was **Concrete**.

With the contingent values created, you tested them by drawing a new sidewalk feature on the map. When you changed **Sidewalk Owner** to **City owned** but **Pavement Type** was left as **Unknown**, ArcGIS Pro indicated an error was present in the data because the contingent values you created were being violated. You then changed **Pavement Type** to **Concrete**, and the error disappeared.

We enabled many advanced geodatabase functionalities that can help improve the quality of the data in your GIS database and track productivity. This included enabling editor tracking, enabling GPS metadata, adding attribute rules, and setting up contingent values. These powerful tools are only found in geodatabases and are not supported by other GIS storage formats such as shapefiles. This makes storing your data in a geodatabase compelling.

These capabilities can be used individually or combined with domains, subtypes, and topologies to ensure data integrity and to increase editing efficiency. They also work with other Esri applications/solutions such as ArcGIS Online, Field Maps, and Survey123 so that, no matter how your data is accessed, collected, or edited, the same limits and rules apply to everyone.

10
Validating and Editing Data with Topologies

GIS has the amazing capability to help us solve real-world problems. ArcGIS Pro includes a wealth of tools that allow users to perform all kinds of analyses to answer questions, find solutions, and determine patterns. However, like most computer systems, the results are only as good as the quality of the data the analysis is based on. The adage garbage in, garbage out applies.

So, how do we ensure our GIS data is clean and accurate? One way is to make use of topology. Topology is a model of how features are related to one another spatially. Are they adjacent, connected, or coincident? This should not be confused with topography, which is how the ground changes in elevation. ArcGIS Pro allows you to use two types of topologies: geodatabase and map.

A geodatabase topology is an item you create within a geodatabase, which allows you to apply rules to the data that is part of the topology. This might include rules such as polygons must not overlap, lines must not intersect, or points must be inside a polygon. These rules are then used to validate the data to locate any features that violate the rules. ArcGIS Pro includes tools that allow you to fix the errors that are found. To make use of geodatabase topologies, you must store your data in a geodatabase, the feature classes must be in the same feature dataset, and you must have a standard or advanced license.

In addition to being able to validate data with geodatabase topology, ArcGIS Pro provides tools to help you edit data so that you don't introduce new errors into your data. These topology edits allow you to edit multiple features so that whatever their existing spatial relationship happens to be, it is maintained after the edit. This can not only keep your data clean but also greatly increase your editing efficiency because you can edit multiple features with a single edit operation.

Map topologies are not as complex as geodatabase topologies. They are temporary, only existing during a given ArcGIS Pro session when you have a map open. Map topologies do not allow you to validate your data to find errors. What they allow you to do is perform edits to existing data, using topology edit tools, so that the edited features maintain their current spatial relationships.

Map topologies are not only limited to data in a geodatabase or in a single feature dataset. Geodatabase feature classes and shapefiles can all be included in a single map topology. Another advantage of the map topology is that it works at all licensing levels, so you can use them even with just a Basic license.

In this chapter, you will work through the following recipes:

- Creating a new geodatabase topology
- Validating spatial data using a geodatabase topology
- Correcting spatial features with topology tools
- Editing data with a map topology

Creating a new geodatabase topology

A geodatabase topology is an item that's stored in a geodatabase, just like a feature class or table. They can only exist and function within a geodatabase. Other formats, such as shapefiles, do not support geodatabase topologies. Like feature classes and tables, you must create the topology. This will include giving it a name, setting tolerances, adding rules, and adding participating feature classes. To create a new geodatabase topology, a couple of things must be true.

First, you can only create a geodatabase topology within a feature dataset. Second, only feature classes stored in the feature dataset containing the topology can participate in that topology. Third, participating feature classes must not be part of another topology or geometric network. A feature class can only participate in one topology or geometric network at a time. Lastly, you must have a Standard or Advanced license of ArcGIS Pro to create a new topology.

In this recipe, you will create a new geodatabase topology in the Trippville GIS database that will ultimately be used to validate the parcel data for the city.

Getting ready

This recipe will require you to have a Standard or Advanced license for ArcGIS Pro. You will not be able to complete it with just a Basic license. If you do not have a Standard or Advanced license, you can request a free trial license from Esri at https://www.esri.com/en-us/arcgis/products/arcgis-pro/trial. Trial licenses are valid for 21 days.

You also need to ensure you have installed the sample data for this book. It is also recommended that you complete all the recipes in *Chapter 1, ArcGIS Pro Capabilities and Terminology*, before starting this recipe. This will ensure you have the foundational skills needed to complete this recipe.

How to do it…

Now that you have a general understanding of what a geodatabase topology is, you will work through the process required to create one in ArcGIS Pro:

1. To get started, you need to launch ArcGIS Pro, as you have done in previous recipes.
2. In the ArcGIS Pro start window, click on **Open another project**, as shown in the following screenshot:

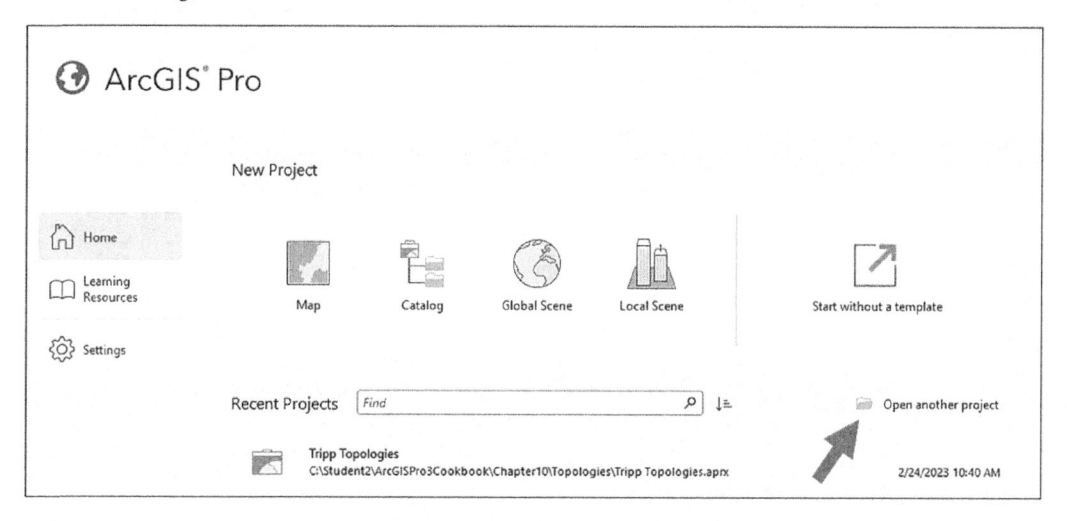

Figure 10.1 – Open another project

3. Expand the **Computer** option in the left panel of the **Open Project** window and select **This PC**.
4. Navigate to `C:\Student\ArcGISPro3Cookbook\Chapter10\Topologies` and select the `Topologies.aprx` project. Click **OK** to open it.

   The project will open in the **Catalog** view. Currently, this project does not contain any maps. Before you create your new topology, you need to verify a couple of things – you need to ensure that your geodatabase contains a feature dataset and that the **Parcels** feature class, which you will validate later, is within a feature dataset:

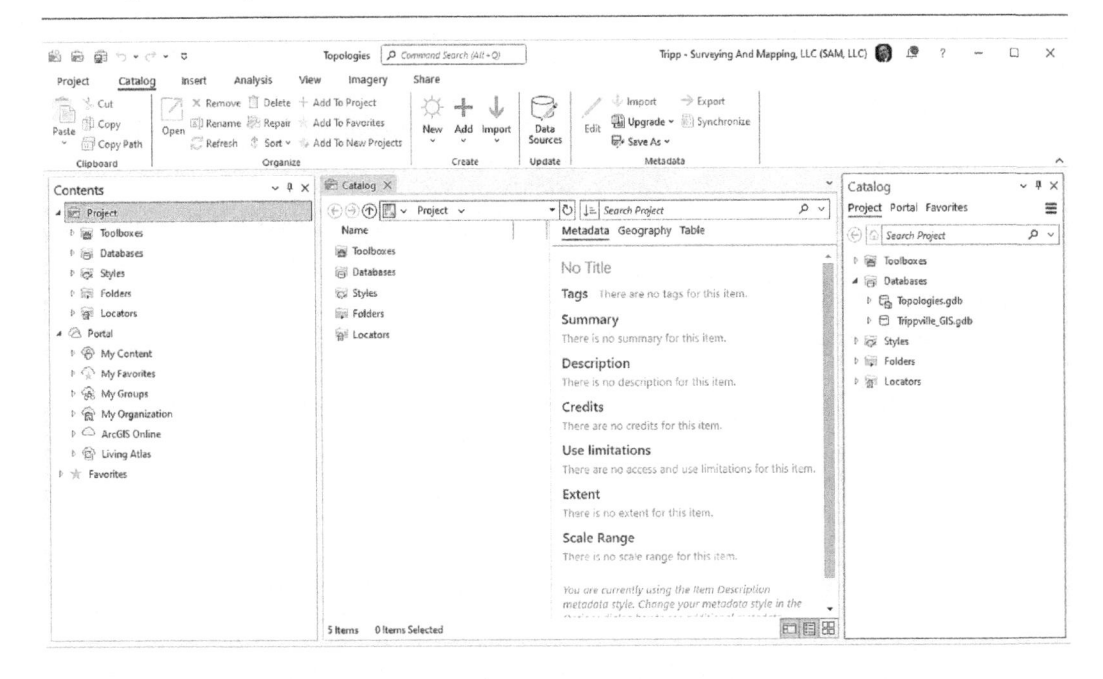

Figure 10.2 – Topologies project open in the Catalog view

5. In the **Contents** pane, click on **Databases** so that the two geodatabases connected to the project are displayed in the **Catalog** view.

6. Double-click on the `Trippville_GIS.gbd` geodatabase so that you can see its contents. Then, answer the following question:

   Question: Does the geodatabase contain any feature datasets, and if so, what are they?

   Answer:

   You should have seen that this geodatabase contains at least three feature datasets: Base, Sewer, and Water. So, that means you can create a topology in this geodatabase. Now, you need to locate the **Parcels** feature class.

7. Double-click on the **Base** feature dataset to reveal its contents.

8. Scroll down to verify that it contains the **Parcels** feature class.

9. Scroll back through the feature dataset to see if it contains any topologies. Then, answer the following questions:

   Question: Is the **Parcels** feature class within the feature dataset?

   Answer:

   Question: Does the feature dataset contain any other topologies?

   Answer:

Now that you have verified that the **Parcels** feature class is stored in the **Base** feature dataset, and there are no other topologies to contend with, you are ready to begin building a new topology.

Creating a new topology

It is now time to work through the process required to create a new topology in the geodatabase. This topology will validate your data to ensure it does not violate the rules you will add:

1. Click on the **Analysis** tab in the ribbon.
2. Click on the **Tools** button in the **Geoprocessing** group on the **Analysis** tab. This will open the **Geoprocessing** pane.
3. Click on the **Toolboxes** tab near the top of the **Geoprocessing** pane, just below the search cell, as shown in the following screenshot. This will open all the system toolboxes connected to the project:

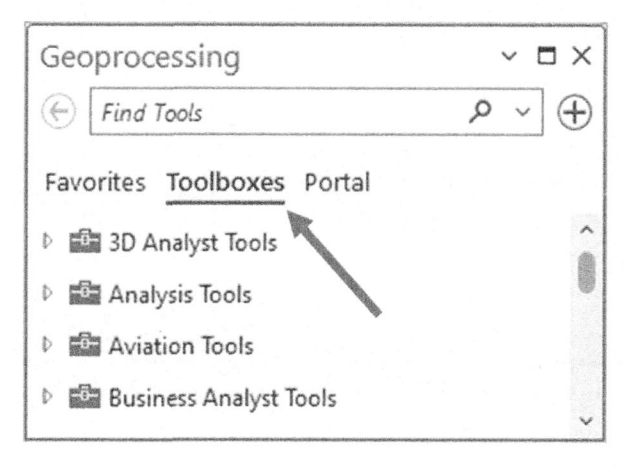

Figure 10.3 – The Toolboxes tab in the Geoprocessing pane

4. Expand the **Data Management Tools** toolbox by clicking on the small arrowhead located to the left of the toolbox's name.
5. Expand the **Topology** toolset located near the bottom of the **Data Management** toolbox.
6. Click on the **Create Topology** tool in the **Topology** toolset. This will open the tool in the **Geoprocessing** pane.
7. Click on the **Browse** button located at the end of the cell for **Input Feature Dataset**.
8. Click on **Databases** in the panel on the left of the **Input Feature Dataset** window that opened. You may need to expand the `Project` folder to see it.
9. Double-click on the `Trippville_GIS.gdb` geodatabase to display its contents.
10. Select the **Base feature** dataset and click **OK**.

> **Tip**
> Do not double-click on the feature dataset. Select it with a single click. Double-clicking will open the feature dataset instead of selecting it. If you accidentally double-click on it, you can click on the back button at the top of the **Input Feature Dataset** window to go back.

11. For **Output Topology**, type `Parcels_Topology`. This is the name of the topology you are creating.

12. Leave **Cluster Tolerance** blank. ArcGIS Pro will calculate a default tolerance. Then, verify that your **Create Topology** tool looks as follows and click **Run**:

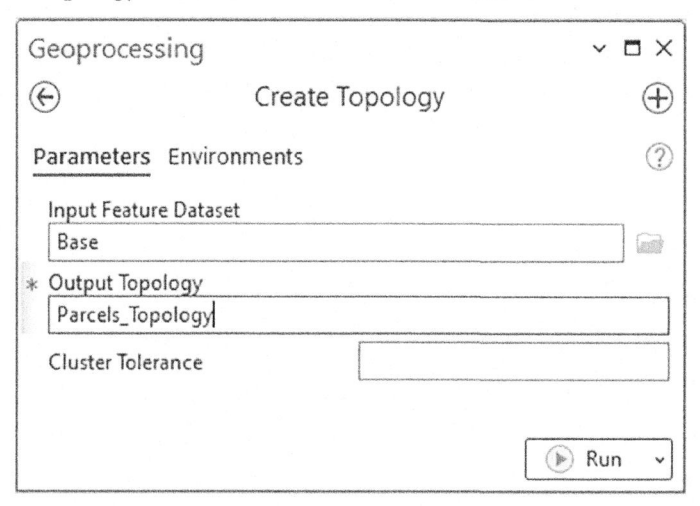

Figure 10.4 – The Create Topology tool with all the parameters completed

> **Note**
> **Cluster Tolerance** is the distance at which ArcGIS Pro believes features should be considered to be in the same location. When you validate the topology, ArcGIS Pro will automatically snap features together that are within the cluster tolerance distance. This means that while validating topology, the data will potentially move. This is both good and bad. It will often clean up data by connecting features that were mistakenly not connected. It will remove small gaps and overlaps between features. However, if your cluster tolerance is set too large, it could snap features together that should not be. So, it is generally recommended to keep the cluster tolerance small. Esri indicates that the best practice is to use the default tolerance calculated by ArcGIS Pro as it is typically small enough to prevent any adverse effects on your data.

13. When the **Create Topology** tool finishes running, close the **Geoprocessing** pane.

In the **Catalog** view, you should now see the parcels topology you just created. The next step is to add the feature class or classes that will participate in the topology. In this recipe, there will be only one feature class: parcels.

Adding the feature class and rules to the new topology

You will now add that feature class to the topology and assign it the rules you want it to validate against:

1. In the **Catalog** view, scroll to **Parcels_Topology**, which you just created. Right-click on it and select **Properties** from the menu that appears:

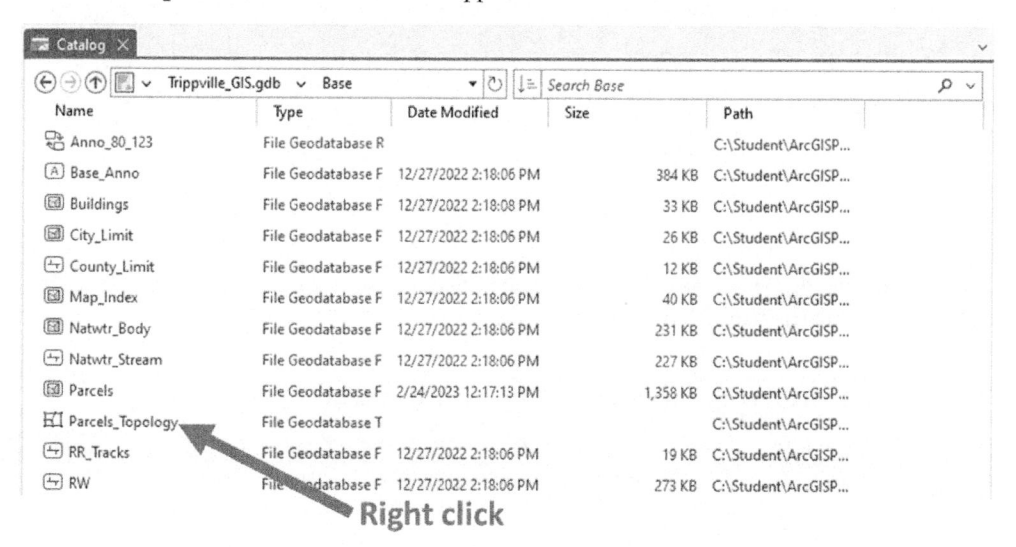

Figure 10.5 – Right-clicking on Parcels_Topology to reveal its menu

2. The **Topology Properties** window should open. Click on the **General** option in the panel located to the left of the window. Review those settings. You will see that the cluster tolerance is set to a very small value.

3. Click on the **Feature Class** option in the left panel of the **Topology Properties** window.

340　Validating and Editing Data with Topologies

4. Click in the box next to the **Parcels** feature class to add it to the topology, as shown in the following screenshot:

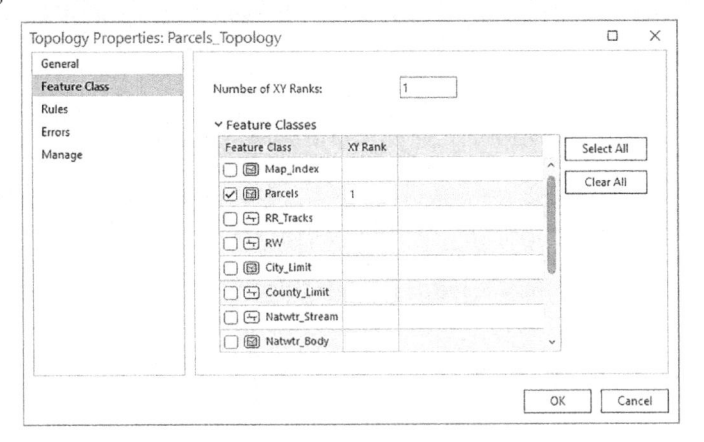

Figure 10.6 – Adding feature classes to a topology

> **Note**
> Multiple topologies can exist in the same geodatabase and feature dataset. However, each feature class can only participate in one topology at a time.

5. Next, click on the **Rules** option in the left panel of the window.
6. Click where it says **Click here to add a new rule.** in the panel on the right-hand side of the window, as illustrated in the following screenshot:

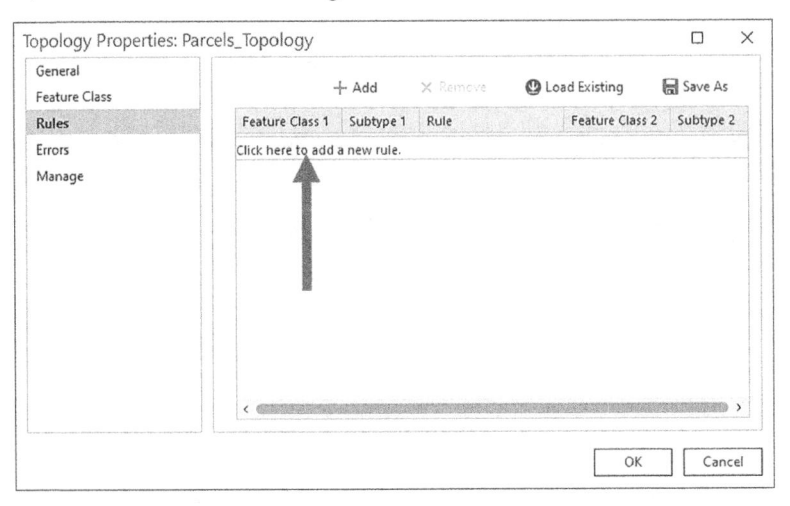

Figure 10.7 – The Click here to add a new rule. option

7. Click on the cell located beneath the column titled **Feature Class 1**. Select **Parcels** from the list that appears. You may need to double-click to get the list to appear. Since you only have one participating feature class in this topology, it should be the only one on the list.

8. Click on the cell located beneath the column titled **Rule**. Select **Must Not Overlap (Area)** from the list of possible rules that appear.

9. Verify that your **Topologies Properties: Parcels_Topology** window looks as follows, then click **OK**:

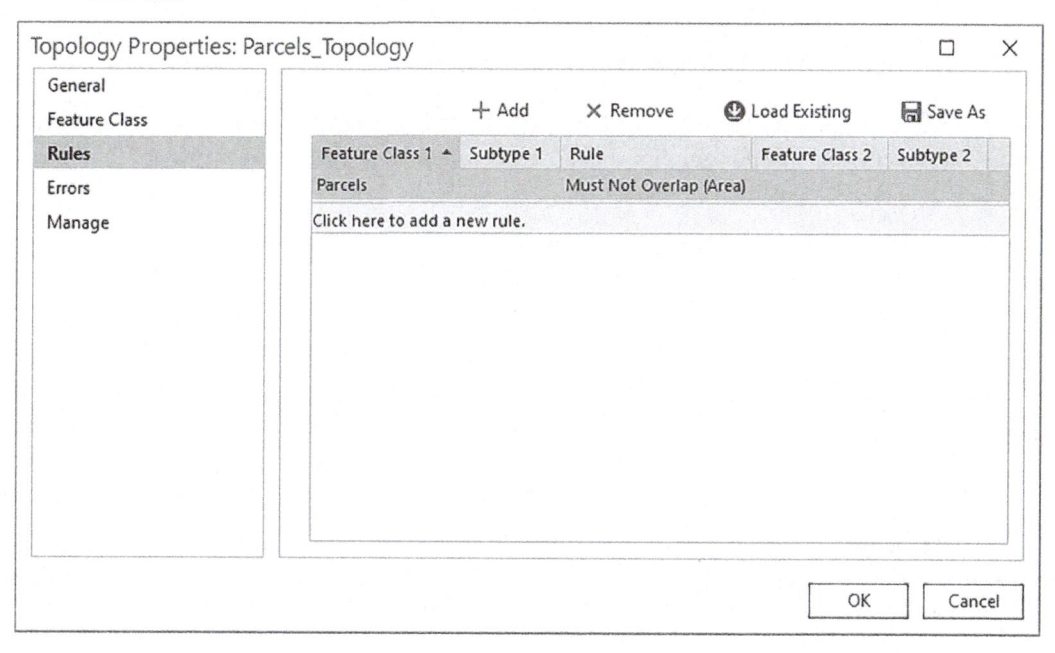

Figure 10.8 – Rule added to topology to validate parcels

> **Note**
> A topology can include many rules. The rules you can apply will depend on the type of feature classes that participate in the topology. For example, the rule you added in this recipe only applies to polygons in a single feature class. For more information about topologies in ArcGIS Pro, as well as the necessary rules, go to `https://pro.arcgis.com/en/pro-app/latest/help/data/topologies/topology-in-arcgis.htm`.

10. Save your project and close ArcGIS Pro.

You have successfully created a new topology in the `Trippville_GIS` geodatabase.

How it works...

In this recipe, you created a new topology using the **Create Topology** geoprocessing tool. You then added the **Parcels** feature class to the new topology from within the **Properties** window. You also added a rule that says parcel polygons must not overlap from within the properties window.

In the next recipe, you will use the topology you just created to validate the polygons in the **Parcels** feature class to locate any that violate the rule you added to **Parcels_Topology**.

Validating spatial data using a geodatabase topology

In the previous recipe, you created a new topology. This will allow you to find errors in the **Parcels** feature class that violate the polygons must not overlap rule you included in the topology. To find any errors, you must first validate the topology.

When you validate a topology, several things occur. First, all features participating within the topology are snapped together if they are within the cluster tolerance. This is done automatically for all features when you validate. Second, the participating feature classes are checked against the rules you have included in the topology. Any areas that include features that violate the rules are identified in the topology so that you can easily locate the errors and fix them.

In this recipe, you will validate your topology to identify any errors in the **Parcels** feature class. To do this, you will create a new map and add the topology, along with the `Parcels` layer, to it.

Getting ready

You will need to have completed the previous recipe to complete this one. You will also need a Standard or Advanced license of ArcGIS Pro. You will not be able to perform this recipe with just a Basic license of ArcGIS Pro. If you only have a Basic license, you can request a trial advanced ArcGIS Pro license from Esri at `https://www.esri.com/en-us/arcgis/products/arcgis-pro/trial`.

How to do it...

You will now work through the necessary steps in ArcGIS Pro to validate the data in the **Parcels** feature class using the topology you created in the previous recipe:

1. To get started, launch ArcGIS Pro and open the `Topologies.aprx` project located in `C:\Student\ArcGISPro3Cookbook\Chapter10\Topologies` using the skills you have learned in past recipes. If you recently completed the previous recipe, this project should be on your recently opened **Project List**.

2. Click on the **Insert** tab in the ribbon.

3. Click on the **New Map** button located in the **Insert** tab. This will create a new blank 2D map in your project. The new map should automatically open.

4. In the **Contents** pane, right-click on **Map** and select **Properties** from the menu that appears, as illustrated in the following screenshot:

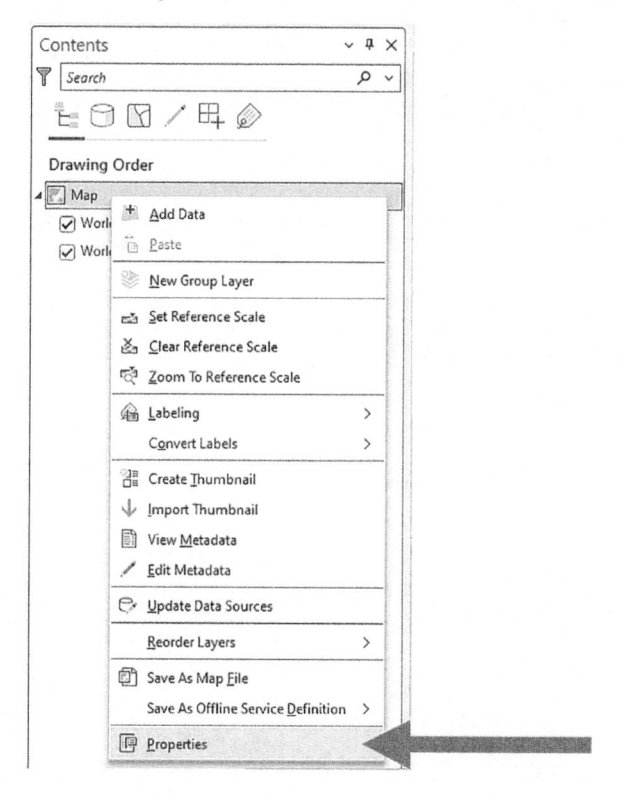

Figure 10.9 – Opening Map | Properties

5. Ensure the **General** option is selected in the left panel in the **Map Properties** window that opened.
6. In the cell for **Name**, type `Parcel Topology` and click **OK**. Where it said **Map** previously, it should now read **Parcel Topology**.
7. Save your project.

You have just added a new map to your project. This map is empty, so you will need to add content to it.

Adding topology and layers to the map

With the project and map open, you will now add the topology and included layer to the new map you just created:

1. In the **Catalog** pane, expand the `Databases` folder by clicking on the small arrowhead located to the left.

2. Expand `Trippville_GIS.gdb` so that you can see its contents.

3. Expand the **Base** feature dataset.

4. Right-click on **Parcels Topology** and select **Add to Current Map** from the menu that appears, as shown in the following screenshot. This will add the topology and the participating feature classes as layers to the map:

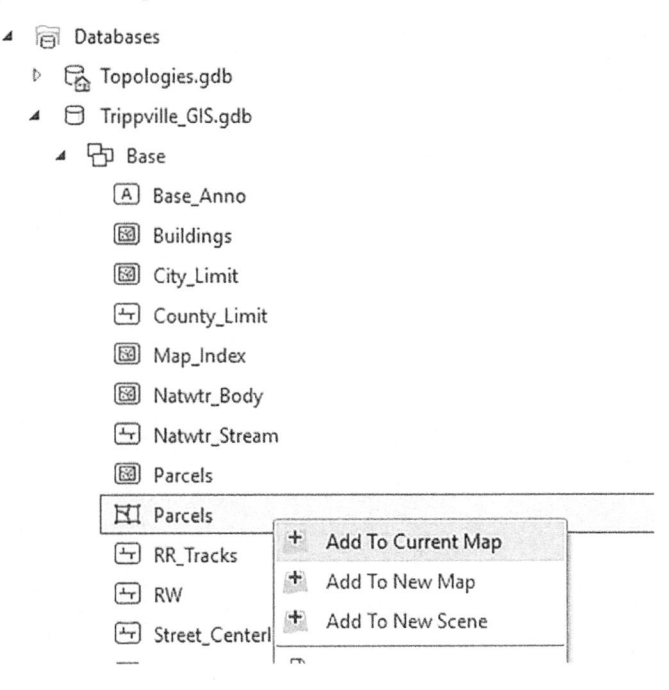

Figure 10.10 – Adding a topology to the current map

5. Right-click on the **Parcels** layer in the **Contents** pane. Select **Zoom to Layer** from the menu that appears. You should now be able to see the parcels for the city of Trippville.

6. In the **Contents** pane, examine the layers listed under **Parcels Topology**. You might need to expand some of them to see all the layers included. Then answer the following question:

Question: What layers are included in the topology?

Answer:

As you can see, a topology includes more than just the rules, participating feature classes, and cluster tolerance. It also includes location errors and **dirty areas**.

7. Turn on the `Dirty Areas` layer in the **Parcels_Topology** group layer in the **Contents** pane. A blue square should appear on the map, as shown in the following screenshot:

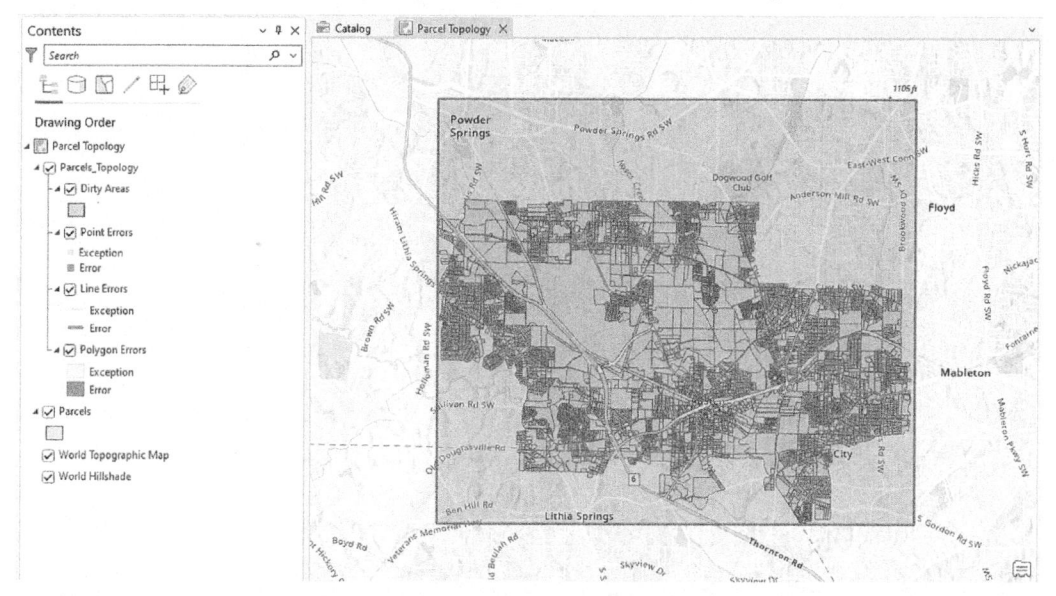

Figure 10.11 – Dirty areas displayed on the map

> **Note**
> **Dirty Areas** are areas in your data that are participating in the topology that you have not validated. In this case, the entire area is considered a dirty area because you have not validated your topology. Just because a dirty area exists does not mean your data has errors or cannot be used. A dirty area is just an area within your data that you have not validated with the rules in your topology because you have either edited the data or just created the topology. Regardless of whether dirty areas exist, or if your data does have errors, you can always use your data.

Validating data using topology rules

Now that the topology has been added to the map and you can see the dirty areas, you need to validate the parcel data using the topology to find any errors that might exist:

1. Click on the **Edit** tab in the ribbon.

2. Click on the drop-down arrow located next to **No Topology** in the **Manage Edit** group on the **Edit** tab and select **Parcels_Topology (Geodatabase)** from the list, as illustrated here:

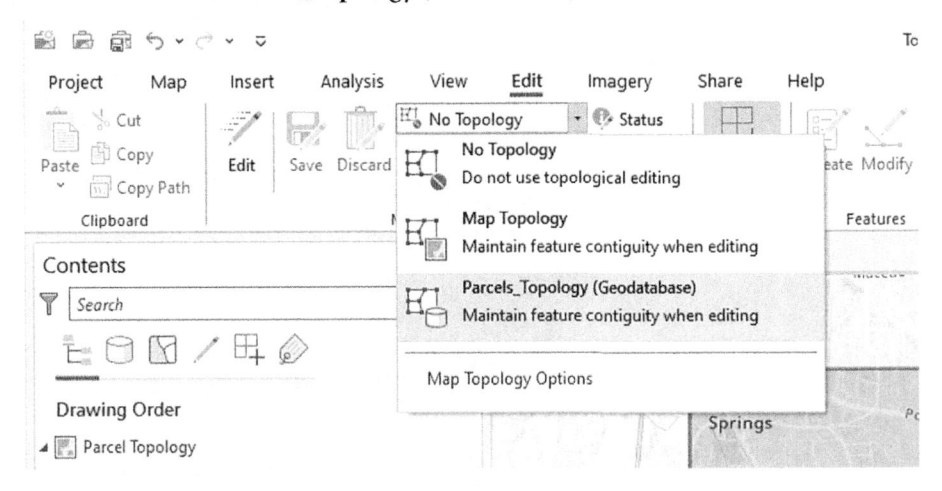

Figure 10.12 – Selecting Parcels_Topology to edit it

3. Now that you have selected the topology you will validate, you need to access the **Validate All** tool. Click on the **Modify** button in the **Features** group on the **Edit** tab. This will open the **Modify Features** pane.

4. Scroll down to the **Validate** toolset. If necessary, expand it so you can see its contents.

5. Since this is the first time you have validated the topology since it was created, select the **Validate All** tool.

> **Tip**
> There are also geoprocessing tools you can use to validate your topology. These are located in the **Data Management** toolbox and the **Topology** toolset. The **Validate Topology** tool does the same thing as the **Validate All** tool.

When you validate the topology, it will look for any parcels that overlap other parcels. Any it finds, it will add those locations to the `Polygon Errors` layer. When the validation is complete, you should see some red polygons on the east side of the `Parcels` layer, which indicates the errors that were found.

6. Close the **Modify Features** pane and save your project.

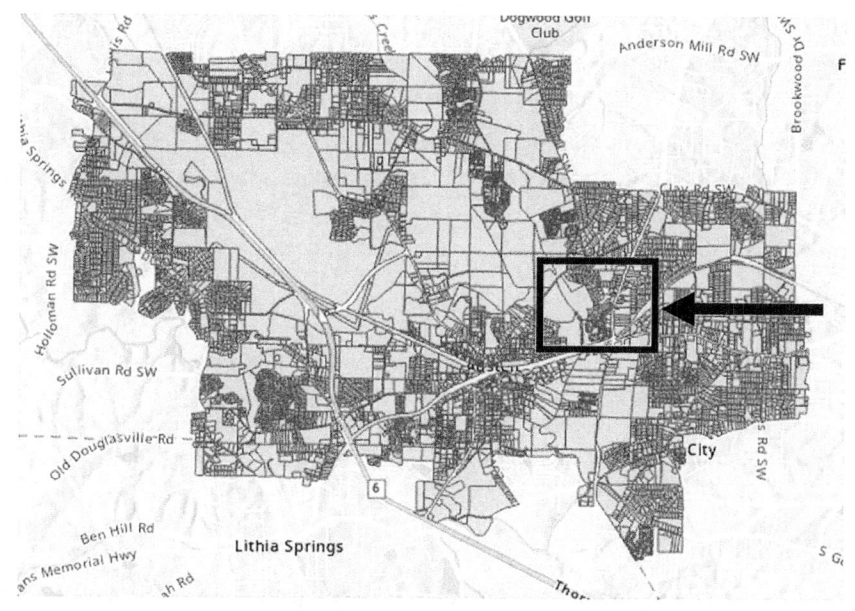

Validating spatial data using a geodatabase topology 347

7. Zoom into the area indicated in the following graphic:

Figure 10.13 – Zoom into this area

Your map should now look as follows. Remember that your colors may be different as ArcGIS Pro assigns a random color to new layers when they are added:

Figure 10.14 – Zooming into the area to see topology errors

8. Let's create a bookmark so that you can easily return to this area later. Click on the **Map** tab in the ribbon.

9. Then, click on the **Bookmarks** button and select the **New Bookmark** option from the list that appears. This will open the **Create Bookmark** window.

10. Name the new bookmark `Parcels Overlap Area to Fix` and type `Area which has parcels that overlap and violate the topology rules` as the description. Then, click **OK**.

> **Note**
>
> As you have seen in previous recipes, bookmarks allow you to quickly zoom into a fixed spatial location. These are very handy for many purposes. Several things I use bookmarks for include printing a map at a specific location and scale, for meetings where I need to highlight specific areas or features, and returning to a specific location for a project or editing. Each map has its own set of bookmarks. There is no limit to the number of bookmarks a project can contain.

11. Save your project and close ArcGIS Pro.

You have now identified areas in the `Parcels` layer where polygons overlap with one another. Next, you will work to clean up those errors.

How it works...

In this recipe, you validated the data contained in the `Parcels` layer using the topology you created in the previous recipe. Once validated, you were able to see where the existing parcel data violated the rule *parcel polygons must not overlap*. Now, you know where the errors are so that they can be fixed.

You did this by using the **Validate All** tool located on the **Edit** tab in the **Modify Features** pane. Validating the data made ArcGIS Pro look through all the parcel polygons and find where there were overlaps between them. This eliminated dirty areas and marked the location of found areas in a `Polygon Error` layer that was part of the topology. With the location of the found errors shown on the map, you were able to zoom into them.

In the next recipe, you will explore methods that can be used to correct the discovered errors.

Correcting spatial features with topology tools

Now that you know where the errors are in your data, you need to fix them. ArcGIS Pro includes specific tools for fixing topology errors. These are generally quick and easy to use for most situations. You may also use standard editing tools to correct errors. In either case, you should always validate your data after completing edits to ensure you have not introduced any new errors into your data.

In this recipe, you will fix several errors you uncovered when validating the topology. You will use the **Error Inspector** tool in ArcGIS Pro to correct these errors.

Getting ready

This recipe continues to build on the work you've done in the previous recipes in this chapter. You will need to have completed all the previous recipes included in this chapter before you can start on this one. As with those previous recipes, you will also need a Standard or Advanced license to complete this recipe.

How to do it...

You will now correct the errors you found when you validated the parcel data using the topology you created. You will use the **Error Inspector** tool in ArcGIS Pro to do this:

1. Launch ArcGIS Pro and open the `Topologies.aprx` project located in `C:\Student\ArcGISPro3Cookbook\Chapter10\Topologies` using the skills you learned in past recipes.
2. In the **Contents** pane, click on the **List by Editing** button.
3. Ensure the `Parcels` layer is enabled for editing.
4. Click on **List by Selection** in the **Contents** pane.
5. Ensure that the `Polygon Errors` and `Parcels` layers can be selected at a minimum.
6. Click on the **Edit** tab in the ribbon.
7. Click on the **Error Inspector** button on the **Edit** tab. This will open the **Error Inspector** pane at the bottom of the interface. Like all panes, this one may be undocked and moved to a different location if desired.

Fixing errors

You will now fix some of the errors the topology has found using tools in the **Error Inspector** pane:

1. Select the first error presented in the pane by clicking on the gray square located to the far left of the first row, as illustrated here:

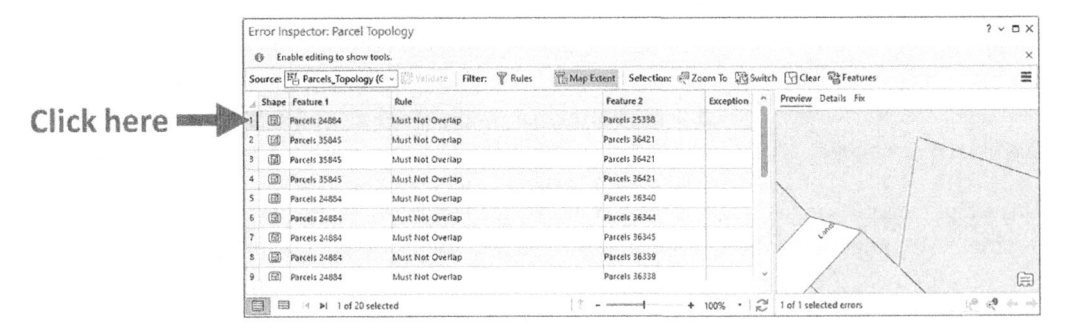

Figure 10.15 – Selecting an error in the Error Inspector pane to fix it

2. Click on the **Zoom To** button located at the top of the **Error Inspector** pane. This will zoom you into the location of the selected error so that you can see it in greater detail.

> **Tip**
> In the **Error Inspector** pane, you can also use the **Preview** window to see the specific error you are interested in correcting. From that window, you can also see additional details about the error in question and access tools to fix it.

3. In the **Preview** window located on the right-hand side of the **Error Inspector** pane, click on the **Fix** tab, as shown in the following screenshot:

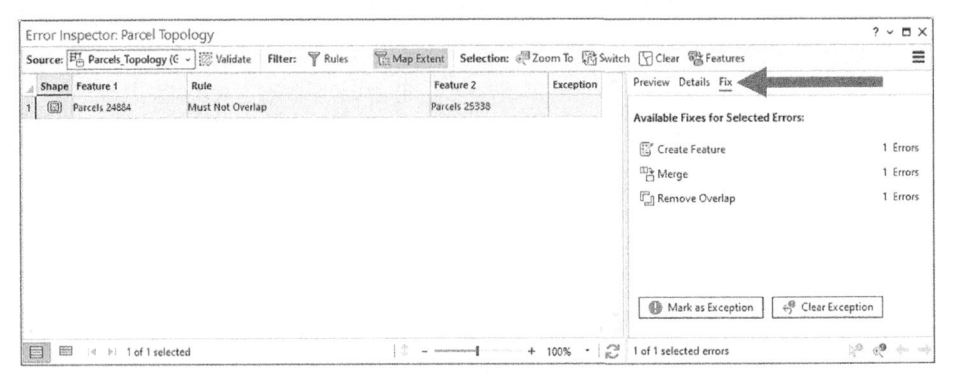

Figure 10.16 – Selecting the Fix tab in the Error Inspector pane

4. Select **Merge** from the options that appear when you click on the **Fix** tab.
5. The **Merge** window should appear. Select **Parcels 24884 (Largest)** and click on the **Merge** button.

   By merging, the overlapping area was subtracted from the other parcel you did not select and was assigned specifically to the parcel you did select. For this recipe, you chose to merge the overlapping area with the largest parcel for the sake of expediency. Normally, you would take time to research and determine which parcel the overlapping area belonged to or if one of the other recommended fixes would apply.

6. A new dirty area should appear on the map because you just edited the `Parcels` layer but have not validated it. Click on the **Validate** button located at the top of the **Error Inspector** pane. This will remove the dirty area and allow you to see if any new errors have appeared due to your edit.

> **Tip**
> You should validate your data regularly when using topologies to ensure it remains clean. How often you validate is up to you, but here are some general guidelines: Beginner users should probably validate after every edit or correction. More experienced users should validate before they save edits to ensure they are not saving erroneous data back to the geodatabase. Remember, you can always undo up until the last time you saved. Once you save, there is no going back.

7. Click the **Save** button in the **Manage Edits** group on the **Edit** tab.

With that, you have successfully fixed a topology error in the data, making your parcel data cleaner and more correct. But as you can see, several more require your attention.

Fixing multiple errors at the same time

In the **Error Inspector** pane, it is possible to resolve multiple errors at the same time. Let's see how that works:

1. Click on the **Map** tab in the ribbon. Then, click on **Bookmarks** and select the **Parcels Overlap Area to fix** bookmark to zoom the map back into this area.
2. In the **Error Inspector** pane, locate the three errors for **Feature 1** relating to **Parcels 35845**.
3. Click on each error, one at a time. Watch both the map and the **Preview** window on the right-hand side of the **Error Inspector** pane. You should be able to see that each of these errors is related to the same two parcels.

> **Tip**
> You can zoom in to get a closer look at each error in the **Preview** window on the **Error Inspector** pane. Simply use the scroll wheel on your mouse to zoom in or out within the **Preview** window.

4. Click on the small square located next to the top record for the errors related to **Feature 1 Parcels 35845**. Holding your *Shift* key down, click on the last of the three records relating to **Parcels 35845**. All three errors should now be selected, as illustrated in the following screenshot:

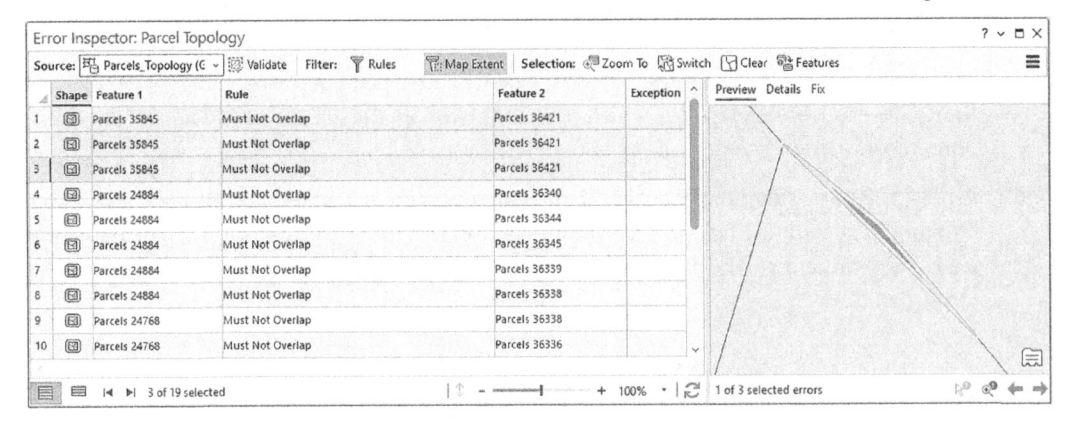

Figure 10.17 – Multiple records selected in the Error Inspector pane

5. In the **Error Inspector** pane, click on the **Fix** tab in the right panel. Then, select **Remove Overlap** from the options that appear in the panel. This will remove the overlapping areas from each impacted parcel polygon, creating a single adjacent boundary between them.

You just fixed three topological errors with a single edit. Using these tools to fix errors in your data not only makes your data cleaner but can also be very efficient as well.

6. Validate your data using the **Validate** button at the top of the **Error Inspector** pane. If no new errors are found for the area you just fixed, save your edits.

7. Using the same processes, you have just fixed the remaining errors. Feel free to try different corrective tools in each case to see how they work. Make sure you validate after each corrective action, as well as save your edits as you go.

8. Save your project and close ArcGIS Pro.

As you have seen, the use of geodatabase topologies can greatly increase the overall accuracy of your GIS data and help you locate where errors may exist. The result is improved data quality and better analysis results. One thing to remember when you want to use geodatabase topologies is that you must have a Standard or Advanced ArcGIS Pro license.

How it works...

In this recipe, you fixed the errors you found in the `Parcels` layer when you validated the data using the topology you created in previous recipes. You fixed these errors using tools provided in the **Error Inspector** pane. This pane allowed you to quickly and easily select one or more errors, preview the location of those errors, zoom the map into the specific error, and provide access to recommended tools to fix the error.

Editing data with a map topology

You have now seen how powerful geodatabase topologies can be to help you find and correct errors in your data. However, to use geodatabase topologies, your data must be stored in a geodatabase, and you must have a Standard or Advanced license. So, what do you do if you want to maintain data that is stored in a shapefile so that you do not introduce topological errors, or if you only have a Basic license? This is where a map topology can be used.

As mentioned in this chapter's introduction, map topologies are temporary and do not allow you to validate data using rules. What they do allow you to do, though, is edit data using topology tools so that the existing spatial relationships are maintained. This keeps you from introducing any new errors into your data. Map topologies also have the advantage of allowing topology editing tools to be used on multiple data formats, including shapefiles, web services, and geodatabase feature class-based layers at the same time. Being able to edit multiple layers with single edits, regardless of format, can greatly increase your efficiency. Map topologies also do not require all the geodatabase feature classes to be in the same feature dataset within the geodatabase. So, while map topologies do not allow you to validate your data like geodatabase topologies do, they do offer greater data flexibility.

In this recipe, you will use a map topology to edit data in multiple layers with a single edit. A road has been widened to include a turn lane. You need to adjust several layers, including city limits, parcels, rights of way, and voting districts, to reflect the newly widened road. Several of these layers are in the Trippville GIS geodatabase you have been working with. However, the voting districts layer is stored in a shapefile because the voter registration system, which needs the layer, cannot read a geodatabase.

Getting ready

This recipe can be completed using all license levels of ArcGIS Pro. It is not required that you complete any previous recipes to complete this one, though it is recommended that you complete the recipes found in *Chapter 1, ArcGIS Pro Capabilities and Terminology*, before starting this recipe to ensure you have the foundational skills required to complete this recipe. You also need to ensure you have installed the sample data for this book.

How to do it...

Follow these steps to enable a map topology and use topology editing tools to update multiple layers at the same time:

1. Start ArcGIS Pro and open the `Topologies.aprx` project located in `C:\Student\ArcGISProCookbook\Chapter5\Topologies` using the skills you learned in past recipes.

2. If they are still open, close all panes except for the **Contents** and **Catalog** panes. This will free up your screen area and make it easier to work.

3. In the **Contents** pane, right-click on the **Parcels_Topology** group layer and select **Remove** from the menu that appears, as shown in the following screenshot:

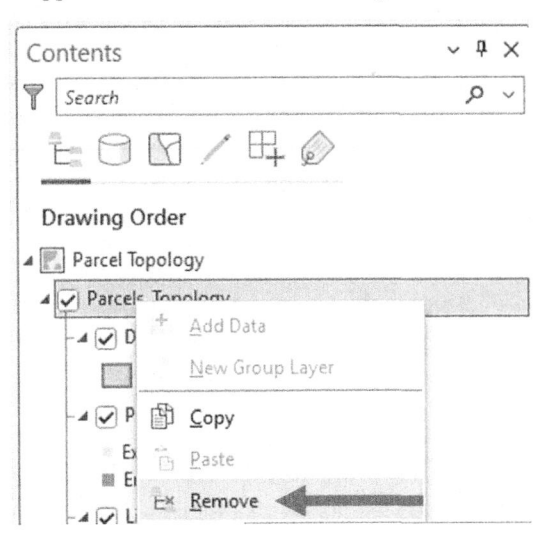

Figure 10.18 – Removing the Parcels_Topology group layer from the map

4. Save your project.

5. Click on the **Add Data** button on the **Map** tab in the **Layer** group in the ribbon.

6. Click on the **Databases** folder located under **Project** in the left panel of the **Add Data** window.

7. Double-click on the `Trippville_GIS.gdb` geodatabase.

8. Double-click on the **Base** feature dataset.

9. Select the **City_Limit** feature class. Then, while holding down the *Ctrl* key, select the **RW** feature class. Then, click **OK** to add these new layers to your map.

10. In the **Contents** pane, ensure you are viewing **List by Drawing Order**. It is the first button on the top left of the pane.

11. Click on the **Symbol** patch, located below the `City_Limit` layer in the **Contents** pane, to open the **Symbology** pane.

12. Ensure you have **Gallery** selected at the top of the pane. Then, scroll down and select the symbol named **Black Outline (2 Pts)** located in **ArcGIS 2D Style**.

> **Tip**
> At the top of the **Symbology** pane, you should see a *Search* function. This allows you to find the symbol you are looking for using keywords or tags. You could type `Black outline 2 pts` into the search to locate the symbol or related symbols.

13. Drag the `City_Limit` layer to the top of the layer list. Then right-click on the `City_Limit` layer and select **Zoom to Layer** from the menu that appears.
14. If desired, change the symbology for the `Parcels` and `RW` layers so that you can easily see them.
15. Save your project.

You have now configured the existing layers in the map so that they will be ready to edit. However, there is another layer you need to add before you start editing: voting districts.

Adding a new layer using a layer file

Follow these steps to add the voting districts layer to your map using a layer file:

1. In the **Catalog** pane, expand the `Folders` folder so that you can see its contents.
2. Expand the `Topologies` folder so that you can see its contents. You should see several folders and files, including `Topologies.gdb`, `Topologies.tbx`, `Voting_Districts.lyrx`, and `Voting Districts.shp`, among others.
3. Right-click on the `Voting_Districts.lyrx` file and select **Add To Current Map**.

> **Note**
> You just added a layer to a map using a layer file. Layer files store configuration settings for a layer so that, when added, they are automatically displayed based on those settings. Layer files can include settings for symbology, labeling, definition queries, scale-visibility ranges, and more. This allows you to standardize layer displays between multiple maps and projects within your organization.

Your map should now look as follows. Your symbology will most likely be different, depending on the colors ArcGIS Pro assigned when you added the layers and any adjustments you made to the symbology. Remember, ArcGIS Pro assigns a random color to newly added layers:

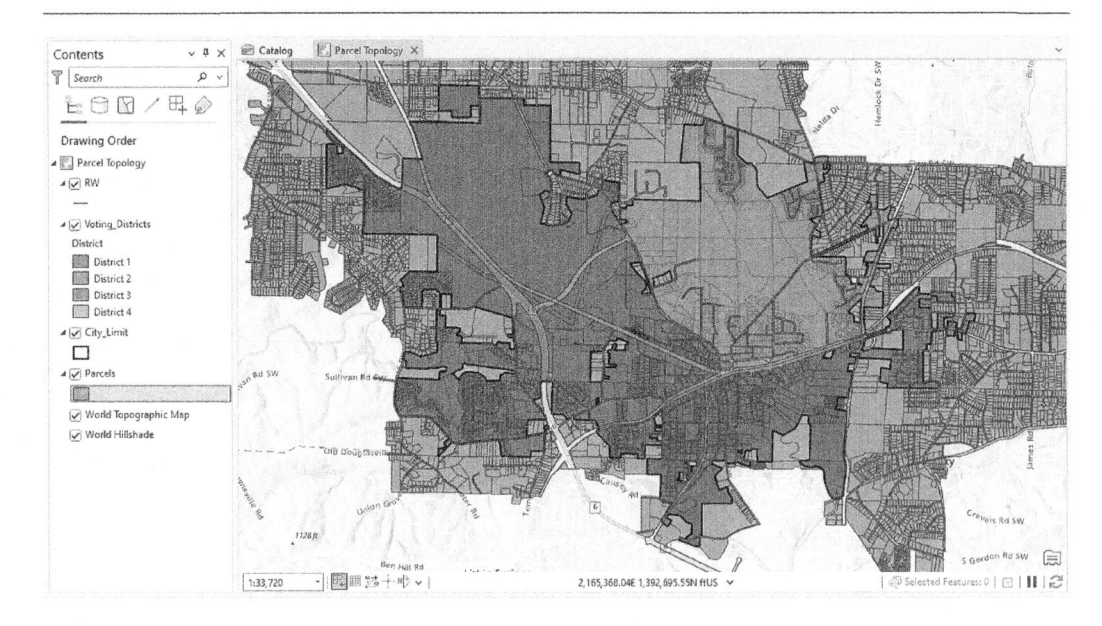

Figure 10.19 – Map with voting districts added

4. Click on the **Map** tab and activate the **Explore** tool.
5. Using the **Explore** tool, zoom into the general area shown in the following screenshot. You do not need to be exact – just try to match the general area:

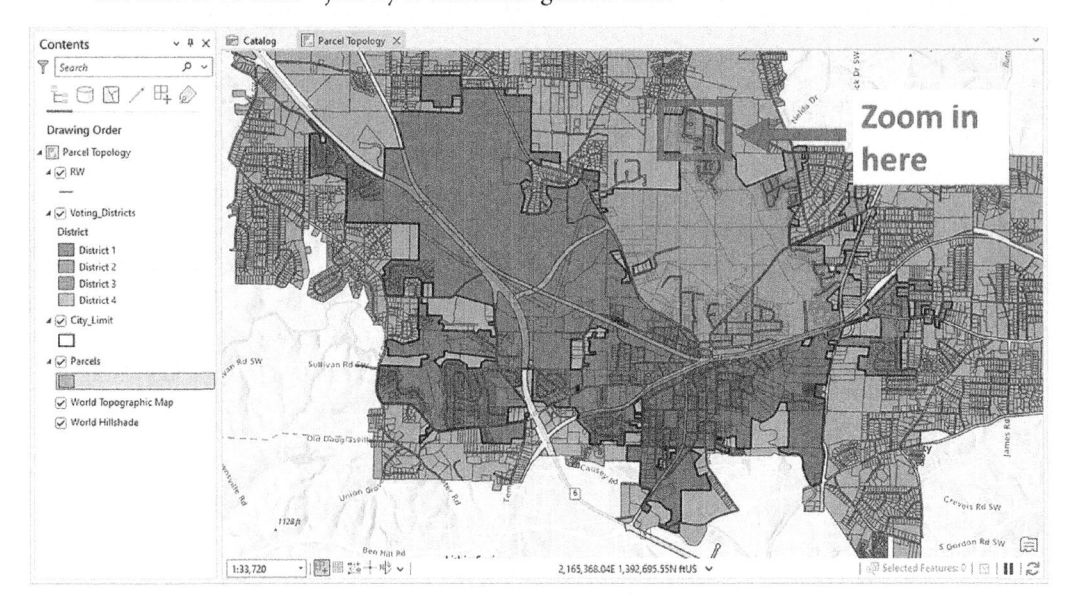

Figure 10.20 – Zooming into the indicated area

6. Save your project to ensure you do not lose any work you have done so far.
7. Click on the **List by Selection** button in the **Contents** pane. Make sure all layers are selectable.
8. Click on the **List by Snapping** button in the **Contents** pane and ensure all layers are snappable.
9. Click on the **List by Editing** button in the **Contents** pane and ensure all layers are editable.

Editing data using a map topology

You are now ready to edit data. You will use a map topology to edit multiple features at one time:

1. Click on the **Edit** tab in the ribbon.
2. Click on the topology drop-down list on the **Edit** tab and select **Map Topology**, as shown in the following screenshot:

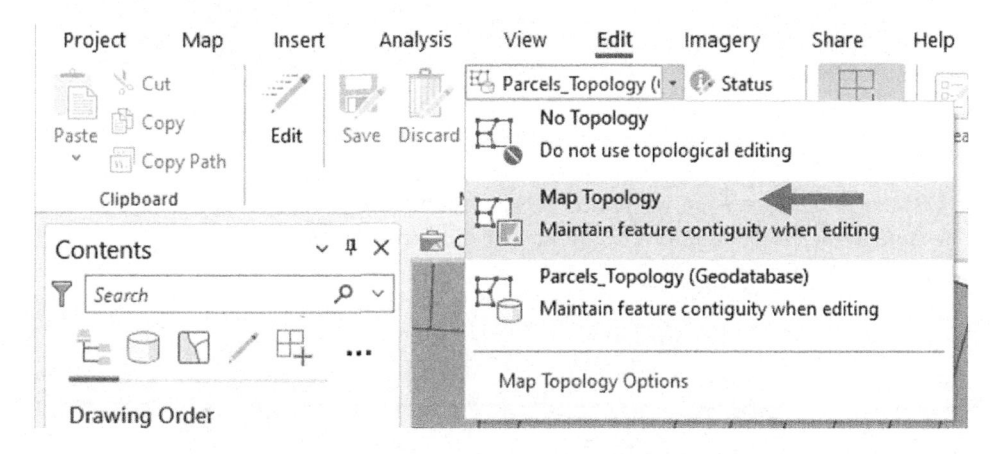

Figure 10.21 – Selecting Map Topology on the Edit tab

You are now ready to begin editing the layers to reflect the new turn lane that has been added to the entrance of Sweetwater Valley Road. This is a new right-turn lane, allowing cars to safely decelerate before turning, and cars continuing down Clay Road to pass turning cars.

3. Click on the **Select** tool in the **Selection** group on the **Edit** tab.

4. On the map, click and draw a selection rectangle, as illustrated in the following screenshot:

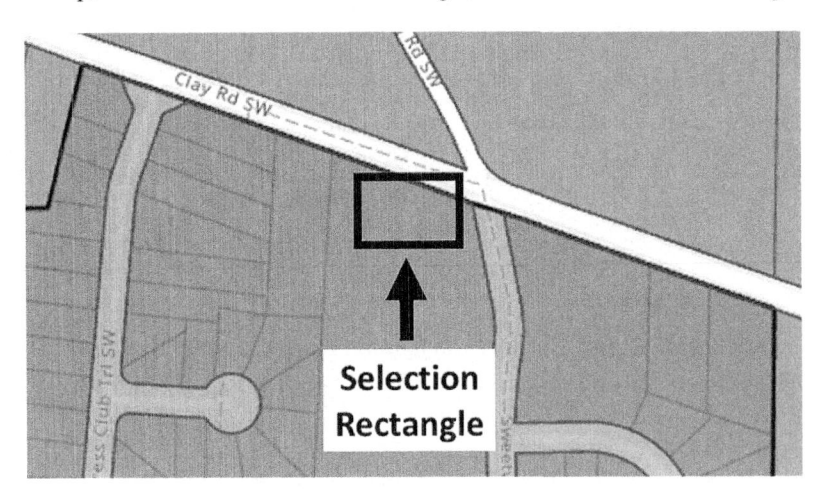

Figure 10.22 – Drawing a selection rectangle

5. Click on the **List by Selection** button in the **Contents** pane. Verify that you have selected one feature in each layer, as illustrated in the following screenshot:

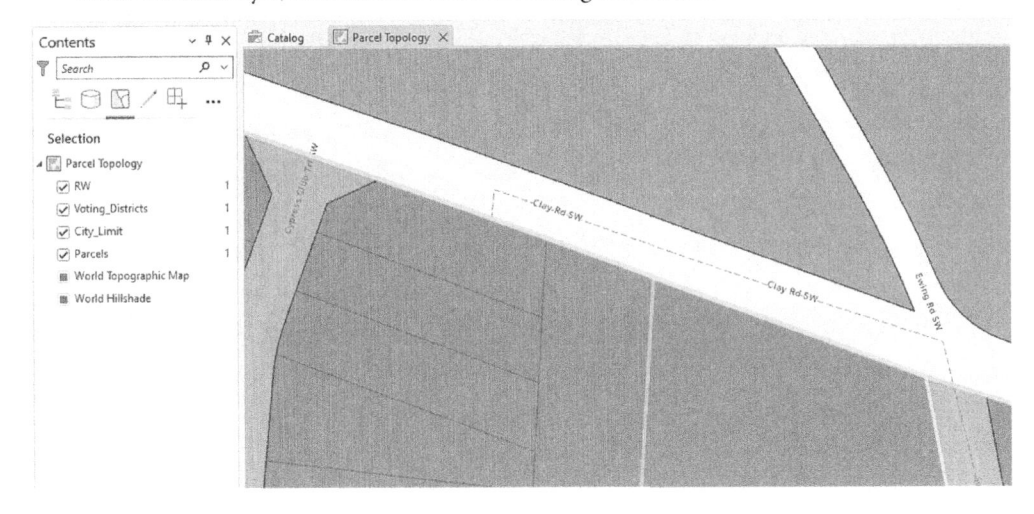

Figure 10.23 – Verifying the selection for the map topology

6. Select the **Reshape** tool in the **Tools** group on the **Edit** tab. This will open the **Modify Features** pane.

7. Click on the northwestern corner of the parcel, located to the west of Sweetwater Valley Road, as illustrated here:

Figure 10.24 – Location to start reshaping features

8. Then, move your mouse pointer in a southerly direction, along the western boundary of the parcel, and right-click. Select **Direction/Distance...** from the context menu that appears, as shown in the following screenshot:

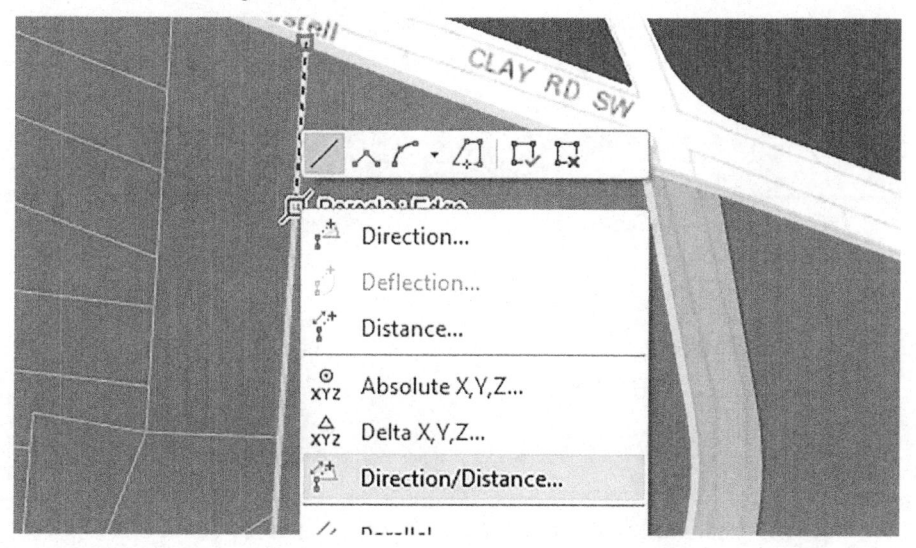

Figure 10.25 – Selecting the Direction/Distance... option to reshape selected features

9. Set **Direction** to S3-42-19W QB and **Distance** to 50.00 FtUS. You may need to press the *Enter* key after entering the distance to apply the values you entered.

360   Validating and Editing Data with Topologies

10. Move your mouse pointer to the northern boundary of the parcel and right-click. Select **Parallel** from the context menu that appears, as illustrated in the following screenshot:

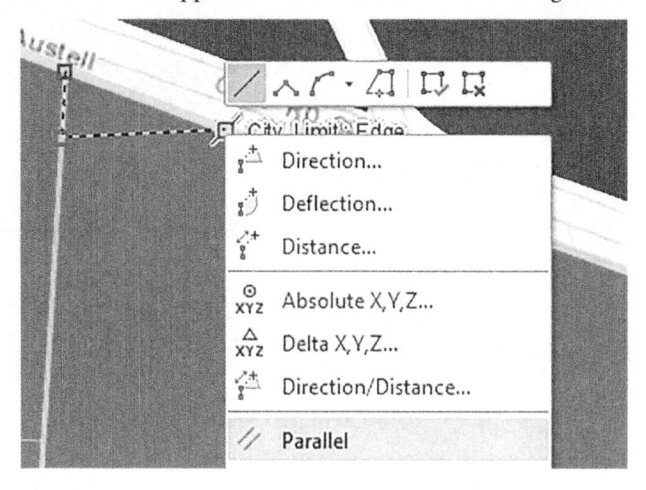

Figure 10.26 – Using a parallel line to reshape selected features

11. Your mouse pointer should now be locked in a direction parallel to the parcel boundary. Move your mouse pointer along that parallel course until it is snapping to the eastern right of way for Sweetwater Valley Road. Click at that location, as shown in the following diagram:

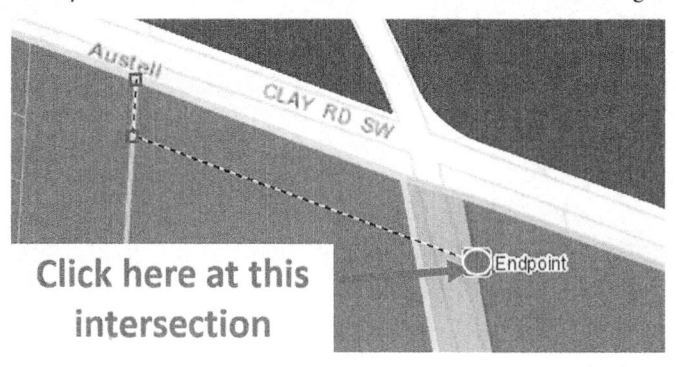

Figure 10.27 – The next location to reshape selected features

12. Move your mouse pointer until it snaps to the intersection of the eastern right of way for Sweetwater Valley Road and the southern right of way for **Clay Rd**, as shown in the following diagram. Double-click when you get to the indicated location to finish your sketch:

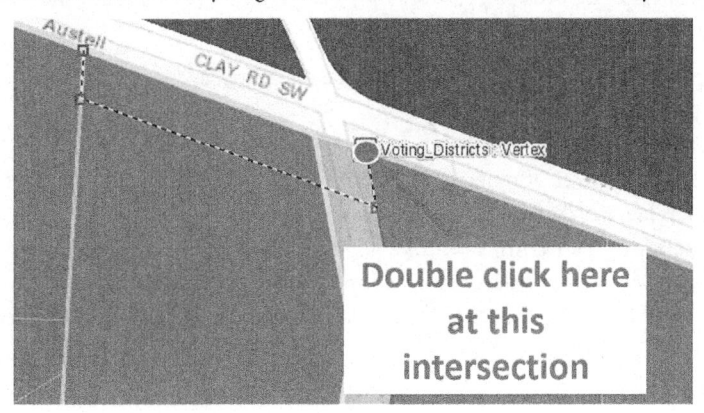

Figure 10.28 – Finishing the sketch to edit selected features using a map topology

Your map should now show the new area for the new turn lane that was installed, as shown in the following screenshot. This new turn lane required you to edit multiple layers. This included all the layers that were on your map. You could have edited each layer individually. However, by using the map topology, you were able to edit features in four different layers with a single edit. This included layers that were stored in different data formats as well.

To learn more about using topologies, go to https://pro.arcgis.com/en/pro-app/latest/help/editing/introduction-to-editing-topology.htm:

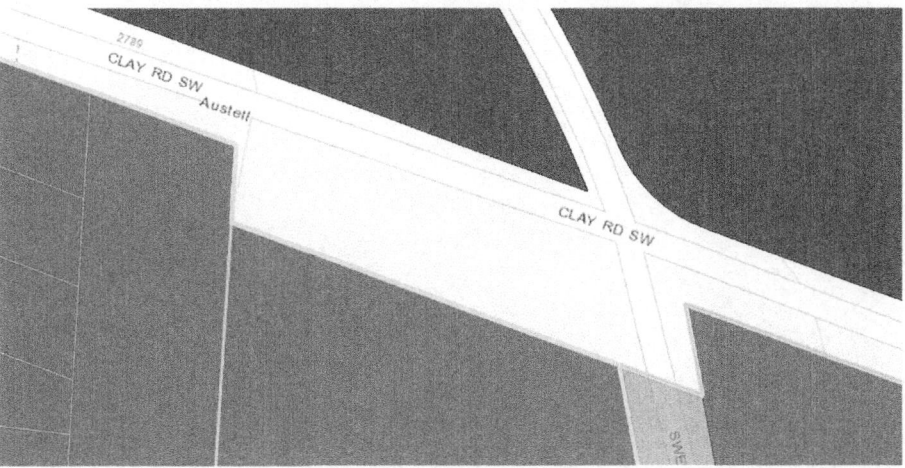

Figure 10.29 – Editing the results using a map topology

13. Once you've verified that your edits were completed successfully, save your edits by clicking on the **Save** button in the **Manage Edits** group on the **Edit** tab.
14. Close the **Modify Features** pane to free up your screen area and reduce the computer resources being used by ArcGIS Pro.
15. Save your project to ensure all your project changes are saved.
16. Close ArcGIS Pro.

You have now seen the power of map topology. It allows you to edit multiple features at one time, regardless of which layer or layers they may be in, so long as they are coincident. This can greatly increase the efficiency of your editing. Unlike geodatabase topologies, you can use map topologies at any license level of ArcGIS Pro. The primary limitation of a map topology is the inability to assign spatial rules and validate your data using those rules, as you can with a geodatabase topology.

How it works...

In this recipe, you were able to edit features in multiple layers with a single edit using a map topology. To do this, you needed to ensure all the layers that needed to be updated were added to your map. Then, you made sure they were all editable, selectable, and snappable. Once you had done that, you went to the **Edit** tab and chose to use a map topology from the topology drop-down list. By selecting **Map Topology**, ArcGIS Pro allowed you to use topology editing tools and techniques to simultaneously edit multiple features in different layers so that they maintained their existing spatial relationships. In this recipe, you reshaped features in the `RW`, `Voting_Districts`, `City_Limits`, and `Parcels` layers all at the same time.

11
Converting Data

As you may have noticed by now, GIS data comes in many formats. There are geodatabases, shapefiles, **Computer Aided Drafting (CAD)** files, DBF files, rasters, spreadsheets, and more. ArcGIS Pro allows you to use all of these. However, there are limits to what ArcGIS Pro allows you to do with some of these formats. Some of these are read-only, such as CAD files or spreadsheets. Others you can read and edit, such as shapefiles and geodatabases. Also, if you have data spread across multiple locations, formats, and files, it can make it difficult to use.

If you are going to be editing and analyzing data, it is recommended that you consolidate all the required data into a single format. For ArcGIS Pro, the primary storage format is a geodatabase. Shapefiles are generally considered a secondary option. To consolidate your data sources, you may need to convert and combine data from one or more sources into a single source. ArcGIS Pro provides several methods and tools to accomplish this. Which one works best will depend on several factors, including the following:

- The existing data format
- The target data format
- How much data needs to be converted and the number of files
- Are you performing a wholesale conversion, or do you need to select specific features?
- Are you trying to combine multiple files into one?
- Are you importing into the GIS or exporting out to use in another application?

All these considerations will help to determine which conversion method and tools will work best for your situation.

In this chapter, you will learn several different methods and tools that will allow you to convert and combine data within ArcGIS Pro. You will work through the following recipes:

- Converting shapefiles to a geodatabase feature class
- Merging multiple shapefiles into a single geodatabase feature class

- Exporting tabular data to an Excel spreadsheet
- Importing an Excel spreadsheet into ArcGIS Pro
- Importing selected features into an existing layer

Converting shapefiles to a geodatabase feature class

Shapefiles are one of the most common GIS data formats you will encounter during your career in GIS. Just about every GIS-enabled application can read, import, export, and edit a shapefile. This includes applications such as QGIS, AutoCAD Map 3D, Civil 3D, GeoMedia, and many others. This has resulted in shapefiles becoming a common format for sharing GIS data with multiple platforms. It is not uncommon to receive data from engineering, planning, and surveying firms in a shapefile format.

In this recipe, you will receive a shapefile containing the location of several stormwater drainage structures, which were collected by a local surveying firm. You will integrate this shapefile into a geodatabase for the City of Trippville.

Getting ready

This recipe does require the sample data to be installed on the computer. While it is not required to have completed any previous recipes, it is recommended that you at least review *Chapter 1, ArcGIS Pro Capabilities and Terminology*, to ensure you have the proper skills required to successfully complete this recipe. You can complete this recipe with any ArcGIS Pro licensing level.

How to do it...

Now, you will work through the steps needed to import a shapefile into a geodatabase as a new feature class:

1. You will need to launch ArcGIS Pro to begin this recipe.
2. Open the `Convert Single Shapefile` project located in the `C:\Student\ArcGISPro3Cookbook\Chapter11\Convert Single Shapefile` folder.

    The project will open with the **Trippville Stormwater** map visible. This is an existing map that was obviously created by someone else. You will notice if you look at the **Contents** pane that the map does not contain any stormwater layers. You will add the shapefile you have received from the local surveying firm.

3. In the **Catalog** pane, click on the small arrowhead located next to the `Folders` folder so that you can see its contents. Then, expand the `Convert Single Shapefile` folder so that you can see its contents.
4. You should see the `Stormwater_Structures.shp` file in the folder you just expanded. Before you use this file, you should examine its metadata to ensure it is the one you are looking for.

5. Right-click on the `Stormwater_structures.shp` file and select **View Metadata** from the menu that appears, as shown in the following screenshot:

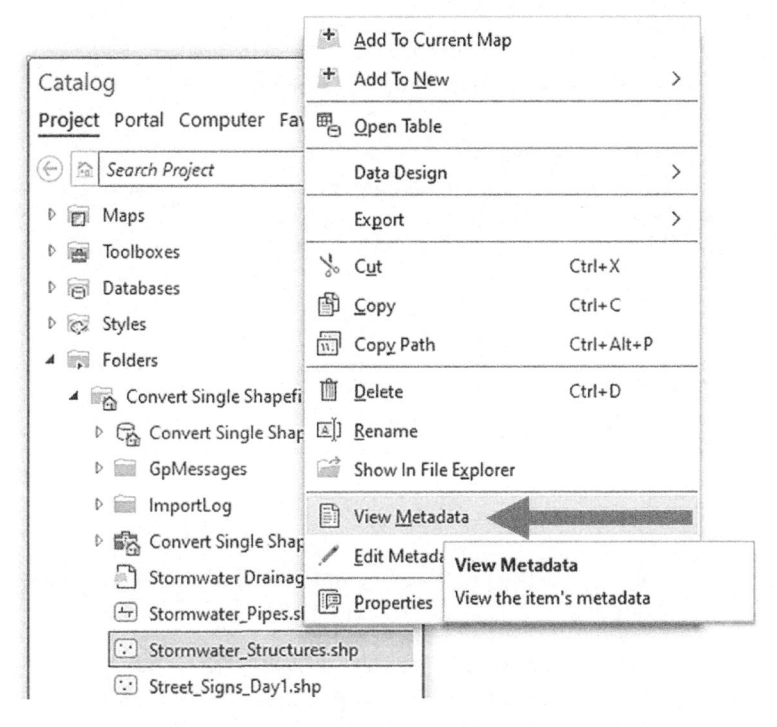

Figure 11.1 – Viewing metadata for a shapefile

> **Tip**
> ArcGIS Pro uses different-colored icons to identify data formats. Shapefiles are shown with a green icon, CAD files with a blue icon, and geodatabases with a gray icon.

The **Catalog** view should open, displaying the metadata for the shapefile you selected. This will provide you with basic information concerning the shapefile.

> **Note**
> Metadata is data about your data. It provides information about how the data was created, when it was created, how often it is updated, what its intended purpose is, and more. It is considered best practice to create and maintain metadata for all GIS datasets. However, it is not uncommon to find data without metadata. This is especially true of shapefiles since they must be manually created and maintained. ArcGIS will automatically create at least some metadata for geodatabase feature classes and tables.

6. Take a moment to read over the metadata for the shapefile. When done, close the **Catalog** view by clicking on the **x** symbol in the tab entitled **Catalog** at the top of the view area. This should return you to the map.

7. Now, it is time to add the shapefile to the map. Right-click on the `Stormwater_Structures.shp` file and select **Add to Current Map** from the menu that appears, as illustrated here:

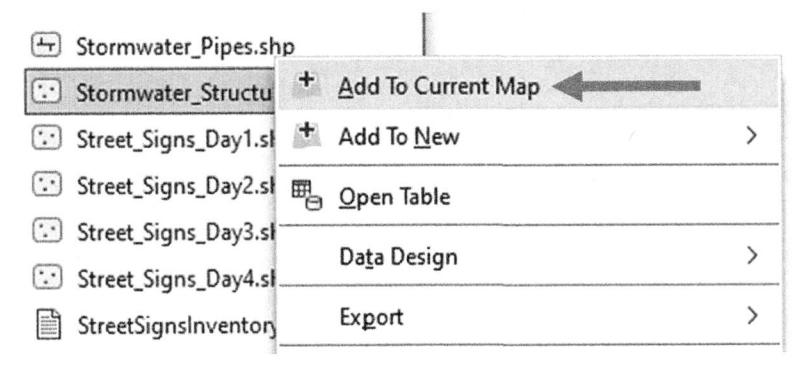

Figure 11.2 – Adding a shapefile to the current map

The `Stormwater_Structures` layer should now appear in your **Contents** pane. Next, you will review the attribute data associated with the layer you just added.

8. Right-click on the `Stormwater_Structures` layer in the **Contents** pane. Select **Attribute Table** from the menu that appears. This will open the attribute table for the shapefile below the map view unless you have moved the table window to another location. Then, answer the following questions:

Question: What fields are located in the attribute table for the shapefile?

Answer:

Question: How many structures/features are stored in the shapefile?

Answer:

The table window is dockable and undockable, so, you can move it wherever you wish, including to another monitor. Unlike ArcMap, ArcGIS Pro allows you to open each table in its own window, providing greater viewing flexibility.

You should see only four fields in the attribute table: `FID`, `Shape`, `Id`, and `Typ`. `FID` and `Shape` are default fields for a shapefile. The `Id` and `Typ` are **user-defined fields** (**UDFs**). Default fields are created and maintained by the software that created the data. UDFs are those created and maintained by users. The default fields and UDFs will vary from application to application and format to format.

The `Typ` field identifies the type of structure. **CB** is a **catch basin**, **AD** is an **area drain**, and **DI** is a **drop inlet**. These are all common stormwater drainage structures.

After viewing the metadata and the attribute fields, you now have a better understanding of the data you will be converting. It is always good to make sure you know the data you are working with before you start using it to perform analysis or create maps.

9. Close the table window by clicking on the **x** symbol in the tab labeled `Stormwater_Structures`. Now, the map should fill the entire view area once more.
10. In the **Catalog** pane, expand the `Databases` folder so that you can see its contents.
11. Expand the `Trippville_GIS.gdb` geodatabase by clicking on the small arrowhead located to the left of the database name.
12. Look at the contents of the geodatabase. You should notice that there are three feature datasets included: `Base`, `Sewer`, and `Water`.

> **Note**
>
> Feature datasets serve several purposes within a geodatabase. First, they help you better organize related data. Second, they ensure all data stored within them shares a common spatial reference, which means they are all in the same coordinate system. This allows the data to take part in topologies and networks that are used to ensure data quality and provide additional functionality.

You will notice there is not one for stormwater.

Creating a new feature dataset

Before you convert the `Stormwater_Structures` shapefile into a geodatabase feature class, you will create a new feature dataset for the stormwater system. This will allow you to easily find the data in the geodatabase later and allow you to create a utility network in the future if you need to:

1. In the **Catalog** pane, right-click on the `Trippville_GIS.gdb` geodatabase. In the menu that appears, go to **New** and select **Feature Dataset**, as illustrated in the following screenshot. This will open the **Create Feature Dataset** tool in the **Geoprocessing** pane:

368   Converting Data

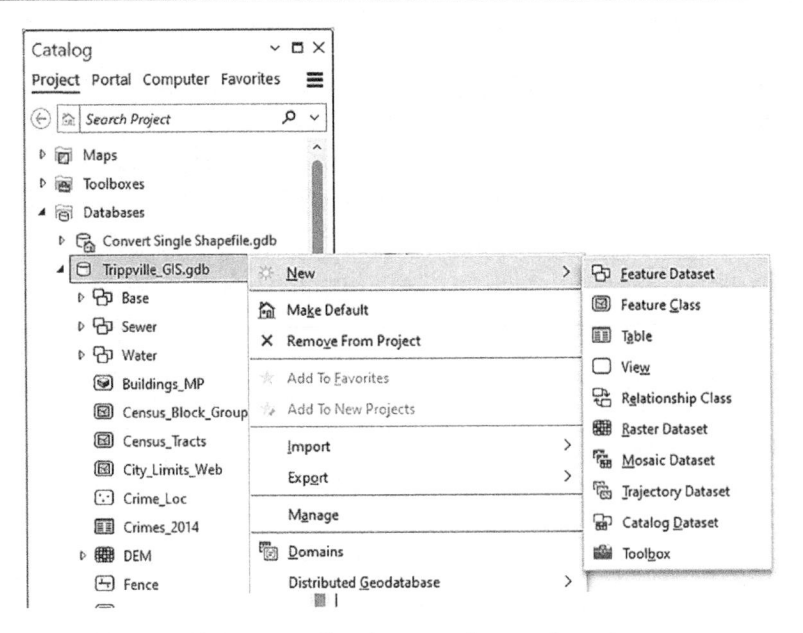

Figure 11.3 – Creating a new feature dataset

2. The **Output Geodatabase** parameter should automatically be set to `Trippville_GIS.gdb` because it was the one you selected and right-clicked on. For the **Feature Dataset Name** parameter, type `Stormwater`.

3. Under the **Coordinate System** option, click on the small drop-down arrow located on the far right side of the cell. Select **Street Rights of Way** from the list, as shown here:

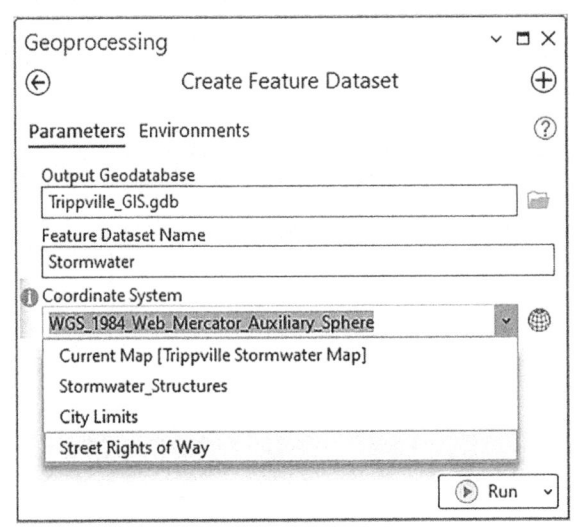

Figure 11.4 – Selecting the coordinate system for the new feature dataset

The coordinate system should now be set to **NAD_1983_StatePlane_Georgia_West_FIPS_1002_Feet**, which is the coordinate system the `Street Rights of Way` layer is in.

4. Verify that your **Create Feature Dataset** tool looks like the following screenshot and click **Run** if it does:

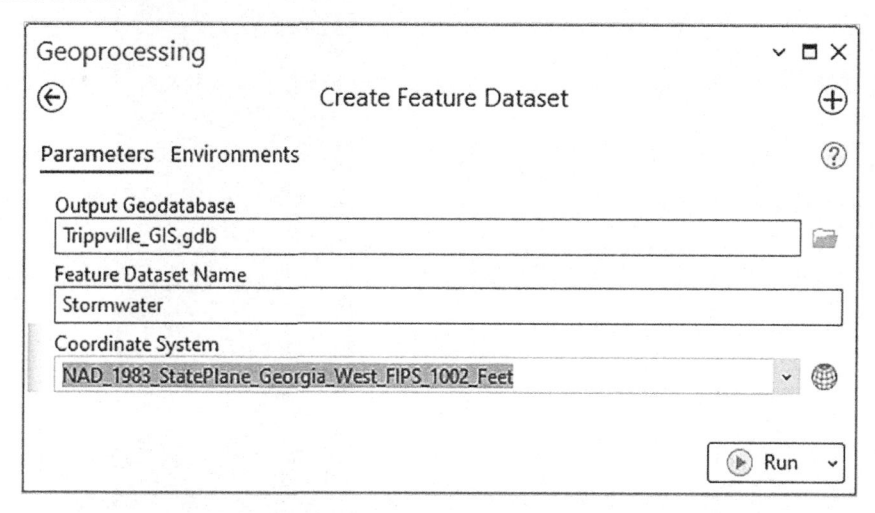

Figure 11.5 – Create Feature Dataset tool with all parameters defined

When the tool finishes, you should see the new feature dataset that you just created appear in the `Trippville_GIS.gdb` geodatabase. You will convert the shapefile containing the stormwater structures to a new feature class that will be stored in the feature dataset you just created.

5. Close the **Geoprocessing** pane once the **Create Features Dataset** tool finishes and the new feature dataset has been created.

You now have created a feature dataset that will house stormwater data in the geodatabase.

Exporting the shapefile to the geodatabase feature class

Next, you will export the shapefile data into the feature dataset you just created. This will add that data to the geodatabase so that it is accessible just like all other utilities.

1. In the **Contents** pane, right-click on the `Stormwater_Structures` layer. In the menu that appears, go to **Data** and select the **Export Features** option, as illustrated in the following screenshot. This will open the **Export Features** tool:

370　Converting Data

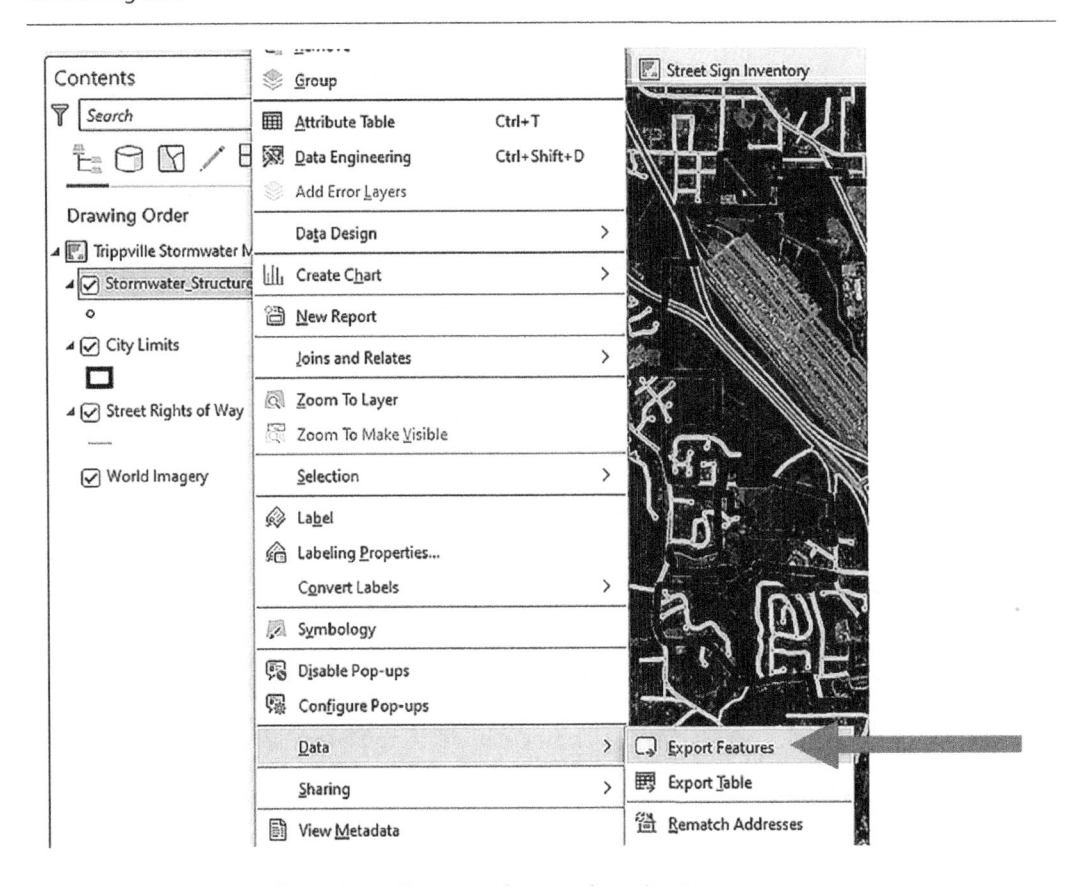

Figure 11.6 – Exporting features from the Contents pane

2. The **Input Features** parameter should automatically be set because you right-clicked on a specific layer to access the tool. Click the **Browse** button located on the far right side of the **Output Feature Class** parameter.

3. In the **Output Feature Class** window that opens, select the Databases folder located under Project in the left panel of the window. Double-click on the Trippville_GIS geodatabase in the right panel of the window.

4. Double-click on the Stormwater feature dataset you created earlier.

5. In the **Name** cell located near the bottom of the **Output Feature Class** window, type Stormwater_Structures and click **Save**.

6. Verify that the **Export Features** tool looks like the following screenshot and click the **OK** button:

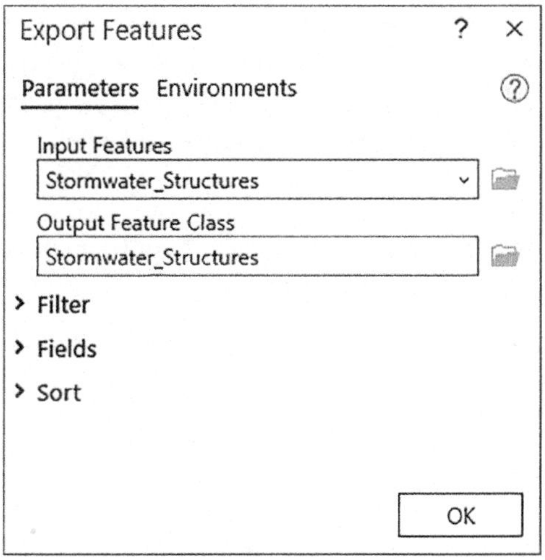

Figure 11.7 – Export Features tool with parameters completely filled in

If the tool ran successfully, a new layer should appear on your map. Now, you will verify that ArcGIS Pro converted the shapefile successfully.

7. Close the **Geoprocessing** pane by clicking on the small **x** symbol located in the upper-right corner of the pane.

8. In the **Contents** pane, click on the **List by Source** button. It should be the second one from the left and looks like a cylinder.

9. You should see that the map now contains two `Stormwater_Structures` layers. Pay attention to the location where these two layers are stored. Then, answer the following question:

    Question: Where are the two `Stormwater_Structures` layers stored?

    Answer:

    As you can see, one of the `Stormwater_Structures` layers is stored in the `Trippville_GIS` geodatabase. The other is stored in the `Convert Single Shapefile` folder. The one in the geodatabase is the one you just converted. The other is the original shapefile. The **Export Features** tool created a new feature class in the geodatabase and left the source shapefile alone. Now that you have verified that it did create the feature class in the geodatabase, you need to verify that it converted all the features.

10. Right-click on the `Stormwater_Structures` layer, which is stored in the `Trippville_GIS.gdb` geodatabase, and select **Attribute Table** from the menu that appears.

11. Look at the bottom of the table window to see the total number of records/features contained in the layer, then answer the following question:

    Question: Does it match the number of features that were in the shapefile? (Look back to earlier in this recipe when you opened the attribute table for the shapefile and answered the question of how many features were stored in it if you don't remember.)

    Answer:

    Both the shapefile and geodatabase feature class should contain the same number of records/features if the **Export Features** tool worked correctly.

12. Right-click on the `Stormwater_Structures` shapefile layer in the **Contents** pane and select **Remove** from the menu that appears. You no longer need the shapefile layer in your map.

13. Click on the **Save Project** button located in the **Quick Access** menu at the top left of the ArcGIS Pro interface.

You have now successfully imported a shapefile into a geodatabase as a new feature class.

How it works...

In this recipe, you converted two different shapefiles that represented features that were part of the stormwater collection system in the City of Trippville. You did this using two different methods.

In the first method, you started by creating a new feature dataset in the `Trippville_GIS` geodatabase to contain the stormwater data you needed to convert. Then, you added the shapefile containing the `Stormwater_Structures` layer to your map. Next, you right-clicked on that layer and opened the **Export Features** tool from the menu. With the tool open, you specified the new feature class you would create when you ran the tool. Lastly, you clicked **OK** to run the tool. The **Feature Export** tool created a new feature class in the `Trippville_GIS` geodatabase and the `Stormwater` feature dataset. This new feature class was named `Stormwater_Structures`, which you specified and contained all the data that was originally the shapefile.

The second method made use of the **Feature Class to Geodatabase** tool. You accessed this tool by right-clicking on the `Stormwater` feature dataset in the `Trippville_GIS` geodatabase and selecting the **Import** and **Feature Class(es)** options from the menu. This opened the **Feature Class to Geodatabase** conversion tool. You then specified the input shapefile you wished to import into the geodatabase and ran the tool. This imported the shapefile into the geodatabase as a new feature class with the same name as the shapefile.

There's more...

As with most things you can do in ArcGIS Pro, there is more than one way to accomplish a task. Converting a shapefile to a geodatabase feature class is no exception. For example, you could have accessed the **Export Features** tool directly from the **Conversion Tools** toolbox and the **To Geodatabase** toolset in the **Geoprocessing** pane.

There is also a third way that you can convert shapefiles to a geodatabase feature class. You can use the **Feature Class to Geodatabase** geoprocessing tool as well. Let's look at how that works.

You have received a shapefile containing stormwater drainage pipes from the same local surveying firm. You want to convert those into the feature dataset that you created earlier in this recipe:

1. In the **Catalog** pane, expand the contents of the `Trippville_GIS.gdb` geodatabase if needed. It may still be expanded from your previous work.
2. Right-click on the `Stormwater` feature dataset. Select **Import** and **Feature Class(es)**. This will open the **Feature Class to Geodatabase** conversion tool in the **Geoprocessing** pane:

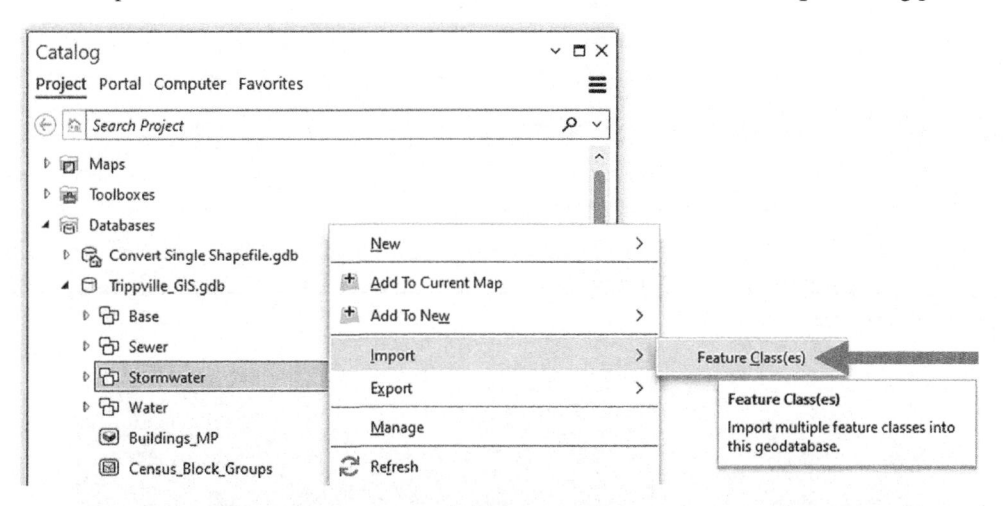

Figure 11.8 – Accessing the Import menu and Feature Class(es) option

The **Output Geodatabase** parameter should already be set in the tool to `Stormwater`.

3. Click on the **Browse** button at the far right end of the **Input Features** parameter.
4. In the **Input Features** window, select the `Folders` folder in the left panel under **Project**. A `Convert Single Shapefile` folder should appear in the right panel of the window. Double-click on the `Convert Single Shapefile` folder.
5. Click on `Stormwater_Pipes.shp` and click **OK**.

6. Verify that your **Feature Class to Geodatabase** tool looks like the following screenshot and click **Run**:

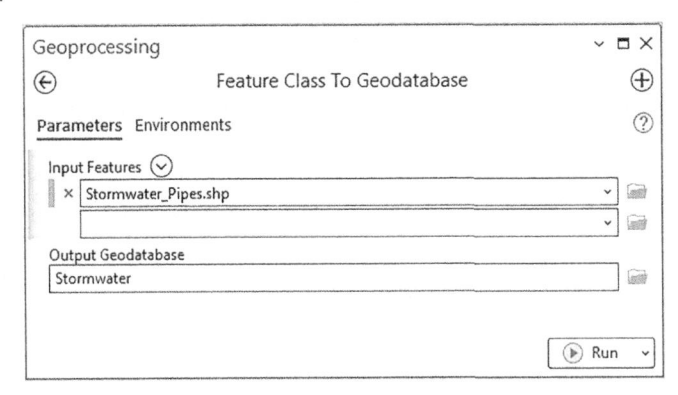

Figure 11.9 – Feature Class to Geodatabase conversion tool

If the tool is successful, a new `Stormwater_Pipes` feature class should appear in the `Stormwater` feature dataset in the `Trippville_GIS` geodatabase. Next, you will add the new feature class to your map so that you can view the data you just converted.

7. Close the **Geoprocessing** pane containing the **Feature Class to Geodatabase** tool.

8. In the **Catalog** pane, right-click on the `Stormwater_Pipes` feature class you just created and select **Add to Current Map** from the menu that appears.

9. A new `Stormwater_Pipes` layer should appear on your map. Right-click on it in the **Contents** pane and select **Zoom to Layer** so that you can see the data better, as shown in the following screenshot:

Figure 11.10 – Stormwater pipes shown on the map

You can now see the stormwater pipes on the map, so they have been successfully imported into the `Trippville_GIS` geodatabase from the original shapefile provided by the surveying company.

> **Tip**
> You may need to adjust the symbology for the new `Stormwater_Pipes` layer so that you can see it. Remember — ArcGIS Pro randomly assigns symbology to new layers when they are added to the map.

10. Save your project and close ArcGIS Pro.

So, you have now seen two different tools you can use to convert shapefiles into geodatabase feature classes. These tools are not just limited to converting shapefiles. They can also convert other spatial data formats, including DWGs, DGNs, DXFs, and more as well.

Merging multiple shapefiles into a single geodatabase feature class

You now know how to convert a single shapefile into an individual geodatabase feature class. What if you have several shapefiles that you need to merge into a single common geodatabase feature class? Can you do that in ArcGIS Pro?

Of course you can. This is not an uncommon workflow. You often need to combine multiple shapefiles or geodatabase feature classes into a single feature class. Combining multiple sources into one makes working with the data easier. For example, you are working on a regional transportation study, and you receive road data from three counties. If you wished to analyze traffic flow or calculate the total length of roads by type for the region, it would be much easier to do so if all the data were in a single feature class. Maybe you are responsible for maintaining data for a 911 center that services both the city and county. The city and county assign addresses for new construction and update the address point file for their respective jurisdiction. You would need to combine these two files for use in the 911 software. These are just two possible examples of situations where you would want to combine multiple shapefiles or geodatabase feature classes into one.

In this recipe, you will merge three different shapefiles into a single common geodatabase feature class. You have had three field crews out collecting the location of street signs as part of a sign inventory the Roads Superintendent asked you to create. Each crew has collected the sign location data into a shapefile. So, you now have three separate shapefiles that need to be merged into a single geodatabase feature class. This will make the data much easier to manage and work with in the future.

Getting ready

This recipe does require the sample data to be installed on the computer. You do not need to complete any previous recipes, though it is recommended that you review *Chapter 1, ArcGIS Pro Capabilities and Terminology*, plus the previous recipe in this chapter, to ensure that you have a basic understanding of the skills required to complete this recipe. This recipe can be completed with any ArcGIS Pro licensing level.

How to do it...

Now, you will merge three different shapefiles into a single geodatabase feature class using the **Merge** geoprocessing tool:

1. Start ArcGIS Pro and open the `Convert Multiple Shapefiles.aprx` project located in `C:\Student\ArcGISProCookbook\Chapter11` using the skills you have learned in past recipes.

   The project should open with a **Street Sign Inventory** map. If you look in the **Contents** pane, you will see three `Street_Signs` layers. These are the three shapefiles collected by your field crews. Take a moment to investigate these three layers so that you know what you have to work with.

2. Right-click on the `Street_Signs_Day1` layer in the **Contents** pane and select **Attribute Table** from the menu that appears. Take a moment to investigate the attribute table and then answer the following questions:

   Question: What type of shapefile is this layer?

   Answer:

   Question: Which attribute fields does it contain?

   Answer:

3. Repeat this process for the other two `Street_Signs` layers. Compare the answers to the previous two questions with each of these layers. Note any similarities or differences.

   As you have just found out, all three layers are point features and have a common attribute table structure or schema. While a common schema is not always required when merging multiple fields together, it does often help. One thing that must match to merge multiple layers together is the feature type. You can only merge points with points, lines with lines, and polygons with polygons.

Merging data

Now that you have verified that the feature types match and the attribute fields are the same, you are ready to merge the three layers together to create a single one:

Merging multiple shapefiles into a single geodatabase feature class   377

1. Close the attribute tables for the three `Street_Signs` layers. This will free up some screen space.
2. Click on the **Analysis** tab in the ribbon.
3. Click on the **Tools** button in the **Geoprocessing** group in the **Analysis** tab. This will open the **Geoprocessing** pane.
4. Click on the **Toolboxes** tab in the **Geoprocessing** pane, as illustrated:

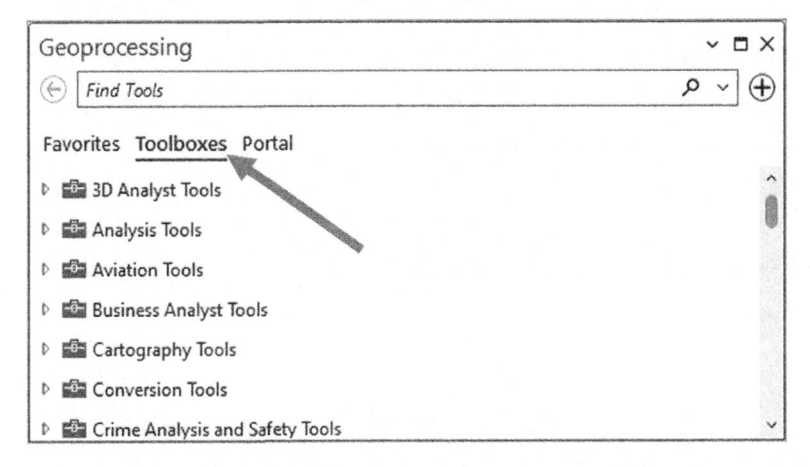

Figure 11.11 – Selecting Toolboxes in the Geoprocessing pane

5. In the **Search** bar at the top of the **Geoprocessing** page, type **Merge** and press your *Enter* key to locate the **Merge** tool.
6. Select **Merge (Data Management Tools)** from the list that is presented. It should be near the top, as shown in the following screenshot:

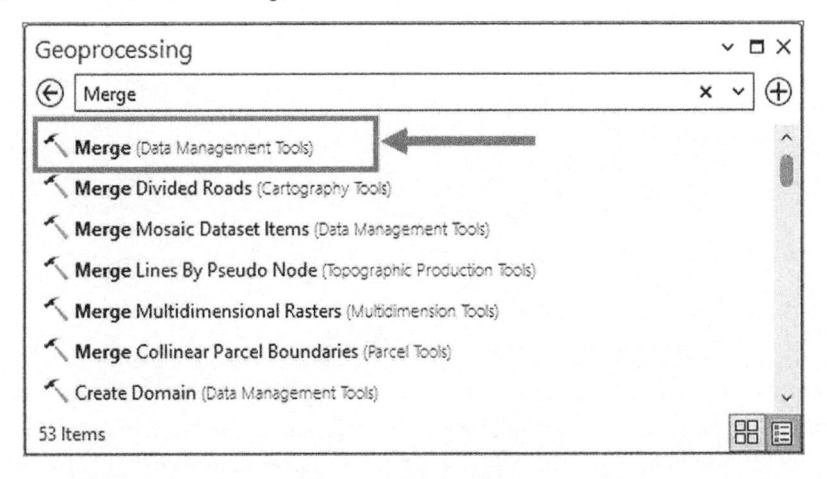

Figure 11.12 – Searching for and selecting the Merge geoprocessing tool

The **Merge** geoprocessing tool should now be open in the **Geoprocessing** pane. This is the tool you will use to combine the three individual shapefiles.

7. Under **Input Datasets**, click on the small drop-down arrow and select `Street_Signs_Day1`. When you do that, another input box should appear below it, allowing you to add another input.
8. Click on the small drop-down arrow at the end of the second input box that appeared and select `Street_Signs_Day2`.
9. Repeat that process to add `Street_Signs_Day3` as the last input for the **Merge** tool.
10. Click the **Browse** button on the right side of the **Output Dataset** box.
11. In the **Output Dataset** window that appears, select **Databases** from the panel on the left side of the window.
12. Double-click on the `Trippville_GIS` geodatabase in the panel on the right side of the window.
13. In the **Name** cell, type `StreetSigns` and click **Save**.
14. Verify that your **Merge** tool looks like the following screenshot. If it does, click **Run**:

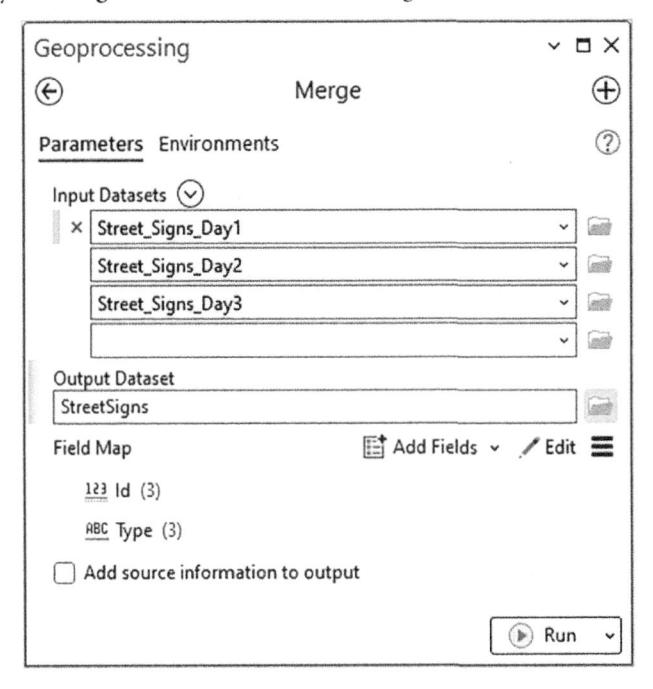

Figure 11.13 – Merge geoprocessing tool with parameters filled in.

When the tool runs successfully, you should see a new `StreetSigns` layer added to your map. This new layer should cover all the points in the original three layers, but notice they still exist.

Verifying results

The **Merge** tool, as with most geoprocessing tools, creates new data and does not alter the original input data. But let's verify that just to be sure.

1. Click on the **List by Source** tab located in the **Contents** pane. It is the button that looks like a gray cylinder.
2. Locate all `StreetSigns` layers that are visible on the map. Note where their source is stored and answer the following questions:

    Question: Where is the `StreetSigns` layer that you just created being stored?

    Answer:

    Question: Where are the three `Street_Signs` shapefiles being stored?

    Answer:

    Question: Are these the same?

    Answer:

    As you can now see, the `StreetSigns` layer is referencing a geodatabase feature class that is being stored in the `Trippville_GIS` geodatabase, whereas the shapefiles for the `Street_Signs` shapefiles are stored in the `Convert Single Shapefile` folder. These are two very different locations. Because we don't see a red exclamation point located beside any of the layer names, we know the data sources still exist and none have been deleted.

    > **Tip**
    > Any time that you see a red exclamation point located just before the layer name in the **Contents** pane, you know that layer has lost the connection to its source data.

3. Right-click on each of the three `Street_Signs` layers and select **Remove** from the menu that appears. This removes them from the map to reduce clutter and confusion. Removing a layer from the map does not delete the source data.
4. Right-click on the `StreetSigns` layer and select **Attribute Table** from the menu that appears.
5. Review the attribute table and compare it with what you saw when looking at the attribute tables for the shapefiles. Then, answer the following question:

    Question: How does the attribute table for the `StreetSigns` layer compare to the ones for the three shapefiles?

    Answer:

6. Save your project.

As you can see, the attribute table for the new `StreetSigns` layer contains the same exact fields as the original three shapefiles. Also, the number of records equals the sum of the total records from the three shapefiles. The **Merge** geoprocessing tool combined the content, both spatial and attributes, of the three shapefiles into a single geodatabase feature class.

How it works...

In this recipe, you used the **Merge** geoprocessing tool to combine the contents of several shapefiles into a single geodatabase feature class. To do this, you first examined the contents and attributes of the shapefiles you wished to combine that represented several different days of collection by field crews. Once you determined that the shapefiles had a similar structure and information, you went to the **Analysis** tab in the ArcGIS Pro ribbon so that you could access the geoprocessing tools. Next, you clicked on the **Tools** button to open the **Geoprocessing** pane. There, you clicked on the **Toolboxes** tab and used the search function to locate the **Merge** tool.

With the **Merge** tool now open, you filled in the required parameters, including listing the three shapefiles representing the three days of field collection as inputs to the tool. You also designated the name of the new geodatabase feature class that the tool created when it ran. Once the tool was complete and the new feature class generated, you took time to review the results to verify it did include all the data from the three input shapefiles.

There's more...

The **Merge** geoprocessing tool works great if you want to combine multiple files or feature classes into a new single feature class. What if you want to add new data from a shapefile or geodatabase feature class to an existing layer of information? For example, your field crews go out and locate additional street signs. You would not want to create another layer. Instead, you would want to add the new signs to the existing layer. How would you do that?

This is where the **Append** geoprocessing tool comes into play. The **Append** tool will add new records from an input table, file, or feature class to an existing one. Let's see it in action so that you can gain a better understanding of how it works. You will append a new shapefile that contains additional sign locations to the `StreetSigns` layer you just created:

1. Click on the **Map** tab in the ribbon.
2. Click on the **Add Data** button in the **Layer** group on the **Map** tab.
3. Click on the `Folders` folder located under **Project** in the panel on the left side of the **Add Data** window.
4. Double-click on the `Convert Single Shapefile` folder in the right-side panel.
5. Select the `Street_Signs_Day4.shp` file and click **OK** to add the shapefile as a layer to the map.

6. Right-click on the `Street_Signs_Day4` layer you just added to your map, and select **Zoom to Layer** from the menu that is presented. This will allow you to more easily see the layer you just added. As you can see, this layer contains data for more signs located on the west side of the city.
7. Right-click on the `Street_Signs_Day4` layer and select **Attribute Table** to open the table for that layer.

> **Tip**
> You can also press the *Ctrl* and *T* keys at the same time to open the attribute table for the selected layer in the **Contents** pane.

8. Review the attribute table contents and note the number of features/records in the table. Use that information to answer the following question:

   Question: How many records/features does the `Street_Signs_Day4` shapefile contain?

   Answer:

9. Open the attribute table for the `StreetSigns` layer and note the number of records/features it contains. Use that information to answer the following question:

   Question: How many records/features does the `StreetSigns` layer contain?

   Answer:

10. Close both attribute tables you have opened.
11. Save your project.

Running the Append tool

You are now ready to run the **Append** tool to add the additional data to the existing `StreetSigns` layer you created using the **Merge** tool previously.

1. Click on the **Analysis** tab in the ribbon.
2. Click on the **Tools** button located in the **Geoprocessing** group on the **Analysis** tab.
3. In the **Geoprocessing** pane that appears, click on the **Toolboxes** tab, as shown in the following screenshot:

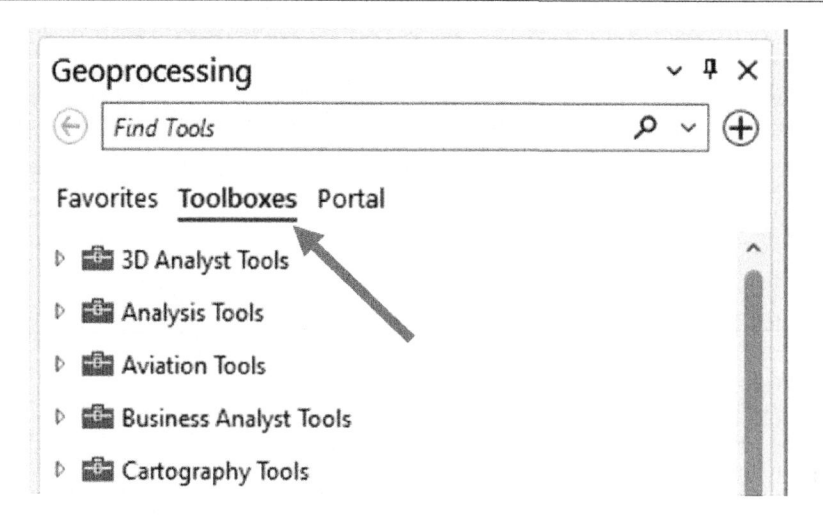

Figure 11.14 – Geoprocessing pane and selecting Toolboxes

4. Locate and expand the **Data Management Tools** toolbox.
5. Expand the **General** toolset, as shown in the following screenshot, and locate the **Append** tool:

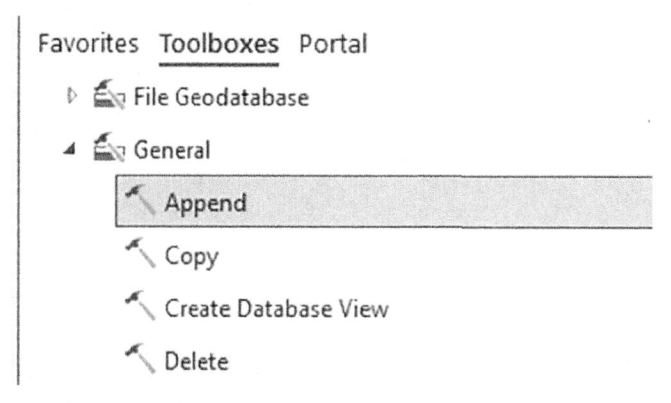

Figure 11.15 – Locating the Append tool in the General toolset

6. Click on the **Append** tool to open it in the **Geoprocessing** pane.
7. Click on the small drop-down arrow located to the right of the **Input Datasets** cell and select Street_Signs_Day4.
8. Click on the small drop-down arrow located to the right of the **Target Dataset** cell and select StreetSigns from the list presented.

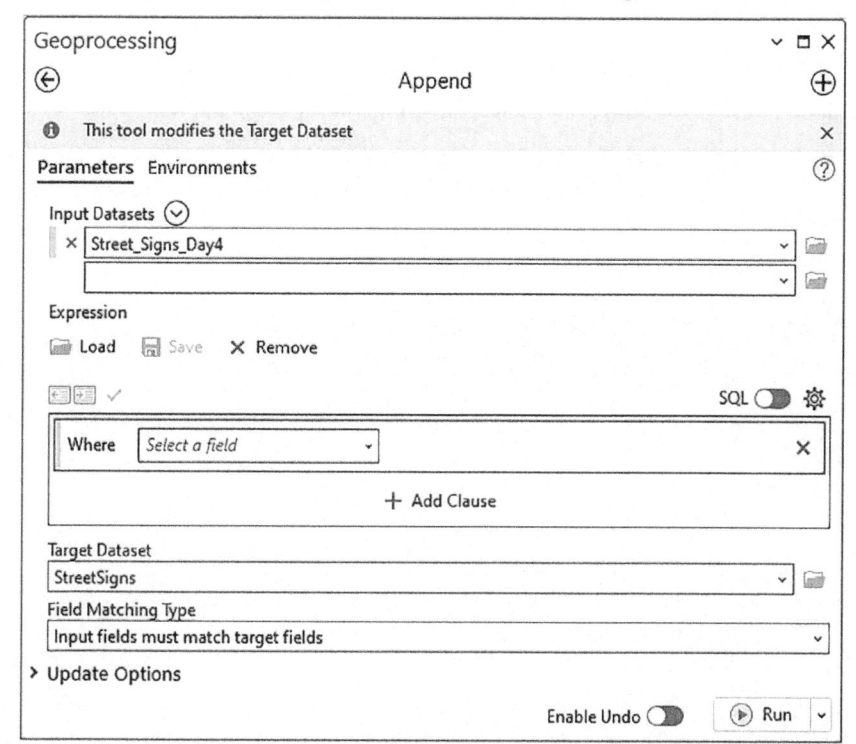

9. Verify your **Append** tool is set up as indicated in the following screenshot, and click **Run**:

Figure 11.16 – Append tool parameters complete

Notice that no new layers were added to your map, unlike when you used the **Merge** tool. This is because no new layers were created. The features from the `Street_Signs_Day4` shapefile have been added to the existing `StreetSigns` shapefile. Let's verify that is what happened.

10. In the **Contents** pane, select the **List by Draw Order** button. It is the first one on the left-hand side.
11. Turn off the `Street_Signs_Day4` layer by clicking on the box to the left of the layer name so that it no longer contains a checkmark.

You should now see a copy of the features that were in the shapefile displayed in the `StreetSigns` layer. This shows that the features from the shapefile were copied to the geodatabase feature class.

Verifying the results

You will now check the attributes to ensure they were copied from the shapefile successfully to the `StreetSigns` layer:

1. In the **Contents** pane, select the **List by Selection** button and ensure the `StreetSigns` layer is set as selectable.

2. Right-click on the `Street_Signs_Day4` layer and select **Zoom to Layer**. This ensures that you are zoomed in on the area covered by this layer.

3. Click on the **Map** tab in the ribbon.

4. Click on the **Select** button in the **Selection** group on the **Map** tab.

5. On your map, click on the lower-right side, as indicated in the following screenshot. Continue to hold your mouse button down and drag your mouse pointer to the northwest until you reach the approximate location shown, creating a rectangle that will allow you to select all the features copied from the shapefile:

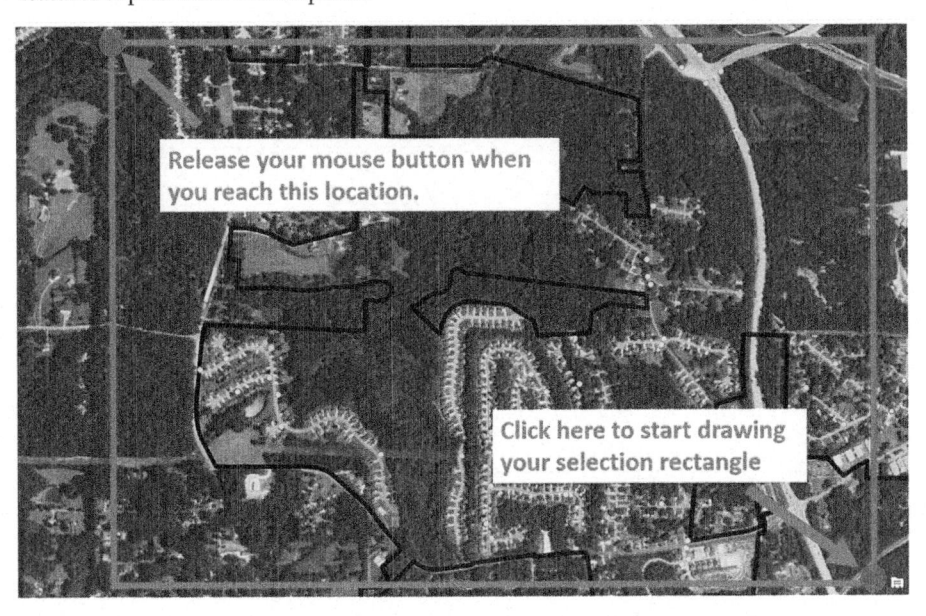

Figure 11.17 – Drawing selection rectangle

6. Right-click on the `StreetSigns` layer and select **Attribute Table**.

7. Click the **Show Selected Records** button located at the bottom of the attribute table. This filters the table so that only the selected features are visible.

8. Note the total number of records. Compare that to the number of records in the table before you used the **Append** tool. Use that information to answer the following questions:

   Question: Do the selected records all include attribute values?

   Answer:

   Question: How does the total number of records in the table compare to the number before you ran the **Append** tool?

   Answer:

As you can see, the attribute values were also copied from the shapefile to the new geodatabase feature class. The new total number of records in the `StreetSigns` layer should equal the sum of the number of records in the shapefile plus the pre-**Append** tool record count for the `StreetSigns` layer. The **Append** tool simply added records/features from the shapefile to the `StreetSigns` layer.

9. Close the attribute table and **Geoprocessing** pane if it is still open.
10. Save your project and close ArcGIS Pro.

As you have now experienced, the **Merge** and **Append** geoprocessing tools are great for combining multiple datasets into one. This can reduce the total number of datasets you must manage and make performing analysis much easier.

Exporting tabular data to an Excel spreadsheet

Many times, the people we work with are not GIS professionals or do not have access to GIS software. They may request data in another format. One of the most requested formats is a spreadsheet. Spreadsheets are something just about everyone with a computer can work with. They can run calculations, summarize data, create graphs, and more.

While you cannot export spatial data to a spreadsheet, ArcGIS Pro will allow you to export tabular data to a spreadsheet-compatible format. This includes Microsoft Excel XLS, XLSX, and CSV files. You can export attribute and standalone database tables to spreadsheets with ArcGIS Pro. As with most other operations, there is more than one way to do it.

In this recipe, you will learn two different methods for converting tables to spreadsheet-compatible formats. The roads superintendent has asked you to provide him with a copy of the street sign inventory you have been working on so that they can get a general count of all signs by type and compare it to the list of signs they have been keeping in a notebook. They also want you to get him a spreadsheet with all roads located within the city. You will use a different method for each of these tasks.

Getting ready

This recipe requires that you have completed the previous recipe. It also requires you to have a spreadsheet application installed, such as Microsoft Excel or Open Office. This recipe can also be completed with all license levels of ArcGIS Pro.

How to do it...

Now that you have a better understanding of the way you might need to export tabular data from your GIS to a spreadsheet or other format, you will work through the steps required to perform this task:

1. You will first need to start ArcGIS Pro and open the `Convert Multiple Shapefiles.aprx` project located in `C:\Student\ArcGISPro3Cookbook\Chapter11` using the skills you have learned in past recipes.

2. Locate the `StreetSigns` layer in the **Contents** pane.

3. Right-click on the `StreetSigns` layer and go down to **Data**. Then, select **Export Table**, as illustrated in the following screenshot. This will open the **Copy Rows** tool in the **Geoprocessing** pane:

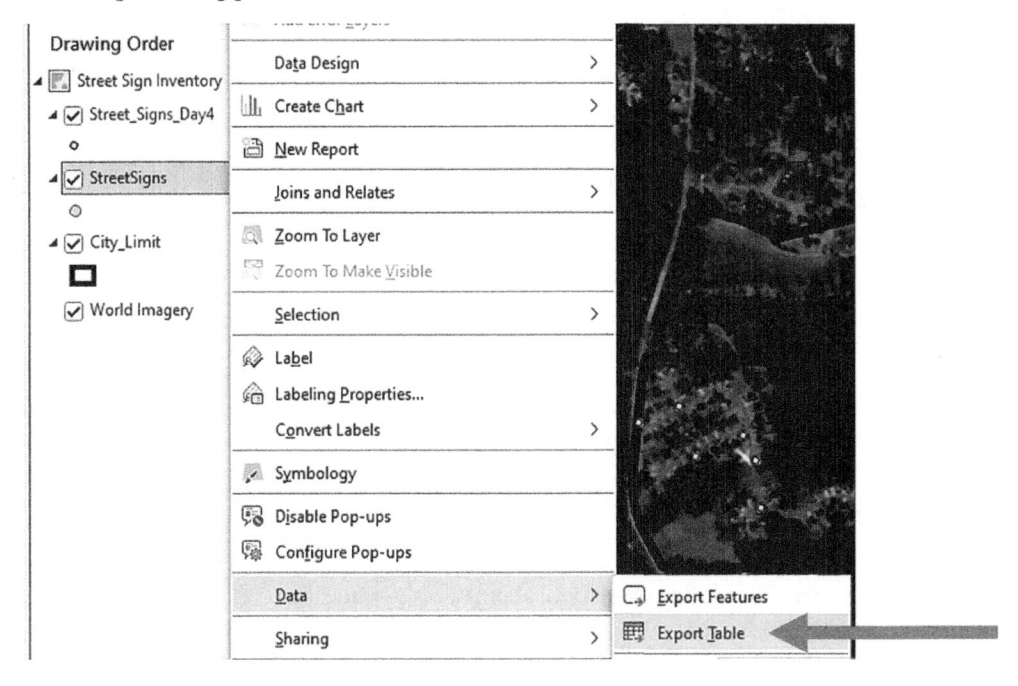

Figure 11.18 – Accessing the Export Table tool

4. The **Input Rows** parameter should already be set to the `StreetSigns` layer since you right-clicked on it to access the tool. Now, click on the **Browse** button located to the right of the **Output Table** window.

5. Click on the `Folders` folder located under **Project** in the left panel of the **Output Table** window.

6. Double-click on the `Convert Single Shapefile` folder in the right panel.

7. In the **Name** cell, type `StreetSignsInventory.csv` and click the **Save** button at the bottom of the window.

8. Verify that your **Export Table** tool looks like the following and click **OK**:

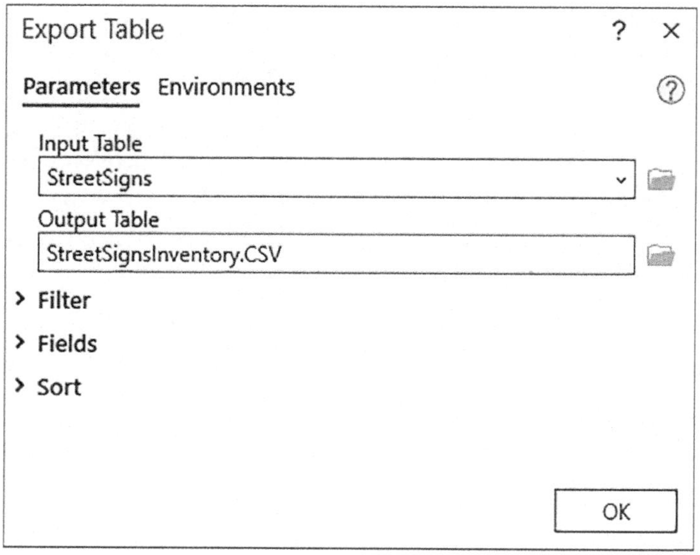

Figure 11.19 – Export Table tool with parameters complete

> Note
>
> The **Export Table** tool can be used to convert multiple types of tabular data to other formats. Input for this tool can include geodatabases, shapefiles, dBase tables, Microsoft Excel spreadsheets, CSV files, and TXT files. Outputs can be geodatabase standalone tables, dBase tables, and CSV and TXT files. For outputs that will result in individual files, you must specify the desired file extension as part of the output table name. To learn more about the **Export Table** tool, go to `https://pro.arcgis.com/en/pro-app/latest/tool-reference/conversion/export-table.htm`.

Remember – if you have features or records selected in a layer or table, the **Export Table** tool will only export the selected items to the new output table. When the **Export Table** tool is complete, you should see a new table appear in the **Contents** pane. This is the CSV file you just created.

Verifying your results

Let's see if it will open in Microsoft Excel or the spreadsheet application you have installed:

1. Open your **File Explorer**, depending on your operating system. There is typically an icon on the taskbar located at the bottom of your monitor that will open **File Explorer**. The icon looks like a file folder sitting on a stand.

Converting Data

> **Tip**
> File/Windows Explorer is not the same as Internet Explorer. Internet Explorer is a web browser. File/Windows Explorer allows you to browse the contents of your computer, network, and connected drives.

2. In the directory tree located in the panel on the left of the **File Explorer** window, locate **This PC** and expand it so you can see its contents.

3. In the left panel of the **File Explorer** window, click on the `C:\` drive. It is normally named either `Local Drive`, `Local Disk`, or `OS`.

4. In the right panel, locate the `Student` folder and double-click on it.

5. Double-click on the `ArcGISPro3Cookbook` folder to open it.

6. Double-click on the `Chapter11` and `Convert Single Shapefile` folders to open them.

7. Scroll down until you find the `StreetSignsInventory.csv` file and double-click to open it.

8. Take a moment to look over the contents of the file. You should see three columns. The first row should contain the field names from the attribute table. The remaining rows contain the data from the attribute table. It should look similar to the following:

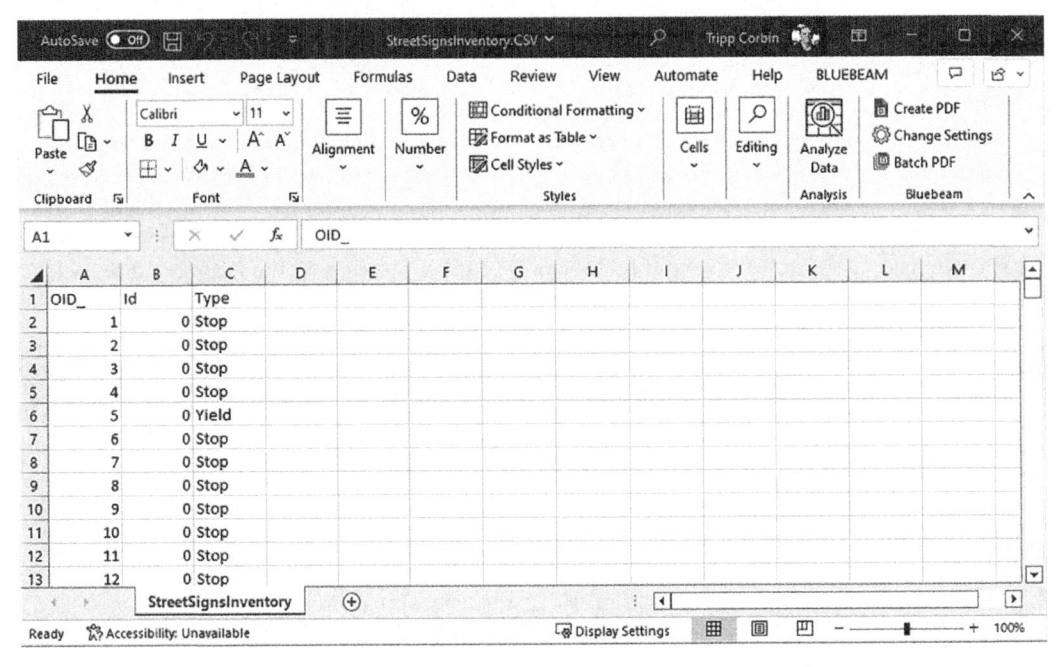

Figure 11.20 – Results of Export Table tool viewed in Excel

> **Note**
> The OID field might not show in the export data in some versions of ArcGIS Pro.

9. Once you are done reviewing the CSV file, close it. If asked to save it, do not.

You have just converted the attribute table for the `StreetSigns` layer to a CSV file that you opened as a spreadsheet. So, you have completed the first task. Now, you need to export the street-centerline attributes.

Preparing to export street centerlines to Excel

Exporting street centerline data to an Excel spreadsheet will actually require you to prepare the data before you export it because, as you will see, your street centerline data actually extends beyond the city limits:

1. Return to ArcGIS Pro. If you accidentally closed ArcGIS Pro, please refer back to the *step 1* of this recipe to reopen the correct project.
2. Close the **Export Table** tool if it is still open.
3. In the **Catalog** pane, expand the `Databases` folder so that you can see its contents.
4. Next, expand the `Trippville_GIS.gdb` geodatabase and the `Base` feature dataset.
5. Drag and drop the `Street_Centerlines` feature class into the map view.

   As you can see, the data extends outside the city limits of Trippville. The roads superintendent only wants data for portions of the roads inside the city limits, so you will need to clip the data.

6. Click on the **Analysis** tab in the ribbon.
7. Click on the **Clip** tool located in the **Tools** group in the ribbon. This will open the **Clip** tool in the **Geoprocessing** pane.
8. Click on the drop-down arrow for the **Input Features** parameter located at the far right side of the cell. Select `Street_Centerlines` from the list.
9. Repeat this process for the **Clip Features** parameter, selecting the `City_Limit` layer.

390   Converting Data

10. Accept the default name for the **Output Feature Class** parameter. Verify that your **Clip** tool looks like the following screenshot and click **Run**:

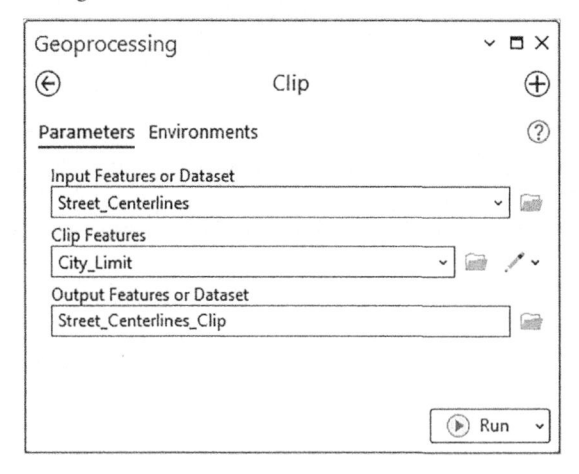

Figure 11.21 – Clip tool with completed parameters

When the **Clip** tool completes, a new layer will appear in your map that contains only the portion of roads within the city limits of Trippville.

Exporting street centerlines to an Excel spreadsheet

With the street centerlines clipped to those inside the city limits, you are ready to export those to an Excel spreadsheet:

1. Close the **Geoprocessing** pane and save your project.
2. Click on the **Analysis** tab in the ribbon to ensure it is active. Next, click on the **Tools** button located in the **Geoprocessing** group.
3. Click on the **Toolboxes** tab at the top of the **Geoprocessing** pane, as illustrated in the following screenshot:

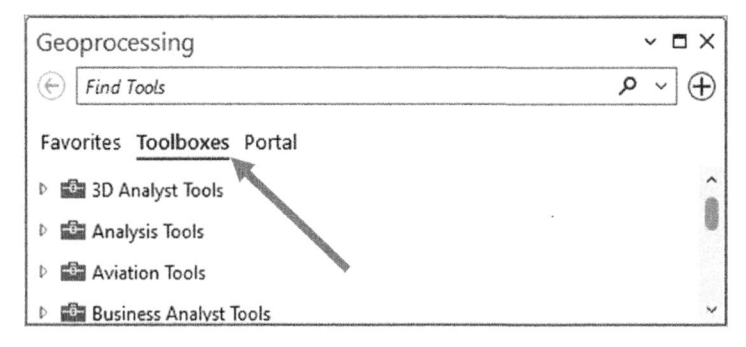

Figure 11.22 – Selecting Toolboxes tab in Geoprocessing pane

4. Locate and expand the **Conversion Tools** toolbox in the **Geoprocessing** pane.
5. Expand the **Excel** toolset to reveal the **Excel to Table** and the **Table to Excel** Python script tools.

> **Tip**
> You may wonder how you can tell these are Python script tools. It is because of the icon located next to the tool. Scrolls, as displayed next to these two tools, indicate Python script tools. Hammers indicate system tools, such as the **Clip** tool you used earlier in this recipe. A group of connected multi-colored diamonds indicates a model tool created with `ModelBuilder`.

6. Select the **Table to Excel** tool. This will open the tool in the **Geoprocessing** pane.
7. Click the drop-down arrow for the **Input Table** parameter and select `Street_Centerline_Clip` from the list.
8. For the **Output Excel File** parameter, click the **Browse** button located on the far right side.
9. In the left panel of the **Output Excel File** window that just opened, locate **This PC** and click on it.
10. In the right panel, locate the `C:\` drive and double-click on it. The `C:\` drive is often named `Local Drive`, `Local Disk`, or `OS`.
11. Navigate to the `C:\Student\ArcGISPro3Cookbook\MyProjects` folder.
12. In the **File Name** box located near the bottom of the **Save As** window, type `Street_Centerlines_Trippville` and click **Save**.
13. Verify that your **Table to Excel** tool looks like the following screenshot and click **Run**:

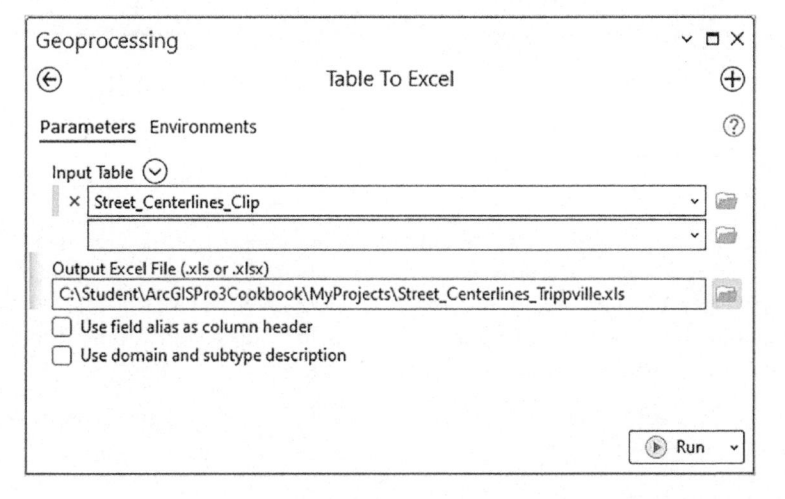

Figure 11.23 – Table to Excel tool with parameters completed

If the tool runs successfully, it should create a new spreadsheet in the folder you designated. Now, you will verify the results.

Verifying the results

You will now open the spreadsheet you just created in Excel to verify it was successfully generated:

1. Open **File Explorer**, as you did earlier in this recipe, and navigate to `C:\Student\ArcGISPro3Cookbook\MyProjects`.

2. Locate the `Street_Centerlines_Trippville.xls` file that you just created and double-click on it to open the file.

3. Take a moment to examine the spreadsheet you created. It should look like the screenshot that follows. If you desire, open the attribute table for the `Street_Centerlines` layer in ArcGIS Pro and compare the two:

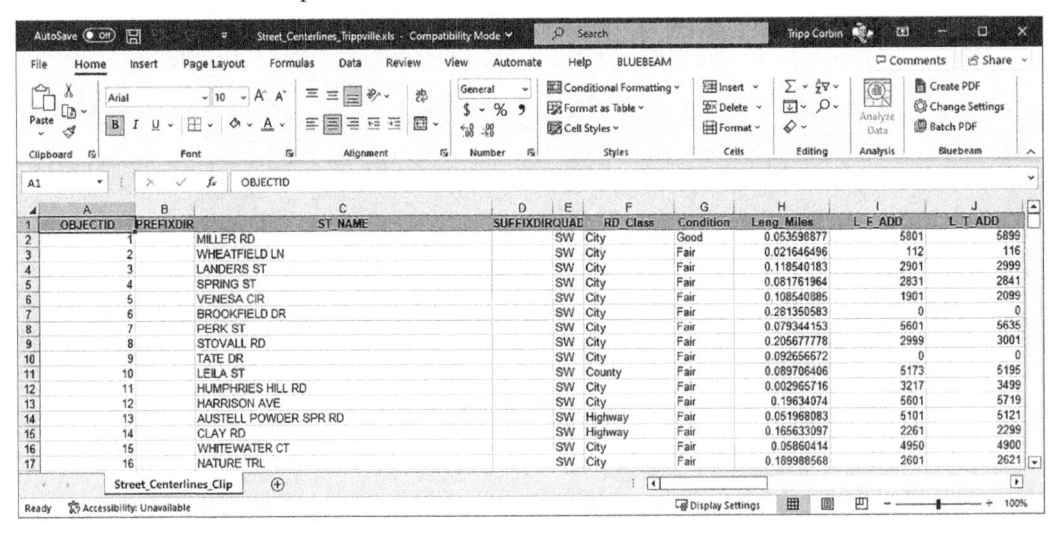

Figure 11.24 – Results of the Table to Excel tool

Your resulting spreadsheet will include several more columns than what is shown in the screenshot, but this should provide you with something to help validate your results. The resulting sheet is too big to be easily seen in its entirety within the limits of this book.

When you are done examining the contents of the spreadsheet, close it and your **File Explorer** window.

4. Return to ArcGIS Pro and save your project.

5. Close ArcGIS Pro.

You have now used two methods to export data from GIS to a format that can be opened by applications such as Microsoft Excel. This capability further expands the usefulness and flexibility of your GIS. It opens up the ability to integrate with other non-spatial applications.

How it works...

In this recipe, you were asked to provide two spreadsheet-compatible files to the roads superintendent: one that contained data from a recently completed street sign inventory and another containing all street centerlines within the city. You used two different methods to accomplish this.

For the street signs, you created a CSV file that you exported from the `StreetSigns` layer in your GIS. You did this by right-clicking on the layer in the **Contents** pane and accessing the **Export Table** tool via the context menu that appeared. You then verified the input was the `StreetSigns` layer and specified the output of the tool as `StreetSignsInventory.csv`, which was created in the `C:\Student\ArcGISPro3Cookbook\MyProjects` folder. Specifying the CSV file extension ensured the output was compatible with spreadsheet applications.

For the street centerlines, you first needed to clip the `Street_Centerlines` layer to the extent of the `City_Limits` layer. This was required because the `Street_Centerlines` layer extended outside of the city limits. You used the **Clip** tool found on the **Analysis** tab and in the **Tools** group to accomplish this task. Next, you used the **Table to Excel** tool to create the desired spreadsheet containing all street segments located in the city. You accessed this tool via the **Analysis** tab and the **Tools** button, which opened the **Geoprocessing** pane. From that pane, you located the **Conversion Tools** toolbox and the Excel toolset that contained the **Table to Excel** tool. You then selected the `Street_Centerlines_Clip` layer as the input and the output as `Street_Centerlines_Trippville.xls`. The results of the **Table to Excel** tool provided the second file the superintendent was asking for.

Importing an Excel spreadsheet into ArcGIS Pro

You have learned how to export data from GIS formats to a spreadsheet within ArcGIS Pro. But what if you need to go the other way? People use spreadsheets to store all kinds of information. This is because they are quick and easy to create and are very versatile. So, what does it take to bring one into your GIS using ArcGIS Pro?

When working with spreadsheets in ArcGIS Pro, remember – simpler is always better. ArcGIS Pro allows you to view XLS and XLSX files. It displays the spreadsheet as a database table. You can view, query, and link a spreadsheet. The linking is limited to using a join or a relate, which you learned about in *Chapter 3, Linking Data Together*.

There are some limitations when trying to use a spreadsheet in ArcGIS Pro. ArcGIS Pro treats a spreadsheet like a database table. So, it applies database limitations to the spreadsheet. This includes the following:

- The values in the first non-empty row in the spreadsheet become the field names.
- Each column must have a unique name.
- Each column name cannot contain special characters. Underscores (_) are allowed.
- The field names (values in the first row) are limited to 64 characters.
- No value in a cell can exceed 255 characters.
- Each field must start with an Alpha character. It cannot start with a number.

If the spreadsheet fails to meet any of these limitations, ArcGIS Pro will not allow you to view the spreadsheet. This means that you may need to customize or simplify the spreadsheet for it to work in ArcGIS Pro. Also, ArcGIS Pro will not allow you to edit a spreadsheet.

If you need to edit the data in a spreadsheet or make it a permanent part of your GIS data, you will need to convert it. ArcGIS Pro has a tool that converts the spreadsheet into a database table. This can be a dBase or standalone geodatabase table.

In this recipe, you will view a spreadsheet containing inspection information in ArcGIS Pro. This will start with you reviewing the contents of the spreadsheet in Excel or another spreadsheet application to ensure that it does not violate some of the limitations. Then, you will open it in ArcGIS Pro.

Getting ready

It is recommended that you complete all the recipes in *Chapter 3*, *Linking Data Together*, before you work on this recipe. *Chapter 3*, *Linking Data Together*, will provide you with a better understanding of how to work with database tables in ArcGIS Pro. You will also need to have Microsoft Excel or a similar spreadsheet application installed. This recipe can be completed with all license levels of ArcGIS Pro.

How to do it...

Now that you understand some of the limitations ArcGIS Pro has when working with a spreadsheet, you will work through the steps to import data from a spreadsheet into a geodatabase so that it can be manipulated in ArcGIS Pro.

1. Open the **File Explorer** application in Windows. It should be an icon that looks like a file folder on your taskbar.
2. In the panel on the left, expand the contents of **This PC** or **My Computer**, depending on your operating system.

3. Locate and expand the C:\ drive, which is typically labeled as OS, Local Disk, or Local Drive.

4. Click on the Student folder so that its contents appear in the right-hand side panel.

5. Double-click on the ArcGISPro3Cookbook folder.

6. Double-click on the Chapter11 folder.

7. Double-click on the Inspections2016to2017.xls file to open it. The spreadsheet should open in your spreadsheet application such as Microsoft Excel. It should look like the screenshot that follows:

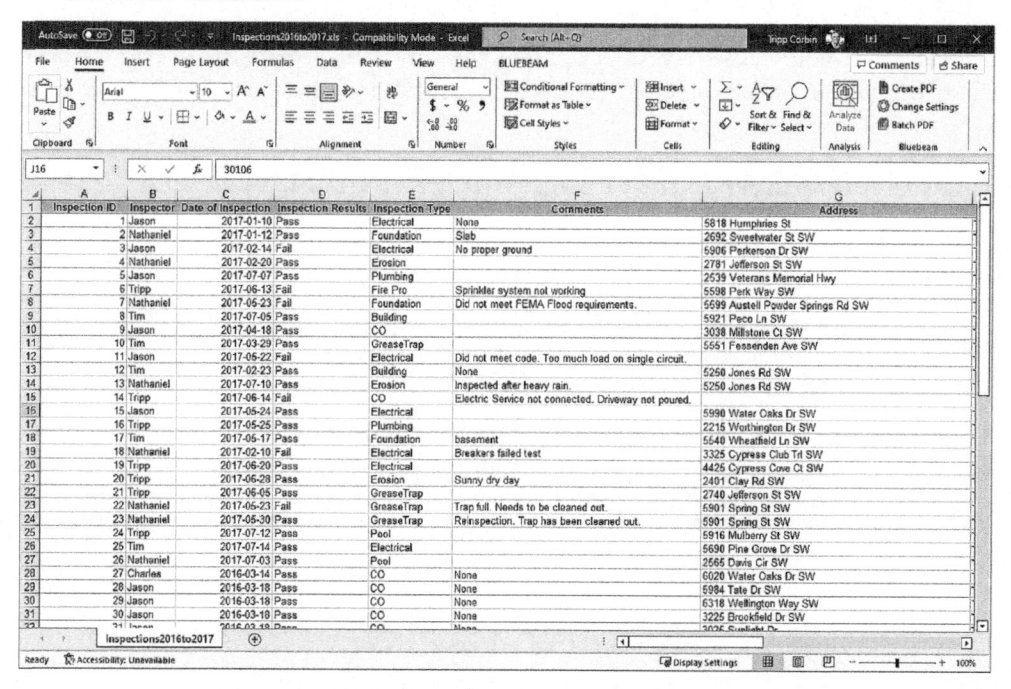

Figure 11.25 – Inspection spreadsheet open in Excel

8. Review the spreadsheet, comparing it to the limitations listed at the start of this recipe. Then, answer the following questions:

    Question: What are the values in the first row that will become the column names when you open the spreadsheet in ArcGIS Pro?

    Answer:

    Question: Is each value in the first row unique?

    Answer:

Question: Do the values in the first row contain any special characters?

Answer:

Question: Do any of the values in the first row exceed 64 characters?

Answer:

Question: Do any of the values in any cell exceed 255 characters?

Answer:

As you can see, this is a simple spreadsheet. It does not contain a lot of title information or complicated equations, which can cause issues in ArcGIS Pro. As a matter of fact, it meets most of the restrictions required by ArcGIS Pro. However, several of the values in the first row do violate the restrictions because they contain special characters (spaces). You will need to adjust these before you can bring the spreadsheet into ArcGIS Pro.

Fixing the spreadsheet so that it successfully imports into a geodatabase

You will now fix issues you have identified with the column headers so that the spreadsheet successfully imports into a geodatabase.

1. You need to change the value in cell **A1**. Replace `Inspection ID` with `Insp_ID`.
2. Replace `Date of Inspection` in cell **C1** with `Date_Insp`.
3. Replace `Inspection Results` in cell **D1** with `Insp_results`.
4. Replace `Inspection Type` in cell **E1** with `Insp_Typ`.
5. Verify that you have properly replaced the values in row 1, and then save the spreadsheet and close it:

	A	B	C	D	E	F
1	Insp_ID	Inspector	Date_Insp	Insp_results	Insp_Typ	Comments
2	1	Jason	2017-01-10	Pass	Electrical	None
3	2	Nathaniel	2017-01-12	Pass	Foundation	Slab
4	3	Jason	2017-02-14	Fail	Electrical	No proper ground
5	4	Nathaniel	2017-02-20	Pass	Erosion	
6	5	Jason	2017-07-07	Pass	Plumbing	
7	6	Tripp	2017-06-13	Fail	Fire Pro	Sprinkler system not working
8	7	Nathaniel	2017-05-23	Fail	Foundation	Did not meet FEMA Flood requirements.
9	8	Tim	2017-07-05	Pass	Building	
10	9	Jason	2017-04-18	Pass	CO	
11	10	Tim	2017-03-29	Pass	GreaseTrap	
12	11	Jason	2017-05-22	Fail	Electrical	Did not meet code. Too much load on single circuit.
13	12	Tim	2017-02-23	Pass	Building	None
14	13	Nathaniel	2017-07-10	Pass	Erosion	Inspected after heavy rain.
15	14	Tripp	2017-06-14	Fail	CO	Electric Service not connected. Driveway not poured.
16	15	Jason	2017-05-24	Pass	Electrical	
17	16	Tripp	2017-05-25	Pass	Plumbing	
18	17	Tim	2017-05-17	Pass	Foundation	basement

Figure 11.26 – Inspection spreadsheet with updated column names to meet ArcGIS Pro requirements

Joining the spreadsheet data to an existing layer

You will now join the data from the spreadsheet with an existing layer in a map:

1. Start ArcGIS Pro and open the `Import Spreadsheet.aprx` project located in `C:\Student\ArcGISPro3Cookbook\Chapter11` using the skills you have learned in past recipes. The project should open with the **City of Trippville** map.

2. You will now add the spreadsheet you just edited to ArcGIS Pro. Click on the **Add Data** button in the **Layer** group on the **Map** tab in the ribbon.

3. In the panel on the left side, expand the **Computer** option so that you can see its contents.

4. Expand the **This PC** option in the left panel and then click on `Local Disk (C:\)` so that its contents appear in the right panel.

5. Scroll down in the right panel until you see the `Student` folder. Double-click on the `Student` folder.

6. Double-click on the `ArcGISPro3Cookbook` folder.

7. Double-click on the `Chapter11` folder.

8. Double-click on the `Inspections2016to2017.xls` file. Select the `Inspections2016to2017$` worksheet and click **OK**. The worksheet should appear as a standalone table in the **Contents** pane.

9. Right-click on the `Inspections2016to2017$` standalone table and select **Open** from the menu that appears. The worksheet opens as a table and should look very familiar. Notice the fields of information this workshop contains:

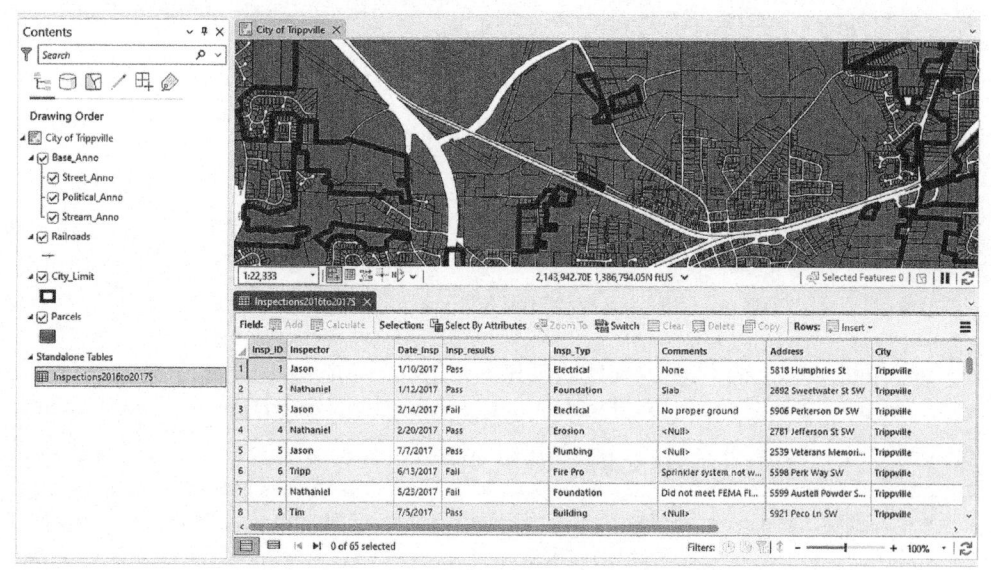

Figure 11.27 – Inspections 2016 to 2017 spreadsheet open in ArcGIS Pro

398　Converting Data

If you look at the list of layers on the map, you will see an `Inspections` layer. This layer shows the locations of inspections referenced in the worksheet you just brought into ArcGIS Pro. These locations were captured at the time the inspections were performed, but the city uses a separate system to log the inspection results, which is why the data is not already in the GIS. Now, you will examine that layer.

10. In the **Contents** pane, right-click on the `Inspections` layer and select **Attribute Table** from the menu that appears.

    As you can see, this table is very simple. It only contains three fields, `ObjectID`, `Shape`, and `InspID`. This means the layer only stores the location of each inspection that was performed. If you were to join it with the worksheet you added, then you would know a lot more about each inspection. This would in turn increase your ability to use the data in ArcGIS Pro. Now, you will join the worksheet to the `Inspections` layer.

11. In the **Contents** pane, right-click on the `Inspections` layer, select **Joins and Relates**, and then **Add Join**, as illustrated in the following screenshot. This will open the **Add Join** tool in the **Geoprocessing** pane:

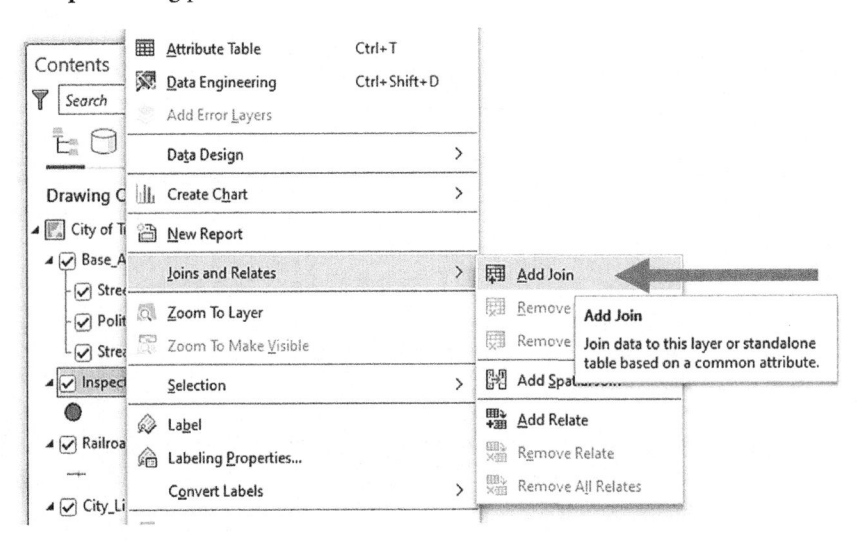

Figure 11.28 – Accessing the Add Join tool from the menu

12. The **Input Table** parameter should already be assigned with `Inspections`. Click the drop-down arrow for the **Input Join Field** parameter and select `InspID` from the list presented.

13. The **Join Table** parameter should also be automatically assigned as the `Inspections2016to2017$` worksheet because it is the only standalone table in your map. Click on the drop-down arrow for the **Join Table Field** parameter and select `Insp_ID`.

14. Verify your **Add Join** tool looks like the following screenshot and click **Run**:

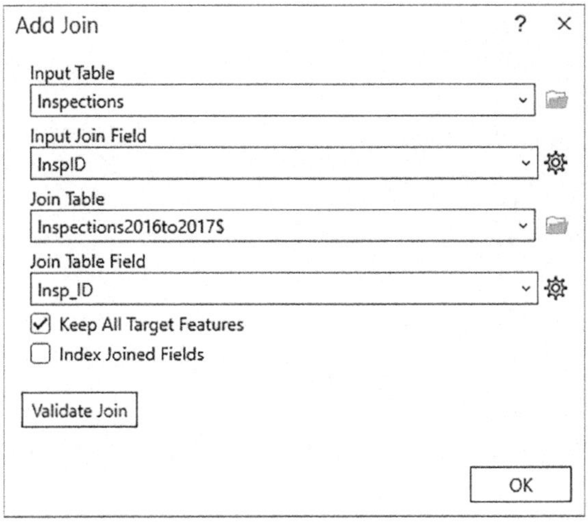

Figure 11.29 – Add Join tool with completed parameters

15. Look back at the **Inspections** attribute table that originally contained only three fields. Now, you should see that the information from the worksheet has been added to that table. You may need to reopen the attribute table to see your results. ArcGIS Pro may have closed it.
16. Close the **Geoprocessing** pane and save your project.

Now that you have joined the worksheet data to the `Inspections` layer, you can use that additional information to label, symbolize, and query, just as if the additional information were part of the layer to start with. The one thing you cannot do is edit the information that comes from the worksheet in ArcGIS Pro. That requires you to convert it.

Exporting data to a new feature class

The method you use to convert data will depend on how you intend to use the data from the spreadsheet. If you want to keep the spreadsheet data in a separate table, you can use the **Excel to Table** tool. If you wish to permanently join the data to the layer, you can export the joined layer to a new feature class. You will do both now:

1. Right-click on the `Inspections` layer and go to **Data** in the menu that appears. Then, select the **Export Features** option, as illustrated in the following screenshot. This will open the **Export Features** tool in the **Geoprocessing** pane:

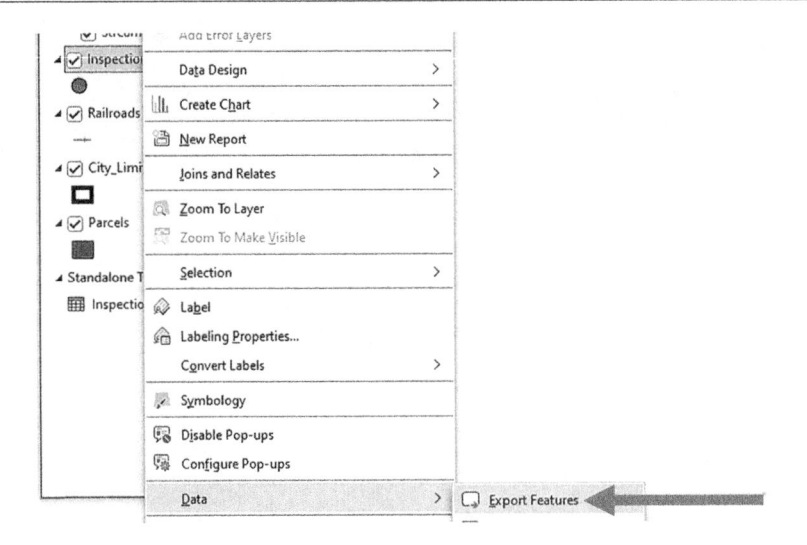

Figure 11.30 – Accessing the Export Features option from the menu

2. The **Input Features** value should be automatically populated. Click the **Browse** button at the end of the **Output Features Class** cell.

3. Select the Databases folder under **Project** in the left panel of the **Output Feature Class** window.

4. Double-click on the Trippville_GIS.gdb geodatabase in the right panel of the window. Then, in the **Name** cell near the bottom of the window, type Inspections_more_data and click the **Save** button.

5. Verify that your **Export Features** tool looks like the following screenshot and click **Run**:

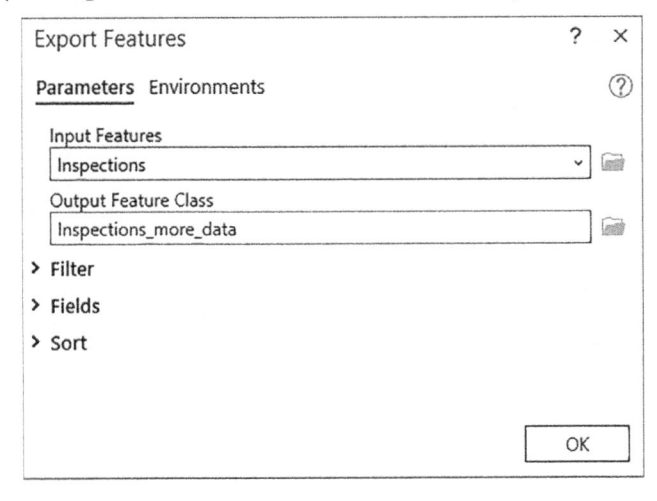

Figure 11.31 – Export Features tool with parameters complete

6. Close the **Geoprocessing** pane.

   You will now verify the results of the **Export Features** tool to ensure it contains all the information you expect.

7. A new `Inspections_more_data` layer should appear in the **Contents** pane. Open the attribute table for this new layer and answer the following question:

   Question: What attribute fields does the layer you created contain?

   Answer:

   As you can see, the new layer contains all the information that was in the attribute table for the `Inspections` layer and the `Inspections2016to2017$` worksheet standalone table. Now that all that information is part of a single layer, you can edit all the data.

8. Save your project using the **Save** button on the **Quick Access Toolbar**.

The ability to import and use data from outside sources greatly increases the flexibility of your GIS. It can provide you with access to a plethora of new information to include in your analysis.

How it works...

You have just learned how to bring an Excel spreadsheet into ArcGIS Pro and join it to a layer so that you can access information. To do this, you first had to examine the spreadsheet to ensure it met the ArcGIS Pro requirements. While looking at the spreadsheet in Excel or another spreadsheet application, you determined that several column names contained special characters such as spaces, which are not compatible with ArcGIS Pro. So, you edited column names in the first row to comply with ArcGIS Pro limitations. Then, you added the spreadsheet to a map in ArcGIS Pro as a standalone table. Next, you examined the attribute table for the `Inspections` layer to see which attribute fields it contained and to identify a key field that would allow you to join it to the spreadsheet. Lastly, you created a join between the `Inspections` layer and the `Inspections2016to2017$` spreadsheet, allowing you to see additional information for each inspection.

While a join allows you to see information associated with each inspection location, you are not able to edit that information in ArcGIS Pro because it is in a spreadsheet. So, you exported the joined data to a new geodatabase feature class. To do that, you opened the **Export Features** tool by right-clicking on the `Inspections` layer, which had the join you created. You then went to **Data and Export Features** from the menu that appeared. You filled out the parameters for the tool and ran it. This created a new feature class in the `Trippville_GIS.gdb` geodatabase that contained the combined information from the `Inspections` layer and the `Inspection2016to2017$` spreadsheet.

There's more...

You just joined a spreadsheet to the `Inspections` layer and then exported that to a new feature class that contains the combined information from both the joined sources. But what if you wanted the `Inspections` layer and the data from the spreadsheet to remain separate but still be editable from ArcGIS? How would you do that?

In that case, you would want to convert the spreadsheet to a true standalone database table. This would allow the two to remain separate, but they could still be joined as needed, or maybe even a relationship class could be set up between them. Keeping them separate would have other advantages as well. First, you could easily replace the standalone table with a new copy of the spreadsheet as needed if your organization wanted to continue to input data through the spreadsheet. Being a standalone table, you could share just the table with others for editing and updating via web forms or mobile applications. These are just a few of the advantages.

Let's look at how you would export the spreadsheet to a standalone table in your GIS database:

1. Click on the **Analysis** tab in the ribbon.
2. Click on the **Tools** button in the **Geoprocessing** group on the **Analysis** tab. This will open the **Geoprocessing** pane once again.
3. If required, click on the **Toolboxes** tab located near the top of the **Geoprocessing** pane.
4. Locate and expand the **Conversion Tools** toolbox.
5. Locate and expand the **Excel** toolset in the **Conversion Tools** toolbox.
6. You should see two tools, **Excel to Table** and **Table to Excel**, as illustrated in the following screenshot. You used the **Table to Excel** Python tool in the *Importing an Excel spreadsheet into ArcGIS Pro* recipe. Now, you will use the **Excel to Table** tool. Click on the **Excel to Table** tool. This will open the tool in the **Geoprocessing** pane:

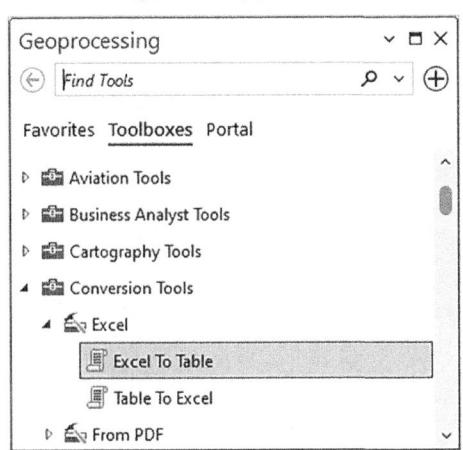

Figure 11.32 – Excel toolset expanded to display available tools

7. Click on the **Browse** button for the **Input Excel File** parameter. This will open a new window that allows you to navigate to the Excel file you will convert.
8. In the panel on the left, locate and expand **This PC** or **My Computer**, depending on your operating system.
9. Select C:\, which is typically labeled OS, Local Disk, or Local Drive. The contents of the C:\ drive should appear in the panel on the right.
10. Scroll down until you see the Student folder, then double-click on the Student folder.
11. Double-click on the ArcGISPro3Cookbook folder.
12. Double-click on the Chapter11 folder.
13. Select the Inspections2016to2017.xls file and click the **OK** button. Do not double-click on the Inspections2016to2017.xls file. That will open its contents and not select it, as desired. This will cause the tool to ultimately fail.
14. Click the **Browse** button for the **Output Table** parameter.
15. Click on the Databases folder in the left panel of the **Output Table** window, located under **Project**. Then, double-click on the Trippville_GIS.gdb geodatabase.
16. In the **Name** cell, near the bottom of the **Output Table** window, type Inspections_2016_2017. Then, click the **Save** button.
17. Click the drop-down arrow for the **Sheet** parameter and select Inspections2016to2017 from the list that appears.

> **Note**
> If you are importing a spreadsheet that contains multiple sheets, you can use the **Sheet** parameter to specify which sheet you wish to import as a new table into your GIS database.

18. Verify your **Excel to Table** tool looks like the following screenshot and click the **Run** button:

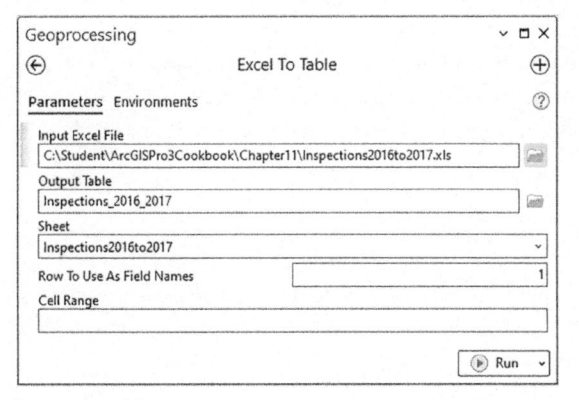

Figure 11.33 – Excel to Table tool with completed parameters

When the tool is complete, a new standalone table will appear in the **Contents** pane. This is the new table that you just created.

Removing and adding joins

Now, you will remove the join between the `Inspections` layer and the spreadsheet and replace it with a join to the new table:

1. Right-click on the `Inspections` layer and go to **Joins and Relates** in the menu that appears. Select **Remove Join**, as illustrated in the following screenshot. This will open the **Remove Join** tool in the **Geoprocessing** pane:

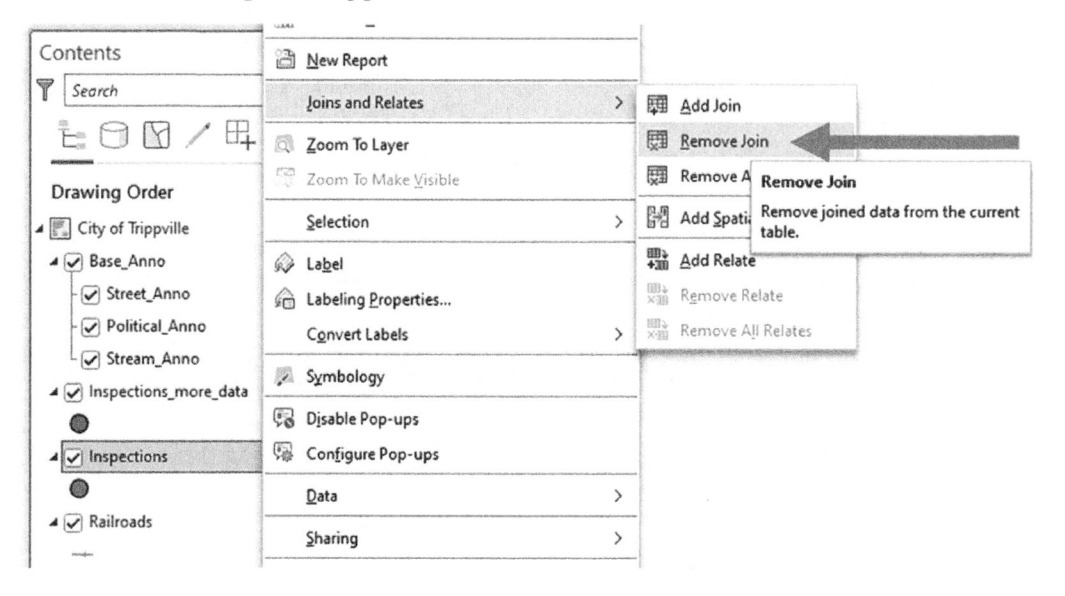

Figure 11.34 – Accessing Remove Join tool from the menu

2. The **Layer Name** or **Table View** parameter should already be populated. For the **Join** parameter, select the drop-down arrow and select `Inspections2016to2017$` from the list. Verify your **Remove Join** tool matches the following screenshot and click the **OK** button:

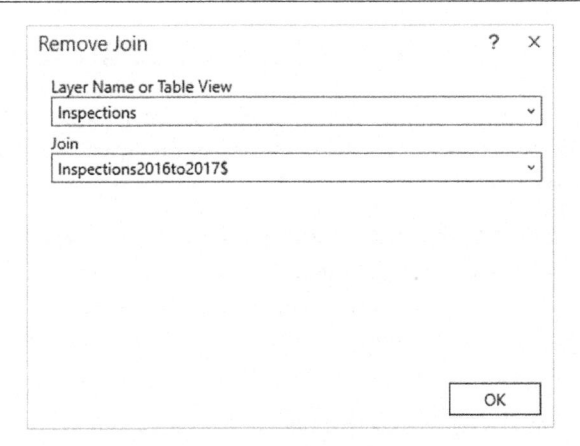

Figure 11.35 – Remove Join tool with parameters filled in

> **Note**
> You can join or relate multiple tables to a single layer or table if needed within ArcGIS Pro. However, be careful how many you join or relate as it can have a negative impact on performance and become difficult to manage.

3. Using the skills you learned in *Chapter 3, Linking Data Together*, and earlier in this recipe, join the Inspections_2016_2017 standalone table to the Inspections layer.
4. Right-click on the Inspections2016to2017$ standalone table and select **Remove**. This will remove the table from the map but does not delete the file.
5. Open the attribute table for the Inspections layer and verify the data from the Inspections_2016_2017 is joined.
6. Close the **Geoprocessing** pane and any open tables that you are viewing.
7. Save your project and close ArcGIS Pro.

You have now successfully imported data from an Excel spreadsheet into a geodatabase as a new standalone table and joined it to an existing layer. With the data imported into your geodatabase, you can use ArcGIS Pro and other Esri applications to update and maintain the table, which you could not do with the data if it remained in the spreadsheet.

Importing selected features into an existing layer

Throughout this chapter, you have learned various methods for importing entire datasets into your geodatabase or exporting from your geodatabase to other formats. What if you only want to import a few selected features or records?

By default, the geoprocessing tools you have already used, such as **Export Features**, **Merge**, **Append**, and **Excel to Table**, will automatically be limited to selected features or records if you have them selected. However, most of these tools create new feature classes or tables. They do not just add data to an existing dataset. ArcGIS Pro will allow you to copy data from one layer and paste it into another existing layer. This allows you to choose exactly which features you want to convert or import.

In this recipe, you will copy records from a shapefile and paste them into an existing layer that points to a geodatabase feature class. You will do this by first performing a query to select features in the shapefile you wish to import into the existing geodatabase layer. Then, you will copy and paste those features from the shapefile into the geodatabase layer. The shapefile you are working with represents a new subdivision being constructed in the City of Trippville.

Getting ready

You are not required to have completed previous recipes in order to complete this one. This recipe can be completed with all license levels of ArcGIS Pro. You will need to ensure the sample data for the book has been installed. It is assumed that by this point you have completed some recipes from the book or at least have worked with ArcGIS Pro in the past and know how to open an existing project.

How to do it...

Now, you will get the opportunity to work through the tasks required to import selected features from a shapefile into a layer based on a geodatabase feature class.

1. Start ArcGIS Pro and open the `Convert Selected Data.aprx` project located in `C:\Student\ArcGISPro3Cookbook\Chapter11` using the skills you have learned in past recipes.

2. Notice the large parcel in the center of the map view. This is the parcel that is going to become the new subdivision. In the **Contents** pane, select the **List by Draw Order** button. It is the first button on the left side at the top of the pane.

3. Turn on the `New Subdivision Shapefile` layer. Once it is visible, you will be able to see the new subdivision.

4. Click on the **List by Editing** button in the **Contents** pane. The button's icon looks like a pencil.

5. Right-click on the `Street Rights of Way` layer and select **Make this the only editable layer** from the menu that appears, as illustrated here:

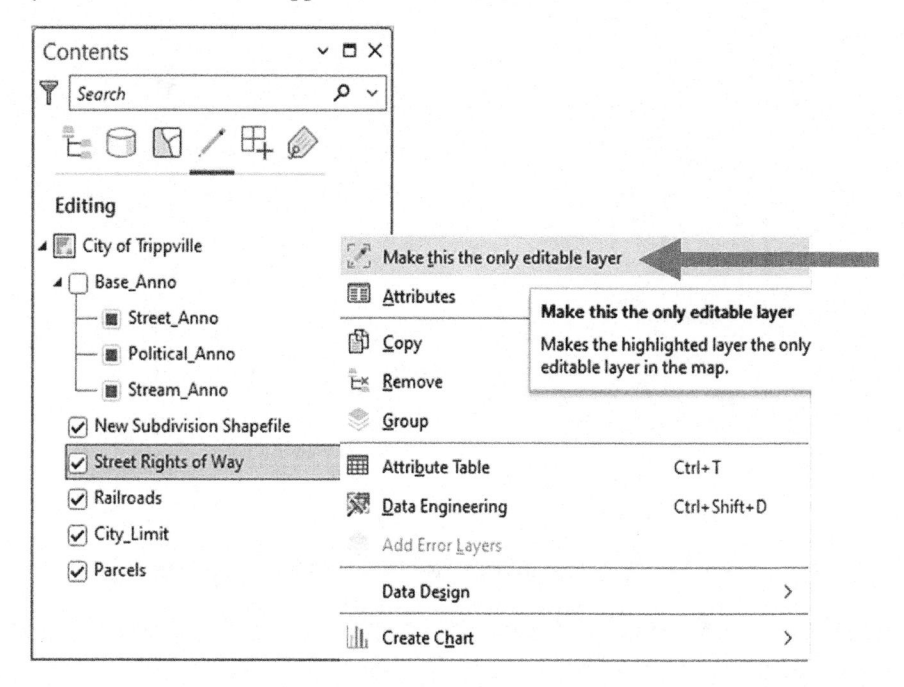

Figure 11.36 – Making the Street Rights of Way layer the only editable one

6. Click on the **List by Selection** button in the **Contents** pane. It is located to the left of the **List by Editing** button. Ensure the `New Subdivision Shapefile` and `Street Rights of Way` layers are the only ones set as selectable.
7. Click on the **Map** tab in the ribbon. Select the **Select by Attribute** tool in the **Selection** group.
8. Click on the drop-down arrow for the **Input Rows** parameter. Select the `New Subdivision Shapefile` layer from the list presented.
9. Ensure the **Selection Type** parameter is set to **New Selection**.
10. Click on the drop-down arrow in the cell located to the right of the word **Where** and select **Layer** from the list.
11. In the cell that appears to the right of **Layer**, ensure it says, **is equal to**. Then, click on the drop-down arrow in the final cell and select `RW` from the list that appears.

12. Verify your **Select by Attributes** window matches the following screenshot and click on the **OK** button:

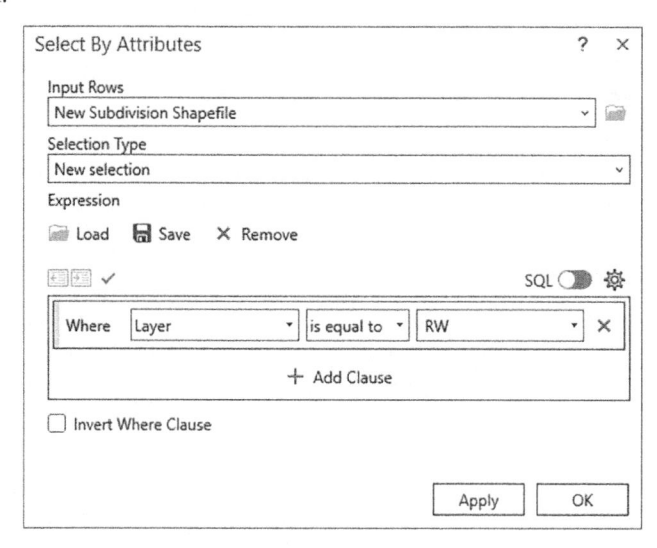

Figure 11.37 – Select by Attributes tool with parameters completed

You should have 258 features selected, which you can verify by looking in the **Contents** pane. You should see the number of selected features next to the `New Subdivision Shapefile` layer. The selected features will be highlighted on the map, as shown in the following screenshot:

Figure 11.38 – Results of Select by Attributes with 258 features selected

Importing selected features into an existing layer   409

The features you want to copy from the shapefile are selected.

Copy and pasting features

Now, you will copy the features you have selected and paste them into the `Street Rights of Way` layer:

1. Click on the **Edit** tab in the ribbon.
2. Click on the **Copy** button located in the **Clipboard** group on the **Edit** tab.

> **Tip**
> If the **Copy** button is grayed out, you may need to enable editing. Click on the **Edit** button so that it is highlighted in *blue*. You will only see this button if you have enabled the **BLAH** option.

3. Click on the drop-down arrow located below the **Paste** tool in the **Clipboard** group on the **Edit** tab and select **Paste Special**, as illustrated in the following screenshot. If the **Paste Special** option is not available, you may need to select the `Street Rights of Way` layer in the **Create Features** pane:

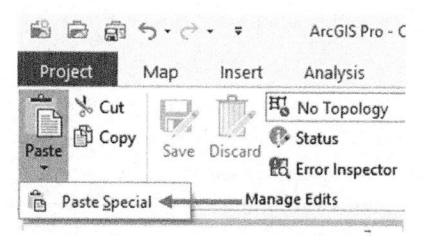

Figure 11.39 – Selecting Paste Special option

4. Set the **Paste into Template** parameter to **Street Rights of Way** using the drop-down arrow, as shown in the following screenshot. Then, click **OK**:

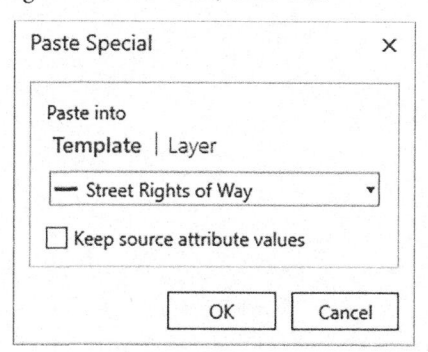

Figure 11.40 – Paste Special tool with parameters complete

5. Look in the **Contents** pane, making sure **List by Selection** is still selected. Notice that there are now no features selected in the `New Subdivision Shapefile` layer. Instead, there are now 258 features selected in the `Street Rights of Way` layer. You just copied the features from one layer and pasted them into another existing layer.

6. Click on the **Clear** button, located in the **Selection** group on the **Edit** tab in the ribbon, to clear your selected features. If needed, close the **Modify Features** window as well.

7. In the **Contents** pane, click on the **List by Draw Order** button and turn off the `New Subdivision Shapefile` layer to verify that you did successfully copy the features to the `Street Rights of Way` layer.

8. Click the **Save** button in the **Manage Edits** group on the **Edit** tab in the ribbon to save the new features you just copied into the `Street Rights of Way` layer. When asked to save all edits, click **Yes**.

9. Save your project and close ArcGIS Pro.

This method will allow you to copy features from layers that reference CAD files (DWG, DGN, or DXF), shapefiles, and geodatabase feature classes to another layer as long as the layer references a shapefile or geodatabase feature class and is the same type of feature (point, line, or polygon).

How it works...

In this recipe, you copied specific features that were in one layer over to another existing layer. You did this by first querying the features in the source layer using the **Select by Attributes** tool. You used this tool to select all features that were assigned an attribute value where the **Layer** field was equal to RW. This resulted in 258 features in the source layer being selected. Next, you went to the **Edit** tab and selected **Copy**. Then, you clicked on **Paste** and selected the **Paste Special** option. Next, you set the **Paste into Template** parameter to **Street Rights of Way** and clicked **OK**. This copied the data from the `New Subdivision Shapefile` layer to the `Street Rights of Way` layer.

12
Proximity Analysis

One of the powers of GIS over other information systems is its ability to analyze data from a spatial perspective – for example, being able to determine how far something is from something else. ArcGIS Pro can perform some amazing spatial analysis. These capabilities are generally broken down into four categories: overlay, proximity, network, and statistical. These can be further specialized as well.

In this chapter, we'll focus on proximity analysis in ArcGIS Pro. Proximity analysis simply involves determining how far or close features are from one another. This could be as simple as selecting features that are within a specified distance from other features, or as complex as calculating the distances between features in one or more layers. Examples of proximity analysis include creating a buffer around selected features to show the area around those features at a set distance. This might involve calculating how far features in one layer are from features in another layer and writing those values to a data table. It might also involve determining which feature in one layer is the closest to a feature in another layer.

In this chapter, you will get some first-hand experience in performing various types of proximity analysis. We will cover the following recipes:

- Selecting features within a specific distance
- Creating buffers
- Determining the nearest feature using the **Near** tool
- Calculating how far apart features are using the **Generate Near Table** tool

Selecting features within a specific distance

For this recipe, the roads superintendent has told you they will be starting a new project along Sloan St. and needs to notify those living and working near the street. They have asked you to generate a list of all the parcels located within 500 feet of Sloan Street so that they can notify the residents of the upcoming work.

In this recipe, you will use the **Select By Location** tool to select all the parcels within 500 feet of Sloan Street, and then export the selected parcels to an Excel spreadsheet to give to the superintendent. Before you can use the **Select By Location** tool, you will need to select all the road centerline segments that make up Sloan Street. You will use the **Select By Attributes** tool to do that.

Getting ready

This recipe requires the sample data for this book to be installed on your computer. It is recommended that you complete the recipes in *Chapter 1, ArcGIS Pro Capabilities and Terminology*, or have some experience working with ArcGIS Pro before you start this recipe to ensure you have the required foundational skills and knowledge needed to complete this recipe. You can complete this recipe with any ArcGIS Pro licensing level.

How to do it...

Follow these steps to perform your first proximity analysis using the **Select By Location** tool to select all the parcels within a specific distance of a selected street:

1. Start ArcGIS Pro and open the `Proximity Analysis.aprx` project located in `C:\Student\ArcGISPro3Cookbook\Chapter12\Proximity Analysis` using the skills you have learned in past recipes.

   The project should open with the **Select in a distance** map. If you look in the **Contents** pane, you will see three layers: `City Limits`, `Street Centerlines`, and `Parcels`.

2. Click on the **Bookmarks** button in the **Navigate** group in the **Map** tab on the ribbon.

3. Select the **Sloan St** bookmark, as illustrated in the following screenshot. This will zoom you into the work area for this recipe:

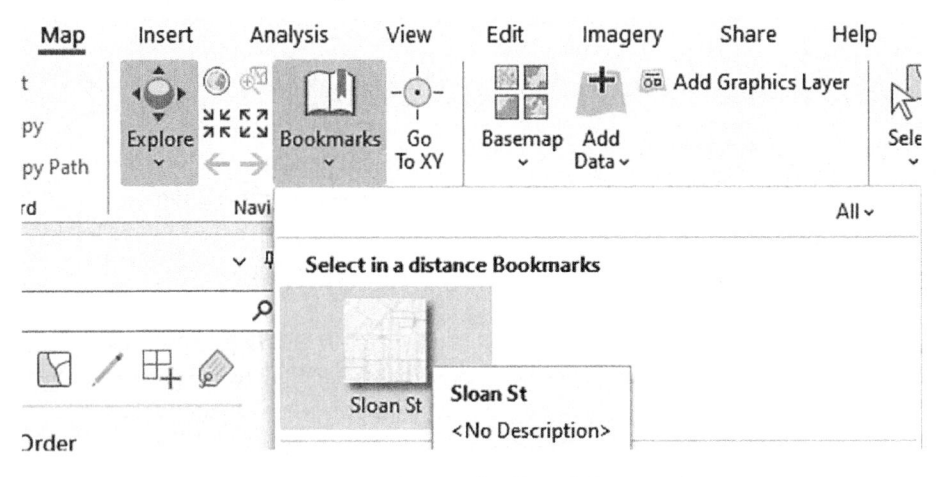

Figure 12.1 – Selecting the Sloan St bookmark

Now, you need to select all the segments that form **Sloan St**. You will use the **Select By Attributes** tool to do this.

4. Click on the **Select By Attributes** tool located in the **Selection** group on the **Map** toolbar. This will open the **Select Layers By Attributes** tool.
5. Click on the drop-down menu for **Input Rows**. Select `Street Centerlines` from the list that appears.
6. Ensure the **Selection** type is set to **New selection**.
7. Set the **Select a field** cell to `ST_NAME` using the drop-down arrow.
8. Ensure the operator cell that appears is set to `is Equal to`.
9. Type `SLOAN ST` into the values cell. It should start to auto-populate as you begin typing.
10. Verify that your **Select By Attributes** tool looks as follows, then click the **OK** button:

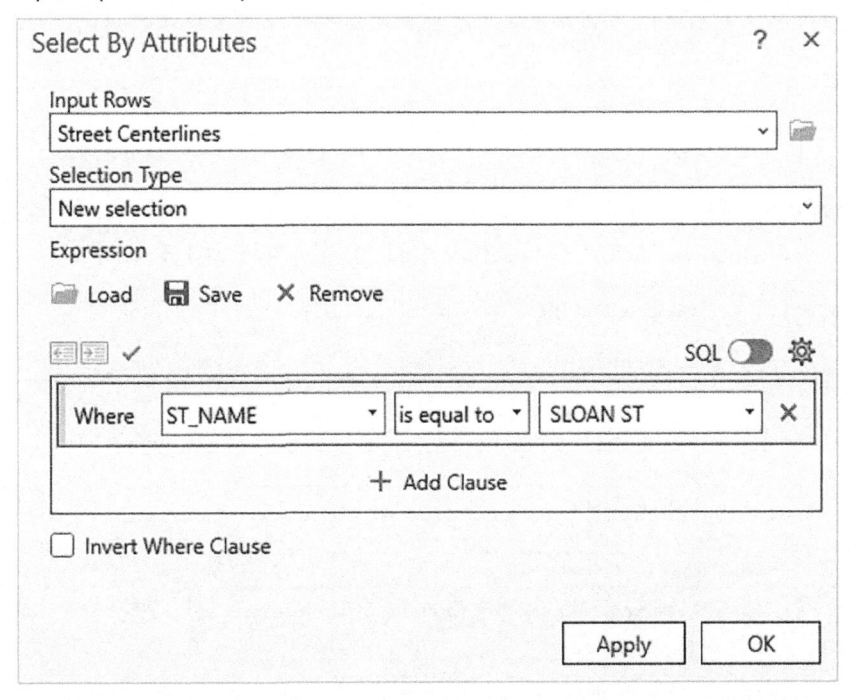

Figure 12.2 – Select By Attributes with completed parameters to select SLOAN ST

If you look at the lower-right corner of the ArcGIS Pro interface, you should see that three features are selected. These are the three centerline segments that make up **SLOAN ST**. Now, you need to select the parcels that are within 500 feet so that you can create the notification list to give to the superintendent.

11. With the Sloan Street segments selected, click on the **Select By Location** tool in the **Selection** group on the **Map** tab in the ribbon.
12. Set **Input Features** to `Parcels` by clicking on the drop-down arrow and selecting the layer from the list.
13. Set **Relationship** to `Within a distance` using the drop-down arrow and select the option from the list.
14. Set **Selecting Features** to `Street Centerlines` from the list presented to you when you click on the drop-down arrow.

> **Note**
>
> This and many other analysis tools will automatically use only the selected features in the **Input Features** or **Selecting Features** parameter area if you have selected features or records in either or both. In previous ArcGIS applications, you would need to verify a separate parameter was enabled to use only the selected features.

15. Type `500` in the **Search Distance** cell and ensure that the units in the cell to the right are set to `US Survey Feet`.
16. Verify that **Selection Type** is set to `New selection`.
17. Review whether your **Select By Location** tool looks as follows and click **OK**:

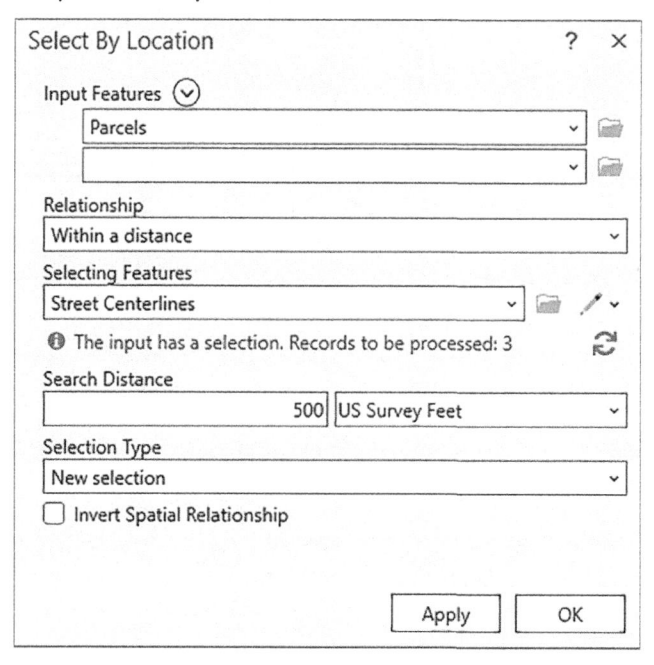

Figure 12.3 – Selecting all parcels within 500 feet of SLOAN ST using the Select By Location tool

> **Tip**
> There's a difference between the **Apply** and **OK** buttons. The **Apply** button runs the tool with the defined parameters and leaves the tool window open for further use, whereas the **OK** button runs the tool with the defined parameters but closes the tool window.

When the tool finishes, you should end up with 107 total selected features. That should include 3 selected street centerline segments and 104 parcels. Now, all that is left is to export the selected parcels to an Excel spreadsheet.

18. Click on the **Analysis** tab in the ribbon. Then, click on the **Tools** button to open the **Geoprocessing** pane.
19. Click on the **Toolboxes** tab in the **Geoprocessing** pane.
20. Expand the **Conversion Tools** toolbox and the **Excel** toolset, as shown in the following screenshot:

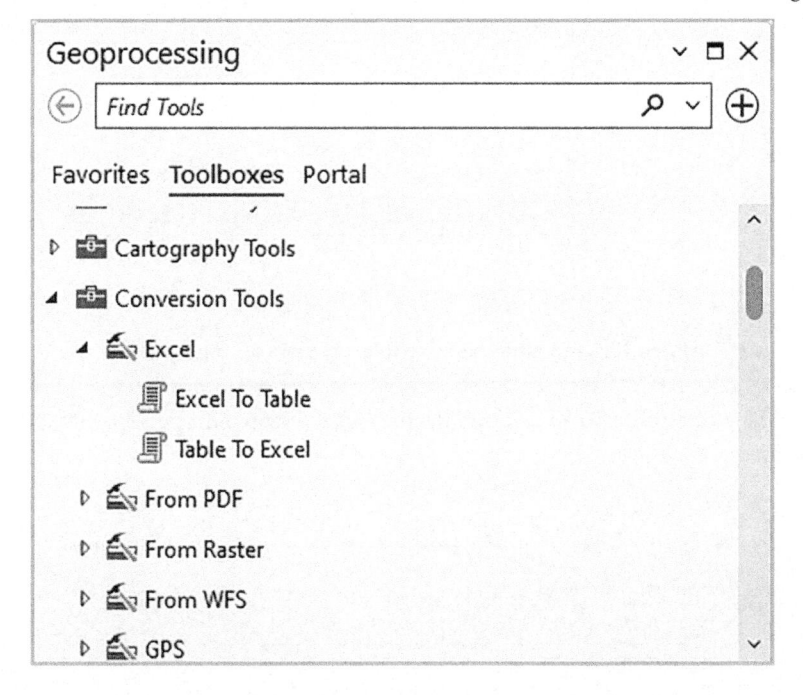

Figure 12.4 – The Conversion Tools toolbox and Excel toolset expanded

21. Click on the **Table To Excel** tool to open it.
22. Set **Input Table** to `Parcels` using the drop-down arrow and select the layer from the list that appears.
23. Click the **Browse** button for **Output Excel File**.

24. In the **Output Excel File** (.xls or .xlsx) window that appears, click on **This PC** or **My Computer** in the panel on the left.
25. In the panel on the right, double-click on the C:\ drive located under **Devices and drives**. It is generally labeled **Local Disk**, **Local Drive**, or **OS**.
26. Navigate to C:\Student\ArcGISPro3Cookbook\MyProjects.
27. Type Parcels_near_SloanST in the **Name** cell, as shown in the following screenshot, then click **Save**:

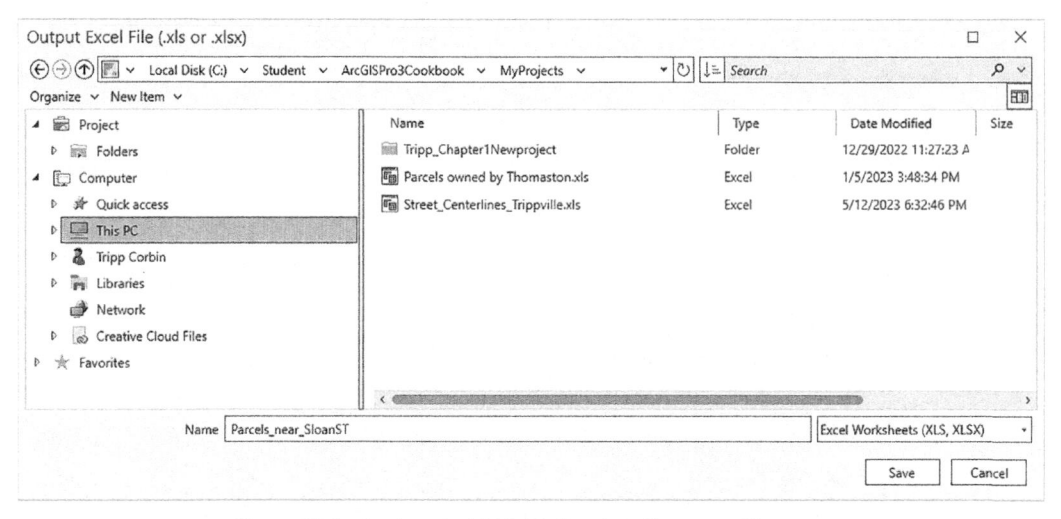

Figure 12.5 – Setting the Table To Excel tool's output filename

28. Verify that your **Table To Excel** tool matches what's shown in the following screenshot and click **Run**:

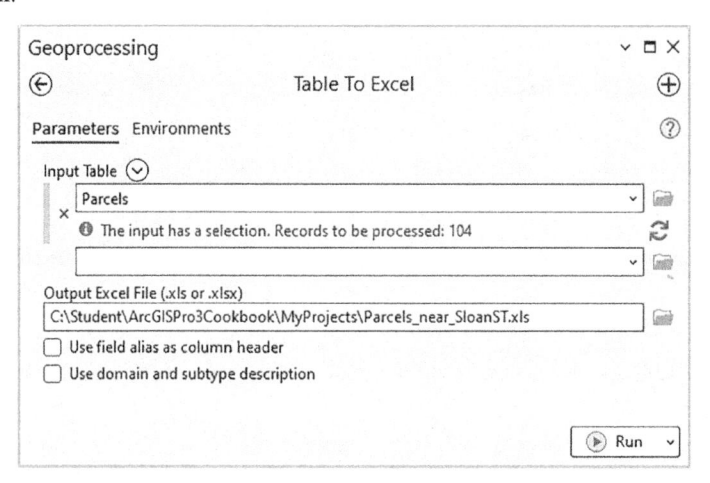

Figure 12.6 – The Table to Excel tool with completed parameters

When the tool is complete, it will export a copy of all the attribute information linked to the selected parcels to a Microsoft Excel spreadsheet.

29. In the **Selection** group on the **Map** tab in the ribbon, click on the **Clear** button to deselect all selected features.
30. Save your project and close ArcGIS Pro.

You have now completed your first proximity analysis workflow in ArcGIS Pro.

How it works...

In this recipe, you needed to create a list that the roads superintendent could use to notify all those who live or work near Sloan Street of an upcoming project. To do this, you had to select all the segments that made up Sloan Street in the GIS database. After, you performed an attribute query using the **Select By Attributes** tool. You then used this selection in the **Select By Location** tool to select all parcels within 500 feet of the selected segments of Sloan Street. Lastly, you exported the selected parcels that were near Sloan Street to a spreadsheet. The roads superintendent will now be able to use this to notify all those near the upcoming road project.

This recipe has not only shown you how to select features that are within a specified distance of other features, but it has also illustrated that analysis is often a multi-step process. It requires you to use a range of tools and skills to arrive at an answer.

Creating buffers

Buffers allow you to see what features are within a specific distance of other features. It generates polygons around the buffered features that provide the visual reference of the specified distance. These polygons can then be used to perform additional analysis or select features that are within, or intersect, the buffers.

In this recipe, you will create a buffer around the creeks and streams in the City of Trippville that will represent a protective zone where special permits are required to perform work. This is due to a recent ordinance passed by the city council to protect water quality in and around the city.

Getting ready

This recipe requires the sample data to be installed on your computer. It is recommended that you complete the recipes in *Chapter 1, ArcGIS Pro Capabilities and Terminology*, as well as the other recipes in this chapter before you start this recipe. This will ensure that you have the skills required to complete the steps in this recipe. You can complete this recipe with any ArcGIS Pro licensing level.

How to do it...

Follow these steps to use the **Buffer** geoprocessing tool to create the boundaries and areas for the protective zones around the creeks and streams in the city:

1. Start ArcGIS Pro and open the `Proximity Analysis.aprx` project located in `C:\Student\ArcGISPro3Cookbook\Chapter12\Proximity Analysis` that you used in the *Selecting features within a specific distance* recipe.
2. In the **Catalog** pane, expand the `Maps` folder. Then, right-click on the `Buffer` map and select **Open** from the menu that appears. This will open a map that contains the layers you need to perform the work you will be doing in the remainder of the recipe.
3. Click on the **Analysis** tab in the ribbon.
4. Click on the **Buffer** tool in the **Tool** group. This will open the **Buffer** tool in the **Geoprocessing** pane.
5. Set **Input Features** to `Creeks and Streams` by clicking on the drop-down arrow and selecting the layer from the list.
6. Accept the default **Output Feature Class**.

> **Note**
> While the input for the **Buffer** tool can be any feature type (point, line, or polygon), the output will always be a polygon.

7. Type `60` into the **Distance** box and select **US Survey Feet** from the **Units** drop-down list located under **Linear Unit**.
8. You will also need to accept the defaults for **Side Type**, **End Type**, and **Method**; here, **Side Type** should be **Full**, **End Type** should be **Round**, and **Method** should be **Planar**.

> **Note**
> You can only access and change these options if you have an Advanced license.

9. Set **Dissolve Type** to **Dissolve all output features into a single feature class** by clicking on the drop-down arrow.

    You must select this option because the protective zone is consistent for all creeks and streams. This will create a single buffer that takes up less space and is easier to manage.

10. Verify that your **Buffer** tool looks as follows, then click the **Run** button:

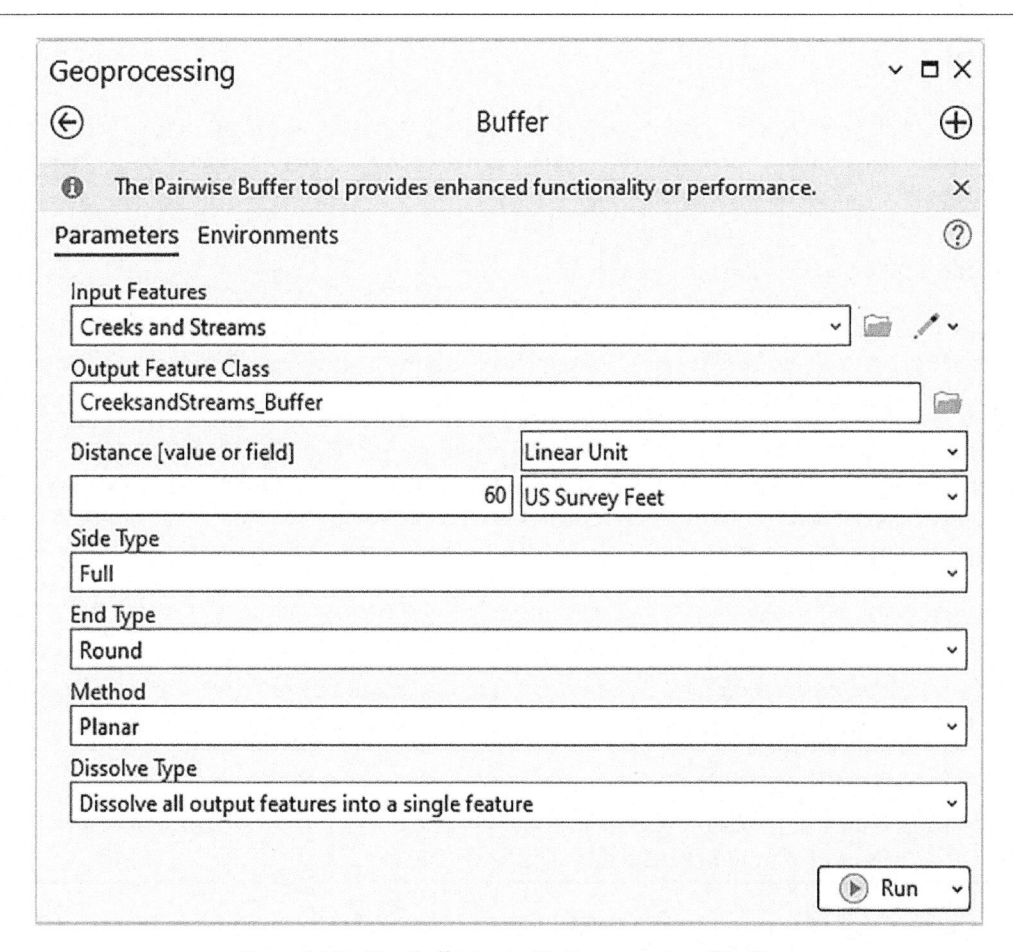

Figure 12.7 – The Buffer tool with its parameters filled in

When the **Buffer** tool completes, a new layer will appear in your **Contents** pane. This is the buffer feature class you just created.

11. Close the **Geoprocessing** pane to free up screen space.
12. Save your project.

You have just created a new feature class in the project geodatabase that represents the protective area around the creeks and streams. You can easily see the areas that are impacted by this new protective zone. You can also use this layer to perform various analyses. For example, you could use the **Select By Location** tool to select all parcels that intersect the protective buffer. This will allow you to know exactly which parcels would need a permit for any work they do near one of the creeks or streams. You could also use the **Union** geoprocessing tool to calculate how much of each parcel is within the protective zone and how much is outside.

Challenge

Using the skills you learned in this recipe and the *Selecting features within a distance* recipe, select all the parcels that are intersected by the new protective zone buffer you just created, then export the selected parcels to an Excel spreadsheet.

There's more...

You have just learned how to create a single buffer around features using the **Buffer** tool. But what if you need to create several buffers at various distances around a set of features or a single layer? You could just use the **Buffer** tool multiple times. However, there is an easier way.

The fire chief for the City of Trippville is working on a fire protection study and needs your help to determine the level of protection provided by the fire hydrants within the city. They need to know how much of each parcel is 100, 200, and 300 feet from each hydrant. You will use the **Multiple Ring Buffer** tool to accomplish this:

1. First, you must clean up the buffer map you have been using. Right-click on the `Natwtr_Stream_Buffer` layer and select **Remove** from the menu that appears.
2. Repeat this process for the `Creeks and Streams` and `Lakes and Rivers` layers.
3. In the **Catalog** pane, expand the `Databases` folder so that you can see its contents.
4. Expand the `Trippville_GIS.gdb` geodatabase.
5. Expand the `Water` feature dataset and right-click on the `fire_hyd` feature class. Select **Add to Current Map** from the menu that appears.
6. Save your project.
7. Click on the **Analysis** tab in the ribbon.
8. Click on the **Tools** button in the **Geoprocessing** group to open the **Geoprocessing** pane.
9. Click on the **Toolboxes** tab near the top of the **Geoprocessing** pane.
10. Expand the **Analysis** toolbox and the **Proximity** toolset.
11. Click on the **Multiple Ring Buffer** tool to open it in the **Geoprocessing** pane. Notice that it has a different icon from all the other tools. This is because this tool is a Python script.

> **Note**
>
> Python is the primary scripting language for ArcGIS. It allows you to create custom tools that help automate tasks. These can then be scheduled to run automatically, at specified times, or run manually. To learn more about creating Python scripts for ArcGIS, you may want to check out *Python for ArcGIS Pro*, by Silas Toms and Bill Parker, at `https://www.packtpub.com/product/python-for-arcgis-pro`.

12. Set **Input Features** to the `fire_hyd` layer you just added by clicking on the drop-down arrow and selecting it from the list.
13. Accept the default value for **Output Feature Class**.
14. Type `100` in the cell under **Distances** and click **Add another**. Another cell should appear.
15. Type `200` in the new cell that appears and click **Add another**. Again, another cell should appear.
16. Type `300` in the new cell that appears and press the *Enter* key.
17. Set **Distance Units** to **Feet** using the drop-down arrow.
18. Accept the default value for **Buffer Distance Field Name**.
19. Set **Dissolve Option** to **Non-overlapping (Rings)**. This will allow the fire chief to know how much coverage they have at each distance.

> **Tip**
>
> **Dissolve Option** determines whether you create rings/donuts or disks. The **Non-overlapping (Rings)** option will generate rings that resemble donuts – that is, circles with their centers cut out. The overlapping area from the other specified distances is cut out. So, in this example, you will get a single ring that extends out to 100 feet. The next ring will cover from 100 to 200 feet. The last ring will cover from 200 to 300 feet.
>
> The **Overlapping (disks)** or **None** option creates solid disks that overlap. Using this recipe as an example, you will have one disk that extends out to 100 feet. You will have a second disk that extends out to 200 feet and overlaps the first disk. The last disk will extend out to 300 feet and overlap the other two disks.
>
> You are not required to set your buffer distances to equal intervals. They can be any distance you wish. To learn more about the **Multiple Ring Buffer** tool, go to `https://pro.arcgis.com/en/pro-app/latest/tool-reference/analysis/multiple-ring-buffer.htm`.

20. Set **Method** to **Planar** using the drop-down arrow.

21. Verify that your **Multiple Ring Buffer** tool matches what's shown in the following screenshot and click **Run**:

Figure 12.8 – The Multiple Ring Buffer tool with completed parameters

The new layer should appear in the **Content** pane, which shows the results of the **Multiple Ring Buffer** tool. Now, let's verify the results.

22. Select the `fire_hyd_MultipleRingBuffer` layer in the **Contents** pane.
23. Click on the **Feature Layer** tab in the ribbon.
24. Click on the **Import** button in the **Drawing** group.
25. Verify that **Input Layer** is set to `fire_hyd_MultipleRingBuffer`.
26. Click on the **Browse** button for **Symbology Layer**. Then, select the **Folders** folder under **Project** in the left panel of the **Symbology Layer** window.
27. Double-click on the **Proximity Analysis** folder in the right panel.
28. Select `Distance_from_Hydrant.lyrx` and click **OK**.

29. Verify that your **Import Symbology** tool looks as follows, then click **OK**:

Figure 12.9 – The Import Symbology tool with completed parameters

The symbology should change for the multiple ring buffer layer you created. Now, you can see each ring for the distances you specified in the **Multiple Ring Buffer** tool, as illustrated in the following screenshot:

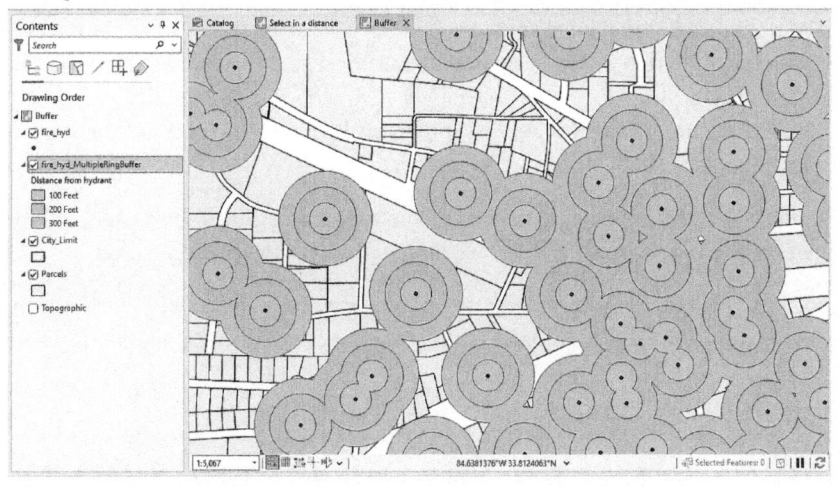

Figure 12.10 – Buffer rings displayed by distance

30. Save your project and close ArcGIS Pro.

With that, you have created a new layer that represents distances around the fire hydrants in the city. This will allow the fire chief to easily determine how much protection the fire hydrants provide to the city's citizens and businesses. Using this data, they can determine if any new fire hydrants should be installed to improve fire protection in the city.

Determining the nearest feature using the Near tool

So far, you have seen how we can use buffers to show areas that are within a specified distance from features. This allows us to see if other features are within that area. We can also use those areas to perform additional analysis. However, these are all general measurements. They do not tell us which feature is the closest to another feature. How do we determine which feature is closest to another feature?

If you have an advanced license for ArcGIS Pro, this is pretty easy. The **Near** tool will determine the closest feature for you. Not only will it determine which feature is closest, but it also has options to calculate the distance, direction, coordinates, and more of the closest features. The **Near** tool works with points, lines, and polygons. You can also specify that it only searches for the closest feature within a specified distance. So, let's put this tool into action.

In this recipe, you will continue to assist the fire chief. They are still working on their fire protection study. They want to know which fire hydrant is closest to each parcel so that the department will know exactly which hydrant they should connect to when responding to a given location. You will use the **Near** tool to calculate which hydrant is closest to each parcel.

Getting ready

You will need to make sure you have an Advanced license for ArcGIS Pro to complete this recipe. You will not be able to do this with a Basic or Standard license. If you are not sure what license level you have or how to determine this, refer to *Chapter 1, ArcGIS Pro Capabilities and Terminology*, the recipe entitled *Determining your ArcGIS Pro license level*. That recipe outlines the steps required to determine your license level if you do not remember.

If you are limited to a Basic or Standard license, then please read through this recipe, and then proceed with the *There's more...* section, which can be completed with a lower license level.

It is also recommended that you complete the other recipes in this chapter before starting this one. This will ensure that you have a good basic understanding of proximity analysis within ArcGIS Pro.

How to do it...

Follow these steps to learn how to use the **Near** tool to locate the fire hydrant that is nearest to each parcel. Remember that the use of this tool requires an Advanced license of ArcGIS Pro:

1. Launch ArcGIS Pro and open the `Nearest Feature.aprx` project located in the `C:\Student\ArcGISPro3Cookbook\Chapter12\Proximity Analysis` folder.

It should open with the **Trippville Fire Protection** map. Verify that this map contains the `Fire Hydrant` and `Parcels` layers at a minimum. It should look as follows. Your scale may be different, depending on the size of your monitor and what panes you have open:

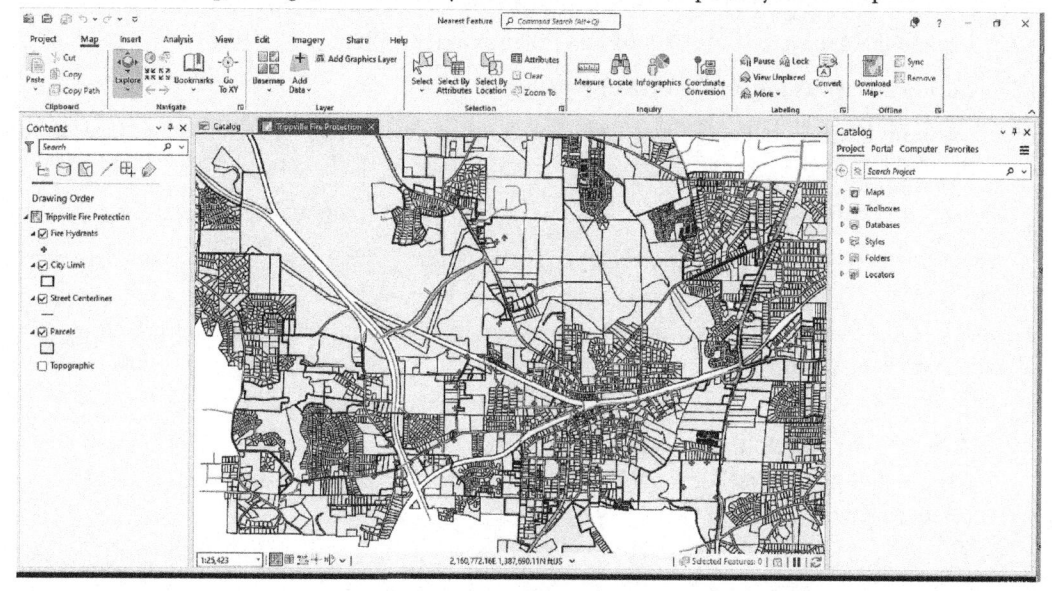

Figure 12.11 – Trippville Fire Protection opened in ArcGIS Pro

2. Right-click on the `Parcels` layer in the **Contents** pane and select **Attribute Table** from the menu that appears.

3. Review the available fields and answer the following question:

   Question: What fields are associated with the parcels in the attribute table?

   Answer:

   As you can see, there are only a few fields associated with the parcels. The attribute table is currently storing **ObjectID**, **Shape**, **LOTNUM**, **STNUMBER**, **STNAME**, **STSUFFIX**, **Shape_Length**, **Shape_Area**, and **Acres**. There is nothing here that tells you which fire hydrant is closest to each parcel. You are about to change that.

4. Close the **Parcels** attribute table by clicking the small **X** next to the table name in the tab at the top of the table view. This will free up more screen real estate.

> **Tip**
>
> If you have multiple monitors, or a really large one, you could also undock and drag the table pane to a different monitor or location. This would free up space for viewing the map but also allow you to continue to view the table. Also, ArcGIS Pro lets you display each table in its own window if you desire, unlike ArcMap, which forces all tables into a single window.

5. Click on the **Analysis** tab in the ribbon, then click on the **Tools** button in the **Geoprocessing** group. This will open the **Geoprocessing** pane.
6. Click on the **Toolboxes** tab at the top of the **Geoprocessing** pane.
7. Expand the **Analysis Tools** toolbox and the **Proximity** toolset.
8. Click on the **Near** tool in the **Proximity** toolset.
9. Set **Input Features** to **Parcels** by clicking on the drop-down arrow and selecting the layer from the list presented.
10. Set **Near Features** to **Fire Hydrants** using the drop-down arrow and selecting it from the list.

> **Tip**
> You can set more than one layer or feature class under **Near Features**. Also, **Input Features** and **Near Features** can be the same if you are looking for the nearest feature in the same layer.

11. Type 300 for **Search Radius** and ensure the units are set to **US Survey Feet**. The fire chief has indicated that any parcel over 300 feet from a hydrant is considered too far from a hydrant to be adequately protected.
12. Click on the **Location** and **Angle** boxes so that a checkmark appears, enabling these options. This will provide additional information for the closest hydrant.

> **Note**
> The **Location** option will add the X and Y coordinates for the nearest hydrant to the **Parcels** attribute table. The **Angle** option will add the angle to the nearest hydrant based on grid north.
>
> It should be noted that the **Near** tool is one of the few geoprocessing tools that edits existing data. In the case of the **Near** tool, it will add new fields to the **Parcels** attribute table since the `Parcels` layer is set for **Input Features**. For more information on the **Near** tool, go to `https://pro.arcgis.com/en/pro-app/latest/tool-reference/analysis/near.htm`.

13. Ensure **Method** is set to **Planar**.
14. Accept all the defaults for **Field Names**.
15. Set **Distance Unit** to **US Survey Feet**.

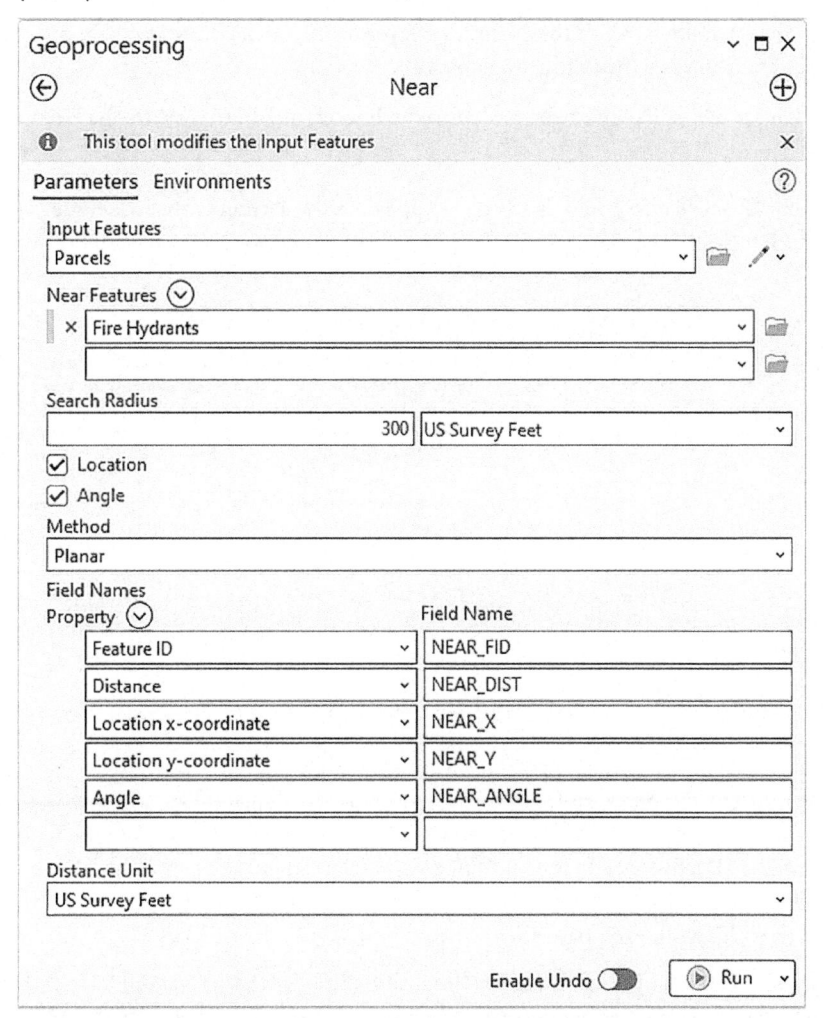

16. Verify that your **Near** tool looks as follows, then click **Run**:

Figure 12.12 – The Near tool with completed parameters

Notice that this time, no new layers were added to your map, unlike with **Buffer** and **Multiple Ring Buffer**. This is because the **Near** tool does not create a new feature class or table. Instead, it adds new fields to the **Input Features** attribute table. Let's see what fields were added to the `Parcels` layer by the **Near** tool.

> **Tip**
> You must also have edit permissions to the **Input Features** layer to use this tool.

17. Close the **Geoprocessing** pane to free up screen space once the tool completes.
18. Open the attribute table for the `Parcels` layer using the same method you did earlier in this recipe. Then, answer the following question:

    Question: What fields have been added to the **Parcels** attribute table by the **Near** tool?

    Answer:

    As you can see, several new fields now appear in the **Parcels** attribute table. This includes **Near_FID**, **Near_Dist**, **Near_X**, **Near_Y**, and **Near_Angle**, as shown in the following screenshot:

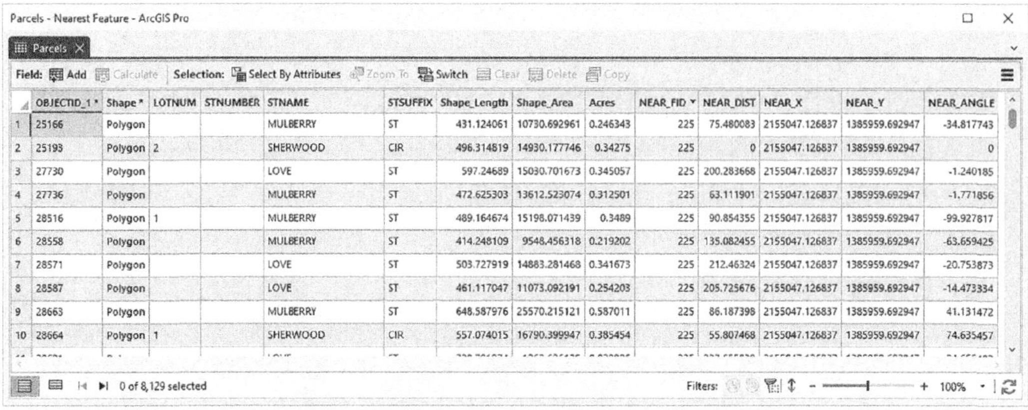

	OBJECTID_1 *	Shape *	LOTNUM	STNUMBER	STNAME	STSUFFIX	Shape_Length	Shape_Area	Acres	NEAR_FID	NEAR_DIST	NEAR_X	NEAR_Y	NEAR_ANGLE
1	25166	Polygon			MULBERRY	ST	431.124061	10730.692961	0.246343	225	75.480083	2155047.126837	1385959.692947	-34.817743
2	25193	Polygon	2		SHERWOOD	CIR	496.314819	14930.177746	0.34275	225	0	2155047.126837	1385959.692947	0
3	27730	Polygon			LOVE	ST	597.24689	15030.701673	0.345057	225	200.283668	2155047.126837	1385959.692947	-1.240185
4	27736	Polygon			MULBERRY	ST	472.625303	13612.523074	0.312501	225	63.111901	2155047.126837	1385959.692947	-1.771856
5	28516	Polygon	1		MULBERRY	ST	489.164674	15198.071439	0.3489	225	90.854355	2155047.126837	1385959.692947	-99.927817
6	28558	Polygon			MULBERRY	ST	414.248109	9548.456318	0.219202	225	135.082455	2155047.126837	1385959.692947	-63.659425
7	28571	Polygon			LOVE	ST	503.727919	14883.281468	0.341673	225	212.46324	2155047.126837	1385959.692947	-20.753873
8	28587	Polygon			LOVE	ST	461.117047	11073.092191	0.254203	225	205.725676	2155047.126837	1385959.692947	-14.473334
9	28663	Polygon			MULBERRY	ST	648.587976	25570.215121	0.587011	225	86.187398	2155047.126837	1385959.692947	41.131472
10	28664	Polygon	1		SHERWOOD	CIR	557.074015	16790.399947	0.385454	225	55.807468	2155047.126837	1385959.692947	74.635457

Figure 12.13 – Results of the Near tool

Those parcels with a -1 value in the fields are not within the 300-foot search radius you specified when you ran the **Near** tool. So, these can easily be identified as not having adequate fire protection. Those parcels that were within 300 feet of a hydrant now have the hydrants feature ID (**Near_FID**), distance to the hydrant (**Near_Dist**), angle to the hydrant (**Near_Angle**), and X, Y coordinates (**Near_X** and **Near_Y**) assigned to each parcel. This allows you to learn which hydrant is closest to each parcel.

Near_FID is the **ObjectID** value for the nearest fire hydrant. You could use those two fields to join, relate, or create a relationship class between the two layers. This would allow you to access even more information.

19. Close the **Parcels** attribute table when you are done reviewing it.
20. Save your project.

There's more...

The **Near** tool requires you to have an Advanced license to determine which feature is the closest. What happens if you only have a Basic or Standard license? Is it possible to determine the closest feature without an Advanced license?

Yes, you can, although it is not as straightforward and does produce a new feature class. You can use the **Spatial Join** tool for this. Let's see how it works. This part of the recipe will assume that you have not been able to complete the recipe up to this point because you did not have an Advanced license:

1. Click on the **Analysis** tab in the ribbon.
2. Click on the **Spatial Join** tool in the **Tools** group on the **Analysis** tab. This will open the **Spatial Join** tool in the **Geoprocessing** pane.

> **Tip**
>
> You can also find this tool in the **Overlay Analysis** toolset in the **Analysis Tools** toolbox. The **Spatial Join** tool allows you to join layers or feature classes together based on many spatial relationships.

3. Set **Target Features** to the `Parcels` layer by using the drop-down arrow and selecting it from the list.
4. Set **Join Features** to `Fire Hydrants`, also using the drop-down arrow.
5. Accept the default **Output Feature Class**, which should be something similar to `Parcels_SpatialJoin`.

> **Note**
>
> Unlike the **Near** tool, the **Spatial Join** tool creates a completely new feature class. It does not simply add new fields to an existing layer. This means that your original `Parcels` layer will remain unchanged.

6. Set **Join Operation** to **Join one to one** using the drop-down arrow. You must use the one-to-one option because only one hydrant should be the closest to each parcel.
7. Ensure the **Keep All Target Features** checkbox is checked. This will ensure the resulting output feature class will contain all the parcels, even if they do not have a fire hydrant nearby.
8. Scroll down to **Match Option** and select **Closest** using the drop-down arrow. This is the spatial relationship that ArcGIS Pro will use to join the data. **Closest** is near the bottom of the list. Take your time and see what other options are available.
9. Set **Search Radius** to **300 US Survey Feet**, as we did when using the **Near** tool. This limits the search for the closest fire hydrant to 300 feet.
10. Leave **Distance Field Name** blank.

11. Verify that your **Spatial Join** tool looks as follows, then click **Run**:

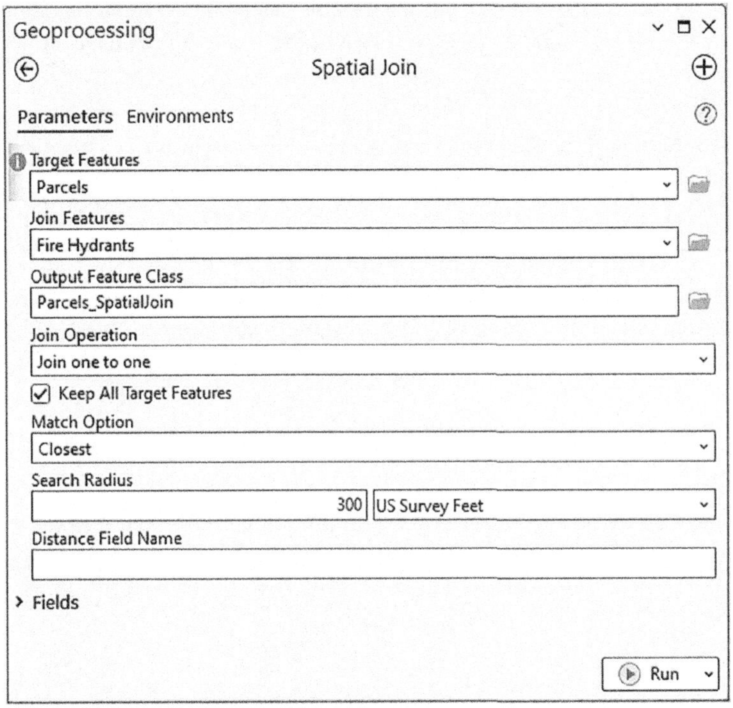

Figure 12.14 – The Spatial Join tool with completed parameters

A new layer should appear in your **Contents** pane. This new layer is the result of the **Spatial Join** tool. Close the **Geoprocessing** pane.

12. Right-click on the new layer that was just added. Select **Attribute table** from the menu that appears.
13. **Attribute table** contains information from both the `Parcels` and `Fire Hydrants` layers. Take a moment to review the results.

> **Note**
>
> The **Join_Count** field tells you how many fire hydrants were joined to the parcels. Because you chose the **Join one to one** option, this number should either be a 1 or a 0. **Target_FID** is the **ObjectID** value for the closest fire hydrant to the parcel. The **FireHyd_ID**, **X**, **Y**, **Picture**, and **Picture 2** fields all come from the closest fire hydrant to the parcel. The remaining fields all came from the `Parcels` layer.

14. Close the attribute table when you are done reviewing it.
15. Save your project and close ArcGIS Pro.

With that, you have successfully joined information from the nearest fire hydrant to each parcel creating a completely new layer using the **Spatial Join** tool.

How it works...

So, you now have two ways to calculate the nearest feature in ArcGIS Pro. The first way is to use the **Near** tool. This tool added new fields to the attribute table of the `Parcels` layer that identified the nearest fire hydrant, as well as the direction and distance to that fire hydrant. The **Near** tool requires you to have an Advanced license of ArcGIS Pro.

The second method was to use the **Spatial Join** tool. This resulted in a new layer being created that contained the combined data for each parcel and the fire hydrant that was nearest to that parcel. Unlike the **Near** tool, the **Spatial Join** tool does create new data.

If you have an Advanced license, you can use either method. If you only have a Basic or Standard license, then you must use **Spatial Join**. But if you can use either, which is best? That depends on what you need to do with the resulting data. **Spatial Join** is best when you are creating data that is intermediate to a large process and will not be needed later. Otherwise, it creates yet another layer you must maintain. The **Near** tool does not create more data for you to manage. It only adds to an existing layer of information that you are already maintaining.

Calculating how far apart features are using the Generate Near Table tool

You now know a couple of different ways to find the closest feature. But what if you want to calculate the distance from one group of features to another group, not just look for the closest? The **Generate Near Table** tool can do this.

The **Generate Near Table** tool will create a new database table that lists how far the features in the input are from those in the near feature layer.

Like the **Near** tool, you have the option to include additional fields for location and angle as well. This tool also requires an Advanced license.

The City of Trippville has several water quality monitoring stations that monitor the quality of the water in the creeks, streams, and rivers in the city. Recently, several of them have been reporting poor or bad water quality. It is believed that this might be due to runoff from an industrial site in the city.

In this recipe, you will use the **Generate Near Table** tool to identify all industrial parcels located near the stations reporting bad water quality. These will be the industrial sites you need to inspect first to check whether they are the cause of the bad water quality.

Getting ready

You will need to make sure you have an Advanced license for ArcGIS Pro to complete this recipe. You will not be able to do it with a Basic or Standard license. If you are not sure what license level you have or how to determine this, refer to *Chapter 1, ArcGIS Pro Capabilities and Terminology*, and the recipe entitled *Determining your ArcGIS Pro license level*. That recipe will tell you how to determine your license level.

If you are limited to a Basic or Standard license, then please read through this recipe, and then proceed with the *There's more...* section, which can be completed with a lower license level.

It is also recommended that you complete the other recipes in this chapter before starting this one. This will ensure that you have a good basic understanding of proximity analysis within ArcGIS Pro.

How to do it...

In this recipe, you will use the **Generate Near Table** tool to create a list of industrial sites located closest to the city's monitoring stations. This table will then form the list of sites the inspection teams need to look at first as possible sources of pollution in the streams, creeks, and rivers:

1. Start ArcGIS Pro and open the `Near Table.aprx` project located in `C:\Student\ArcGISPro3Cookbook\Chapter11` using the skills you have learned in past recipes. The project should open with the Trippville map displayed.

    With the correct project opened and the map displayed, start by selecting the parcels in the city that are industrial-zoned. You will use the **Select By Attributes** tool to do this.

2. Activate the **Map** tab in the ribbon and select the **Select By Attributes** tool in the **Selection** group. The **Select Layer By Attributes** tool will open in the **Geoprocessing** pane.

3. Set **Input Rows** to **Zoning** by using the drop-down arrow and selecting it from the list.

4. Ensure **Selection Type** is set to **New selection**.

5. Set **Select a Field** cell to **ZONING** using the drop-down arrow.

6. Make sure the **Operator** cell is set to **is equal to** and the blank cell that appears to the right of the operator is set to **HI** using the drop-down arrow. HI stands for **Heavy Industrial**.

7. Verify that your query looks as follows, then click the **Add Clause** button:

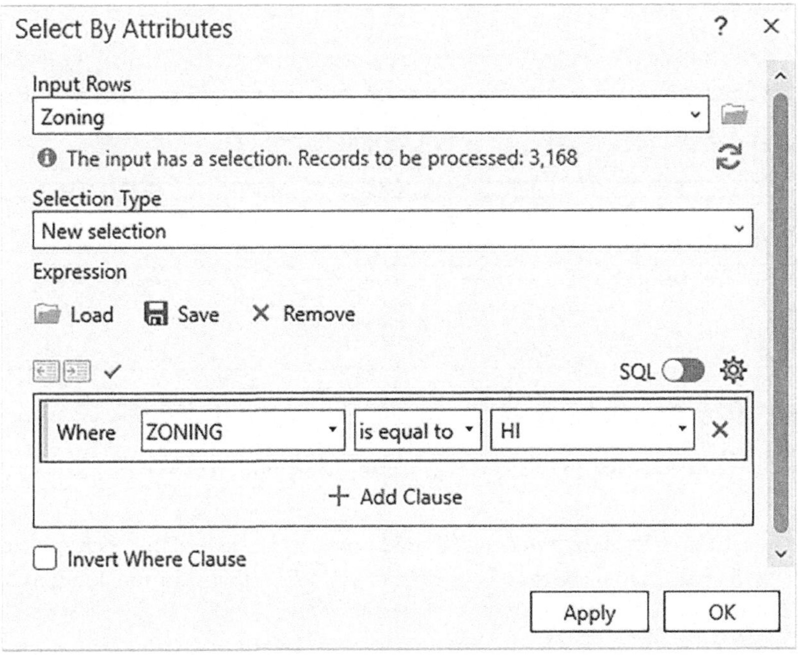

Figure 12.15 – Select By Attributes with a clause to select parcels zoned as HI

Two zoning classifications are used for industrial zoning. You need to add a second clause to account for the second classification.

8. Where it says **And** below the clause you created previously, click the drop-down arrow and select **Or** from the list that appears.

> **Note**
>
> **And** and **Or** are clause connectors. They determine how clauses in a query work together. If you had left the query with **And**, then the selection would have to meet both clauses. Since a parcel cannot be zoned both heavy and light industrial, that means nothing would be selected. **Or** means it only must meet the requirements of one of the two clauses. So, in this case, the query will select all features that are zoned either light or heavy industrial.

9. Set **Select a Field** to ZONING, **Operator** to **is equal to**, and the value to **LI**. Your query should now look like this:

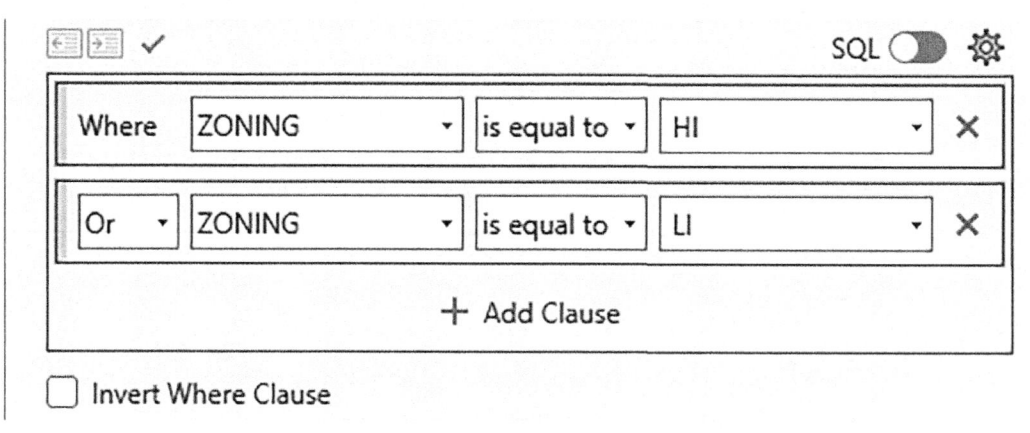

Figure 12.16 – Query to select parcels Zoned HI or LI

10. Your **Select Layer By Attributes** tool should now contain two SQL Where clauses. They will select all the parcels that are zoned either LI or HI. Verify that your tool looks as follows, then click **OK**:

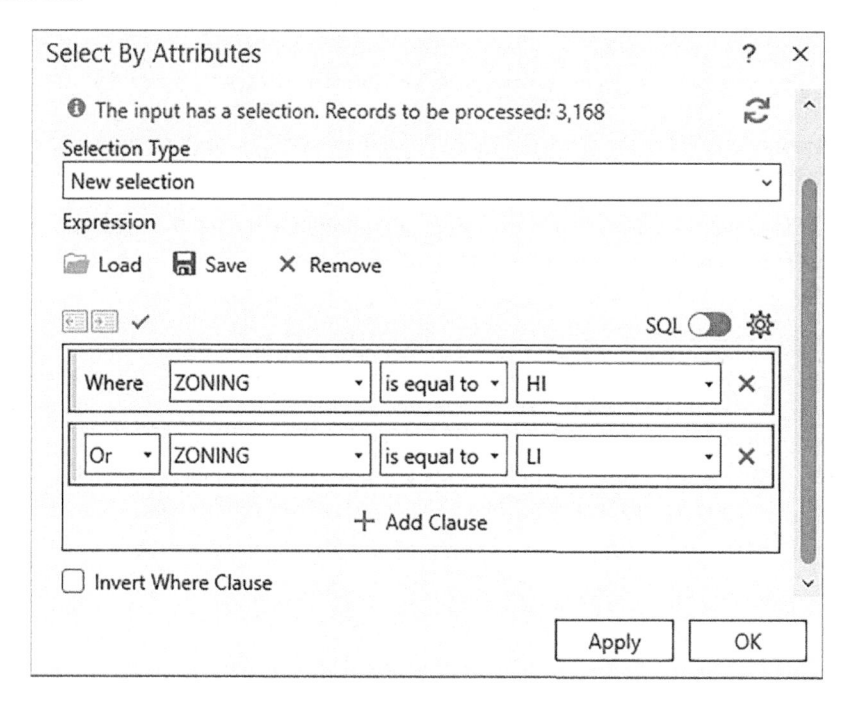

Figure 12.17 – Using Select By Attributes to select all parcels with industrial zoning classification

When you run the query, approximately 151 parcels should be selected within the map. All of them should be zoned either LI or HI. Next, you need to select the water quality stations that are reporting bad water quality. You will do this with another **Select By Attributes** query.

11. On the **Map** tab, click on the **Select By Attributes** button again.
12. If the previous query you created is still showing, click the **X** button to clear it out. You will be creating a new one.
13. Set **Input Rows** to **Water Quality Station** using the drop-down arrow.
14. Ensure **Selection Type** is set to **New Selection**.
15. In the **Where** clause, set **Select a Field** to **Current Status** using the drop-down arrow.
16. Ensure **Operator** is set to **is equal to** and the value cell is set to **Bad**.
17. Verify that your **Select By Attributes** tool looks as follows, then click **OK**:

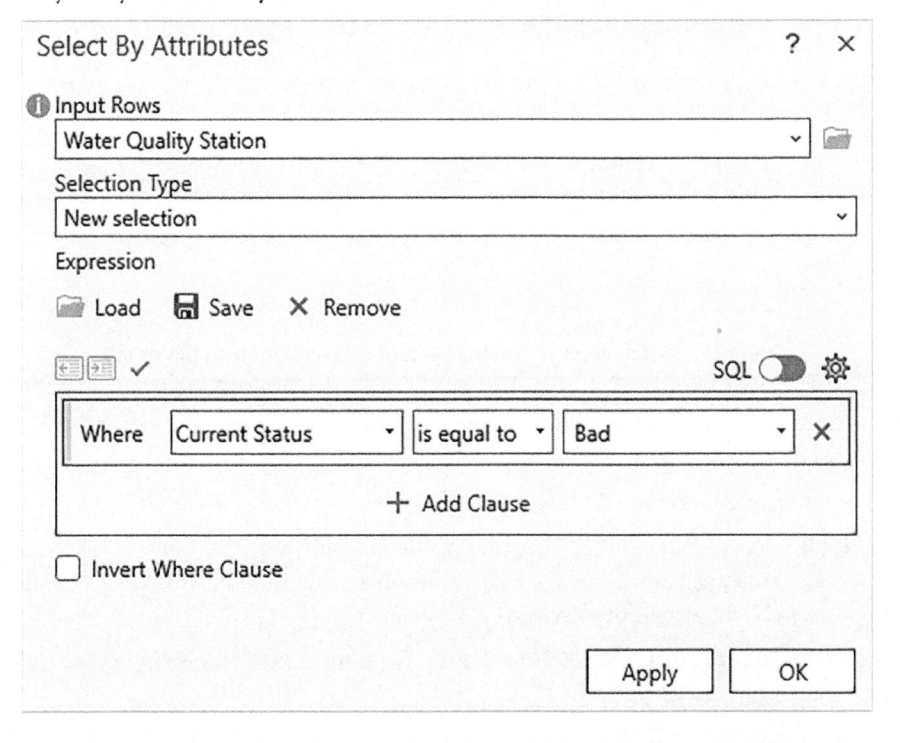

Figure 12.18 – Selecting water quality stations with a current status of Bad

You now have two selection sets. You have features selected in the `Parcels` layer that are zoned for industrial use – LI or HI. You also have all the water quality monitoring stations that have a status of **Bad**. Now, you need to determine which of the industrial-zoned parcels are close to the selected monitoring stations. You will use the **Generate Near Table** tool to do this.

18. In the **Contents** pane, click on the **List By Selection** button. It is the third one from the left. You should now have approximately two water quality stations and 151 zoning polygons selected.

    Now that you have selected the appropriately zoned parcels and water quality stations, you need to create a table that lists which of the selected parcels is near the selected water quality stations. You will limit your search to a 1,500-foot radius around the water quality stations.

19. Click on the **Analysis** tab in the ribbon.
20. Click on the **Generate Near Table** tool in the **Tools** group. If you don't see the tool, click on the arrows located on the right-hand side of the group to navigate up and down the list of tools in the group, as illustrated in the following screenshot:

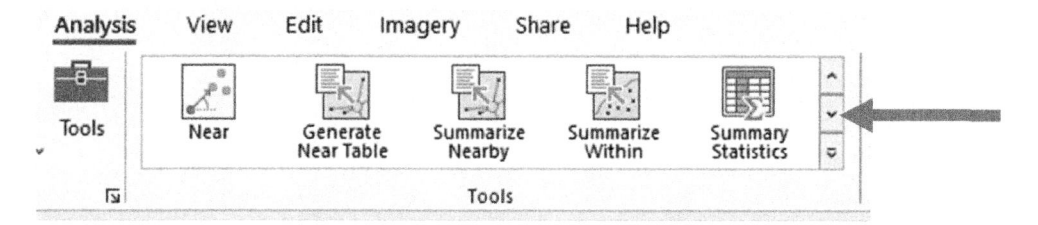

Figure 12.19 – Finding the Generate Near Table tool on the Analysis tab

21. The **Generate Near Table** tool will appear in the **Geoprocessing** pane. Set **Input Features** to **Water Quality Station** using the drop-down arrow.
22. Now, set **Near Features** to **Zoning** using the drop-down arrow on the far right-hand side.
23. Accept the default value ArcGIS Pro generates for **Output Table**. Like most geoprocessing tools, this one is creating a new table. By default, the table is stored in the **Project** geodatabase. This happens to be the **Proximity Analysis** geodatabase.
24. Set **Search Radius** to **1,500 US Survey Feet** by typing 1500 and verifying the units in the adjoining cell.
25. Uncheck **Find only closest feature** if it is checked. You want to find all industrial-zoned parcels within 1,500 feet of the water quality monitoring stations that are reporting bad quality, not just the closest one. Leave **Maximum number of closest matches** set to **0**. This will ensure all matches within 1,500 feet are returned.
26. **Method** should be set to **Planar**.
27. Set **Distance Unit** to **US Survey Feet** using the drop-down arrow.

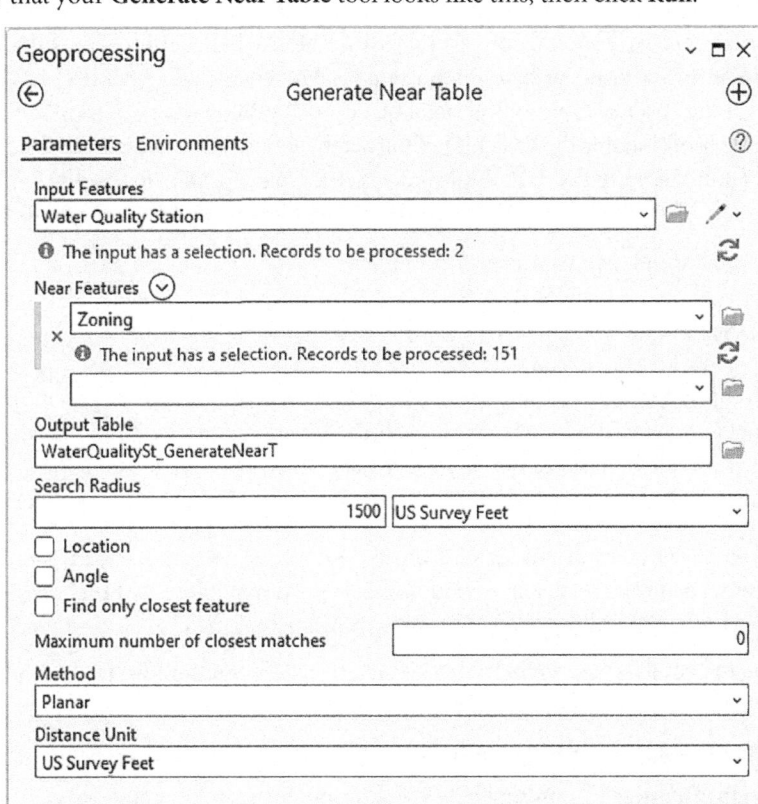

28. Verify that your **Generate Near Table** tool looks like this, then click **Run**:

Figure 12.20 – The Generate Near Table tool with its parameters complete

When the tool finishes, a new standalone table should appear in your **Contents** pane. This table is the results of the tool you just ran. Let's take a look at the results.

29. Close the **Geoprocessing** pane once the tool finishes.
30. Right-click on the new table you just created, that was added to the **Contents** pane. Select **Open** from the menu that appears.

> **Tip**
> As we have mentioned previously, geoprocessing tools, such as the **Create Near Table** tool, only use selected features automatically. Since you selected specific features in the `Water Quality Stations` and `Zoning` layers, this tool only used those selected features to run its analysis.

31. Review the newly created table.

    In the new table, you should see several fields. The **IN_FID** field contains the object ID for **Water Quality Station**. You should see a value of 3 or 4, which matches the object ID for those stations in the attribute table for the water quality stations that have a current status of **Bad**. The **Near_FID** field contains the object ID of the nearby industrial-zoned parcel. Each water quality station should be paired with multiple parcels. The **Near_Rank** field identifies which is closest to each water quality station. This helps you prioritize which parcels you should begin with.

32. Close the table and save your project.

There's more...

So, can you do this without an Advanced license of ArcGIS Pro? The answer is yes. You can use the **Spatial Join** tool to accomplish something very similar. However, the result is not a table – it is a new feature class that will contain many overlapping features. If you just want a table, you can export the records from the results of the **Spatial Join** tool to a new table. Here's how all that works:

1. If you still have the features selected from the `Water Quality Station` and `Zoning` layers, proceed to the next step. If you do not, repeat *steps 2* to *24* of this recipe. You should have 2 water quality stations and 151 zoning polygons selected.

2. Click on the **Analysis** tab in the ribbon. Then, select the **Spatial Join** tool in the **Tools** group. This will open the tool in the **Geoprocessing** pane.

3. Set **Target Features** to **Water Quality Station** using the drop-down arrow.

4. Set the **Join Features** to **Zoning** using the drop-down arrow.

5. Click **Browse** button to set **Output Feature Class**.

6. Select **Databases** from the panel on the left. Then, double-click on the `Proximity Analysis.gdb` geodatabase in the panel on the right.

7. Type `WaterQual_IndustZoned_SpatialJoin` into the **Name** field and click **Save**.

8. Set **Join Operation** to **Join one to many**. This will ensure the result has all the industrial-zoned parcels within 1,500 feet of the water quality stations reporting bad quality.

9. Set **Match Option** to **Within a distance** using the drop-down arrow.

10. Set **Search Radius** to `1500` and the units to **US Survey Feet**.

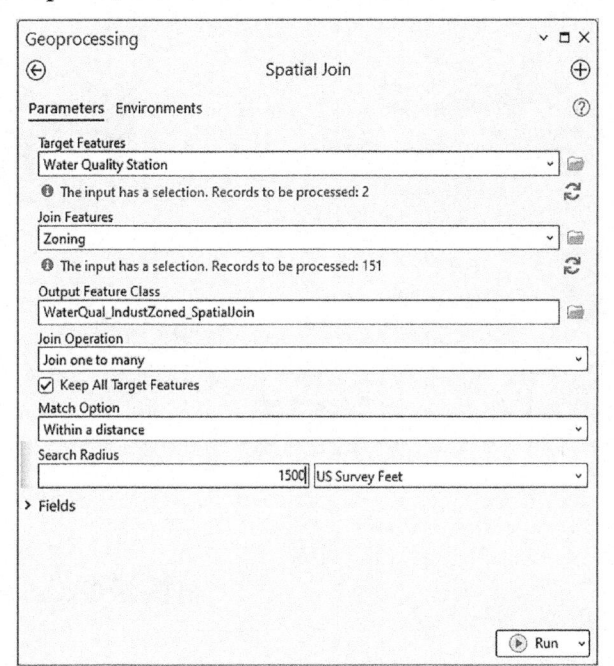

11. Verify that your **Spatial Join** tool matches what's shown here, then click **Run**:

Figure 12.21 – The Spatial Join tool with completed parameters

When the tool completes, a new layer named `WaterQual_IndustZoned_SpatialJoin` will appear in your **Contents** pane. This is the result of the **Spatial Join** tool you just ran. This is a new feature class.

12. Close the **Geoprocessing** pane to declutter your screen. Then, click on the **List by Draw Order** button in the **Contents** pane.
13. Click on the **Map** tab in the ribbon. Then, click on the **Clear** button in the **Selection** group.

You should now be able to see what looks like two new features that overlap the two water quality stations that had bad water quality. You might need to turn off the `Water Quality Station` layer to see the overlapping features. Now, it's time to verify how many features are actually in those locations.

14. Right-click on the new `WaterQual_IndustZoned_SpatialJoin` layer and select **Zoom To Layer**.
15. Select the new `WaterQual_IndustZoned_SpatialJoin` layer in the **Contents** pane.
16. Click on the drop-down arrow located below the **Explore** tool and select **Selected in Contents**. Now, click on one of the two new features. The information popup should appear when you click on one of the features.

17. Look at the lower-left corner of the information popup to see how many features are in the location you selected, as illustrated in the following screenshot:

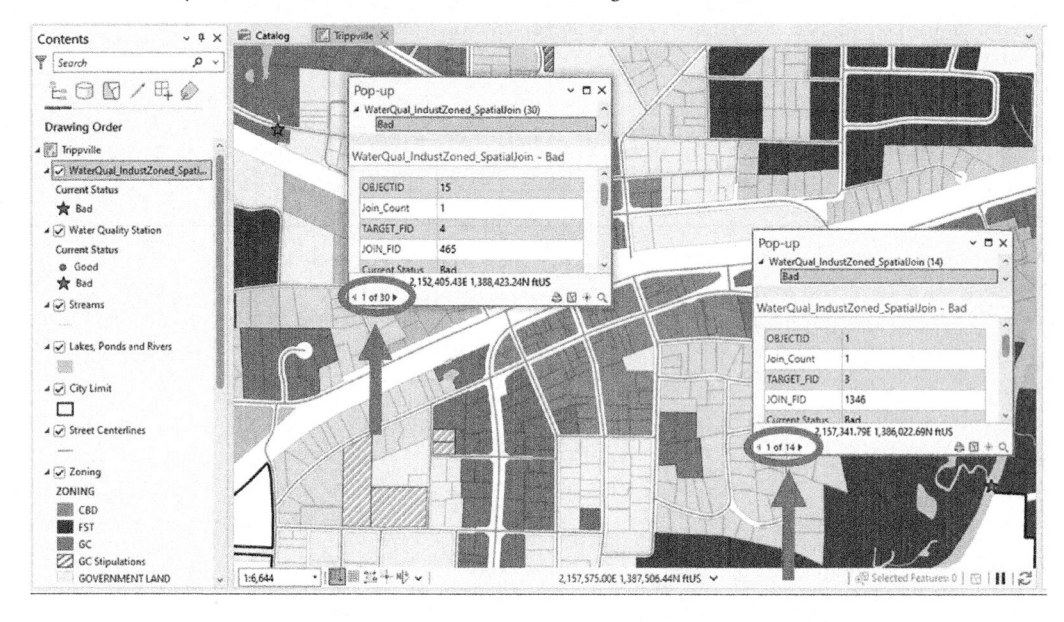

Figure 12.22 – Using the Explore tool to examine the results of the Spatial Join tool

On your map, only one popup will be visible. The two in the preceding screenshot are for illustration purposes only. If you click on the northernmost location, you should have about **30** features, whereas if you click on the southernmost location, you should have about **14** features.

18. Right-click on the `WaterQual_IndustZoned_SpatialJoin` layer and select **Attribute Table** to look at the resulting attributes.
19. Review the table and see if you can identify what the fields mean and where they came from.
20. Close the attribute table and save your project.
21. Close ArcGIS Pro.

As you can see, you have created another new feature class. If you wanted to export this as a standalone table, you could right-click on the layer and go to **Data**. Then, you can select **Export Table**. If you wanted to export it to an Excel spreadsheet, you could use the **Table To Excel** tool you have used in past recipes.

13
Spatial Statistics and Hotspots

If you have completed the previous chapter, you have begun to get an idea of the analytical power ArcGIS Pro brings to the table. You have seen how it can be used to find features that are near one another and calculate those distances in addition to other attributes. But what if you are not looking to find just the closest feature or calculate how far features are from one another? What if you want to determine whether there are any spatial clusters in your data, to see whether there is some pattern or possible point of origin?

This type of analysis is often called **cluster** or **hotspot analysis**. It uses various methods to determine the spatial distribution of data and associated values so that you can see clusters, determine center points and directional distribution, and more. This is done by calculating spatial statistics that can be weighted by other attributes, such as size, total amounts, or counts.

ArcGIS Pro contains over 25 different tools for calculating spatial statistics. These are all stored in the **Spatial Statistics** toolbox and divided between five different toolsets: **Analyzing Patterns**, **Mapping Clusters**, **Measuring Geographic Distributions**, **Modeling Spatial Relationships**, and **Utilities**. All but a couple of the tools in this toolbox work with the core ArcGIS Pro product at all license levels. This means that even with a Basic license, you can perform some pretty amazing analysis.

In this chapter, you will get the opportunity to work with a few of the spatial statistics and hotspot analysis tools that are in ArcGIS Pro.

In this chapter, we will cover the following recipes:

- Selecting features within a specific distance
- Finding the mean center of a geographic distribution
- Identifying the central feature based on geographic distribution
- Calculating the geographic dispersion of data

Selecting features within a specific distance

The police chief has come to you for help. They are working on a grant to help fund more police officers through a neighborhood policing program. They need to determine whether there are areas where crime has been occurring that have experienced more damage or loss than others. If such areas exist, they could include that as justification in their grant proposal.

They have provided you with crime data for the city that covers 2013 and 2014. This data has been geocoded, so you have a point layer that shows the location of the crimes and is attributed to the crime type, the date it occurred, the officer responding, the damage or loss in dollars, and the number of victims involved. The grant requires all numbers to be summarized to the Census Block level. You will have to use the **Spatial Join** tool to aggregate the crime data by Census Block.

In this recipe, you will use the **Hot Spot Analysis** tool to see whether there are any geographic clusters of crimes that have resulted in higher-than-normal damage or loss to the victims. To perform this analysis, you will need to aggregate the point data to the Census Blocks using a spatial join to summarize the crime data.

Getting ready

This recipe requires this book's sample data to be installed on your computer. It is recommended that you complete the recipes in the previous chapter before working on these. It is also recommended that you have completed the *Joining features spatially* recipe in *Chapter 3, Linking Data Together*. This will ensure that you have a better foundational understanding of how to access analysis tools and general workflows. You can complete this recipe with any ArcGIS Pro licensing level.

How to do it...

It is now time for you to perform the analysis requested by the police chief. You will use the **Hot Spot Analysis** tool to do the analysis:

1. Start ArcGIS Pro and open the `Spatial Stats Hotspot.aprx` project located in the `C:\Student\ArcGISPro3Cookbook\Chapter13\Spatial Stats Hotspots` folder using the skills you have learned in past recipes.

   The project should open with the `Union City Crime` map. This map contains the crime locations provided by the police chief, as well as the streets, city limits, parcels, census blocks, and police beat zones. Before we begin with the analysis, we will need to get a better idea of the data we have to work with.

2. Right-click the Crime Locations layer in the **Contents** pane and select the **Attribute Table** option from the menu that appears. Take a moment to explore the table and answer the following questions:

   Question: What fields are in the attribute table for the Crime Locations layer?

   Answer:

   Question: What types of crimes are identified in the attribute table?

   Answer:

   Question: What is the largest amount of loss or damage you can find in the table?

   Answer:

> **Tip**
> You can sort fields in ascending or descending order by right-clicking the field name in the table and selecting the desired sorting, much like you can do in a spreadsheet.

3. Open the attribute table for the Census Blocks layer using the same process. Review its contents and answer the following question:

   Question: What fields are in the attribute table for the Census Blocks layer?

   Answer:

   You now have a better understanding of the data you will be working with. Since the hotspot analysis is looking for spatial clusters within an area, you will generally get better results using polygon inputs than points. So, it is recommended that you aggregate or summarize your point data to get the best results. This means you will need to aggregate the crime locations data into the Census Blocks layer so that you can see how many crimes occurred in each block and the total amount of damage in each block. You will use the **Spatial Join** tool to do this.

4. Close the open attribute tables for the Crime Locations and Census Blocks layers.
5. Click the **Analysis** tab in the ribbon. Then, click the **Spatial Join** tool in the **Tools** group. This will open the tool in the **Geoprocessing** ribbon.
6. Set **Target Features** to **Census Blocks** using the drop-down arrow.
7. Set **Join Features** to **Crime Locations** using the drop-down arrow.
8. Accept the default value for **Output Feature Class**.
9. Set **Join Operation** to **Join one-to-one** using the drop-down arrow and ensure **Keep All Target Features** is checked.

> **Note**
>
> As you may recall from *Chapter 3, Linking Data Together*, you have two options for the join operation: **Join one-to-one** or **Join one-to-many**. The **Join one-to-one** option will aggregate data or sum values if more than one match is found. This is what we want. If the **Join one-to-many** option is selected, the tool will create multiple records for each match it finds between the **Target** and **Join** features. To learn more about the **Spatial Join** tool, go to `https://pro.arcgis.com/en/pro-app/latest/tool-reference/analysis/spatial-join.htm`.

10. Set **Match Option** to **Contains** using the drop-down arrow.
11. Do not set a value for **Search Radius**.
12. Expand the **Fields** option, as shown in the following screenshot:

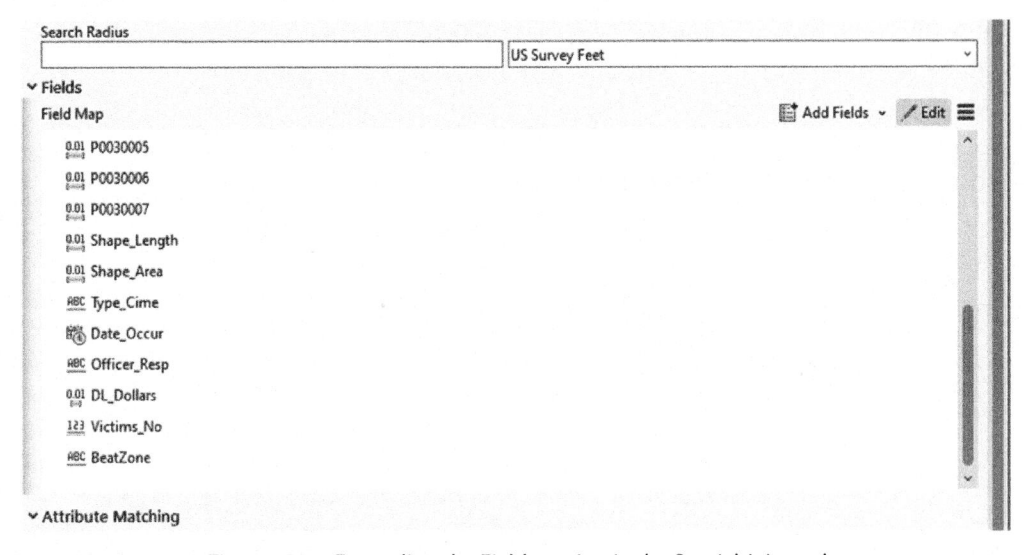

Figure 13.1 – Expanding the Fields option in the Spatial Join tool

13. Click the **Edit** button to open the **Field Properties** window.

14. In the **Fields** panel, scroll down until you see `DL_Dollars` and select it. This is the Damage and Loss field from the Crime Locations layer. As illustrated in the following image, set the **Actions and Source Fields** to `Sum` in the right panel so it adds the total damage from all the crimes that happen within the associated Census Block:

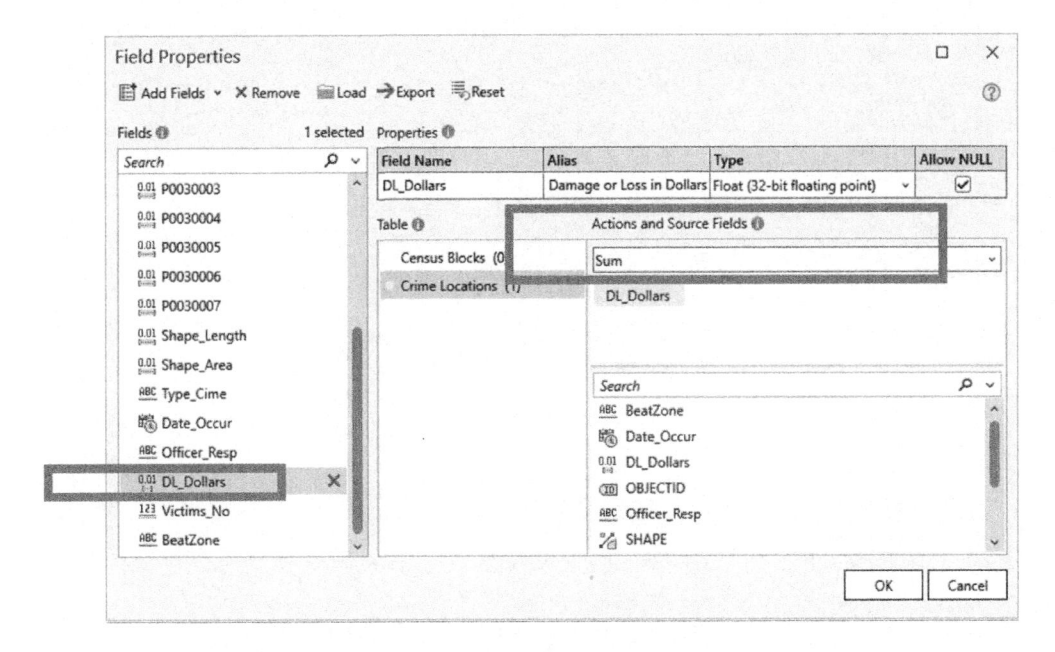

Figure 13.2 – Setting field options in Spatial Join tool

15. Under **Table**, select `Crime Locations`. Then verify `DL_Dollars` appears below the Sum action as show in the image above.

16. In the **Fields** panel of the **Field Properties window**, select `Victims_No` field located below `DL_Dollars`. Set the **Actions and Source Fields** to Sum and verify the `Victims_No` field appears below it. You might need to reselect the `Crime_Locations` under the **Table panel**. Then click the **OK** button in the Field Properties window to apply these settings.

17. Verify that your **Spatial Join** tool looks like the following and click **Run**.

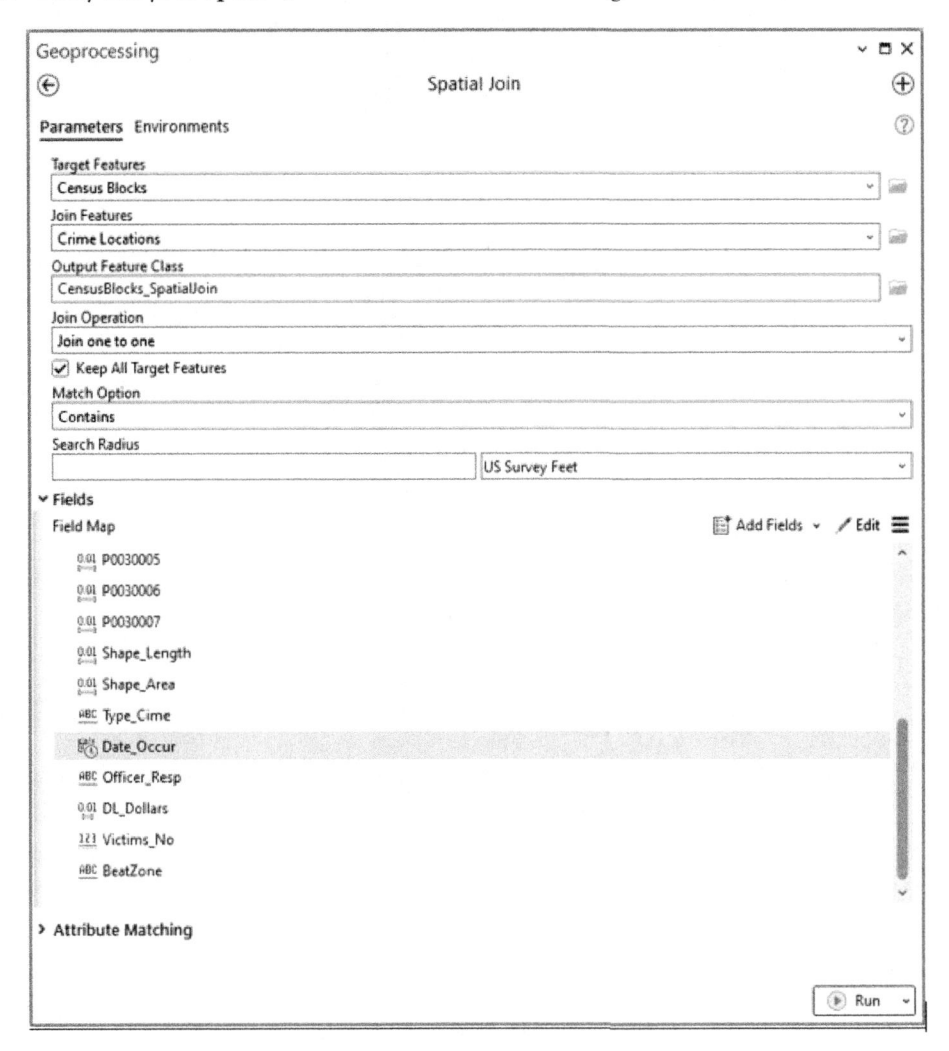

Figure 13.3 – Spatial Join tool with completed parameters

18. Close the **Geoprocessing** tool and save your project.

    A new layer should have appeared in the **Contents** pane. This is the result of the **Spatial Join** tool. It should have aggregated the crime data by Census Block. Let's examine the result to verify this.

19. Right-click the new `Census_Blocks_SpatialJoin` layer you just created and select **Attribute Table** from the menu that appears.

20. Take a moment to scroll through the fields to see which ones are present. Then, answer the following questions:

    Question: What fields does this new layer contain that the original `Census Blocks` layer did not?

    Answer:

    Question: What census block contains the largest amount of damage or loss from crime?

    Answer:

    As you can see, the fields from the **Crime Locations** table have been added to those of the census blocks in the new layer you created. The damage or loss in dollars and the number of victims now represent the totals for all crimes that occurred within the given census block. You should have also seen a new field, **Join_Count**. This provides you with a count of the number of crimes that were contained in each census block.

    Now, it is time to perform hotspot analysis to determine areas where the greatest amount of damage or loss has occurred due to crime. But before we do that, we need to update those Census Blocks that have a null damage value with a 0 to prevent issues with performing our analysis.

21. Click the **Map** tab in the ribbon. Then, click the **Select By Attributes** tool in the **Selection** group.
22. Ensure **Input Rows** is set to the `CensusBlocks_SpatialJoin` layer you just created.
23. **Selection Type** should also be set to **New Selection**.
24. Set the **Select a field** cell to **Damage or Loss in Dollars**.
25. Set **Operator** to **is Null** and click **Apply**. This will select all the census blocks that have no value in the **Damage or Loss In Dollars** field.
26. In the **Census_Blocks_SpatialJoin** attribute table, right-click the **Damage or Loss in Dollars** field and select **Calculate Field** from the menu that appears, as shown in the following screenshot:

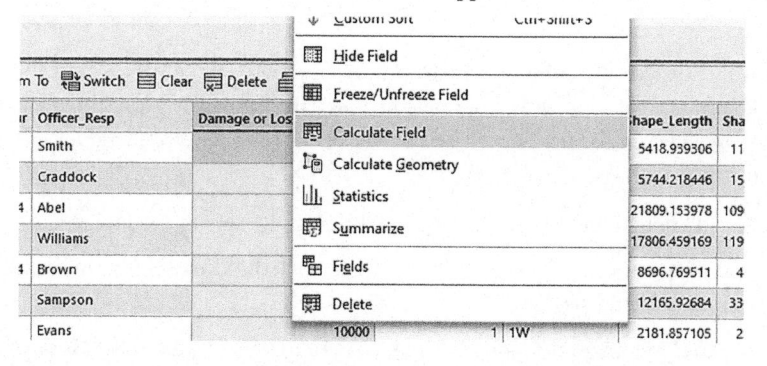

    Figure 13.4 – Selecting the Calculate Field tool

27. This will open the **Calculate Field** tool. Verify that **Input Table** is set to **CensusBlocks_SpatialJoin** and **Field Name** is set to **Damage or Loss in Dollars**.

28. Set **Expression Type** to **Python 3** using the drop-down arrow.
29. In the expression cell located just beneath where it says **DL_Dollars**, and just above the code block area, type **0**. That is the number zero. It is not the letter O. Then, click **OK**:

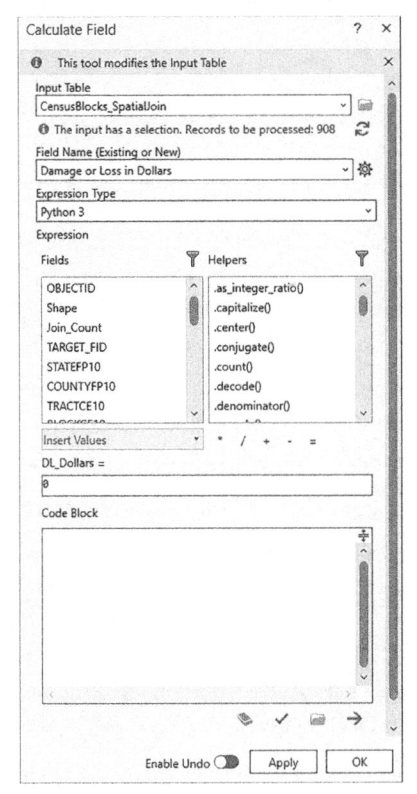

Figure 13.5 – The Field Calculate tool with its parameters completed

You have just used the **Calculate Field** tool to populate the census blocks that did not have any matching crime data with a damage or loss value of 0. The **Hot Spot Analysis** tool does not work well when the field it will be analyzing has missing or null values. Now, it is time to begin hotspot analysis.

30. Close the attribute table for the `Census_Blocks_SpatialJoin` layer.
31. Ensure the **Map** tab is active in the ribbon. Then, click the **Clear** button in the **Selection** group to deselect all selected records.
32. Click the **Analysis** tab in the ribbon. Then, click the **Tools** button in the **Geoprocessing** group to open the **Geoprocessing** pane.
33. Click the **Toolboxes** tab at the top of the pane. Scroll down to the **Spatial Statistics Tools** toolbox. It is near the bottom of the list.

34. Expand the **Spatial Statistics Tools** toolbox. Then, expand the **Mapping Clusters** toolset.
35. Click the **Hot Spot Analysis (Getis-Ord GI*)** Python tool to open it in the **Geoprocessing** pane.
36. Set your **Input Feature Class** to **Census_Blocks_SpatialJoin** using the drop-down arrow.
37. Set **Input Field** to **Damage or Loss in Dollars** using the drop-down arrow. This is the field you will be analyzing.

> **Tip**
> **Input Field** must be a numeric field type. This means it must be a long or short integer, or a float or double field type. It cannot be a text, date, or other field type.

38. To set **Output Feature Class**, click on the **Browse** button and select **Databases** in the left panel of the window that opens under **Project**.
39. Select the **Spatial Stats Hotspots.gdb** geodatabase in the right panel and type `Crime_Hotspots_DamageLoss` into the **Name** cell. Lastly, click the **Save** button.
40. Leave **Conceptualization of the Spatial Relationships** set to the **Fixed distance band**. This setting determines how features are analyzed in spatial relation to one another.
41. Set **Distance Method** to **Euclidean** using the drop-down arrow.
42. Leave **Distance Band or Threshold Distance** and **Self Potential Field** undefined as these are optional parameters.
43. Verify that your **Hot Spot Analysis (Getis-Ord Gi*)** tool looks as follows, then click **Run**:

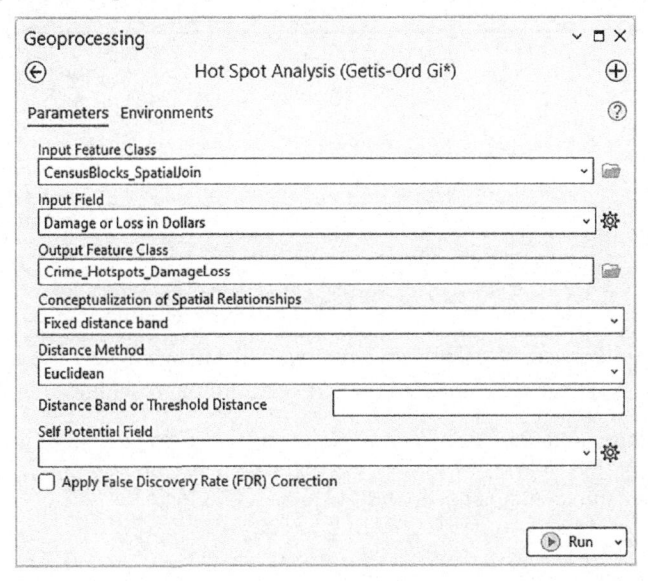

Figure 13.6 – The Hot Spot Analysis tool with its parameters filled in

When the **Hot Spot Analysis** tool completes, a new layer will be added to your map. This layer will automatically be symbolized based on the statistical significance or confidence of the data. The higher the confidence, the hotter or colder it is. In this case, the hotspots are those areas that have the highest level of damage or loss in dollars. The cold spots are those where there was significantly less damage or loss in dollars.

44. Close the **Geoprocessing** pane.
45. In the **Contents** pane, select the **List by Draw Order** button.
46. Drag the new layer you just created so that it is below the Crime Locations layer so that you can see the results concerning the city as a whole. Then, turn off the CensusBlocks_SpatialJoin layer.

The results of your analysis have identified two hotspots or clusters where the amount of damage or loss in dollars is significant compared to the surrounding areas:

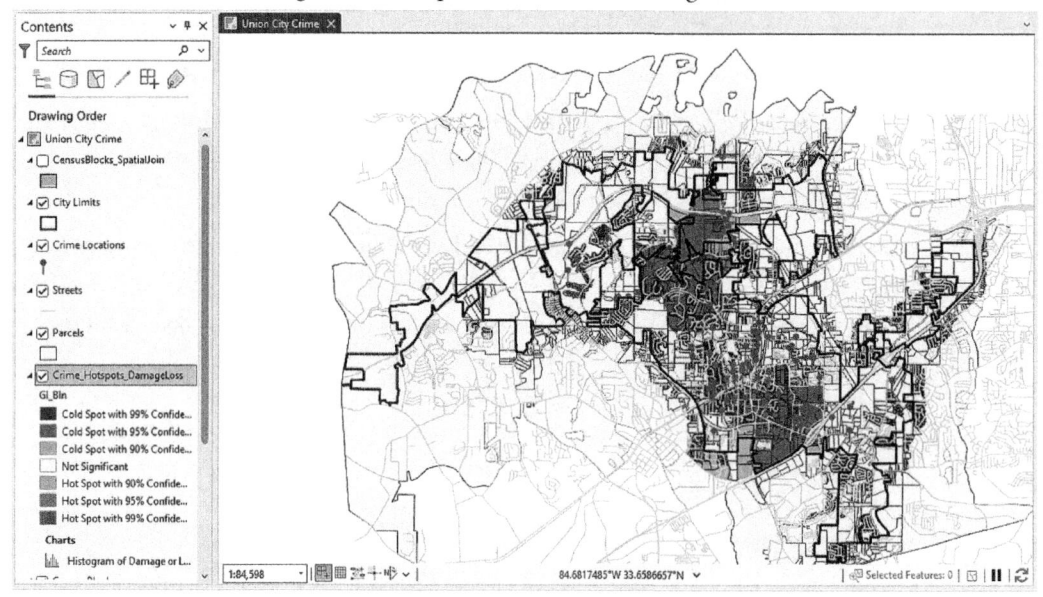

Figure 13.7 – Results of hotspot analysis

As you can see, there are no cold spots. This means that there are now areas that are significantly below the average damage compared to their surrounding neighbors.

47. Save your project and close ArcGIS Pro.

With that, you have completed hotspot analysis using ArcGIS Pro.

How it works...

In this recipe, you performed hotspot analysis to determine if there were clusters of crimes in the city that resulted in higher damage or losses than in other areas. You used the **Hot Spot Analysis** tool to create the results. However, before you could use that tool, you had to aggregate the crime data with Census Blocks to generalize the crime data by area. You used the **Spatial Join** tool to aggregate this data and summarized the total amount of damages or losses that occurred in each Census Block.

Finding the mean center of a geographic distribution

When we are looking for clusters of data or trying to determine the overall geographic distribution of our data, one of the first things many of these tools do is determine a center of mass for the data. From there, it can compare nearby features by looking for clusters, determining area concentrations of data, looking at the directional distribution, and more.

However, finding the center of the geographic distribution of our data can be a powerful analytical tool. This can allow us to strategically locate new facilities, pick a central meeting place, plan a reaction to events, and more. There are three types of centers we can calculate: mean, feature, and median.

The mean center is the easiest to calculate. It is simply the average of all the X and Y coordinates for all the features in the layer you are analyzing. The result is the mean center. This can be useful in tracking movement or shifts over time, such as population shifts from urban to suburban areas. You might also use this to locate a good place to hold a meeting that would be centrally located for all attendees.

The median center is sometimes referred to as the minimum distance center. Simply put, it is the coordinate pair that represents the location closest to all features in a given layer. Areas will automatically be weighted by the number of features present. Those with many features will pull the median center toward them more than those with fewer features. This can be useful in locating new facilities that may rely on other structures for support.

The center feature locates an existing feature that is closest to all other features. This might be helpful if you are looking for a location that would be centrally located to other facilities. For example, let's say you are planning training for firefighters. You want to hold this training at an existing fire station because they have the equipment you need for this. The center feature would allow you to locate the fire station that would be closest to the other stations.

In this recipe, you have been asked to find a central location to hold a conference for a professional organization. You have been provided with a layer showing the locations of all the members. You will use the **Mean Center** tool to determine the geographic mean center of all the members. This will then identify the area you should look to for holding the conference.

Getting ready

This recipe requires this book's sample data to be installed on your computer. You do not need to complete any previous recipes, though it is recommended that you work through the first three chapters of this book, or have similar experience with ArcGIS Pro, to ensure that you have the proper foundational understanding to complete this recipe. You can complete this recipe with any ArcGIS Pro licensing level.

How to do it...

Follow these steps to use the **Mean Center** tool to locate the average center location from a cluster of points:

1. Start ArcGIS Pro and open the `Spatial Stats Hotspots.aprx` project located in the `C:\Student\ArcGISPro3Cookbook\Chapter13\Spatial Stats Hotspots` folder.

2. In the **Catalog** pane, expand the **Maps** folder. Then, right-click the `Member Center` map and select **Open** from the menu that appears. This will open a map that contains the layers you need to perform the work for the remainder of the recipe.

   As you can see, this organization has members located across the US. So, it is important to pick a central location for the conference to reduce travel time and costs for all members.

3. Click the **Analysis** tab in the ribbon. Then, click the **Tools** button in the **Geoprocessing** group to open the **Geoprocessing** pane.

4. Click the **Toolboxes** tab at the top of the pane, scroll down to the **Spatial Statistics Tools** toolbox, and expand it so that you can see its contents:

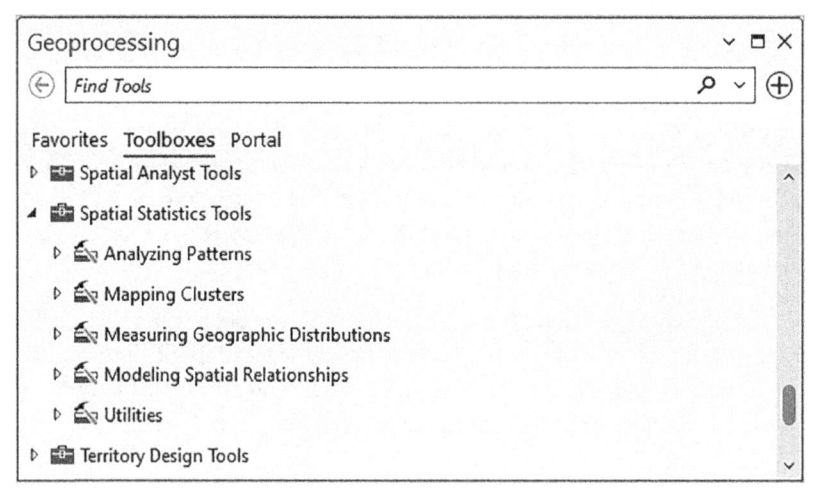

Figure 13.8 – The Spatial Statistics Tools toolbox expanded

5. Expand the **Measuring Geographic Distribution** toolset. Notice the various tools included in this toolset. You should see tools for calculating the various centers that were discussed earlier, along with others.
6. Click the **Mean Center** tool to open it.
7. Set **Input Features Class** to **Members** using the drop-down arrow.
8. Accept the default value for **Output Feature Class**.
9. Leave **Weight Field**, **Case Field**, and **Dimension Field** blank for this analysis.
10. Verify that your **Mean Center** tool looks as follows, then click **Run**:

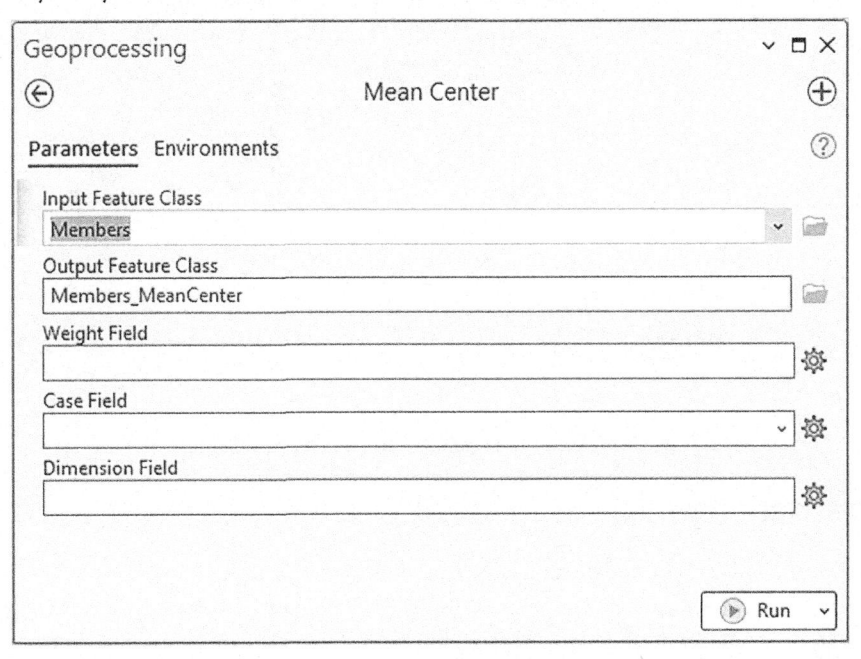

Figure 13.9 – The Mean Center tool with its parameters filled in

The results should produce a point that is located just north of Lebanon, Missouri, and west of Richland, Missouri, as shown in the following screenshot. This represents the mean geographic center of all the members of the organization:

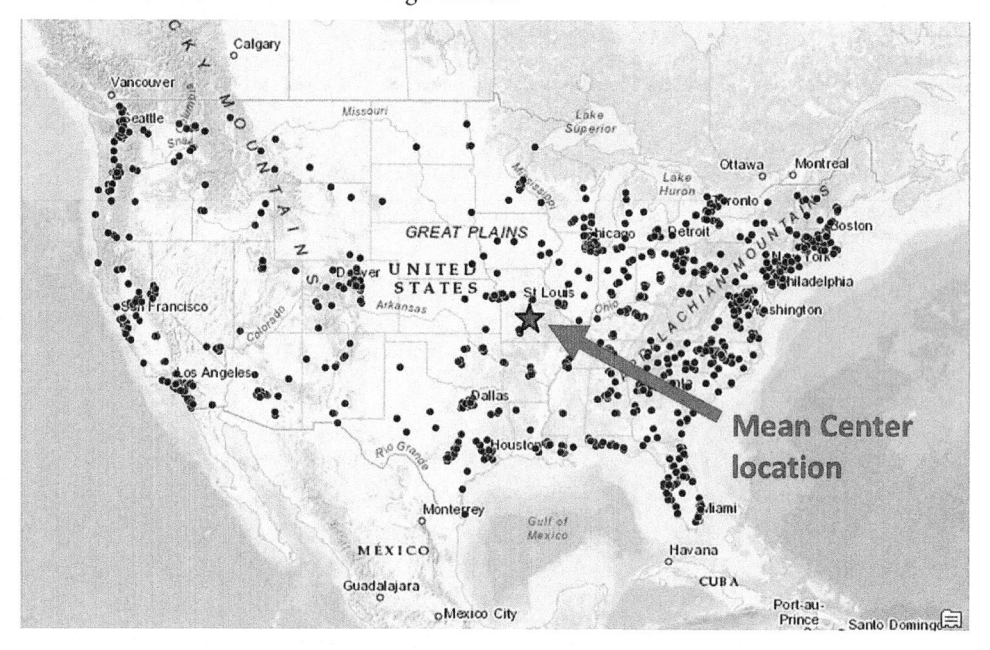

Figure 13.10 – The results of the Mean Center tool

11. Close the **Geoprocessing** pane and save your project.

You have just calculated the mean center for the members of a professional organization to help them pick a location for their next conference. You used the **Mean Center** tool, located in the **Spatial Statistics Tools** toolbox, and the **Measuring Geographic Distributions** toolset to do this.

There's more...

You just calculated the mean center for the members of a professional organization using the **Mean Center** tool. However, when you ran the tool, you left several parameters blank, including **Weight Field** and **Case Field**. Both of these parameters can change the results of the tool and greatly increase its flexibility.

Often, when we are analyzing data, not all the inputs are equal, even in a single layer. Some may need to be given more influence over others. This is certainly true when we are trying to calculate centers. We can reflect this unequal influence by using a weighted field.

When we calculate centers without using a weighted field, as you did with the members, all features are treated the same. No single feature or location carries more weight or influence on the result than another. However, if you use **Weight Field**, some locations receive greater consideration than others when calculating the mean center. This will cause the center to be pulled toward those weighted locations, as illustrated in the following figure:

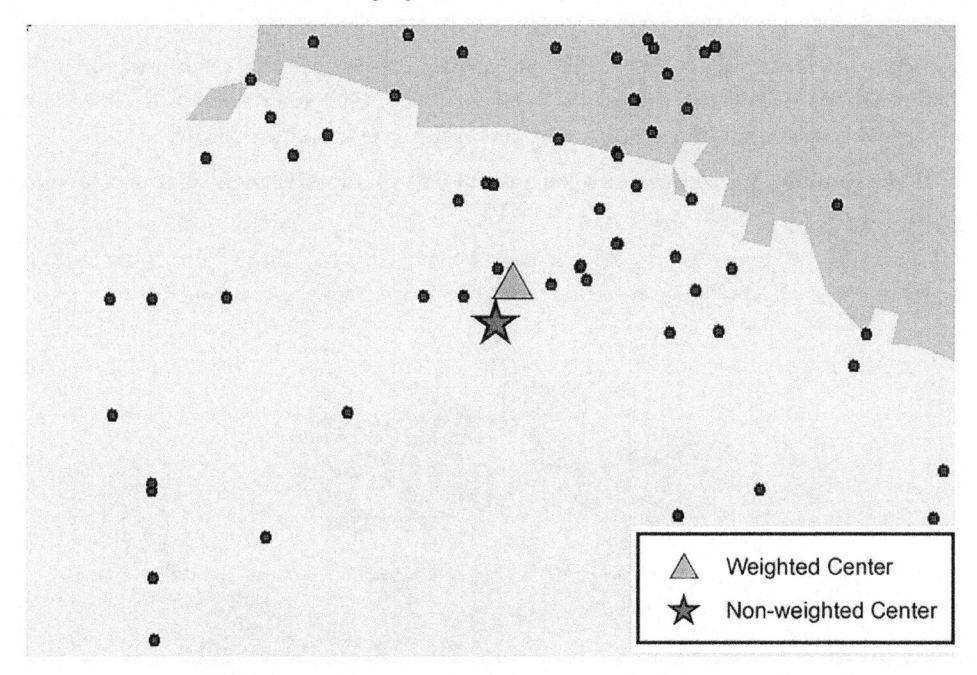

Figure 13.11 – Weighted center versus non-weighted center

The weight each feature carries when running the analysis is based on a field in the attribute table of **Input Feature Class**. This attribute field must be a numeric type field. Text fields cannot be used as weighted fields. Those with a higher number will have more pull when calculating the center than those with a lower number. This will cause the calculated center to be pulled closer to those with a higher weight.

The **Case Field** groups features for analysis based on values found in that field. This means that your results will have more than one center point returned. A center location will be calculated for each grouping of features.

So, let's see how these two fields work. The police chief needs more help with their grant application. They are trying to determine the possible location of new precincts within each of the police beat zones. These zones represent patrol and response areas for the precincts. They need to know the mean center of the crimes that occur in each Zone. They want those crimes with the greatest number of victims to be considered more than those with the least. You will use the same **Mean Center** tool to perform this analysis for the police chief. However, this time, you will provide values for the **Weight Field** and **Case Field** parameters:

1. Ensure the `Spatial Stats Hotspots` project is still open.

2. Click the **Union City Crime** map tab at the top of the map view. The map you used to perform the hotspot analysis in the previous recipe should now be displayed.

3. Right-click the `Crime_Hotspots_DamageLoss` layer and select **Remove** from the menu that appears. Repeat this process for the `Spatial Join` and `Census Blocks` layers as well. This just cleans up the map so that it is easier to read and work with.

4. In the **Contents** pane, make sure you are on **List by Draw Order**. This is the first button on the left at the top of the pane.

5. Turn on the `Police Beat Zones` layer so that it is visible. Turn off the `Census Blocks` layer if needed. Now, you can see the zones the police chief was referring to:

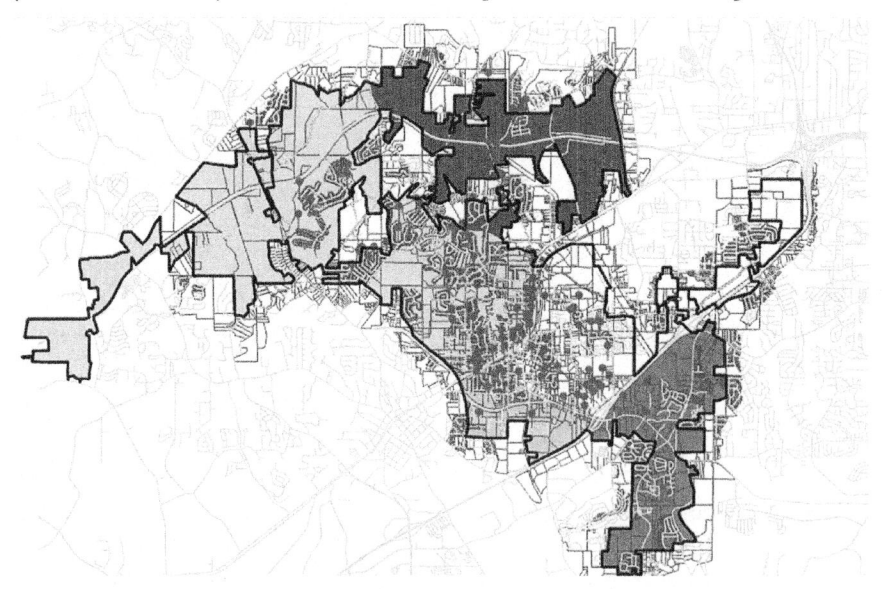

Figure 13.12 – Police beat zones turned on in the map

6. Using the skills you have learned, open the **Mean Center** tool in the **Geoprocessing** pane.

> **Tip**
> If you don't remember how to do this, refer to *steps 3* to *7* when you calculated the mean center of the members earlier in this recipe.

7. Set **Input Feature Class** to **Crime Locations** using the drop-down arrow.

Finding the mean center of a geographic distribution

8. Accept the default value provided for **Output Feature Class**.
9. Set **Weight Field** to **Number of Victims** using the drop-down arrow.
10. Set **Case Field** to **Police Beat Zone**.
11. Verify that your **Mean Center** tool looks as follows, then click **Run**:

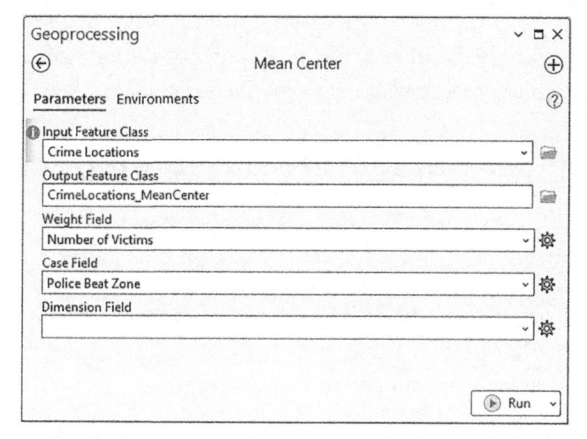

Figure 13.13 – The Mean Center tool with Weight Field and Case Field specified

12. Close the **Geoprocessing** pane.

   You should now see that a center point has been located for each of the police beat zones. This is the geographic mean center of all the crimes that occurred in each zone, weighted by the number of victims:

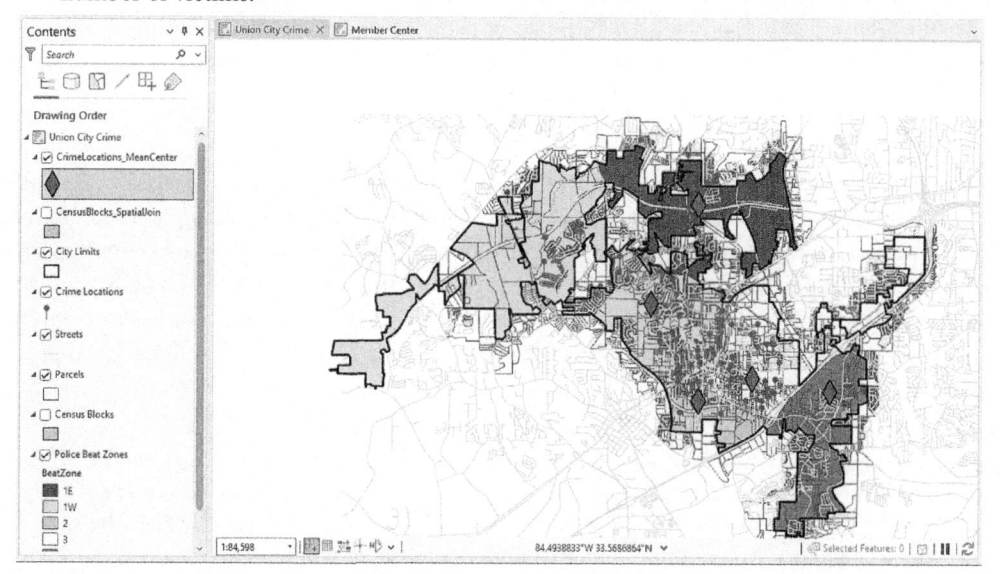

Figure 13.14 – Results of the Mean Center tool with Weight Field and Case Field used

This means the centers were pulled toward those crimes with the greatest number of victims. If you wanted to see the impact **Weight Field** had on the results, you could re-run the **Mean Center** tool without **Weight Field** applied. Depending on the number of victims, the difference between the weighted and non-weighted centers will vary.

13. Save your project and close ArcGIS Pro.

With that, you have successfully calculated the mean center of crimes occurring in each police beat zone weighted by the number of victims impacted by each crime.

Identifying the central feature based on geographic distribution

You now know how to calculate the geographic center for a group of features. As you have seen, this can be very useful to help locate new structures, event sites, and so on. It can also be used to compare two sets of data to see whether there is a shift or movement over time. But what if you need to know what feature within a group is the most central? Can you do that?

You can do this using the **Central Feature** tool. This tool is in the **Spatial Statistics Tools** toolbox and the **Measuring Geographic Distributions** toolset. The **Central Feature** tool identifies the most central feature in a point, line, or polygon in the input feature class. It does this by calculating the distances from the centroid of each feature to every other feature's centroid. Once that has been calculated, the tool selects the feature that has the shortest distance to all other features and copies it to a new output feature class. The shortest distance calculation will honor Z (elevation) coordinates as well if they exist. Like the **Mean Center** and **Median Center** tools, you can also use a weight and case field if desired. This tool works best with a projected coordinate system. The use of a geographic coordinate system can skew the results due to the inconsistency of the units. To learn more about the specifics of how this tool calculates the center feature, go to https://pro.arcgis.com/en/pro-app/latest/tool-reference/spatial-statistics/central-feature.htm.

In this recipe, you will be working for the ACME GIS company that recently merged with another company. This resulted in many new locations being added. The CEO thinks it might be best to move from the corporate headquarters, currently located in the Chicago area, to a more centrally located office. This will help reduce travel time for meetings and other events. So, they have asked you to determine the most central office within the organization. They want you to consider the number of employees at each location as well. The offices with more employees should be given greater consideration in your analysis.

Getting ready

It is recommended that you complete the other recipes in this chapter before starting this one. This will ensure that you have a good foundational understanding of where to access the required tools and basic concepts required to complete this recipe.

Identifying the central feature based on geographic distribution   459

You will also need to ensure the sample data is installed before you continue.

How to do it...

You will now use the **Central Feature** tool to locate the central office for the ACME GIS company:

1. Launch ArcGIS Pro and open the `Spatial Stats Hotspots.aprx` project located in the `C:\Student\ArcGISPro3Cookbook\Chapter13\Spatial Stats Hotspots` folder.

2. In the **Catalog** pane, expand the **Maps** folder. Then, right-click the `ACME GIS Office Locations` map and select **Open**. This will open a map, as illustrated in the following screenshot, that shows the locations of all the offices for the ACME GIS company:

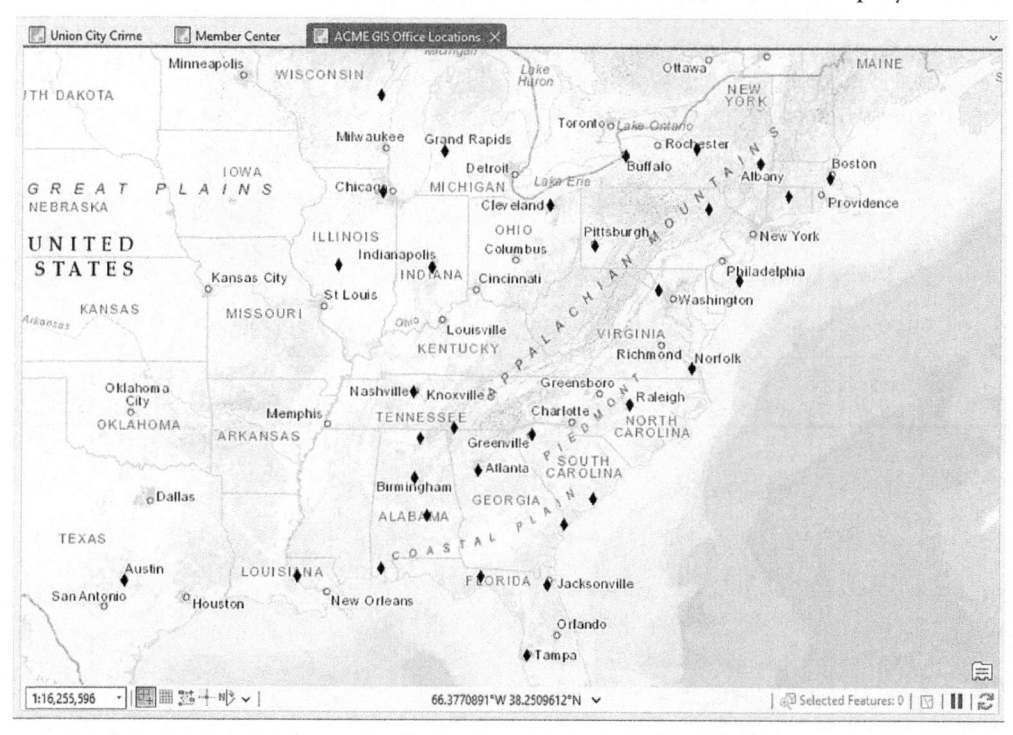

Figure 13.15 – Map showing the ACME office locations

3. As always, you should get to know your data before you use it. So, right-click the **Office Locations** layer in the **Contents** pane and select **Attribute Table** from the menu that appears.

4. Review the available fields and answer the following questions:

   Question: What fields are associated with the office locations in the attribute table?

   Answer:

Question: How many offices does ACME GIS have?

Answer:

Question: Which office has the most staff?

Answer:

As you can see, this table has many attributes for each office. It has fields containing the address information for each location, as well as the number of staff currently employed at the location.

5. Close the **Office Locations** attribute table by clicking the small **X** next to the table name in the tab at the top of the table view.
6. Take a moment to look at the spatial distribution of the offices. See if you can estimate which office is the central feature.

As you can see, many are located along the east coast of the United States. There is a gap along the Appalachian Mountains and Kentucky. This is because ACME GIS mostly had offices in the northeastern part of the United States before it merged with another company. The other company had offices mostly in the southeast of the United States.

7. Click the **Analysis** tab in the ribbon. Then, click the **Tools** button in the **Geoprocessing** group. This will open the **Geoprocessing** pane.
8. Click the **Toolboxes** tab at the top of the **Geoprocessing** pane.
9. Expand the **Spatial Statistics Tools** toolbox and the **Measuring Geographic Distribution** toolset:

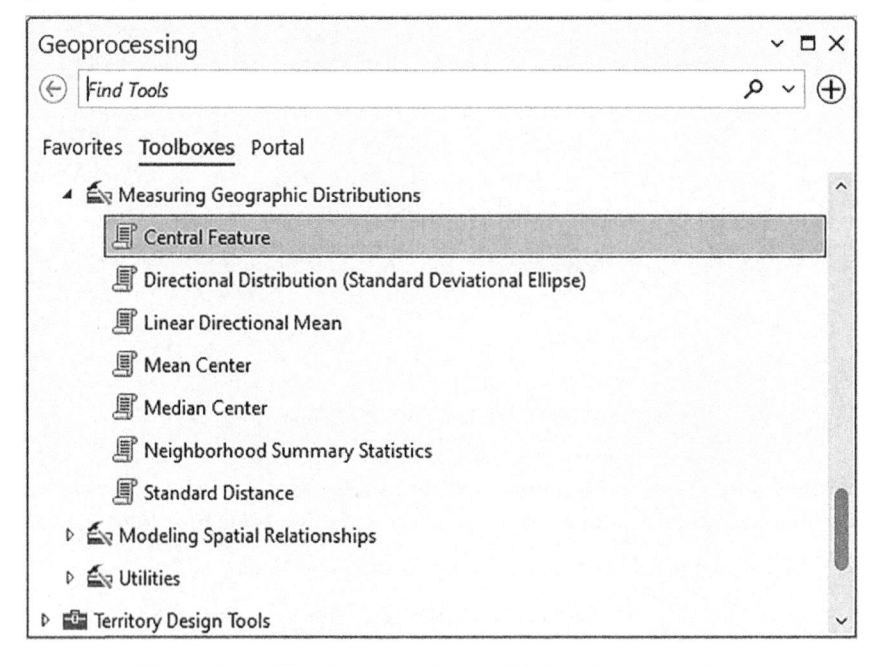

Figure 13.16 – The Measuring Geographic Distributions toolset

10. Click the **Central Feature** tool to open it.
11. Set **Input Feature Class** to **Office Locations** by clicking the drop-down arrow and selecting the layer from the list presented.
12. Type `Central_Office` for the **Output Feature Class**, replacing the default value completely.
13. Accept **Euclidean** for **Distance Method**.
14. For this attempt to locate the central feature, you will not use a **Weight Field** value, as requested by the CEO. So, leave it blank. You will rerun this tool later using the **Staff Number** field to weigh the results. This will allow you to see the difference that **Weight Field** can make to the results.
15. Leave **Self Potential Weight Field** and **Case Field** blank as well.
16. Verify that your **Central Feature** tool looks as follows, then click **Run**:

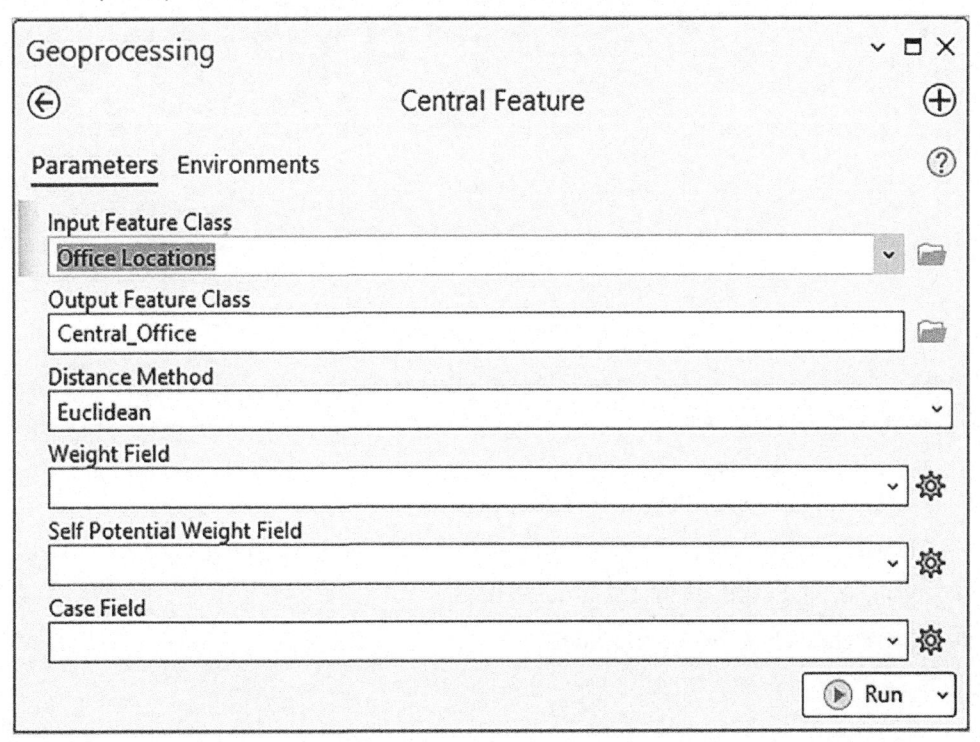

Figure 13.17 – The Central Feature tool with its completed parameters

A new layer will appear in your **Contents** pane, which is the result of the **Central Feature** tool. As you can see, and as illustrated ahead, the office located in Greenville, South Carolina is the most central if you do not use a **Weight Field** value:

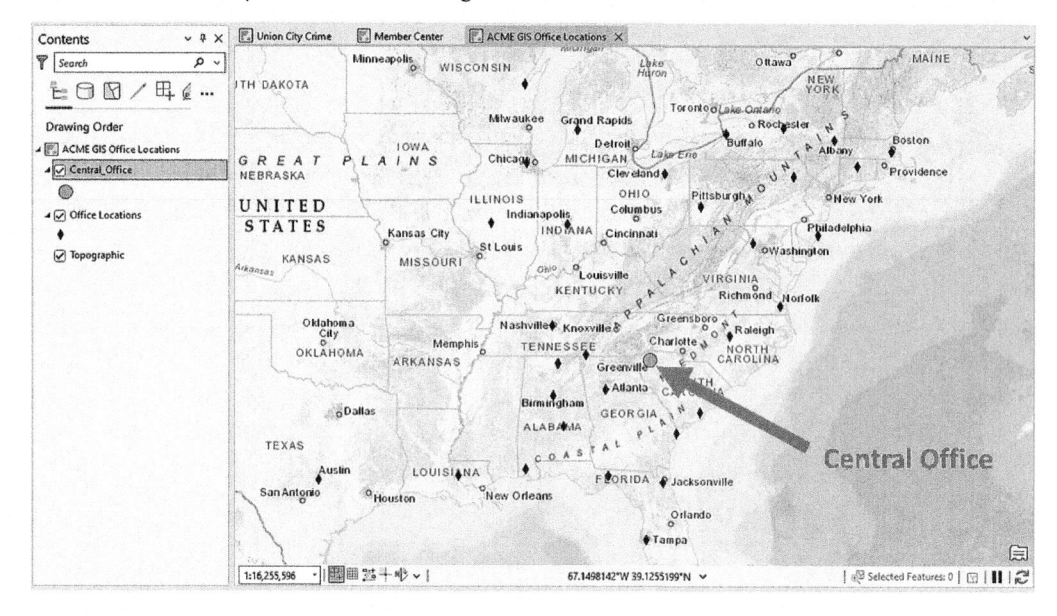

Figure 13.18 – The Central Feature tool's result

Now, let's see what happens when you apply the number of staff in each office to the analysis as the **Weight Field** value.

17. The **Geoprocessing** pane should still be open, with the **Central Feature** tool open. **Input Feature Class** should still be set to **Office Locations**.
18. Change **Output Feature Class** to **Central_Office_Weighted** by typing that into the cell.
19. Again, accept **Euclidean** for **Distance Method**.
20. Set **Weight Field** to **Number of Staff**.

21. Leave the other parameters blank. Verify that your **Central Feature** tool looks as follows, then click **Run**:

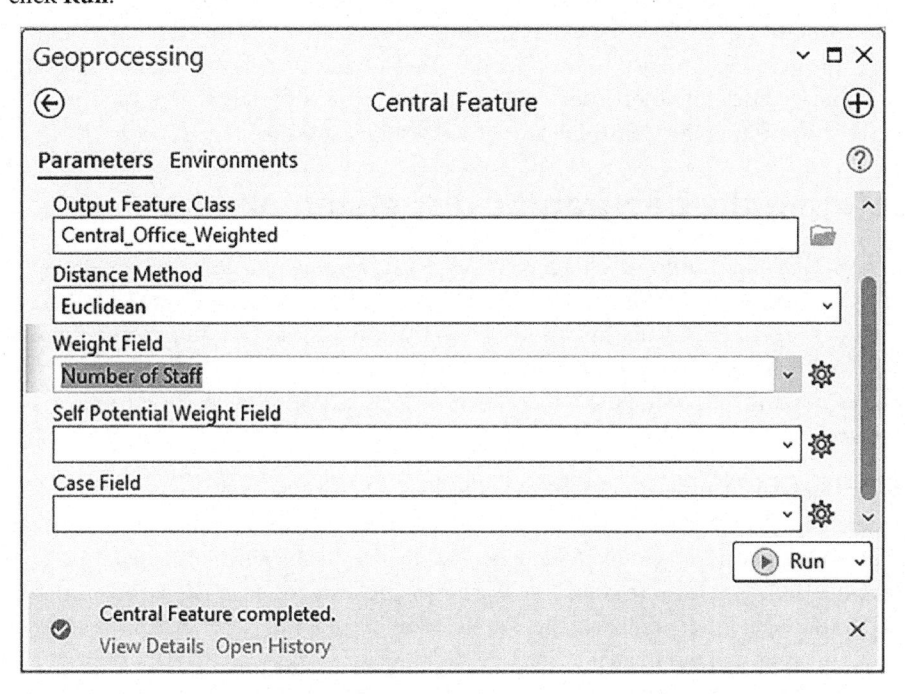

Figure 13.19 – The Central Feature tool using Weight Field

This shows that the northern offices have more staff than those in the south. This difference is great enough to pull the central feature to a completely new location.

22. Save your project and close ArcGIS Pro.

You now know how to locate the central feature within a layer. This can be very useful for several purposes, such as locating a new operation in an existing facility, trying to find the starting point for a spreading fire by comparing sightings, or locating a new facility based on where most people are.

How it works...

In this recipe, you located the central feature in a layer based on the spatial distribution of those features in a single layer. You did this using the **Central Feature** tool. As the name of the tool implies, it locates the feature within a layer that is closest to the geographic center of all the features within the specified input layer.

In the first part of this recipe, you ran the **Central Feature** tool without specifying any of the optional parameters. This treated each feature within the input layer equally. Next, you ran the tool, specifying the use of a **Weight Field** value. You used the **Number of Staff** field as this value. The tool used that field to give greater weight to each feature based on the number of staff working in that location. This meant that those offices with more staff were considered more important than those with fewer staff. This caused the results of the **Central Feature** tool to shift to the northwest.

Calculating the geographic dispersion of data

You can now determine the geographic center and central feature for a grouping of data. As you have seen, this can be a very powerful type of analysis. It can help you seek new locations, determine the focal point of a series of incidents, find the center of mass for a group of features, and more. But what if you need to know where the area of greatest concentration of features is, or how compact or spread out the data is around its geographic center? Such analysis can help you locate clusters of data or see shifts in behavior.

ArcGIS Pro's **Standard Distance** tool, located in the **Spatial Statistics Tools** toolbox, and the **Measuring Geographic Distributions** toolset allow you to do this. It measures the compactness of a distribution around the mean center of a group of features. The smaller the distance calculated, the less the data is dispersed, meaning it is more compact. The larger the distance calculated, the more the data is dispersed, meaning it is less compact. The output from this tool will either be a new polygon feature class for 2D analysis or a multipatch 3D sphere. So, this tool works with 3D data analysis as well. To learn more about how this tool calculates the dispersion of data, go to `https://pro.arcgis.com/en/pro-app/latest/tool-reference/spatial-statistics/standard-distance.htm`.

Because of your help with the police chief's neighborhood policing grant, they were impressed and are back for more help. They believe that there is a shift in crime patterns as the year progresses and the holiday season draws closer. They want to shift patrols from the residential areas to the area around the mall, commercial district, and interstates as the holiday season approaches. However, the city council believes that this might make the citizens unhappy, so they oppose the shift in police resources.

The chief has asked for your help to determine whether there is indeed a change in the crime patterns as they believe and to provide them with something they can take to the city council to make their case. They also want analysis results for each police beat zone.

In this recipe, you will use the **Standard Distance** tool to determine the concentration of crimes for the first half of the year and the last half by beat zone. This will allow you to see whether there is truly a change in the crime patterns as the year progresses by demonstrating whether there is a shift in the central location as well as the area of concentration.

Getting ready

It is recommended that you complete the other recipes in this chapter before starting this one. This will ensure that you have a good foundational understanding of where to access the required tools and basic concepts required to complete this recipe.

You will also need to ensure the sample data is installed before you continue.

How to do it...

Follow these steps to use the **Standard Distance** tool to determine if there has been a shift in crimes within the city:

1. Launch ArcGIS Pro and open the `Spatial Stats Hotspots.aprx` project located in the `C:\Student\ArcGISPro3Cookbook\Chapter13\Spatial Stats Hotspots` folder.

2. Click on the **Union City Crime** map tab at the top of the map view area. If you closed the **Union City Crime** map, in the **Catalog** pane, expand the **Maps** folder, right-click the **Union City Crime** map, and select **Open** from the menu.

3. If needed, turn off all layers in the map except for `City Limits`, `Crime Locations`, `Streets`, `Parcels`, and `Police Beat Zones`. Your map should look like this:

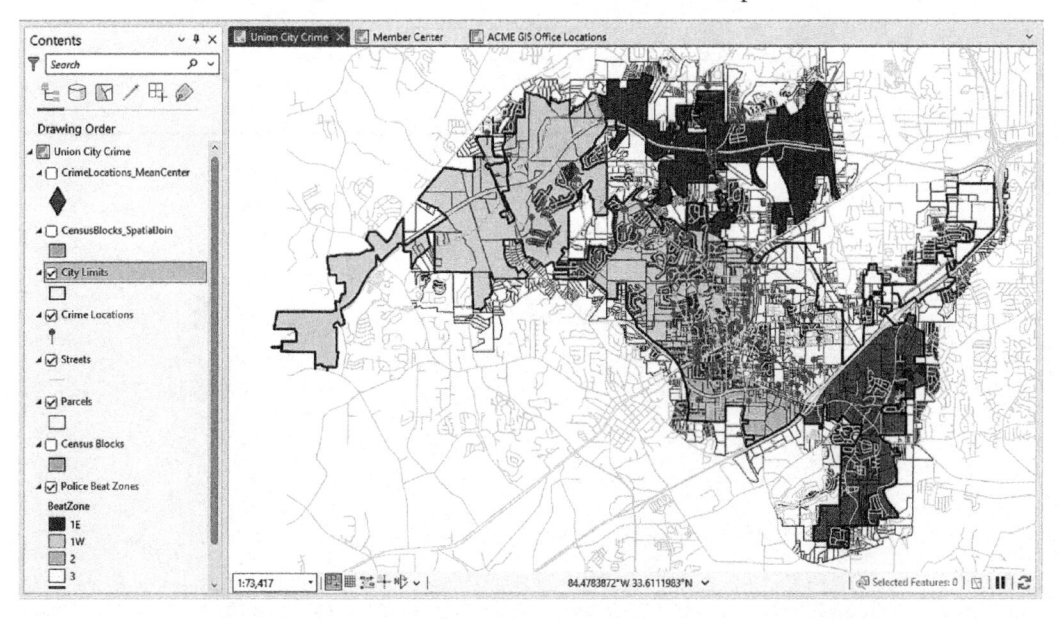

Figure 13.20 – The Union City Crime map with layer visibility set

Next, you will need to add some additional crime data to your map. This data will be part of your analysis.

4. In the **Catalog** pane, expand the Databases folder. Then, expand the **Union City** feature dataset.
5. Right-click on the **Crime_Loc_2022** feature class and select **Add to Current Map**. A new layer should appear in your map and the **Contents** pane. Because it is a point layer, it should be at the top of the layer list.
6. Select the new **Crime_Loc_2022** layer in the **Contents** pane. Then, select the **Feature Layer** tab in the ribbon.
7. Click the **Import** button located in the **Drawing** group. This will open the **Import Symbology** tool.
8. Ensure **Input Layer** is set to **Crime_Loc_2022**. Because you selected that layer in the **Contents** pane earlier, this should be automatically set.
9. Set **Symbology Layer** to **Crime Locations** using the drop-down arrow.
10. All other parameters should remain empty. Verify that your **Import Symbology** tool matches the following and click **OK** to apply the symbology from the Crime Locations layer to the **Crime_Loc_2022** layer:

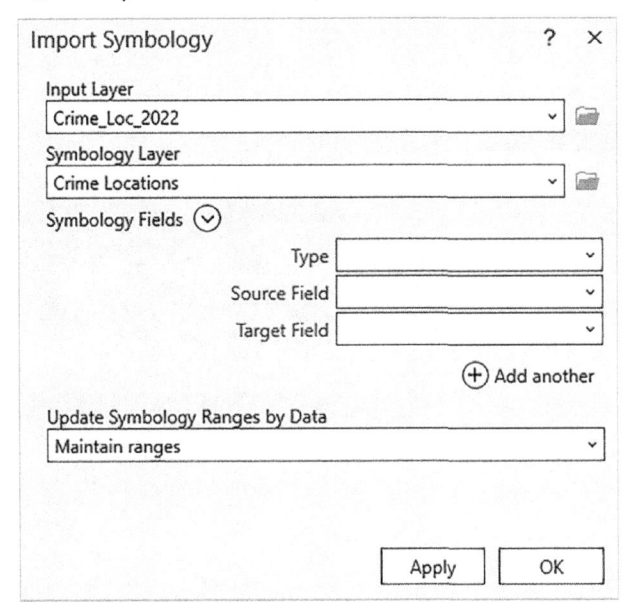

Figure 13.21 – The Import Symbology tool with completed parameters

11. Right-click on the Crime Locations layer and select **Remove** from the menu that appears.

You have now prepared the map for the analysis you need to conduct for the police chief. Now, you can start analyzing the distribution of the crimes for the first half of the year. We will start by selecting all the crimes that occurred on or before June 30, 2022.

12. Select the **Map** tab in the ribbon and then click the **Select by Attributes** button in the **Selection** group.
13. Ensure **Input Rows** is set to **Crime_Loc_2022** using the drop-down arrow if needed.
14. Set **Selection Type** to **New selection**.
15. In the **Where** clause, set the field to **Date_Occur** using the drop-down arrow.
16. Next, set **Operator** to **is On or Before** using the drop-down arrow.
17. Then, set the value to **06/30/2022**. You can either type that in or use the calendar button located to the right of the value.
18. Verify that your **Select By Attributes** window matches what's shown here and click **OK**:

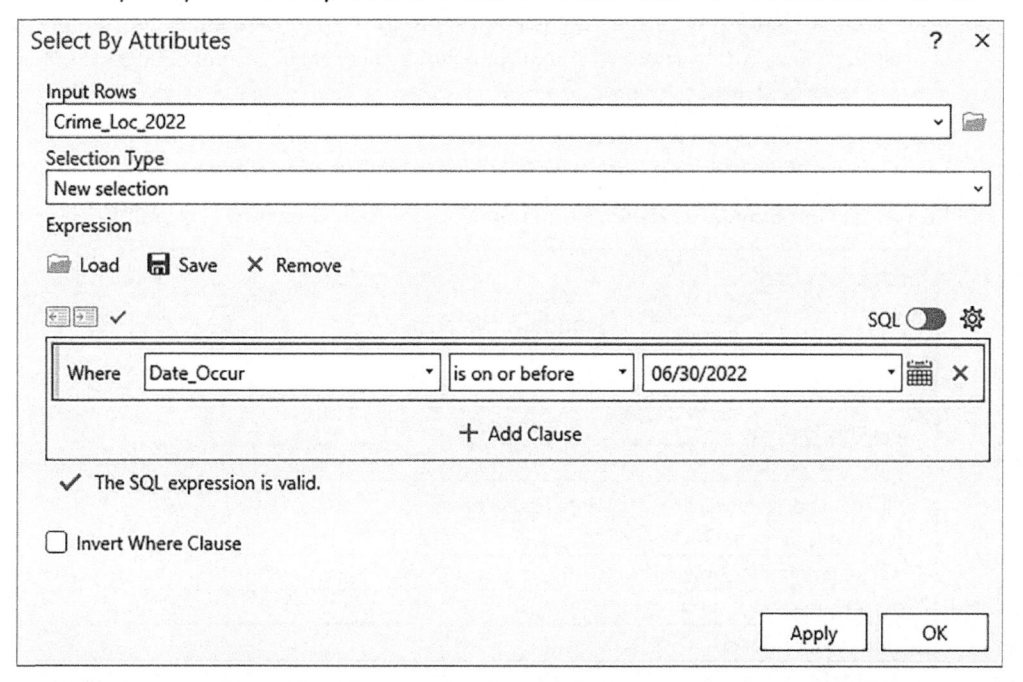

Figure 13.22 – The Select By Attributes query to select all crimes that occurred before 6/30/2022

When you run the query, approximately 70 crimes should be selected within the map. This should be all the crimes that occurred during the first half of the year. You can verify this by opening the attribute table for the `Crime_Loc_2022` layer and setting the table to only display selected records. Then, sort on the **Date_Occur** field.

Now that you have selected the crimes that occurred during the first half of the year, it is time to analyze their geographic distribution.

19. Click on the **Analysis** tab in the ribbon. Then, click the **Tools** button to open the **Geoprocessing** pane. Next, click on the **Toolboxes** tab at the top of the pane.

20. Scroll down through the list of toolboxes until you see the **Spatial Statistics Tools** toolbox and expand it.
21. Expand the **Measuring Geographic Distributions** toolset.
22. Click the **Standard Distance** tool to open it in the **Geoprocessing** pane.
23. Set **Input Feature Class** to **Crime_Loc_2022** using the drop-down arrow. You should see a warning indicating you have features selected in this layer.
24. Set the **Output Standard Distance Feature Class** to **Crime_Distribution_First_Half** by typing that value into the cell.
25. Leave **Circle Size** set to **1** standard deviation.
26. Leave **Weight Field** blank. You do not need to apply a weight to the features for this analysis because the police chief has not told you to apply any greater weight to some crimes over others. They are to all be treated the same.
27. You will need to apply a **Case Field** value so that you can break the analysis down by police beat zone. Set **Case Field** to **Police Beat Zone** using the drop-down arrow.
28. Verify that your **Standard Distance** tool resembles the following, then click **Run**:

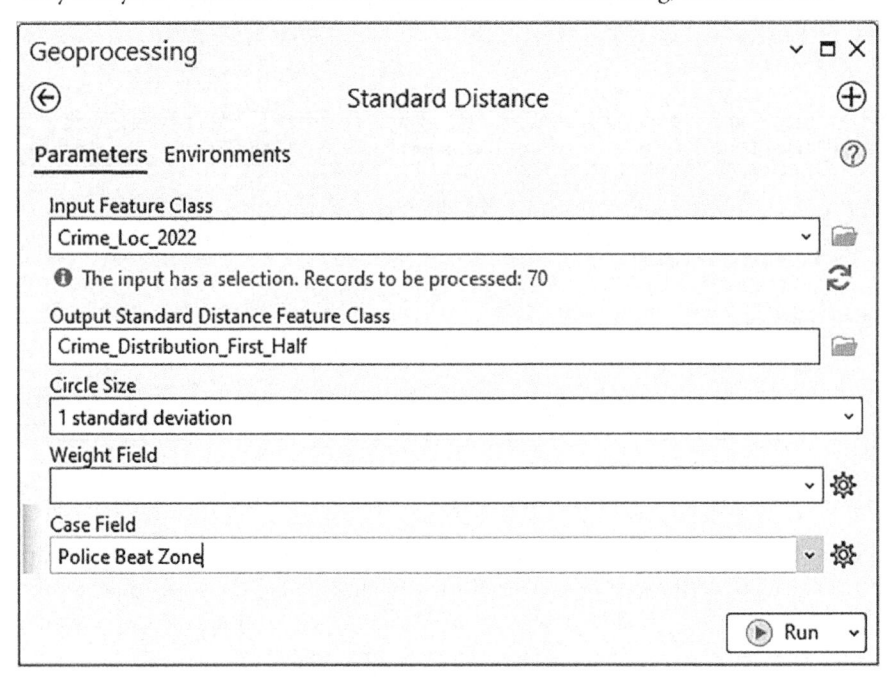

Figure 13.23 – The Standard Distance tool with its parameters filled in

A new layer will appear on your map once the tool is complete. This represents the geographic distribution of the crimes that occurred during the first half of 2022 in the city. The map should look something like this:

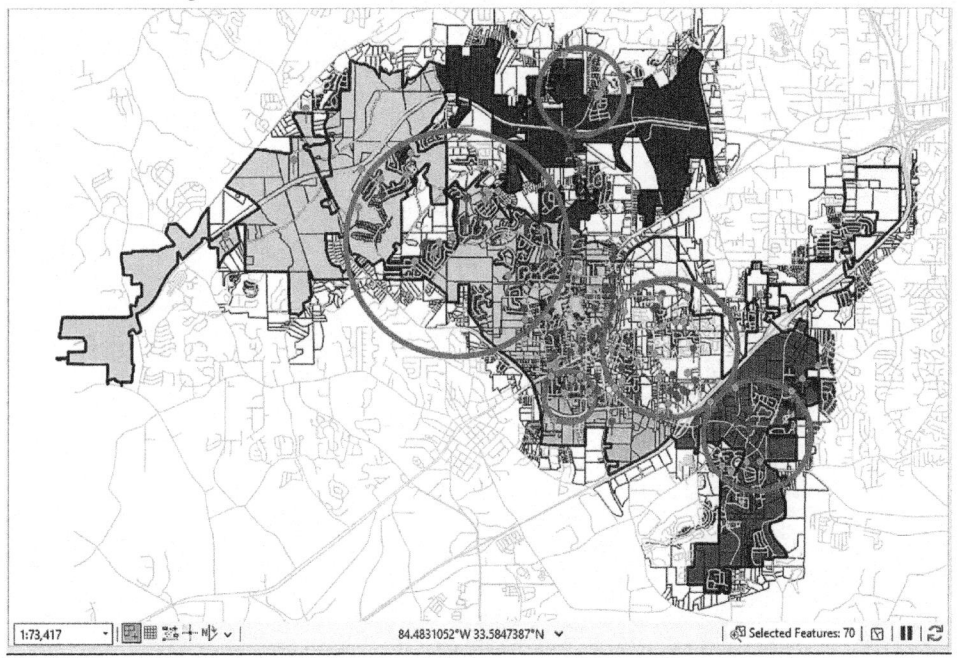

Figure 13.24 – Results of the Standard Distance tool

The circles represent the concentration of the crimes around the mean center for each group by police beat zone. So, you should have five circles. Notice that not all the crimes that occurred in each zone are inside the circle. This is because the circle represents the overall distribution or concentrations of the crimes.

Now that you know the distribution for the first half of the year, it is time to analyze the last half of the year.

29. Close the **Geoprocessing** pane.
30. Select the **Map** tab in the ribbon. Then, click the **Clear** button located in the **Selection** group to deselect all selected features.
31. Click the **Select by Attributes** button in the **Selection** group so that you can build a query to select all crimes that occurred during the last half of 2022.
32. Set **Input Rows** to **Crimes_Loc_2022** using the drop-down arrow.
33. Ensure **Selection Type** is set to **New selection**.
34. Set the **Select a Field** cell to **Date_Occur** using the drop-down arrow.

35. Make sure the **Operator** cell is set to is **is after** and the blank cell that appears to the right of the operator is set to **06/30/2022** by typing in the date or using the **Calendar** option.

36. Verify that your query looks as follows, then click the **OK** button:

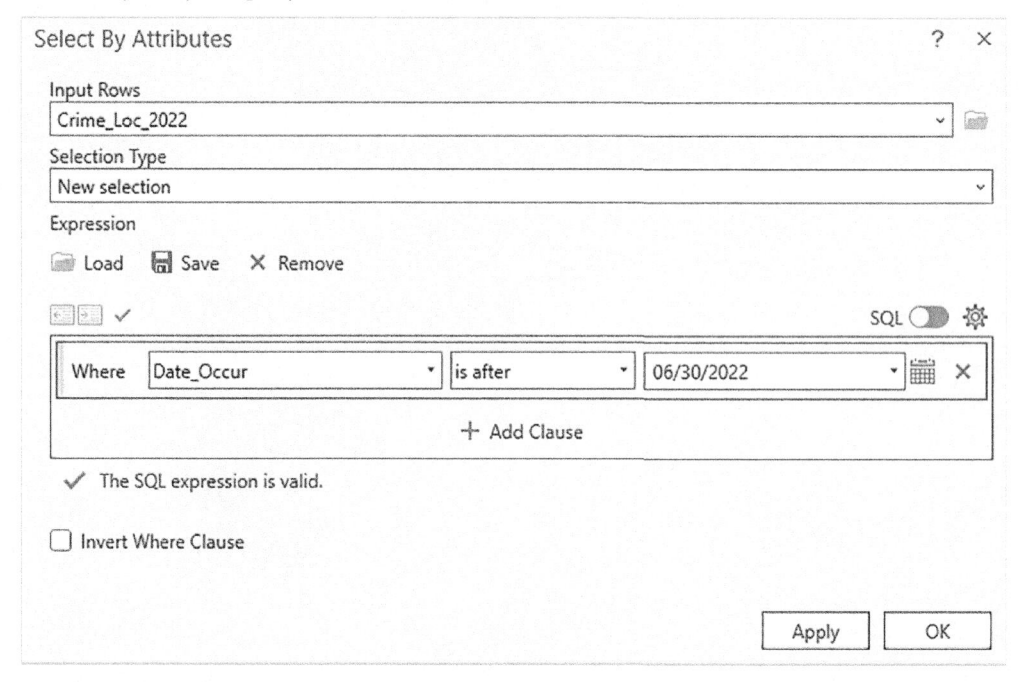

Figure 13.25 – Select By Attributes with a clause to select crimes that occurred after 6/30/2022

Now, you should have all the crimes selected that occurred during the last half of the year. This means you should have 82 crimes selected. You can verify this using the same method you used to verify the previous selection.

37. Click on the **Analysis** tab in the ribbon. Next, click on the **Tools** button. Then, click the **Toolboxes** tab at the top of the pane.

38. Scroll through the list of toolboxes until you see the **Spatial Statistics Tools** toolbox; expand it.

39. Expand the **Measuring Geographic Distributions** toolset.

40. Click the **Standard Distance** tool to open it in the **Geoprocessing** pane.

41. Set **Input Feature Class** to **Crime_Loc_2022** using the drop-down arrow. You should get a warning that indicates you have features selected in this layer.

42. Set **Output Standard Distance Feature Class** to **Crime_Distribution_Last_Half** by typing that value into the cell.

43. Leave **Circle Size** set to **1 standard deviation**.

44. Leave the **Weight Field** blank. You do not need to apply a weight to the features for this analysis because the police chief has not told you to apply any greater weight to some crimes over others. They are to all be treated the same.

45. You will need to apply a **Case Field** value so that you can break the analysis down by **Police Beat Zone**. Set **Case Field** to **Police Beat Zone** using the drop-down arrow.

46. Verify that your **Standard Distance** tool resembles the following, then click **Run**:

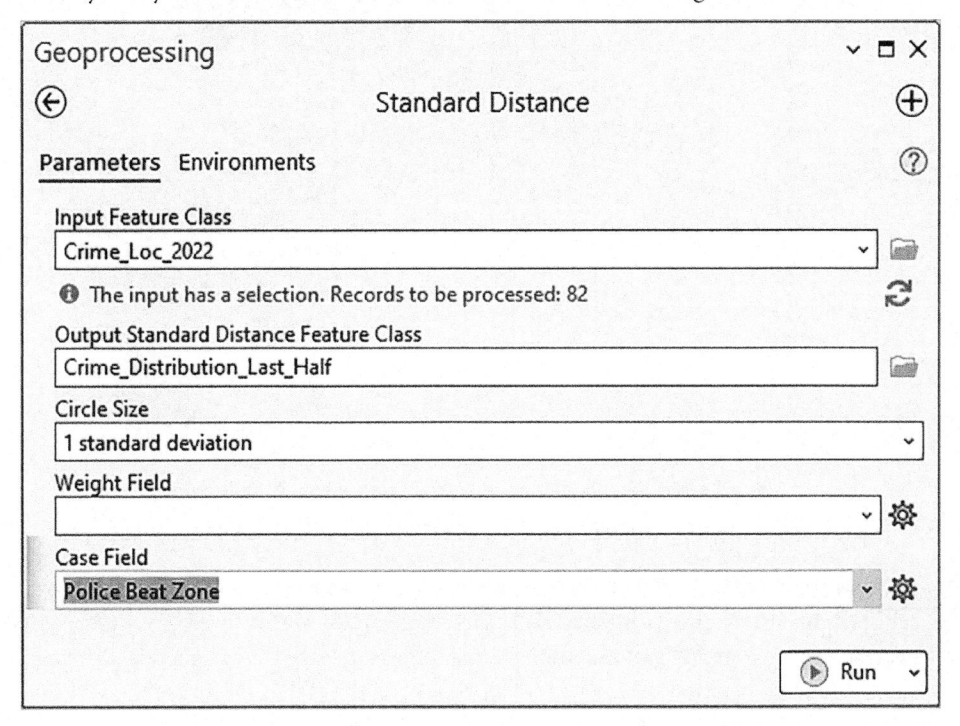

Figure 13.26 – The Standard Distance tool with its parameters complete

47. Close the **Geoprocessing** pane when the tool completes. A new layer will appear on your map.

48. Select the **Map** tab in the ribbon and click on the **Clear** button to deselect all features.

    At this point, you will see the distribution of crimes for the last half of the year and be able to visually compare this to the first half:

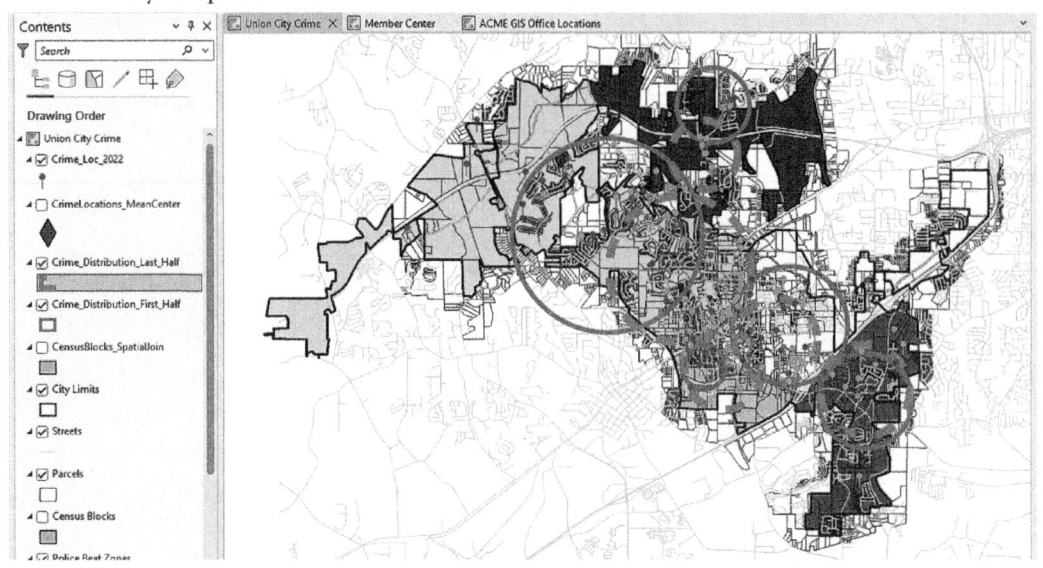

Figure 13.27 – Results of the Standard Distance tool showing the difference between the first and second half of the year

It would appear that the police chief is correct – the data does show that the concentration of crimes shifts toward the commercial district, mall, and interstate area. You can see that the distribution for zone 1W contracts and shifts southeast toward the commercial district located along Roosevelt Highway, which runs through the middle of the city. The same is true for zone 3, which contracts along Shannon Parkway, another major commercial thoroughfare. Zone 2 expands toward both Shannon Parkway and Roosevelt Highway during the last half of the year. The concentration for zone 1E stays about the same size but certainly shifts south toward the intersection of US Highway 29, Scarborough Road, and Stonewall Tell Road. This is another major artery with a lot of commercial development.

49. Save your project and close ArcGIS Pro.

So, as you saw, your analysis supported the police chief's theory. They seem to be correct. You can now provide them with the data they need to take to the city council so that they can make an informed decision about the distribution of police resources.

How it works...

In this recipe, you were able to show the general shift in locations of crimes that occurred in a city between the first and second half of the year. You did this by querying the crimes that occurred during the period of interest using the **Select By Attributes** tool. Then, you calculated areas of concentration in which the crimes occurred every 6 months using the **Standard Distance** tool. This tool created polygons that identified where the concentration of crimes occurred by police beat zone. You then were able to visually compare the results from every 6 months to see there was a shift in the concentration of where the crimes were occurring.

14
3D Maps and Analysis

Since its beginnings, GIS has relied on 2D maps. They were what we created and where we performed our analyses. Recently, this is changing. More and more, we are having to take the third dimension, Z, into account. This is because the world is becoming more crowded. Our infrastructure is stacking up upon itself, both above ground and below. So, the need to track and view data in 3D is increasing. Supporting technologies, such as **Light Detection and Ranging (LiDAR)**, **Unmanned Aerial Vehicles (Drones/UAS/UAV)**, **Building Information Models (BIM)**, Digital Twins, and increasing computer power are also adding to this push.

As you have seen if you have completed previous recipes in this book, ArcGIS Pro is a very powerful tool for visualizing and analyzing data. However, up to this point, we have been working in a 2D environment. While we can do a lot in 2D, it does have limitations because we live in a 3D world. ArcGIS Pro brings new 3D capabilities to the table that we have not had in previous desktop GIS applications. It natively supports the 3D viewing, querying, and limited analysis of data. When you add the 3D Analyst extension, you open up a whole new world of possibilities.

Working in 3D provides much better ways to view and manage our infrastructure. We can begin to see the true spatial relationships between the features we are maintaining, viewing, and analyzing. ArcGIS Pro allows you to create 3D Maps, called **scenes**, using the core application, as illustrated here:

Figure 14.1 – Sample 3D scene from ArcGIS Pro

The preceding scene was created using ArcGIS Pro. You can see the buildings, power poles, light poles, and fire hydrants, all depicted in 3D. This allows you to see things such as whether a building will be in the way of a power line or whether a building will cast a shadow over an area that might prevent certain species of plants from growing. If you add the 3D Analyst extension, then you can begin to calculate the slope of road or pipe, determine whether one building might block the view from another, and more.

Don't limit your 3D thinking or displays to infrastructure. Any data can be displayed in 3D. This opens a whole new way to look at your data, as illustrated in the following maps in *Figure 14.2*:

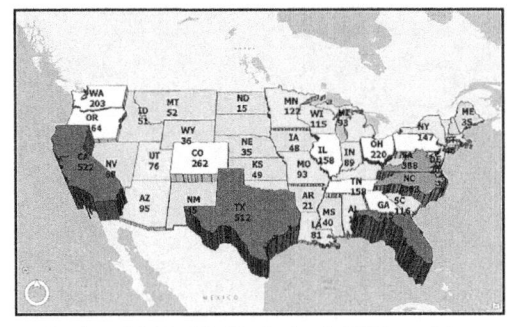

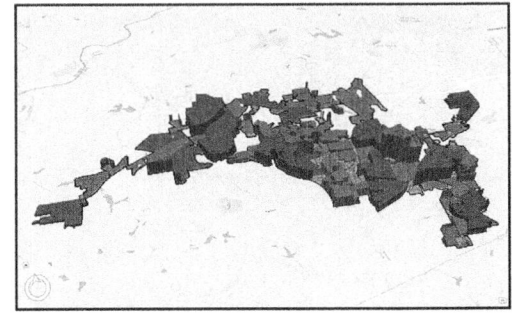

Number of GIS Certified Professionals (GISP) per state   Total Population by Census Block of Union City, GA

Figure 14.2 – More 3D scene examples

As you can see, these two maps have nothing to do with infrastructure. Instead, they are highlighting areas with higher and lower values. The use of 3D makes those areas with higher and lower values stand out visually even more so than using traditional symbology. These maps also illustrate that you can combine traditional categorical and quantitative symbology with 3D to produce even more impactful maps. This capability is included in the core functionality of ArcGIS Pro.

What if I need to do more than visualize data in 3D? What if I need to analyze the data so that I can make use of elevational or height information? Well, we have already hinted at that. Esri has an extension for ArcGIS Pro called 3D Analyst. This extension greatly enhances ArcGIS Pro's ability to work within 3D. The 3D Analyst extension adds a new toolbox to your arsenal that contains 11 toolsets and over 100 tools for creating, converting, and analyzing 3D data. The 3D Analyst extension is not part of the core ArcGIS Pro product. It must be purchased as an additional license. However, it will work with all license levels of ArcGIS Pro. To learn more about the 3D Analyst extension, go to `https://www.esri.com/en-us/arcgis/products/arcgis-3d-analyst/overview`.

In this chapter, you will get an opportunity to explore some of the 3D capabilities of ArcGIS Pro. You will learn how to create a new 3D scene to visualize data in 3D, enable your data so you can not only store *X* and *Y* coordinates but also *Z* coordinates (elevations), and learn how to create 3D features. The recipes in this chapter will include the following:

- Creating a 3D scene
- Enabling your data to store *Z* coordinates (elevations)
- Creating multipatch features from 2D
- Creating 3D features
- Calculating lines of sight
- Calculating the volume of a polygon

Creating a 3D scene

In ArcGIS Pro, a 3D map is called a scene. When creating a new 3D scene, one of the first things you need to determine is what data represents the ground surface. The ground surface becomes the canvas that all 2D layers are draped across. Yes, a 3D scene will include both 2D and 3D layers. Typical 2D layers might include an aerial photo, parcels, political boundaries, and natural water features. These often help put your 3D layers into context. The ground surface can also serve as the starting point for displaying the 3D layers. It may provide the starting elevation for those features. They are then extruded above or below that surface. Esri provides a terrain model that is the default ground surface for any new 3D scene. This model is a web service published through ArcGIS Online. You can also use your own elevation or terrain data. This can include a **Digital Elevation Model (DEM)**, **Triangulated Irregular Network (TIN)**, or other web services.

In this recipe, you are an employee for an engineering firm that is working on a project to assess the feasibility of using an abandoned quarry as a water reservoir. The lead engineer on the project has asked you to create a 3D map of the area that he will use during a presentation to the client. This will help the client and engineer visualize the area as he creates his plan.

Getting ready

This recipe requires the sample data to be installed on the computer. It is recommended that you complete the recipes from *Chapter 1, ArcGIS Pro Capabilities and Terminology*, before starting this recipe. This will ensure you have a better foundational understanding of how to add layers, set symbology, and navigate within a map. You can complete this recipe with any ArcGIS Pro licensing level.

It is also recommended that you verify that your computer hardware is sufficient for working with 3D data. This requires more power than working with traditional 2D Maps. It is recommended that you have at least an 11th generation Intel i5 processor or better with 16 GB or more RAM (32 GB if you don't have a dedicated video card), a dedicated video card with a **graphics processing unit** (**GPU**), and 4 GB video RAM and 12 GB of free hard drive space. If your computer does not meet or exceed these specifications, you may have issues completing this and other recipes in this chapter.

How to do it...

You will now create the scene requested by the lead project engineer. It will include both 2D and 3D layers:

1. Start ArcGIS Pro and click on the **Local Scene** template option located under **New Project** in the **Startup** window, as illustrated in the following figure:

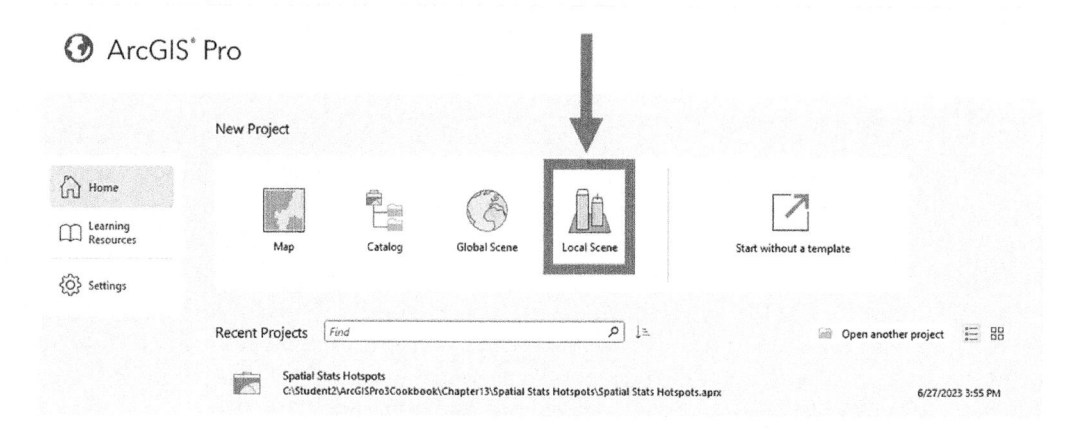

Figure 14.3 – Selecting the new Local Scene template

2. In the **Create a New Project** window that appears, type `3DAnalyst` into the cell located to the right of the **Name** option.

3. Click the **Browse** button at the end of the **Location** cell. Then, in the left panel of the **New Project Location** window, click **Computer**.

4. In the panel on the right, double-click **This PC**.

5. Scroll down and double-click on the `C:` drive. It is sometimes labeled as `Local Disk`, `OS`, or `Local Drive`.

6. Scroll down and double-click the `Student` folder. Then, repeat for the `ArcGISPro3Cookbook` folder.

7. Scroll down and select the `Chapter14` folder. Click on the **OK** button to set the location where your new project will be saved.

> **Tip**
> Remember, for this part of creating a new project, you are selecting the location where the new project will be stored, not what the project is being named. So, you must select an existing folder on a local drive or network share. If you double-click instead of single-clicking to select, you will need to back up one folder. You can do that simply by clicking the **Back** button located in the left corner. It looks like an arrow inside a circle.

8. Verify your **Create a New Project** window matches that in the following figure and click **OK**:

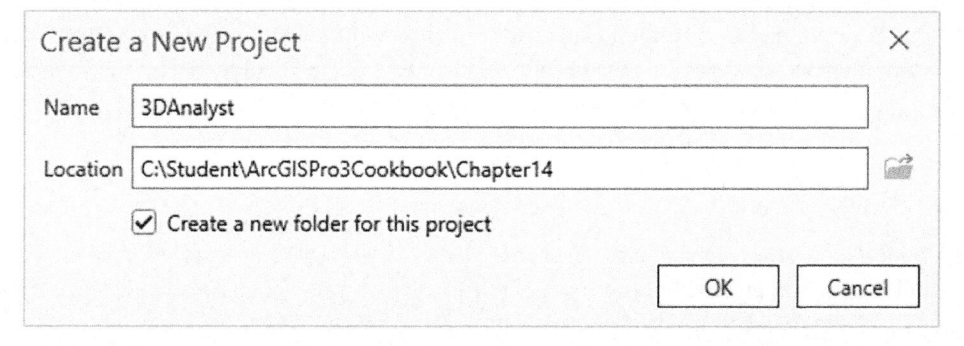

Figure 14.4 – The Create a New Project window with completed parameters

You have just created a new project. This new project contains a single local scene. Now you will need to add data and folder connections to the project that you will need in order to add layers and access data.

1. In the **Catalog** pane, right-click the `Folders` folder and select **Add Folder Connection**.
2. In the panel on the left of the **Add Folder Connection** window, select **Computer**. Then double-click the `C:` drive located in the right panel of the window. It may be labelled as `Local Disk`, `OS`, or `Local Drive`. You may also need to scroll down to see it.
3. Scroll down and double-click the `Student` folder. Then, scroll down and select the `ArcGISPro3Cookbook` folder with a single click. Lastly click **OK**.
4. If needed, expand the `Folders` folder in the **Catalog** pane so you can see its contents.

You should see that two folders are now connected when you created the new project in *step 8* and performed *steps 9–12*. The first is your project folder. This has the same name as the project you just created. The second folder is the connection you just added. If you were to expand the `ArcGISPro3Cookbook` folder connection, you would see all the other folders inside. Adding the connection to that folder allows you to access and use data contained in any of those folders inside this project.

Adding a database connection

Now, you will connect to a couple of databases that you will need in this and future recipes:

1. Right-click the `Databases` folder in the **Catalog** pane and select **Add Database**.
2. Click on the **Folders** option located under **Project** in the left panel of the **Select Existing Geodatabase** window. You may need to expand the `Project` folder to see it.
3. Double-click the `ArcGISPro3Cookbook` folder in the right panel of the window.
4. Scroll down and double-click the `Databases` folder.
5. Select the `AthensQuarry.gdb` geodatabase and click **OK**.

You have just added the connection to the primary database you will be using in this recipe. But since you will be using this project for other recipes in this chapter, you will go ahead and connect to the `Trippville_GIS.gdb` geodatabase that you will need later.

6. Repeat that same process to add the `Trippville_GIS.gdb` geodatabase to the project. The one difference is that you need to select the `Trippville_GIS.gdb` at the end before you click **OK**.

7. Your project should look like that in the following screenshot. The scene may be at a different scale. Save your project to ensure all your work so far is not lost:

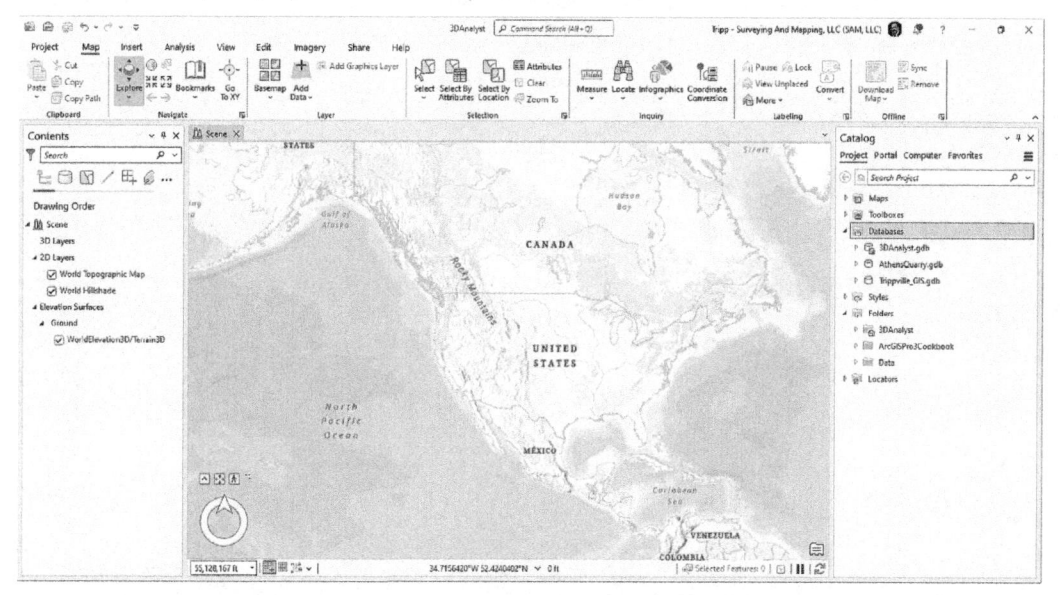

Figure 14.5 – New project with folder and database connections established

You have added the databases and folder connections that you need in this and future recipes to the project.

Configuring the 3D scene

Now you are ready to begin configuring your 3D scene. You will need to first configure some properties for the scene:

1. In the **Contents** pane, right-click on **Scene** and select **Properties** from the menu that appears. This will open the **Map Properties** window for the **Scene**.
2. On the **General** tab, rename the scene from `Scene` to `Quarry Project`. If you accidentally press the *Enter* key and close the **Map Properties** window, reopen it by right-clicking on the map name and selecting **Properties** from the menu.
3. Set the **Display units** to **US Feet** using the drop-down arrow. This project is located in the **State of Georgia**, which is in southeastern United States.
4. Set the **Elevation Units** to the same value, **US Feet**.

5. Verify that your **Map Properties** window looks like that in the following figure and click **OK**:

Figure 14.6 – Scene properties with updated values

Next you will add the elevation surface you want to use for ground surface for this `Quarry Project` scene you are creating. The ground elevation surface serves as the canvas upon which both 2D and 3D layers are overlaid. For this project, you have been provided with a highly accurate digital elevation model for the quarry project area that was created by a local surveyor. You will use this as the ground elevation for this scene.

6. Activate the **Map** tab in the ribbon.
7. Click on the drop-down arrow located below the **Add Data** button, as illustrated in the following screenshot. A long menu should appear when you click on the arrow.

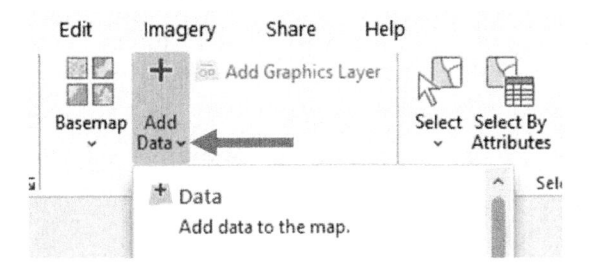

Figure 14.7 – Selecting the Add Data drop-down menu

8. Scroll down the menu that appears and select **Elevation Source Layer**. This will open the **Add Elevation Source Layer** window.
9. Click on the `Databases` folder option under **Project** in the left panel of the window.
10. Double-click on the `AthensQuarry.gdb` geodatabase in the right panel of the window.
11. Select the `quarry_DEM` raster dataset and click OK to add it to your scene. Your scene may automatically zoom in on the extents of the new ground elevation source you just added.

You should now have two elevation sources listed in the **Ground** group for this project. You may also see a warning telling you a datum transformation was applied. This is normal because your DEM and the Esri web service make use of two different coordinate systems and datums.

Figure 14.8 – Quarry Project scene with new ground elevation source added

It is not unusual for a scene to have more than one elevation source. It depends on the area covered by both the project and local elevation sources. ArcGIS Pro treats overlapping elevation sources much like layers in the **Contents** pane. It will use the top one that covers the area in question if two or more of the sources overlap. In this case, it will use the local DEM you just added until you reach the spatial edge where the DEM ends. Then it will make use of the Esri Terrain Model as the source.

12. Double-click on the `Quarry Project` scene at the top of the **Contents** pane. This will open the **Map Properties** window.
13. Click **Coordinate System** in the left panel.

14. Look at the value located in the box below **Current XY**. It should read **NAD 1983 StatePlane Georgia West FIPS 1002 (US Feet)**.

15. Look at the value located in the box below **Current Z**. It should read **NAVD88 height (ftUS)**.

16. If these two values are correct, click the **OK** button to close the window. Then, save your project.

These two values were updated when you added the new `quarry_DEM` ground surface. This was done automatically.

Adding layers to the scene

Now you will start adding layers to the scene you have created and been configuring. This will include both 2D and 3D layers:

1. On the **Map** tab, click on the **Add Data** button.
2. Select **Project** in the left panel of the **Add Data** window. Then, double-click on the `Databases` folder in the right panel.
3. Double-click on the `AthensQuarry.gdb` geodatabase in the right panel.
4. While holding your *Ctrl* key down, select the following feature classes: `Bldg`, `HydroLine`, `HydroPoly`, and `Parcels`. Click the **OK** button once you have the feature classes selected.

The new layers you added should appear in your map as 2D layers, as shown in the next screenshot. You will need to adjust the draw order for the layers and the symbology. Also, keep in mind that the colors you see will most likely be different, as they are assigned by ArcGIS Pro at random:

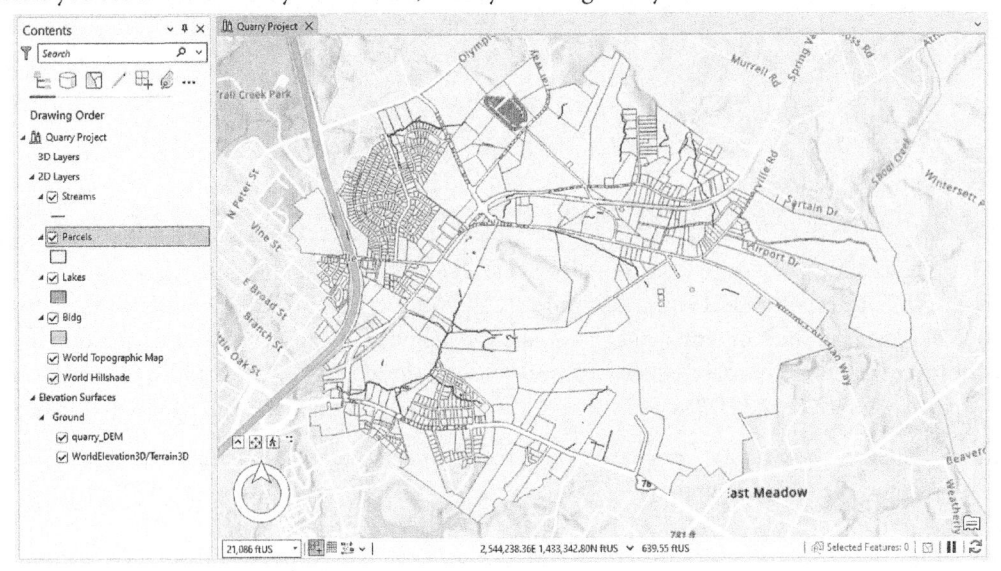

Figure 14.9 – New layers added to the scene

5. In the **Contents** pane, select the **List by Draw Order** button.
6. Select and drag the `Lakes` layer so it is above `Parcels`. `Lakes` is an alias for the `HydroPoly` feature class. Lakes should now be visible on top of the parcels in the scene.

> **Note**
> Aliases can be assigned to many things in ArcGIS Pro, including layers and table field names. An alias provides a more detailed or informative description for the item. In this case the alias `Lake` is more understandable than the feature class name of `HydroPoly`. Most people know what a lake is, so that layer name makes more sense.

7. Select and drag the `Bldg` layer so it is above the `Parcels` layer as well. It can be either above or below the `Lakes` layer.

All layers should now be visible in your map. No layer should be hiding behind another.

Adjusting layer symbology

Now, let's work on the symbology so each layer is easier to interpret and locate in the scene:

1. Click the symbol patch located below the **Streams** layer name in the **Contents** pane, as shown in the following screenshot. This should open the **Symbology** pane:

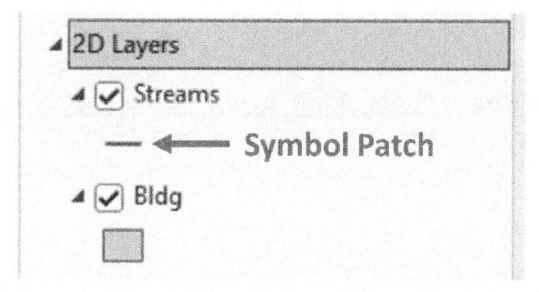

Figure 14.10 – Selecting the symbol patch

2. Ensure you are viewing the **Gallery** by clicking the **Gallery** tab next to the top of the pane. Select the **Water (line)** predefined symbol.
3. Click the symbol patch located below the `Lakes` layer name. Then, select the **Water (area)** predefined symbol in the **Symbology** pane.
4. Click the symbol patch below the `Bldg` layer name. Then, select the **Building Footprint** predefined symbol. If you see more than one, select any one you want to use.
5. Click the symbol patch located below the `Parcels` layer name. Select the **Black Outline (1pt)** predefined symbol.

6. Close the **Symbology** pane.

   Your scene should now look like that in the following screenshot:

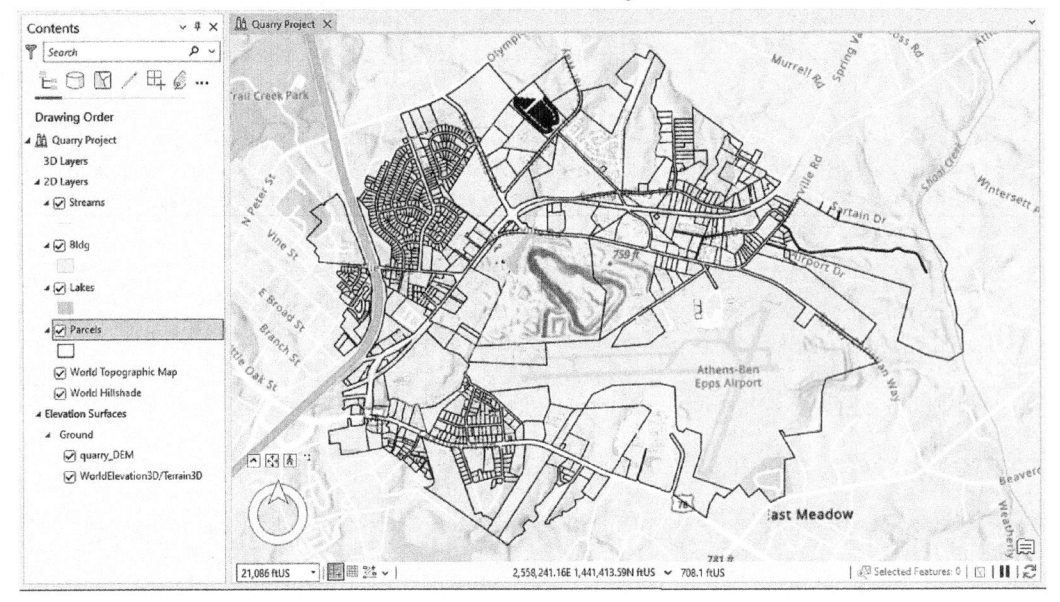

Figure 14.11 – Added layers with symbology configured

So far, you have created a new project that contains a 3D scene. You then configured the source elevation, which is the ground surface for the scene, using a local digital elevation model. Lastly, you added several layers to the scene. These layers are all being displayed as 2D layers, which means they are just being draped across the DEM you specified as the elevation source.

Making a 2D layer 3D

Now it's time to make one of the 2D layers a 3D layer. This is fairly easy to do:

1. Select the `Bldg` layer in the **Contents** pane.

2. Drag and drop it on the **3D Layers** in the list. A box will appear around 3D Layers when your mouse pointer is on top of it. This is the indication to release your mouse button. The `Bldg` layer should now appear under **3D Layers**, indicating it is now a 3D layer. This means you will be able to display the buildings as 3D objects in the scene.

3. Activate the **Feature Layer** tab in the ribbon.

4. Click on the drop-down arrow located below the **Type** button. Select **Min Height** from the menu that appears, as shown in the following screenshot:

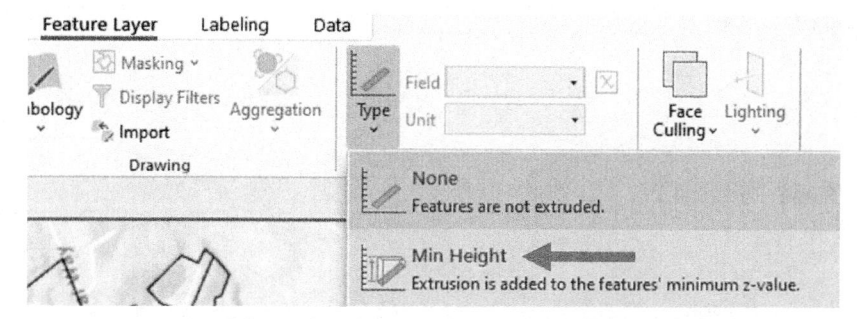

Figure 14.12 – Selecting the Min Height extrusion type

5. In the cell located next to the **Type** option, click the drop-down arrow and select **Estimated Height** from the list, as illustrated in the following screenshot. This is a field from the attribute table for the Bldg layer. Depending on some of your settings, you might also see this listed as **Est_HGT**:

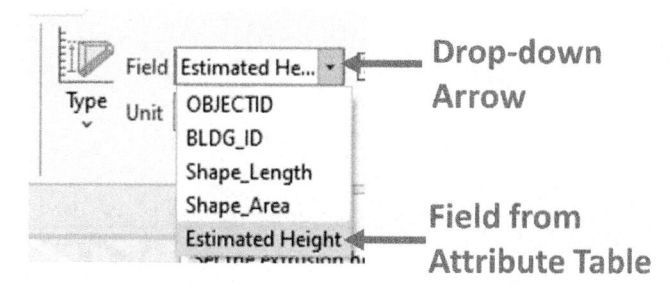

Figure 14.13 – Setting the extrusion field

You have just extruded the building in the Bldg layer based on a field in the attribute table that has each building's estimated height. Let's look at the results of your work so you can see the buildings are indeed 3D.

1. Place your mouse pointer on the southern side of the scene; anywhere on the southern side will work.
2. Press the scroll wheel located in the center of your mouse. While holding it down, push the mouse away from you. This should rotate the scene around the Z axis. Rotate it until you can begin to see the building heights above the ground and 2D layers. Then, release your scroll wheel.
3. Roll your scroll wheel so that when you zoom in closer, you can see the buildings. Verify that they are being shown above the ground and 2D layers.

4. Your scene should look like the following screenshot. Yours might be a little different depending on how far you rotated the scene along the Z axis and how much you zoomed in:

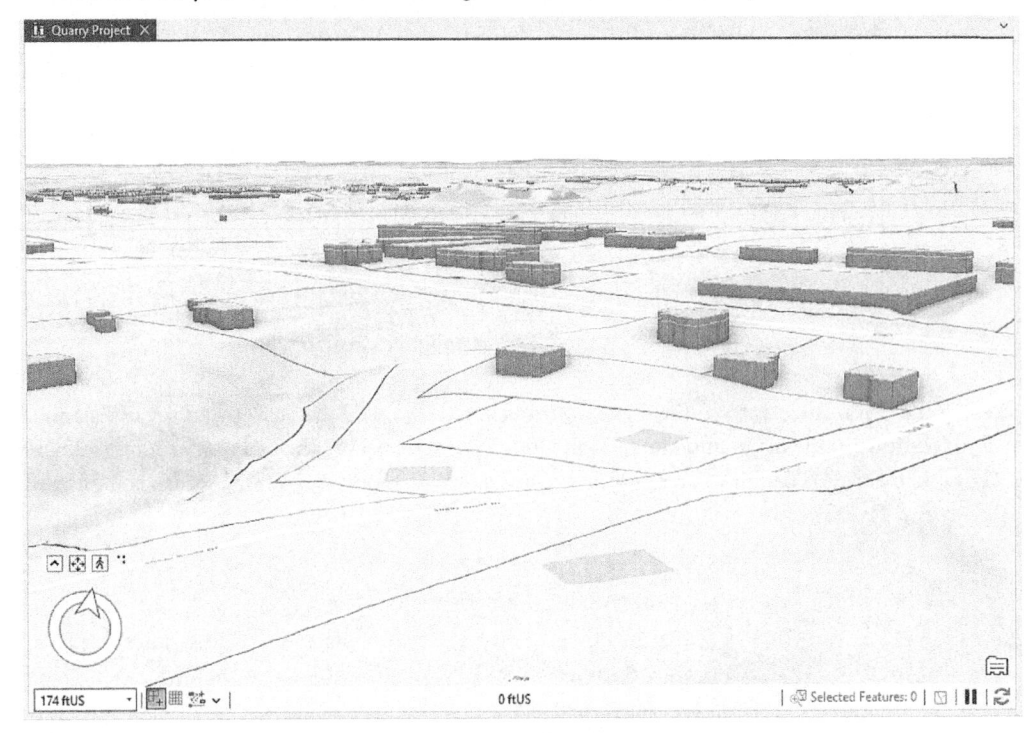

Figure 14.14 – Bldg layer extruded to view in 3D

5. Click the **Map** tab on the ribbon. Then, click the **Basemap** button.
6. Select the **Imagery** basemap from the list provided. This provides a more realistic 3D view of the quarry project area.

> **Tip**
> If you do not see the **Imagery** basemap, it may be that you are not connected to ArcGIS Online or are connected to your organization's portal for ArcGIS. If possible, connect to ArcGIS Online using your login details. If you are not able to, skip to *step 11*.

7. Save your project.
8. Continue to navigate through the scene. Try zooming in and out, panning, and rotating along all axes. Make sure to examine the area around the quarry. You should be able to see all the piles of rock and the quarry pit itself in detail, as shown here:

Figure 14.15 – Area around the quarry showing piles of rock

9. When you are done exploring the scene, right-click the `Bldg` layer in the **Contents** pane and select **Zoom to layer**.
10. Save your project and close ArcGIS Pro.

You have just created a very basic 3D scene using the core functionality found in ArcGIS Pro. What you just did can be accomplished at all license levels of ArcGIS Pro and does not require any extensions.

How it works...

In this recipe you created a new scene that allowed you to look at a project area in 3D with buildings extruded to their estimated heights and aerial photography draped across an accurate digital elevation model. To do this, you first created a new project using the **Local Scene** template. This opened the new project with a new blank local scene.

With the new project and scene open, you next added database and folder connections to the project. These new connections provided access to data resources you used in this recipe to create the scene and will use in later recipes. Next, you configured the scene by setting the display and elevation units to **US Feet** under the scene properties. Then, you added the `quarry_DEM` digital elevation model for the project area as a ground elevation surface by clicking on the drop-down arrow under the **Add Data** button. You did this because it was more accurate than other available elevation data.

Next, you added and configured the layers you needed to display in the scene. This included extruding the `Bldg` layer so that it displayed the builds represented in the layer at their estimated height in 3D using the `Estimated Height` field in the attribute table for the layer.

Enabling your data to store Z coordinates (elevation)

In *Chapter 8, Creating a Geodatabase*, we learned that vector data, points, lines, and polygons store data using the *X* and *Y* coordinates for the features. This determines their location, which is then displayed in a map. We expanded on that in *Chapter 7, Projections and Coordinate System Basics*, where we learned those *X* and *Y* coordinates were referencing locations in a specific, real-world coordinate system that ties our data to the Earth. This allowed us to bring data from all over into our maps so that we could see their spatial relationships. ArcGIS Pro will project data that is in different coordinate systems on the fly, so they are displayed together. However, that only represents two dimensions.

Can you enable data in ArcGIS Pro to store Z, the third dimension? Of course you can. You typically do this when you first create the feature class or shapefile. Some formats, such as an AutoCAD `DWG` and `DXF` files, are always Z-enabled. The 3D Analyst extension for ArcGIS Pro also has tools that allow you to convert existing 2D data to 3D Z-enabled data.

In this recipe, you will continue to work with the data for the quarry project you used in the last recipe. The lead project engineer is now concerned about the nearby airport. He needs to see the runways in relation to the quarry and may need to make calculations concerning runoff from the quarry to those runways. He has acquired a shapefile that contains the location of runways, taxi ways, and parking areas at the airport. However, they are not Z-enabled. You will create a new feature class in the `Athens_Quarry` geodatabase that is Z-enabled. Then, you will import the data from the shapefile into it. Lastly, you will calculate the elevation of the features based on the Digital Elevation Model you selected as the ground elevation in the *Creating a 3D scene* recipe.

Getting ready

This recipe requires the sample data to be installed on the computer. You will need to have completed the previous recipe before starting this one. You will also need to have the 3D Analyst extension to complete this recipe.

To determine whether you have access to the 3D Analyst extension, open ArcGIS Pro. Then, open any project you have previously opened. Click the **Project** tab in the ribbon. Click the **Licensing** option in the left panel of the **Project** pane. This will tell you what license level and extensions you currently have access to, as illustrated in the following screenshot:

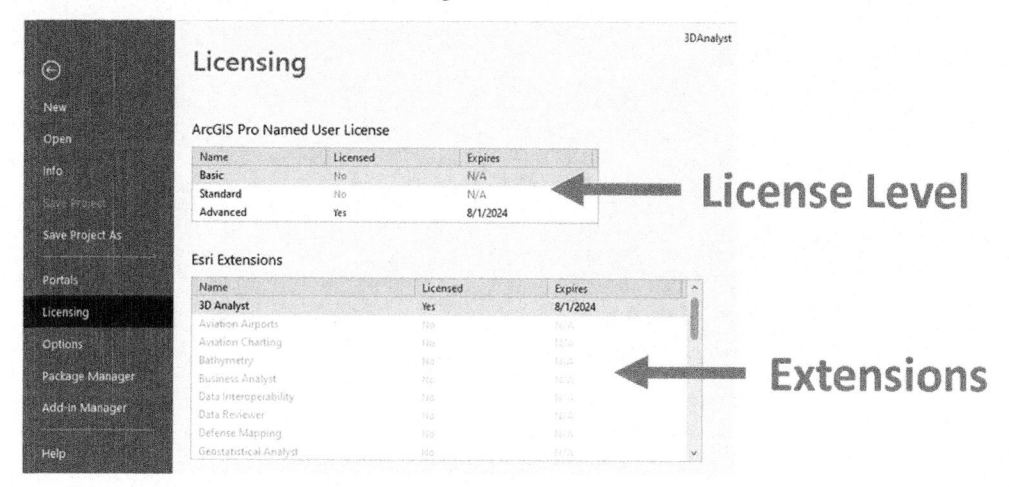

Figure 14.16 – Determining licensing for extensions

If you do not have access to the 3D Analyst extension, first contact your administrator to see whether they can provide one for you. If you do not have a license for this extension, you can request a trial license from Esri at https://www.esri.com/en-us/arcgis/products/arcgis-pro/trial.

This recipe can be completed with any license level of ArcGIS Pro if you have the 3D Analyst extension.

How to do it...

You will now go through the process to enable GIS data to store Z values:

1. Start ArcGIS Pro and open the 3D Analyst project you created in the previous recipe. It should be in your list of recently opened projects.

2. Ensure the QuarryProject scene you created in the previous recipe is open. If not, open it from the **Catalog** pane under the **Maps** folder by right-clicking it and selecting **Open Local View**.

3. In the **Catalog** pane, expand the `Database` folder. Right-click the `AthensQuarry.gdb` geodatabase. Then, go to **New** and **Feature Class**, as shown in the following screenshot. This will open the **Create New Feature Class** tool:

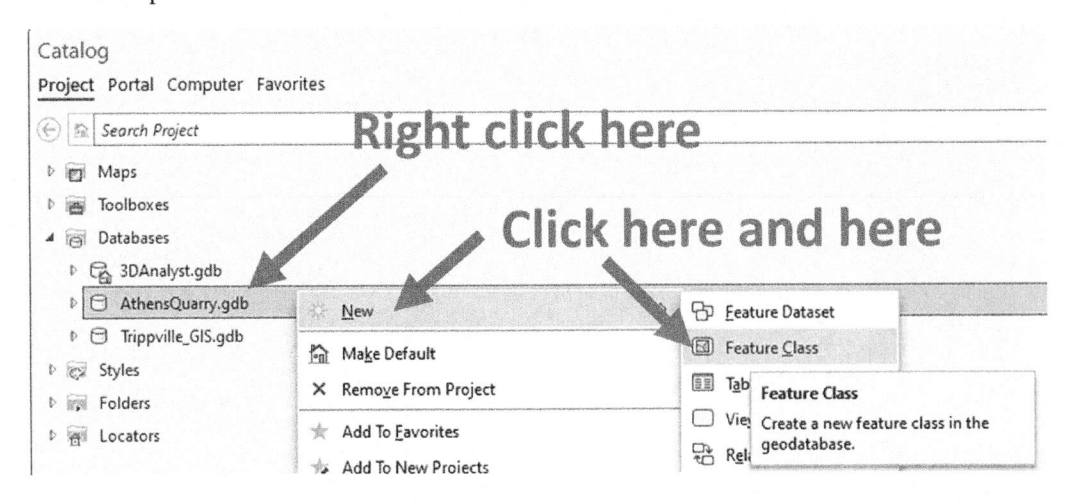

Figure 14.17 – Creating a new feature class

4. For the **Name** of the new feature class, type `Airport_Runways` into the feature class name and leave the **Alias** blank.
5. Verify the **Feature Class Type** is set to **Polygon**. If it is not, select it using the drop-down arrow.
6. Under **Geometric Properties**, ensure **Z Values – Coordinates include Z values used to store 3D data** is enabled.
7. Make sure **Add output dataset to current map** is enabled. Then, click **Next**.
8. The **Fields** tab should appear to allow you to configure attribute fields. Click on the **Import** button so you can import the fields from an existing shapefile.
9. Click on the **Folders** option in the left panel of the **Import** window.
10. Double-click the `ArcGISPro3Cookbook` folder in the right panel of the window. Then, double-click the `SourceData` folder.
11. Select the `Airport.shp` shapefile and click **OK**. If you do not see the `Airport.shp` shapefile, verify that the cell to the right of the **Name** cell is set to **Tables and Raster Catalogs**.
12. You should see three new fields were added to the list of fields. This should include `TYPE`, `Shape_len`, and `Status`, as shown in the following screenshot:

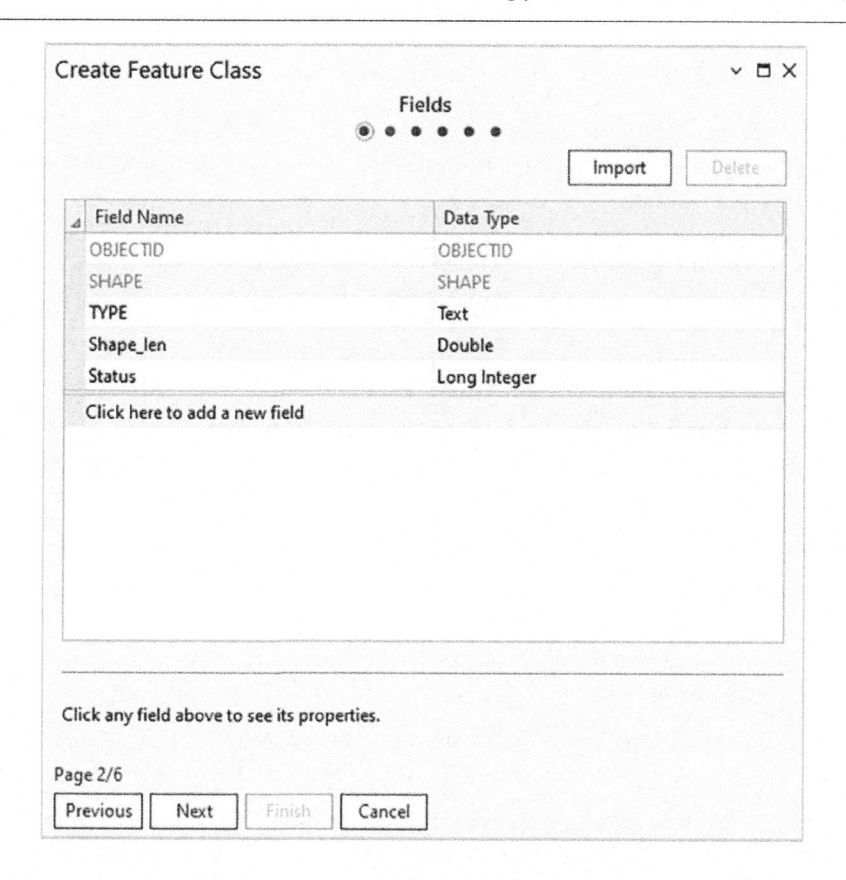

Enabling your data to store Z coordinates (elevation) 493

Figure 14.18 – Importing fields from a shapefile

13. Click the **Next** button in the **Create Feature Class** tool.
14. The **Spatial Reference** options should now be shown. Verify that **Current XY** is set to **NAD 1983 StatePlane Georgia West FIPS 1002 (US Feet)** and **Current Z** is set to **NAVD88 (ftUS)**. Then, click **Next**.
15. These settings have automatically been set because it used the settings from the `Quarry Project` scene you opened.
16. For the **Tolerance** settings, accept the defaults that have been calculated and click **Next**. Then, do the same for the **Resolution**.
17. Lastly, you should see settings for **Storage Configuration**. Ensure **Default** is selected under **Configuration Keyword**. Then, click **Finish** to create the new feature class.

When the tool is complete, a new layer should appear in the `Quarry Project` scene. This new layer references the new feature class you just created. This new feature class will not only store the *X* and *Y* coordinates for all the features it contains but it will allow you to store the *Z* coordinates as well.

Adding data to the new layer with the Append tool

Since this is a brand-new feature class, it is currently empty. You will next import the features that are in the shapefile you used to import the fields from when you created the new feature class. You will use the **Append** tool to do this:

1. Select the **Analysis** tab in the ribbon and click on the **Tools** button.
2. In the **Geoprocessing** pane that opened, click on the **Toolboxes** tab located below the **Find Tools** cell.
3. Scroll down and select the **Data Management** toolbox. Then, expand the **General** toolset.
4. Click on the **Append** tool to open it in the **Geoprocessing** pane.

> **Tip**
> You can also access the **Append** tool by right-clicking on the `Airport_Runways` feature class in the **Catalog** pane and selecting **Load Data** from the menu that appears.

5. Click the **Browse** button for the **Input Datasets**. Then, select **Folders** in the left panel of the **Input Datasets** window.
6. In the right panel of the **Input Datasets** window, navigate to `ArcGISPro3Cookbook\Chapter14\SourceData`. Then, select the `Airport.shp` shapefile and click **OK**.
7. You will not need to build an expression since you will be loading all the data from the shapefile. So, skip down to the **Target Dataset** and set it to the `Airport_Runways` layer you just created using the drop-down arrow.

> **Tip**
> You had to use the **Browse** button to set the **Input Dataset** because the `Airport.shp` shapefile was not already a layer in your map. The drop-down arrows only list existing layers or tables included in the active map or scene.

8. Ensure **Field Matching Type** is set to **Input fields must match target fields**. You can use this option because you imported the fields from the shapefile so they should match.

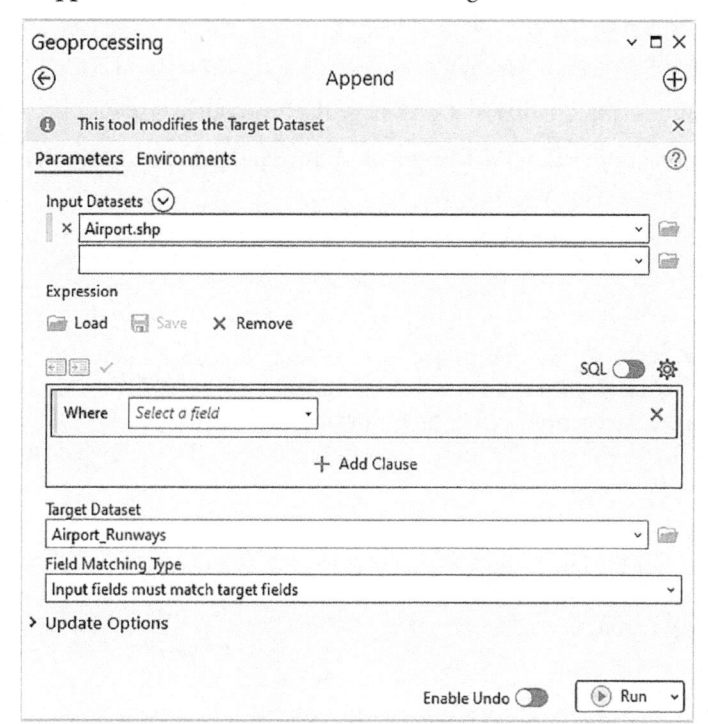

9. Verify your **Append** tool matches that in the following screenshot and click **Run**:

Figure 14.19 – Append tool with completed parameters

When the **Append** tool is complete, it will have copied the features from the shapefile to the `Airport_Runways` feature class. You will not be able to see them because their Z coordinate is still currently set to 0 (zero) because the shapefile was a 2D shapefile. Next, you will need to calculate a Z coordinate for each vertex associated with the features in the new feature class.

Calculating Z values

You will use a tool from the 3D Analyst extension that will calculate the Z value for the vertices based on where it lies in relation to the DEM you set as the ground surface. The 3D Analyst extension is an add-on for ArcGIS Pro and does require a license to use this extension:

1. Click the **Back** button in the **Geoprocessing** pane that has the **Append** tool open. The **Back** button is a small arrow inside of a circle located on the top left side.

2. Locate and expand the **3D Analyst** toolbox so you can see its contents. If you do not see this toolbox, you most likely do not have access to the 3D Analysis extension for ArcGIS Pro.

3. Expand the **3D Features** toolset and then expand the **Interpolation** toolset.

4. Click on the **Update Feature Z** tool to open it. This tool will calculate the Z coordinates of features based on an input surface.

5. Set the **Input Features** to **Airport_Runways** using the drop-down arrow.

6. Set the **Input Surface** to **quarry_DEM** using the drop-down arrow.

7. ArcGIS Pro will automatically determine which **Interpolation Method** it believes is best. Accept the one determined by ArcGIS Pro.

8. **Status Field** will be left blank.

> **Note**
> This field can be used to track which features are not updated with new Z coordinate values. This is typically because the features do not overlap the surface. In this case, all the features in the **Input Features** layers overlap the **Input Surface** for the project area, so there is no need to be concerned. There are parts of some of the runway features that are outside the project area, but those are not a concern.

9. Verify that your **Update Feature Z** tool matches the following screenshot and click **Run**:

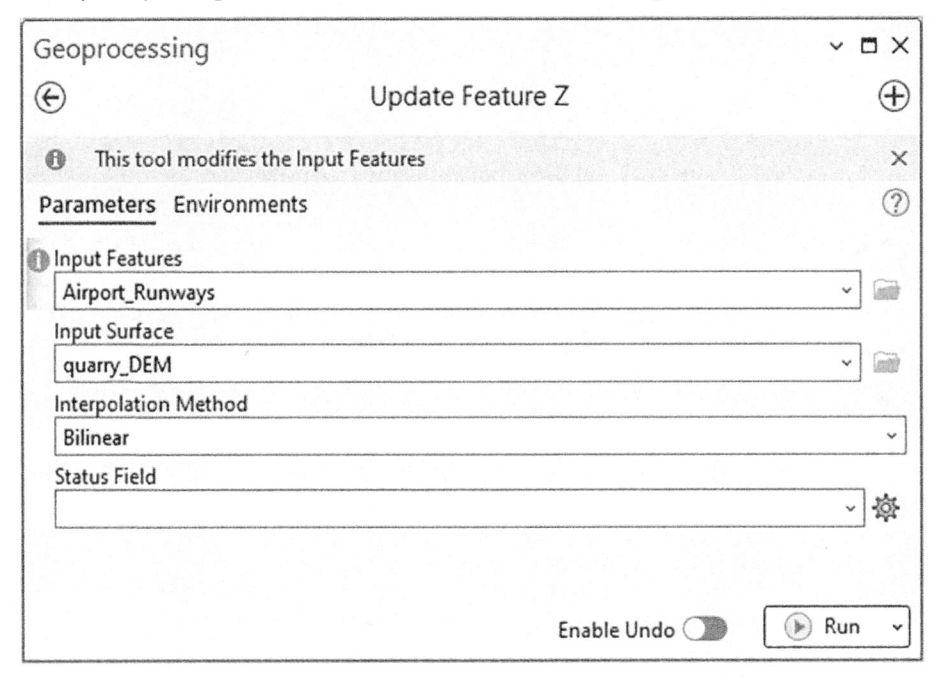

Figure 14.20 – The Update Feature Z tool with completed parameters

Once the tool is complete, you should start to see parts of the `Airport_Runways` layers, as shown in the following screenshot:

Figure 14.21 – Results from Update Feature Z tool

Some runway features may appear hidden or cutoff. This is because they extend past the area covered by `quarry_DEM` that represents the project area. The features that do not overlap `quarry_DEM` were not assigned a Z value and thus cannot be seen. However, since those features are outside the project area, they are not a concern.

You will further verify the success of the tool by using another tool that will add and calculate several attribute fields that are based on the 3D characteristics of the feature. This is the **Add Z Information** tool that is also included in the 3D Analyst extension.

10. Save your project.
11. In the **Geoprocessing** pane that contains the **Update Feature Z** tool you just used, click the **Back** button.
12. If needed, click the **Toolboxes** tab in the **Geoprocessing** pane.
13. Expand the **3D Analyst Tools** toolbox to see its contents.
14. Expand the **Statistics** toolset. Then, click the **Add Z Information** tool to open it.
15. Set the **Input Features** to `Airport_Runways` using the drop-down arrow.
16. Under the **Output Property** options, enable the following options:

    - Lowest Z
    - Highest Z
    - Average Z
    - Average Slope

17. Leave **Noise Filtering** blank. This parameter can be used to filter out values that might skew statistics by being too small or large to the point of being considered outliers.

18. Verify your **Add Z Information** tool looks like that in the following screenshot and click **Run**:

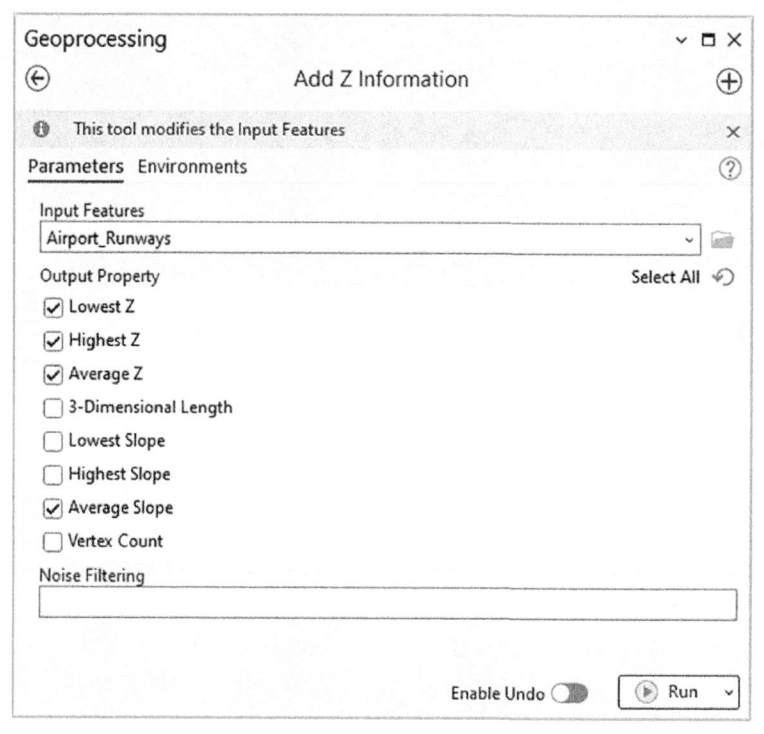

Figure 14.22 – The Add Z Information tool with completed parameters

Fields for the properties you checked in the **Add Z Information** tool should now be added to the attribute table for `Airport_Runways` along with the appropriate values.

Verifying results

Now, to verify that the tools worked, do the following:

1. Close the **Geoprocessing** pane.
2. Right-click the `Airport_Runways` layer and select **Attribute Table** from the menu that appears. Then, review the fields contained in the table.

Hopefully you were able to pick out the four fields that were added to the attribute table for the `Airport_Runways` layer. These correspond to the options you selected in the tool. The values stored in these newly added fields allow you to see that the **Update Feature Z** tool was successful except for the two features that extended beyond the project area.

3. Close the `Airport_Runways` attribute table.
4. Save your project and close ArcGIS Pro.

You have now successfully enabled GIS data so that it will store elevations as a Z coordinate in addition to X and Y coordinates as well as import data from another file and update the Z coordinates using an existing **digital elevation model (DEM)**.

How it works...

You have just created a new Z-enabled feature class and imported data from a 2D shapefile into it. You did this by creating a new feature class in the `AthensQuarry.gdb` geodatabase from the **Catalog** pane. This launched the **Create Feature Class** tool wizard. This tool then took you through the process and parameters needed to create a new feature class in a geodatabase. To allow this new feature class to store Z values, you made sure that option was enabled in the wizard. Once the new feature class was created, you imported data from an existing 2D shapefile into it using the **Append** geoprocessing tool.

You then used the **Update Feature Z** to calculate the Z coordinates for all the vertices for the features in the new feature class you created. Lastly, you then used the **Add Z Information** tool to verify that the Z coordinates had been successfully added.

Creating multipatch features from 2D

In working through the recipes in this chapter, you have learned how to display 2D data in 3D and how to create basic Z-enabled layers. All of these methods for working in 3D open the door to a wealth of capabilities within ArcGIS Pro, from display to analysis. However, they all still have limitations. Extruding 2D data to display in 3D does allow us to see those features and their relationships in three dimensions compared to other features, but we are not able to easily calculate volumes or locate a position vertically within the extruded feature.

Adding Z coordinates to a point, line, or polygon allows us to place it in the correct space, but again, they still only form a single plane. What do we need to create a solid shape (meaning something that has volume)? There is a more advanced 3D data format that is supported in ArcGIS Pro that allows for this. It is called a **multipatch**.

A multipatch is a true 3D object. It is constructed using a series of planes or polygons that are drawn in a 3D environment. They are connected to form a 3D feature. The planes or polygons that form the feature are also called **patches**, hence the term multipatch. ArcGIS Pro is not the only application that allows for the display and creation of multipatch features. Most **computer-aided drafting and design (CADD)** software also can be used to create multipatch features. Because of the complexity of a multipatch feature, you are often able to show more detail than what you can accomplish with simple extrusion. Multipatch feature classes also increase the number of analysis tools you can use from the 3D Analyst extension for calculating line of sight and more.

The city planner for the City of Trippville wants to incorporate line-of-sight requirements into the development regulations for the city. He will want to see whether new buildings will adversely impact the views of key elements, such as parks and rivers, from existing buildings in the city. Many of these types of calculations require the use of multipatch features. In this recipe, you will convert the building footprints from extruded buildings to a new multipatch feature class.

Getting ready

To complete this recipe, you will need access to the 3D Analyst extension for ArcGIS Pro. You will need to have completed the other recipes in this chapter and ensure the sample data is installed before you continue. This recipe can be completed with all license levels of ArcGIS Pro.

How to do it...

You will now work through the process to create a new multipatch feature class representing the buildings. These will be based on existing 2D building footprint polygon features:

1. Start ArcGIS Pro and open the 3DAnalyst.aprx project you used in the previous recipes in this chapter. This project should be included in your list of recently opened projects.

2. In the **Catalog** pane, right-click the Maps folder and select **New Local Scene**. A new scene should open in the view area.

> **Note**
>
> Scenes will either be global or local. Global scenes are used for large area 3D maps where the curvature of the earth can impact the display and analysis of the data. Local scenes are for smaller areas such as a city, township, or county. In these cases, the curvature of the earth has less impact due to the scale of the 3D map. Local scenes most often make use of a projected coordinate system, whereas global scenes use a geographic coordinate system. For more information about global and local scenes, go to http://pro.arcgis.com/en/pro-app/help/mapping/map-authoring/scenes.htm.

With the new scene created, you next need to add a ground elevation source that you have for the area. This one is more accurate than the World Elevation 3D service provided by Esri that is used by default.

3. Activate the **Map** tab in the ribbon.
4. Click on the drop-down arrow below the **Add Data** button to expose the menu.
5. Click on the **Elevation Source Layer** option located near the bottom of the menu that appeared.
6. In the left panel of the **Add Elevation Source Layer** window, select Databases.
7. Double-click on the Trippville_GIS.gdb geodatabase in the right panel of the window.

8. Select **DEM** and click **OK** to add it as an **Elevation Surface** in your new scene. If it was added successfully, your scene should automatically zoom in on the general area covered by the DEM you added, as shown in the following screenshot:

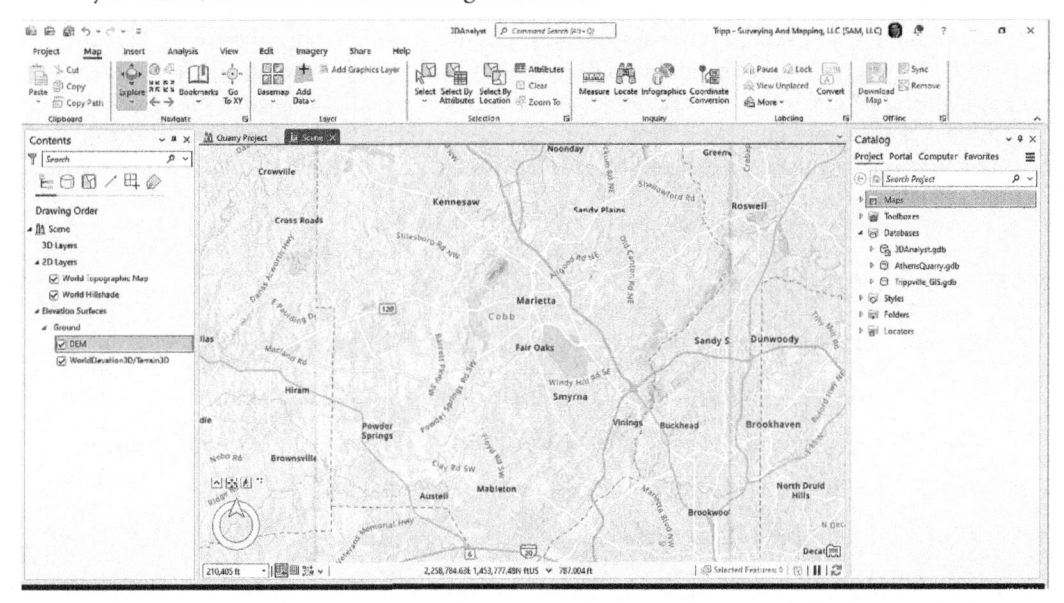

Figure 14.23 – New scene with new ground elevation source added

9. In the **Contents** pane, right-click on **WorldElevation3D/Terrain3D** ground elevation surface and select **Remove**.

When you remove **WorldElevation3D/Terrain3D**, the scene display area will be clicked to the area covered by the DEM surface you just added. This is because a scene requires a ground surface for any location or area to display it. The DEM you added is for the area representing the City of Trippville. Your scene will be limited to that area.

10. Now, rename the scene to something that is more descriptive than the default name of **Scene**. In the **Contents** pane, double-click on **Scene** to open its properties.
11. Click on the **General** option in the left panel of the **Map Properties** window.
12. Set the **Name** to `Trippville Buildings` and **Elevation Units** to **US Feet**. Then, click **OK** to apply the changes.

13. Save your project.

> **Tip**
> Just a reminder: you should save your projects to avoid possibly losing your work. Working with 3D scenes requires a lot more resources than 2D maps. This makes them more prone to crashing and failure. You might also want to enable the **Create a backup when the project has unsaved changes** option. This is located on the **Project** tab under **Options|General|Project Recovery**. This can allow you to quickly recover if your system crashes and you have not saved changes you have made to the project.

Adding layers to the scene

The new scene is ready to begin adding layers. You will add a few layers to the new scene now:

1. Click the **Add Data** button in the **Map** tab in the ribbon.
2. Click the `Databases` folder on the left panel of the **Add Data** window under **Project**.
3. Double-click the `Trippville_GIS` geodatabase in the right panel.
4. Double-click the **Base** feature dataset.
5. Select the **Buildings** and `Parcels` feature classes. Hold your *Ctrl* key down to select both. Click **OK** once both are selected.

The two layers are added as 2D layers to your scene, as illustrated in the following screenshot:

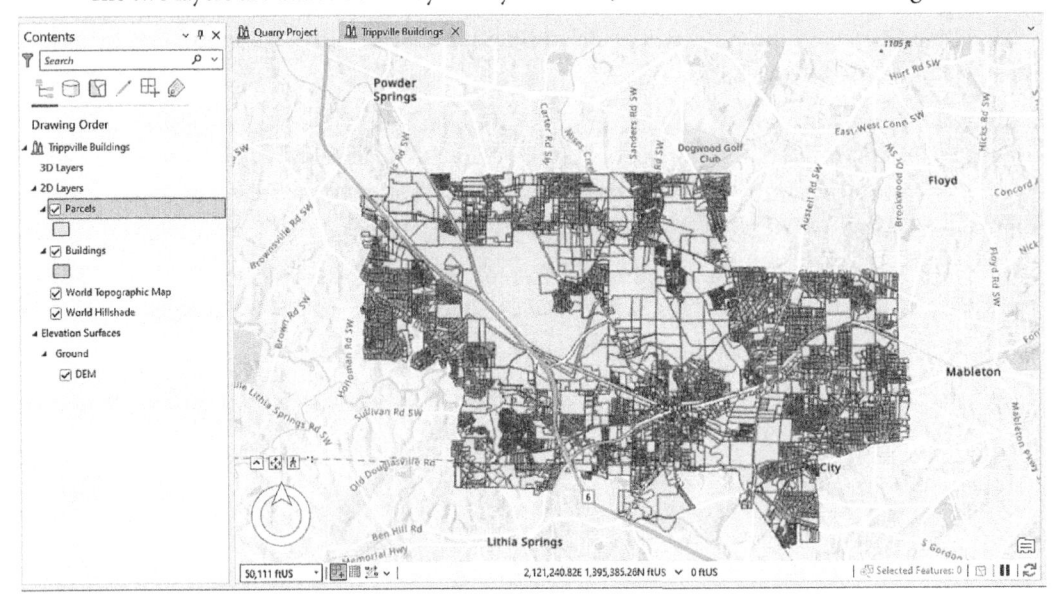

Figure 14.24 – Two new layers added to scene

Now, we will work on adjusting the symbology for both layers.

1. Click on the **List by Drawing Order** button in the **Contents** pane.
2. If needed, drag the Buildings layer so it is above Parcels in the **Contents** pane. When doing this, make sure the Buildings layer stays in the **2D Layers** group.
3. Click the small symbol patch located below the Parcels layer in the **Contents** pane.
4. In the **Symbology** pane that appears, ensure you are looking at the **Gallery**. The **Gallery** contains preconfigured symbols. Scroll down until you find the **Black Outline (1pt)** symbol and select it.
5. Click the small symbol patch located below the **Buildings** layer in the **Contents** pane.
6. Scroll down through the list of symbols until you come to the dark gray **Building Footprint** symbol that is in **Category Schema3**. Select that symbol for your buildings layer.

> **Tip**
> If you hover your mouse pointer over a symbol in the **Symbology** pane, you will see information about the symbol, including its name, the style in which it is stored, and its category. Also at the top of the **Symbology** pane is a **Search** function that you can use to locate predefined symbols based on a keyword. For example, you could have searched Building and it would have returned all symbols that are associated with buildings, including the one you were looking for.

7. Close the **Symbology** pane.
8. Right-click the **Buildings** layer in the **Contents** pane and select **Zoom to Layer**. Your scene should now look very similar to the following screenshot. The scale may be different, depending on the size of your monitor and your resolution:

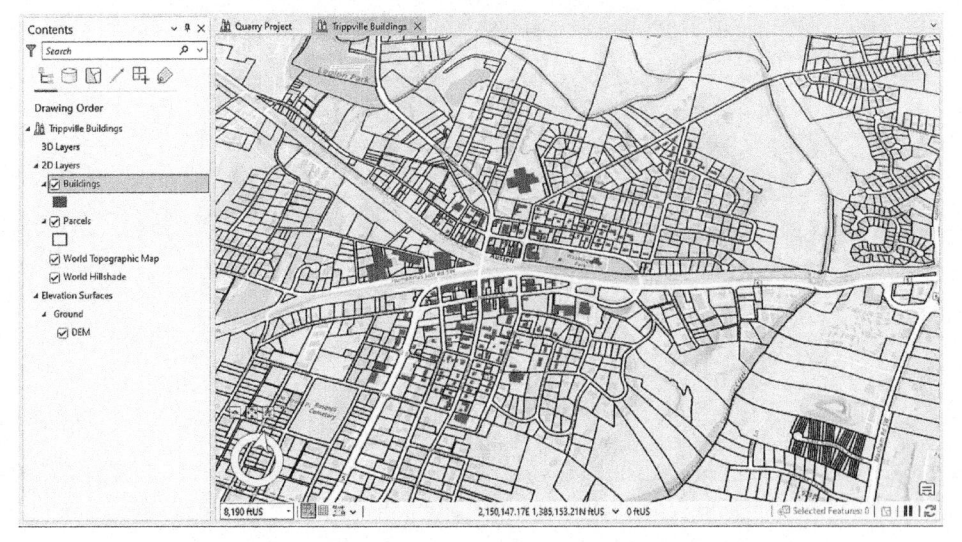

Figure 14.25 – Symbology configured for the two added layers

Making a layer 3D and extruding features

With the layers now added to the scene, you will make the Buildings layer 3D. Then you will extrude the features in the layer, so they display with a height based on attribute values in the layer's attribute table:

1. Drag and drop the **Buildings** layer from the **2D Layers** up to the **3D Layers**.
2. Save your project.
3. Click the **Feature Layer** tab on the ribbon.
4. Click the drop-down arrow below **Type** in the **Extrusion** group. Select **Max Height** from the list presented.
5. Ensure the **Unit** is set to **US Feet**.
6. In the **Field** cell located to the right of the **Type** button, click the drop-down arrow and select the `Estimated Height` field from the list. This will extrude the buildings based on their estimated height stored in the attribute table for the layer.
7. Using the skills you learned in the first recipe of this chapter, *Creating a 3D map*, zoom in on the downtown area and rotate your scene so it looks similar to the following screenshot:

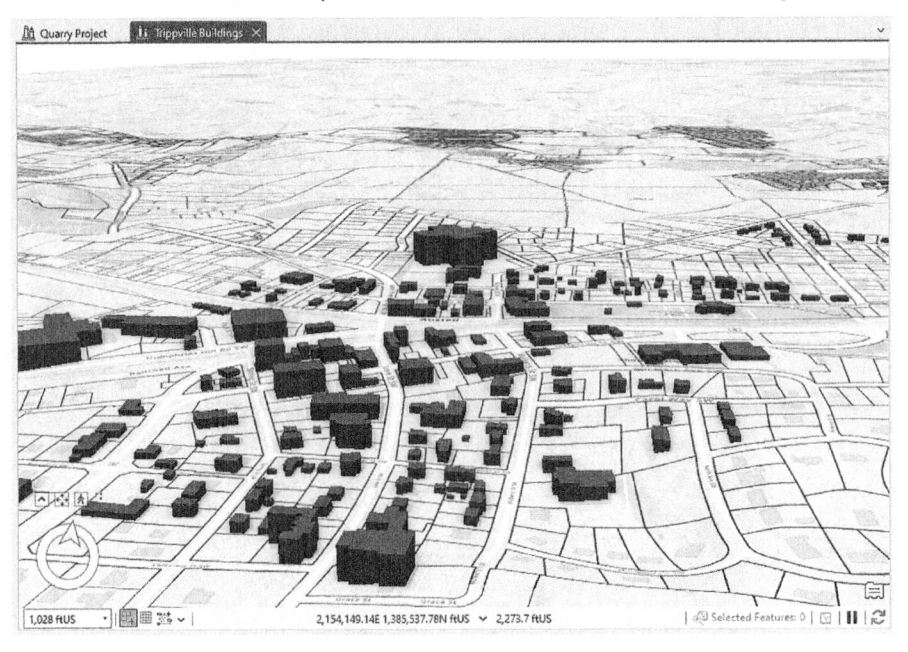

Figure 14.26 – Buildings successfully extruded

You can now see the buildings extruded to their estimated height. While they look 3D, they are still just 2D features that are being artificially expanded along the Z axis to appear in 3D.

Converting to multipatch

Now you will convert them to 3D multipatch features:

1. Click the **Analysis** tab in the ribbon.
2. Click the **Tools** button to open the **Geoprocessing** pane.
3. Click the **Toolboxes** tab in the **Geoprocessing** pane.
4. Expand the **3D Analyst Tools** toolbox.
5. Expand the **Conversion** toolset and select the **Layer 3D to Feature Class** tool.
6. Set the **Input Feature Layer** to **Buildings** using the drop-down arrow.
7. Click the **Browse** button for the **Output Feature Class**.
8. Click the `Databases` folder under **Project** in the left panel of the **Output Feature Class** window. Double-click the `Trippville_GIS.gdb` geodatabase.
9. Type `Buildings_MP` for the **Name**. Then, click the **Save** button.
10. Leave the **Grouping Field** parameter blank. The **Buildings** layer does not contain a field that could be used for this parameter.
11. Ensure the **Disable Color and Texture** option is not selected.
12. Verify your **Layer 3D to Feature Class** tool looks like that in the following screenshot and click **Run**:

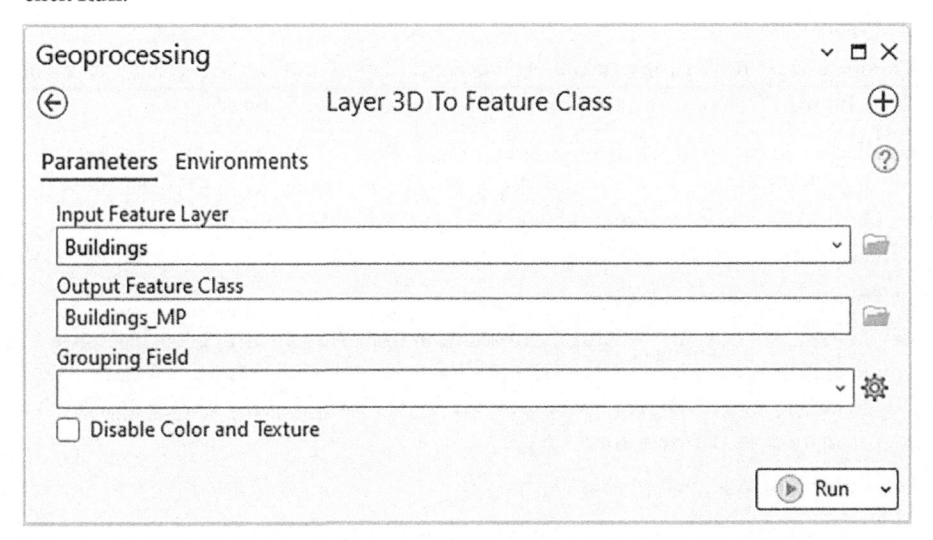

Figure 14.27 – The Layer 3D To Feature Class tool with completed parameters

Verifying results

You will now verify the results of the **Layer 3D to Feature Class** tool. You will check to see if it did indeed create a new multipatch layer:

1. Turn off the original **Buildings** layer you added at the beginning of this recipe so the Buildings_MP layer is the only visible 3D layer to confirm your conversion was successful. You should still see 3D buildings on the map. These are the multipatch features you just created.
2. Right-click the Buildings_MP layer and select **Attribute Table**. Review the values in the Shape field.

As you can see, the new layer is a multipatch shape. It is no longer a polygon. Now let's see an example of the increased capability a multipatch allows you to have. We will use a tool you are already familiar with: **Add Z Information**.

3. Close the attribute table for the Buildings_MP layer.
4. Click on the **Analysis** tab in the ribbon and then the **Tools** button.
5. Open the **Add Z Information** tool in the **3D Analyst Tools** and **Statistics** Toolset, using the skills you have learned.
6. Set the **Input Features** to the **Buildings_MP** layer. Then, look at the **Output Property** options that are available. Make a note of the available options so you can compare them later.
7. Now set the **Input Features** to the **Buildings** layer and notice the **Output Property** options available for this layer.
8. Compare the **Output Property** options that are available between the two different input layers. Note those differences and what they might mean for future analysis.

Each input layer allows you to calculate some common values, including lowest Z, highest Z, average Z, lowest slope, highest slope, and average slope. When you set the Buildings_MP as the **Input Features** for the **Add Z Information** tool, you are presented with a few options that are not available for the Buildings layer. You can have the tool calculate the volume of each building and the total surface area.

9. If you wish, you can run the **Add Z Information** tool using the Buildings_MP layer as the input and whatever options for the **Output Property** you want, just so you can see the results. If you do, close the **Geoprocessing** pane when it is complete. If you do not want to try this step, simply close the **Geoprocessing** pane.
10. Save your project and close ArcGIS Pro.

You have just successfully created a multipatch feature class from existing 2D data.

How it works...

In this recipe, you created a new multipatch feature class that contained true 3D features. You did this by adding a 2D layer, in this case `Buildings`, and extruded them so they displayed in 3D using an attribute field that contained the estimated height for each building.

Then, using the **3D Layer to Feature class** tool, which is part of the 3D Analyst extension for ArcGIS Pro, you generated a new multipatch feature class that contained true 3D features representing the buildings.

Creating 3D features

So far, you have learned how to create 3D features from existing 2D features. The first method you learned was creating a new Z-enabled feature class and importing existing 2D features into it. Then, you update the Z coordinates based on an elevation surface. Next, you learned how to convert extruded features into a new multipatch feature class. So, how do you create new 3D features from scratch?

Like most things in ArcGIS Pro, there are several methods you can use. You can create new 3D features in either a 2D map or a 3D scene. You can then specify a specific Z coordinate or have them automatically inherit the ground surface elevation.

In this recipe, you will create a few new 3D features. These are based on requests from the director of planning for the City of Trippville. For the first, he wants to flesh out the 3D view of the city so that it includes fences. So, you will begin digitizing fences. You will start with a 2D map and then move to the 3D scene. Next, the director is concerned about the construction of a new proposed building. He believes it will block the view to one of the main parks in the city from an existing building. So, he needs you to establish a few observation points that you will use in a later recipe to establish sight lines and determine visibility.

Getting ready

You must have completed the other recipes in this chapter before starting this one. This will ensure that you have the required project, maps, scenes, and data to complete this project. It will also confirm that you have a good foundational understanding of where to access the required tools and basic concepts required to complete this recipe. You will also need to ensure the sample data is installed before you continue.

How to do it...

You will now get to work through the process of creating new 3D features from scratch:

1. Start ArcGIS Pro and open the `3DAnalyst.aprx` project you created in the first recipe, *Creating a 3D scene*, and have used in all other previous recipes in this chapter. This project should be included in your list of recently opened projects.

2. Add an existing 2D map that contains the layers you need. Right-click the **Maps** folder in the **Catalog** pane and select **Import** for the menu.

3. In the **Import** window, expand the **Computer** folder in the left panel.

4. Double-click on **This PC** in the right panel of the Import window.

5. Then navigate to `C:\Student\ArcGISPro3Cookbook\Chapter14` in the right panel of the window.

6. Select the `TrippvilleMap.mapx` file and click **OK**. This is a map file. Importing this file will create a new 2D map in your project.

   A new 2D map should be added to your project. You should now see this new map in your map view area of the ArcGIS Pro interface, as shown in the following screenshot:

Figure 14.28 – Trippville Map imported into your project

Before you can edit 3D data in a 2D map, you need to ensure it has an elevation surface assigned. Next, you will make sure the new 2D map is using the DEM for the City of Trippville.

1. Save your project.

2. Double-click on the **Trippville** map located at the top of the **Contents** pane just above all the layers in the map to open the **Map Properties** window.

3. Set the **Display units** to **US Feet** using the drop-down arrow. Then, click **OK** to apply this change.

4. As you have done in past recipes, activate the **Map** tab and click on the drop-down arrow below the **Add Data** button.

5. Select the **Elevation Source Layer** option from the menu list that appears.
6. In the left panel of the **Add Elevation Source Layer** window, select the `Databases` location beneath **Project**.
7. Double-click the `Trippville_GIS.gdb` geodatabase in the right panel of the window.
8. Select **DEM** and click **OK**. You should now see **Elevation Surface** appear at the bottom of the layer list in the **Contents** pane.

You have prepared your 2D map so that it allows you to edit 3D data by ensuring it has an **Elevation Surface** assigned to it.

Creating a new 3D feature

Sometimes it is easier to edit 3D data in a 2D map because you are more comfortable with those tools. Now, we will digitize a fence:

1. Click the **Map** tab to activate it in the ribbon.
2. Click the **Bookmark** button and select the **New Fence** bookmark. This will take you to the area that has two fences that you will digitize.

Figure 14.29 – A zoomed-in view of the indicated parcel

3. Click the **List by Editing** button at the top of the **Contents** pane. It is the one that resembles a pencil.
4. Right-click on the `Fence` layer and select **Make this the only editable layer**.
5. Click the **Edit** tab in the ribbon. If you see an **Edit** button in the ribbon, click on it so it is highlighted in blue to allow for data edits. You will not see that button if you have not enabled the **Enable and disable editing from the Edit tab** under the **Options** for ArcGIS Pro located in the **Project** tab. You do not need to enable this option to continue with this recipe.

Since the `Fence` layer is Z-enabled, you need to tell ArcGIS Pro where the Z coordinate will come from when you draw a new fence. You will need to enable and configure **Z Mode** to do this.

6. Click on the **Z Mode** button to enable it. The button should be highlighted in blue once it is enabled.
7. Next, click the small arrow located below the **Z Mode** button and select **Surface** from the small menu that appears. This should make `Ground` appear in the cell located to the lower right of the button. This means the Z value for each vertex will be determined based on the ground surface elevation where you click.

> **Tip**
> ArcGIS Pro 3.2 appears to have an issue with reading the ground surface as an option to select for **Z Mode**. To work around this issue, you will need to add the DEM a second time by right-clicking on **Elevation Surfaces** and selecting **Add Elevation Surface**. Then, right-click on **Surface**, which has now appeared, and select **Add Elevation Source Layer**. Lastly, navigate to and select DEM in the `Trippville_GIS.gdb` geodatabase.

> **Information**
> Selecting **Constant** allows you to manually enter a constant Z value that will be applied to all new Z-enabled features you create.

8. Once more on the **Edit** tab in the ribbon, select the **Create** button to open the **Create Features** pane.
9. Select the **Fence** feature template in the **Create Features** pane. Then, select the **Line** tool located below the template.
10. Digitize the new fence by clicking on the locations shown in the following screenshot. Double-click or press the *F2* key to finish drawing the new feature.

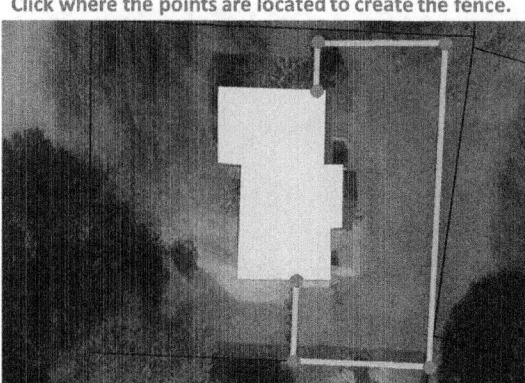

Figure 14.30 – Digitizing a new fence

11. Click the **Attributes** button located in the **Selection** group on the **Edit** tab. This will open the **Attribute** pane. You need to set the height of the fence.
12. In the **Height** cell, type 6 and press the *Enter* key to show it is a six-foot-tall fence. Then, click the **Apply** button.

> **Tip**
> If you have enabled the **Auto Apply** function located in the bottom-left corner of the **Attributes** window, you will not need to **Apply** edits to the attributes.

13. Click the **Geometry** tab in the **Attribute** pane and look at the values. Pay close attention to the Z values. These were automatically populated using the DEM.

As you can see, even though you were creating the Z-enabled fence features in a 2D map, ArcGIS Pro was still able to determine a value for the Z coordinate as well as the X and Y. It did this by comparing the location you clicked to the ground surface and interpolating the Z value for that location.

14. Close the **Attributes** pane.
15. Click the **Save** button in the **Manage Edits** group on the **Edit** tab to save your edits. When asked to **Save all edits**, click **Yes**.
16. Click the **Clear** button in the **Selection** group to deselect the new fence you just created.

You have just created a 3D feature in a 2D map. This can be easier than doing so on a 3D scene because your computer does not have to expend the resources required to render the data in 3D. That makes your system respond faster and reduces the chances that it might have issues.

Creating a new feature in a Scene

You will now see how to create a new 3D feature in a Scene.

1. Click the `Trippville Buildings` scene you have worked with in previous recipes. It should be at the top of the view area.

> **Tip**
> If you closed the `Trippville Buildings` scene, you could reopen it from the **Project** pane. Simply locate the scene in the `Maps` folder in the **Project** pane. Then, right-click and select **Open Local View**.

2. Activate the **Map** tab in the ribbon. Then, click the **Basemap** button in the **Layer** group and select the **Imagery** basemap.
3. Click the **Add Data** button in the **Layer** group on the **Map** tab.

4. Click the **Computer** in the left panel of the **Add Data** window. Then, double-click on the **This PC** option in the right panel of the window.

5. Next, scroll down and double-click on the `C:` drive, which may be labeled as `Local Disk`, `OS`, or `Local Drive`. Then, navigate to the `C:\Student\ArcGISPro3Cookbook\Chapter14` folder.

6. Select the `Fence_3D.lyrx` layer file and click **OK**. The fence layer should appear in your scene with the symbology already configured.

7. Click the **Bookmarks** button located on the **Map** tab and select the **New Fence** bookmark. This will zoom you into the same area you were just working with in your 2D map. It is possible you may be zoomed in too closely to see things clearly. If this happens, use your scroll wheel on your mouse to zoom out until you can see the new fence you created.

8. Click the **Explore** tool and pan the scene to the northwest until you see the house with the pool located to the northwest of the parcel with the fence you just drew.

9. Zoom in on the house with the pool until your view looks similar to the following screenshot:

Figure 14.31 – Zooming in on the house with the pool in the backyard

10. Click on the **Edit** tab in the ribbon.
11. Enable **Z Mode** and click on the drop-down arrow to see what options you are allowed to select.

As you can see, the surface options are grayed out. This is because in a 3D scene, the ground surface is automatically used to determine the Z coordinates for any new features. You do not need to set this when creating new Z-enabled features in a scene like you do in a 2D map.

12. Disable **Z Mode**.

13. The **Create Features** pane should still be open from when you created the first fence in the 2D map. If not, click the **Create** button to open it.

    Select the **Fence** feature template and digitize in the new fence, as illustrated in the following screenshot. I recommend starting on the northwest point where the fence meets the building then proceeding clockwise around the fence. Remember to double-click on the last point or press *F2* after you digitize the ending point.

Figure 14.32 – Digitizing fences located around the pool in the backyard

You have just created a new 3D feature. It has elevations or Z coordinates for the vertices along the ground. That is just the start of creating a full 3D feature. It is still missing some information, which you will add next.

Adding height to 3D features

The fence may not be visible in the scene after you complete it. This is because it has no height. It is running along the ground with all the other non-extruded layers. You will fix that now:

1. Click the **Attribute** button in the **Selection** group on the **Edit** tab. This will open the **Attributes** pane.

2. In the **Height** cell, type 8 to make the height of the new fence display eight feet. Click the **Apply** button if it is visible. You should now see the fence in the scene.

3. Close the **Attributes** and **Create Features** panes to free up screen space.

4. Click the **Clear** button in the **Selection** group in the ribbon.

5. Click the **Save** button in the **Manage Edits** group to save the new fence you just created. Again, when asked if you want to save all edits, click **Yes**.

6. Activate the **Map** tab in the ribbon and click on the **Explore** tool.

7. Using the skills you learned in previous recipes, rotate the view so you can see the new 3D fence you just created. It should look like that in the following figure:

Figure 14.33 – The new fence around the pool

8. Save your project.

You can see that the fence you created is being displayed in 3D using the Z coordinates for the starting elevation and extrusion to show the height. You are off to a good start with the request from the planning director to create and populate a new fence layer. However, he really wants to know whether that proposed new building will create a problem for the existing buildings by blocking their view of the park.

Creating new 3D observation points feature class

So, you will move on to conducting a line-of-sight analysis, the first part of which is establishing observation points from the existing buildings. These must be 3D points. You will start by creating a new Z-enabled point feature class. Then, you will digitize new observation points from the existing buildings the Planning Director is concerned about. In the next recipe, you will complete the analysis:

1. In the **Catalog** pane, expand the `Databases` folder so you can see its contents.

2. Right-click the `Trippville_GIS.gdb` geodatabase and go to **New**. Then, select **Feature Class**, as illustrated in the following screenshot:

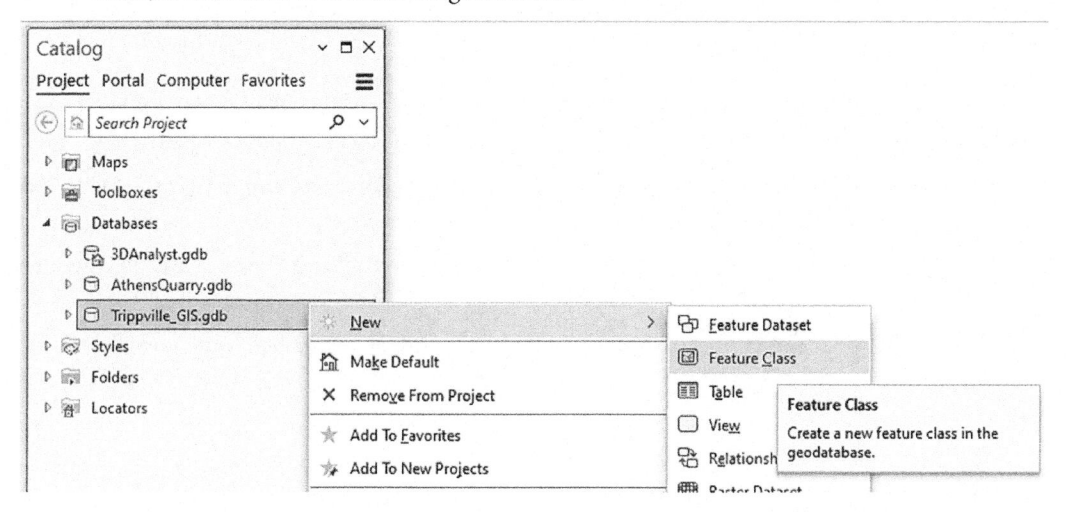

Figure 14.34 – Creating a new feature class from the Catalog pane

3. In the **Create Feature Class** tool, the **Name** should be `Observ_PT`. Provide an **Alias** of **Observation Points**.

4. Set the **Feature Class Type** to **Point** using the drop-down arrow.

5. Under **Geometric Properties**, ensure **Z Values – Coordinates include Z values used to store 3D data** is enabled.

6. Make sure **Add output dataset to current map** is also enabled. Then, click the **Next** button.

7. You should now be on the **Fields** part of the tool. Since you will only be using these locations for simple analysis, you do not need to define any additional fields. Click the **Next** button to continue.

8. Verify that **Current XY** is set to **NAD 1983 StatePlane Georgia West FIPS 1002 (US Feet)** and **Current Z** is set to **NAVD88 height (ftUS)**. These should be automatically set based on the Trippville buildings scene. Click **Next** to continue.

9. Accept the calculated values for **XY Tolerance** and **Z Tolerance** by clicking **Finish**.

10. Save your project

When the **Create Feature Class** tool is complete, you should see a new `Observation Points` layer appear in your **Contents** pane. This new layer references the new feature class you just created.

Digitizing the new point features

Now you need to digitize some points into the new layer that represent the observation locations the **Planning Director** is concerned about:

1. Now you need to zoom in on the proposed building location. You will do this by selecting the proposed building and then zooming in on it. In the **Map** tab, click the **Select By Attributes** button.
2. Set the **Input Rows** to `Buildings` using the drop-down arrow and ensure the **Selection Type** is set to **New Selection**.
3. Under **Expression** next to **Where**, set the **Select Field** cell to **Building Status**. Leave the **Operator** set to `is equal to`. Set the **Value** cell that appears to `Pr - Proposed`.
4. Verify that your **Select By Attributes** tool looks like the following screenshot and click **OK**:

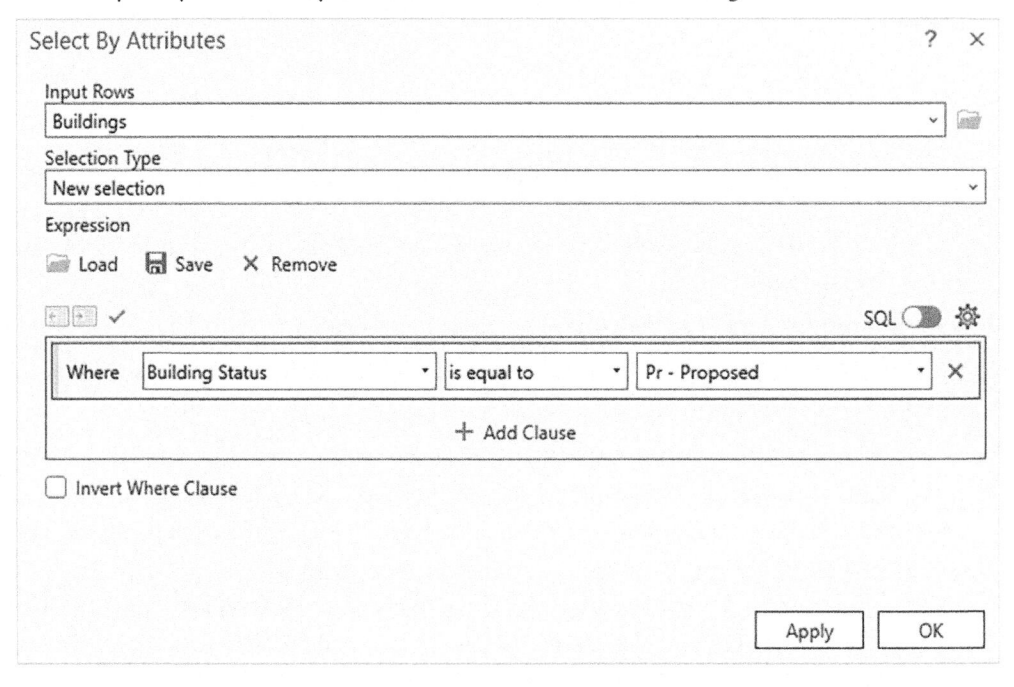

Figure 14.35 – Selecting Proposed Buildings with the Select By Attribute tool

5. Right-click the `Buildings` layer in the **Contents** pane. Then, go to **Selection** and select **Zoom to Selection**, as illustrated in the following screenshot:

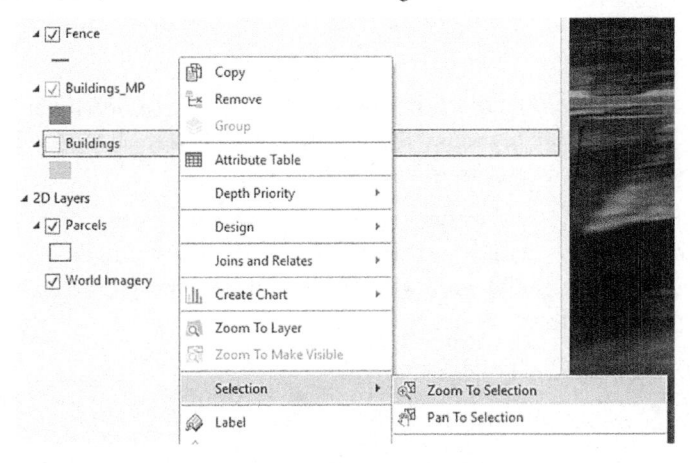

Figure 14.36 – Zooming in on the selected building

The scene will zoom in on the location of the proposed building that is causing all the concern. Your scene should now look like that in the following screenshot. Depending on the size of your monitor and the size of your ArcGIS Pro application, yours may look slightly different:

Figure 14.37 – The scene zoomed in on the selected building

> **Note**
> You can still query and zoom in on features in a layer that have been turned off, so they are not visible.

6. Using skills you have learned previously, add the `Parks` feature class located in the `Trippville_GIS.gdb` geodatabase to the scene. This will allow you to see the location of Washington Park, which is located to the northeast of this building.

7. Using the **Explore** tool on the **Map** tab, zoom and rotate the `Trippville Buildings` scene so that you can see the area surrounding the proposed building. Something like the following screenshot should work well. You may need to drag the `Parks` layer to the `2D Layers` group in order to see it:

Figure 14.38 – Scene zoomed in so you can see the selected building and park

The proposed building is highlighted in blue just to the left of the center. Washington Park is the green area to the northeast of the proposed building. Now, you need to digitize observation points for the buildings directly south and west of the proposed building.

8. In the **Contents** pane, click the **List by Snapping** button. Then, right-click on the `Buildings_MP` layer and select **Make this the only snappable layer**.

9. Click the **List by Editing** button in the **Contents** pane. Right-click the Observation Points layer and select **Make this the only editable** layer.
10. Click the **Edit** tab in the ribbon. If you see the **Edit** button, click on it to enable editing. This button will be highlighted in blue when enabled. Depending on what ArcGIS Pro options you have enabled, you may not see this button.
11. Click the **Create** button to open the **Create Features** pane.
12. Ensure **Snapping** is turned on and it is set to **Edge**, **Vertex**, **End Point**, and **Intersection**.
13. Select the Observation Points feature template in the **Create Features** pane.
14. Click the locations shown in the following screenshot. You may need to zoom in and rotate the scene to get the exact locations:

Figure 14.39 – Locations of the observation points

15. Click the **Map** tab in the ribbon and activate the **Explore** tool.
16. Rotate your scene so you can see the location of the new points you just digitized. They should be snapped to the tops of their associated buildings.
17. If they are, click the **Edit** tab in the ribbon. Then, click the **Save** button to save the new points you just created. If asked if you want to save all edits, click **Yes**.
18. Close the **Create Features** pane.
19. Save your project and close ArcGIS Pro.

You have just successfully created new 3D features including points and lines.

How it works...

In this recipe, you created several new 3D features. You created two new fence features and four new observation points.

To create the new fence features, you used a couple of methods. The first fence you created you did in a normal 2D map. As you learned, you can still create 3D features in a 2D map. You need to ensure you have a surface added to the map to do this. You then needed to enable the **Z Mode** option and select a surface. This then allows the new fence to inherit the Z coordinate from the location you clicked using the ground surface as a reference. The next fence feature you created was in a scene. This provided a 3D display environment to create the 3D feature and meant the new feature automatically inherited the Z from the surface without needing to enable the **Z Mode** option. You then specified a height for each fence as an attribute that was then used to extrude them, so they displayed at their true height above the ground.

Then, you created a completely new Z-enabled feature class to store observation points that you will use in the following recipe to analyze the ability to see Washington Park from those locations. The Z coordinates for those new points were determined by the corner of the buildings you snapped to.

Calculating lines of sight

You've learned several ways to create new 3D features. Now it is time to investigate some analyses you can perform with these features that you cannot do with 2D data. A common analysis is to determine lines of sight. This has many uses. It can be used by property appraisers to determine whether a property has a good line of sight to a key feature, such as a river or ocean, which might increase its value. It is used by the police to establish security perimeters for special events. It might be used to help plan a parade route. So, as you can see, there are many uses for such an analysis.

In this recipe, you will use the observation points you created in the *Creating 3D features* recipe to determine whether there are lines of sight to Washington Park or whether the proposed building will block the view of the park entirely. This will be one of the determining factors the Planning Director will use when deciding whether to approve the new building. This will require the use of several geoprocessing tools from the 3D Analyst extension for ArcGIS Pro.

Getting ready

You must have completed the other recipes in this chapter before starting this one. This will ensure you have the required project, maps, scenes, and data to complete this project. It will also confirm you have a good foundational understanding of where to access the required tools and basic concepts required to complete this recipe.

You will also need to ensure the sample data is installed before you continue and access the 3D Analyst extension for ArcGIS Pro.

How to do it...

You will now use analysis tools in the 3D Analyst extension for ArcGIS Pro to calculate lines of sight from the observation points you created in the *Creating 3D features* recipe to the nearby park.

1. Start ArcGIS Pro and open the 3DAnalyst.aprx project you used in the previous recipes in this chapter. This project should be included in your list of recently opened projects.
2. Click the **Analysis** tab in the ribbon. Then click the **Tools** button in the **Geoprocessing** group.
3. Click the **Toolboxes** tab in the **Geoprocessing** pane. Then, expand the **3D Analyst Tools** toolbox.
4. Scroll down to the **Visibility** toolset and expand it so you can see its contents.
5. Select the **Construct Sight Lines** tool to open it. This tool does require you to have a license for the 3D Analyst extension.
6. Set the **Observation Points** to the Observation Points layer using the drop-down arrow.

> **Tip**
> If you get a warning indicating the layer has a selection, activate the **Map** tab and click on the **Clear** button to deselect all features. Then, click on the **Refresh** button next to the warning in the **Construction Sight Lines** tool.

7. Set the **Target Features** to the Parks layer using the drop-down arrow.
8. For the **Output**, click the **Browse** button and navigate to the Trippville_GIS.gdb geodatabase using the skills you have learned from previous recipes. Name the new output feature class Park_Sight_Lines and click **Save**.
9. Ensure both the **Observer Height Field** and **Target Height Field** are set to Shape.Z.
10. Set the **Sampling Distance** to 200.
11. Verify that **Sampling Method** is set to the default value of 2D Distance.

12. Verify your **Construct Sight Lines** tool looks like that in the following screenshot and click **Run**:

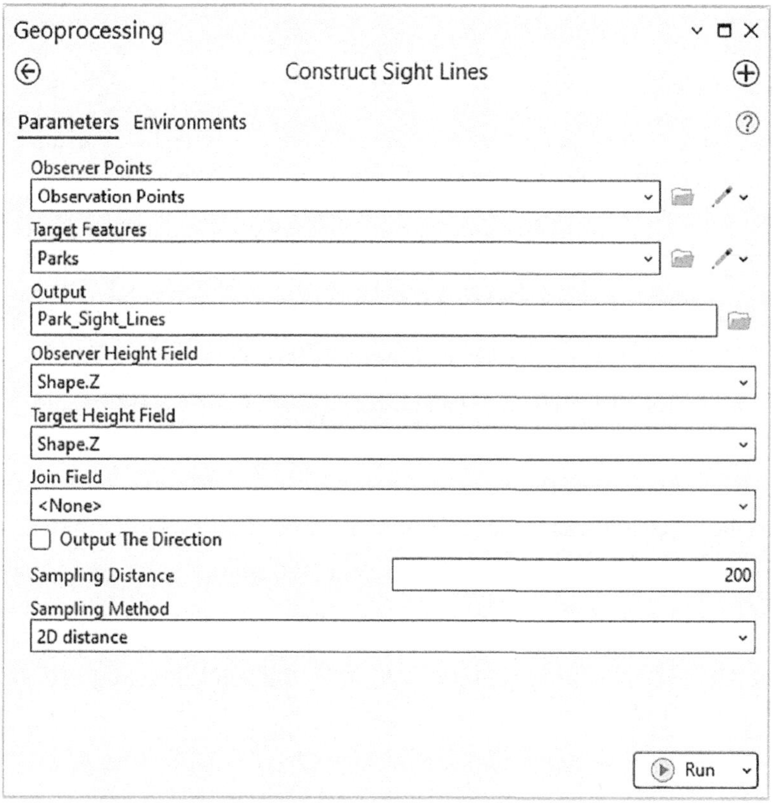

Figure 14.40 – Construct Sight Lines tool with completed parameters

A new layer appears in your scene that represents every possible sight line from the observation point to the park using the interval you specified in the **Sampling Distance**. This tool ignores any possible obstructions.

Determining whether there is an obstruction to the line of sight

Next, you will use another tool that will investigate each of the sight lines you created to see whether they are blocked by the ground surface or other obstructions, such as buildings:

1. In the **Contents** pane, turn off the Park_Sight_Lines layer you just created. You do not need to see the layer to use it for further analysis.
2. Close the **Geoprocessing** pane.

> **Tip**
> Turning off layers that you do not need to see for the final results of any analysis process you are pursuing can improve overall performance and reduce map clutter.

3. Activate the **Analysis** tab in the ribbon. Then, click on the **Tools** button.
4. Click on the **Toolboxes** tab and expand the **3D Analyst Tools** toolbox.
5. Select the **Line of Sight** tool in the **Visibility** toolset.
6. Click the **Browse** button for the **Input Surface**. Navigate to the `Trippville_GIS.gdb` geodatabase and select DEM. Click **OK** once the DEM is selected.
7. Set **Input Line Features** to `Park_Sight_Lines` using the drop-down arrow.
8. Set **Input Features** to `Buildings_MP` using the drop-down arrow.

> **Note**
> **Input Features** is a layer that contains additional features you want the tool to consider that might block the line of sight along the sight lines. This must be a multipatch feature class layer.

9. Name your **Output Feature Class** `Buildings_to_Park_LOS`.
10. You can leave **Output Obstruction Point Feature Class** blank. This would create a new point feature class that shows exactly where the views are obstructed. For this analysis, we are not interested in those exact positions.
11. Verify that your **Line of Sight** tool looks like that in the following screenshot and click **Run**:

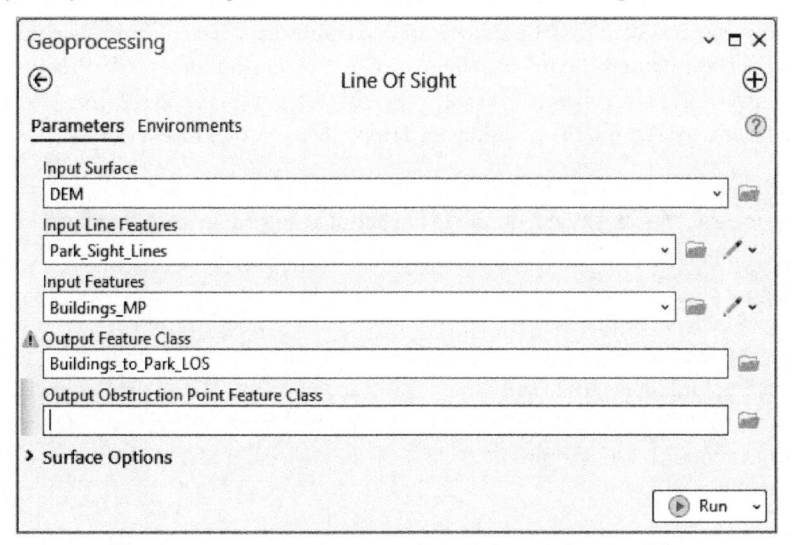

Figure 14.41 – Line of Sight tool with completed parameters

When the tool is complete, you will see a new layer added that has identified which sight lines are visible and which are not, as shown in the following screenshot. Remember, this tool takes the ground as well as the buildings into account when determining what is visible and what is not:

Figure 14.42 – Line of Sight analysis results

You can now easily see which sight lines are being blocked by the proposed building and which are not. This is the information the Planning Director needed.

12. Close your **Geoprocessing** pane.
13. Save your project and close ArcGIS Pro.

Now you have performed your first 3D analysis to determine lines of sight from specified points to an existing feature. This required you to have the 3D Analyst extension for ArcGIS Pro. In addition, you had to create several 3D layers before you were able to perform this analysis. This included creating a multipatch layer representing the buildings in Trippville and observation points.

You could have performed this same analysis in a 2D map. ArcGIS Pro can still perform 3D analysis in a 2D map. However, the results will not be as impactful or informative when displayed in a 2D map.

How it works...

In this recipe, you determined how and where a new proposed building will impact the views of a local park from existing buildings. To do this, you require access to the 3D Analyst extension for ArcGIS Pro. This extension includes a host of different tools for analyzing and creating 3D data above what is included in the core ArcGIS Pro product.

To start, in the previous recipe, you first created observation points that were located in the top corners of existing buildings where those corners faced Washington Park. Next, you use the **Construct Sight Lines** tool from the 3D Analyst extension to generate lines that represented lines of sight from the observation points to various locations of the park boundary. Lastly, you used the **Line of Sight** tool to analyze which of those lines of sight would be obstructed by the new proposed building if it was constructed.

Calculating the volume of a polygon

Another common 3D analysis task is to calculate the volume of the area covered by a polygon feature. This is done by overlaying the polygon feature across a surface and then calculating the volume that exists between the plane created by the polygon features and the surface they overlay.

In this recipe, you will return to the quarry scene you worked on earlier. The project engineer has determined the optimum full pool elevation and created a polygon layer for you. He wants you to determine the total volume of water needed to fill the quarry to this level in gallons.

You will use the **Polygon Volume** tool that is part of the 3D Analyst extension to accomplish part of this analysis. You will then use the **Calculate Field** tool to calculate the total gallons required to fill the quarry.

Getting ready

You must have completed the other recipes in this chapter before starting this one. This will ensure you have the project, maps, scenes, and data required to complete this project. It will also confirm you have a good foundational understanding of where to access the tools and basic concepts required to complete this recipe.

You will also need to ensure the sample data is installed before you continue and access the **3D Analyst** extension for ArcGIS Pro.

How to do it...

You will now use the **Polygon Volume** tool that is part of the 3D Analyst extension to calculate the total volume of water needed to fill the old quarry with water:

1. Start ArcGIS Pro and open the 3DAnalyst.aprx project you used in the previous recipes in this chapter. This project should be included in your list of recently opened projects.
2. Close all the Trippville maps and scenes. If needed, reopen the Quarry Project scene.

> **Tip**
> Remember, you can right-click a map or scene in the **Catalog** pane to open it.

3. Now you need to add the `Full Pool` layer created by the project engineer. Click the **Add Data** button located on the **Map** tab in the ribbon.

4. Navigate to the `C:\Student\ArcGISPro3Cookbook\Chapter14` folder and select the `Full_Pool_Level.lyrx` layer file. Click **OK**. A new layer should have been added to your scene that shows the water level of the quarry when it is filled, as shown in the following screenshot:

Figure 14.43 – The Full Pool layer added to scene

5. Activate the **Analysis** tab in the ribbon. Then, click the **Tools** button.

6. Next, click the **Toolboxes** tab in the **Geoprocessing** pane and expand the **3D Analysis Tools** toolbox. Then, expand the **Area and Volume** toolset.

7. Click the **Polygon Volume** tool to open it. This tool does require a license for the **3D Analyst** extension.

8. Click the **Browse** button for **Input Surface**. Then, expand the **Computer** folder in the left panel of the **Input Surfaces** window.

9. In the right panel, scroll down and double-click on the `C:` drive, which might be labeled `Local Disk`, `Local Drive`, or `OS`.

10. Then, navigate to the `Student\ArcGISPro3Cookbook\Databases` folder. Select `quarry_tin` and click **OK**.

> **Note**
> For this tool, the input surface must be a TIN, LAS, or terrain dataset. It will not accept a DEM surface as the input.

11. Set **Input Polygons** to `Full Pool Level` using the drop-down arrow.

12. Set **Height Field** to **Z_Mean** using the drop-down arrow.
13. Ensure the **Reference Plane** is set to `Calculate below the plane`. The plane it is referring to is the one created by **Input Features**.
14. Accept the default values for **Volume Field** and **Surface Area Field**.
15. Verify that your **Polygon Volume** tool looks like that in the following figure and click **Run**:

> **Tip**
> If you see warning indicators next to **Volume Field** and **Surface Area Field**, it indicates that those fields already exist. This will not stop the tool from running. It will just skip the need to create the specified fields and overwrite any values those existing fields contain.

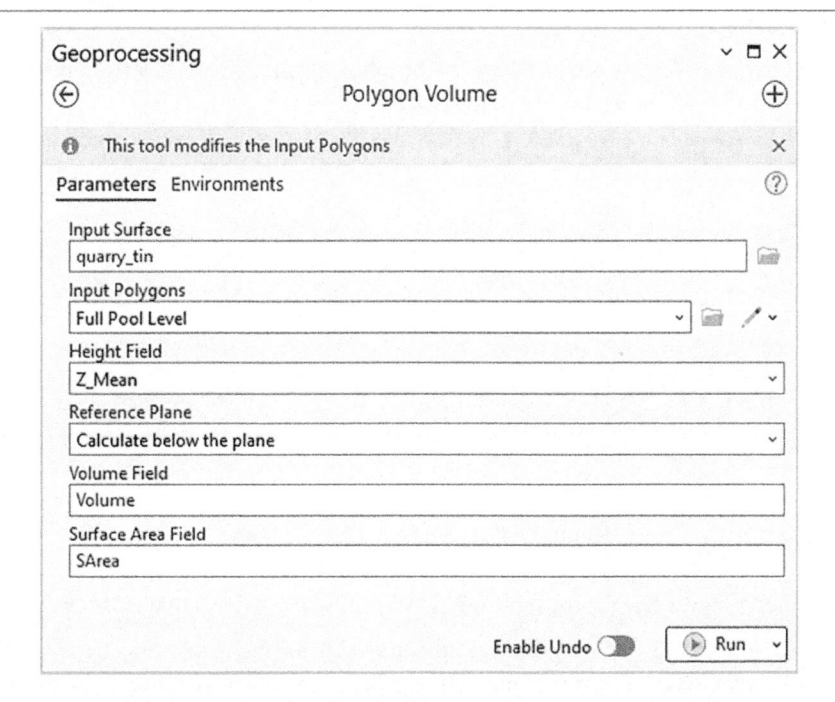

Figure 14.44 – Polygon Volume tool with completed parameters

16. When the tool is complete, it will have created two new fields in your attribute table: `Volume` and `SArea`. It will have then calculated values into those two new fields. You will examine those results next.
17. Close the **Geoprocessing** pane.
18. Right-click **Full Pool Level** and select **Attribute Table**. When the table opens, you should see the two fields that were added: `Volume` and `SArea`.

The volume is in cubic feet because your horizontal and vertical units for the layer are set to **US Feet**. The engineer needs to know the volume in US gallons.

Calculating volume in gallons

You will now create a new field to contain the volume in gallons and use the **Calculate Field** tool to populate it:

1. In the **Table** pane, click the **Add Field** button. It is located in the top-left of the pane.

2. Under the **Field Name** value, type Gallons, then press your *Tab* key to move over the **Alias** column and leave it blank. It will automatically default to the field name you typed in.

3. Next, click on the **Data Type** column and set it to **Float** using the drop-down arrow. This will allow it to store decimal values.

4. Verify your table design looks like that in the following figure and click the **Save** button in the **Changes** group on the **Fields** tab in the ribbon:

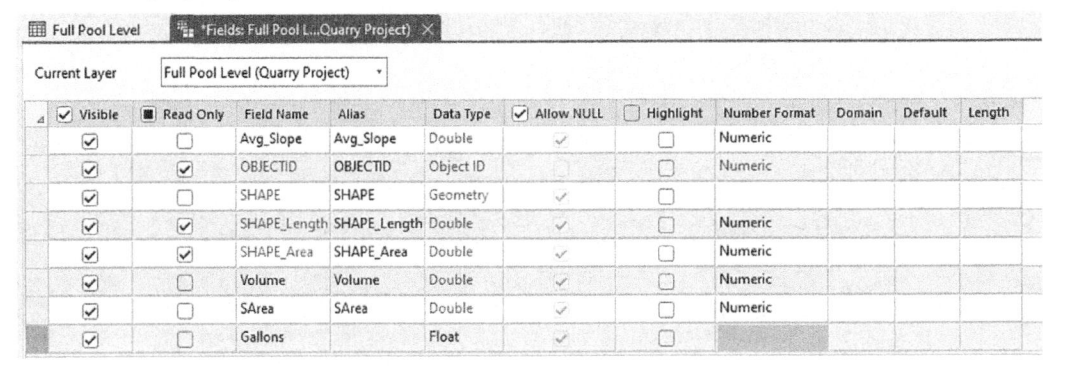

Visible	Read Only	Field Name	Alias	Data Type	Allow NULL	Highlight	Number Format	Domain	Default	Length
✓	☐	Avg_Slope	Avg_Slope	Double	✓	☐	Numeric			
✓	✓	OBJECTID	OBJECTID	Object ID		☐	Numeric			
✓	☐	SHAPE	SHAPE	Geometry	✓	☐				
✓	✓	SHAPE_Length	SHAPE_Length	Double	✓	☐	Numeric			
✓	✓	SHAPE_Area	SHAPE_Area	Double	✓	☐	Numeric			
✓	☐	Volume	Volume	Double	✓	☐	Numeric			
✓	☐	SArea	SArea	Double	✓	☐	Numeric			
✓	☐	Gallons		Float	✓	☐				

Figure 14.45 – New field added to table

5. Close the **Design View** for the table but make sure to leave the actual table open.

6. You have just added a new field to the attribute table for the Full Pool Level layer. This new field will be used to store the number of gallons that will be required to fill the quarry. You will use the **Calculate Field** tool to populate a value into the field you just created.

7. Right-click the Gallons field you just added in the table view and select **Calculate Field**. The **Calculate Field** tool will open in the **Geoprocessing** pane.

> **Tip**
> If the **Calculate Field** tool is greyed out so you are not able to select it, that is most likely because you have adjusted the editing options for ArcGIS Pro. You will need to go to the **Edit** tab in the ribbon and click on the **Edit** button to enable data editing. Once you do that, you should be able to access the **Calculate Field** too.

Calculating the volume of a polygon   529

8. Ensure the **Input Table** is set to `Full Pool Level`. This should have been done automatically.
9. Ensure the **Field Name (Existing or New)** is set to `Gallons`.
10. Verify that the **Expression Type** is set to **Python 3**.
11. Locate the cell under `Gallons =` and type the expression `!Volume! * 7.48052` into the cell.
12. Verify that your **Calculate Field** tool matches that in the following screenshot and click **OK**:

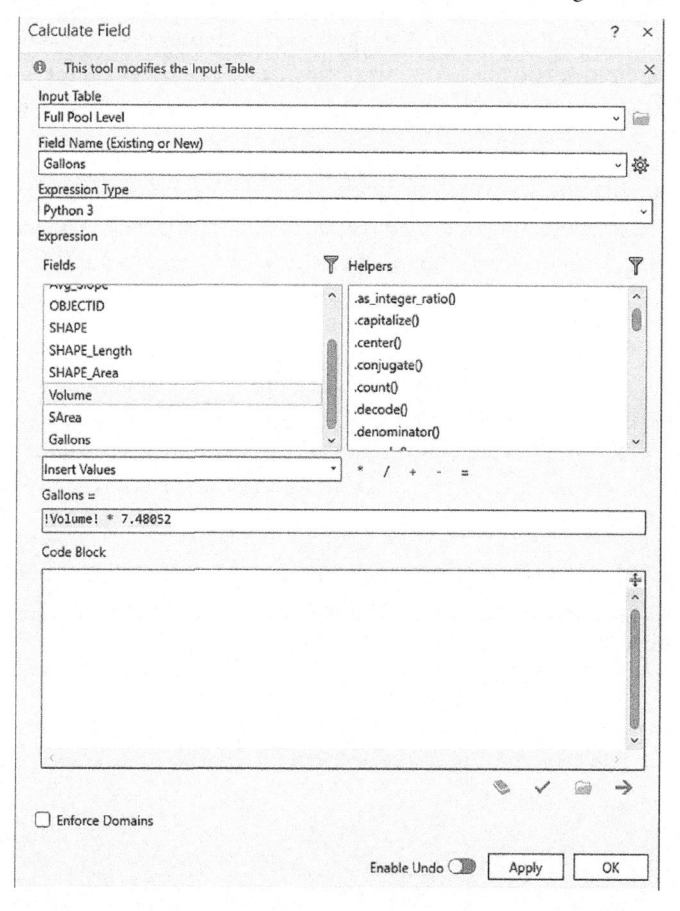

Figure 14.46 – The Calculate Field tool with completed parameters

13. The `Gallons` field you created should now have a number populated into it. This is the total number of gallons the quarry will hold to reach the fill line designated by the engineer and represented by the `Full Pool Level` polygon.
14. Close the `Full Pool Level` attribute table and save your project.
15. Close ArcGIS Pro.

You have just calculated the volume for an area covered by a polygon. This required you to have two datasets. You needed an elevation surface. The **Polygon Volume** tool you used required the surface to be in TIN, Terrain or LAS formats. The other dataset you needed was a polygon that formed the plane that was used to overlay the surface.

How it works…

You have successfully calculated the volume of a polygon that represents a new reservoir that is being built. To do this, you needed two datasets. One of the datasets represented the elevation surface for the area in question. The other dataset was the polygon that represented the full level of the reservoir being built.

You used the **Polygon Volume** tool in the 3D Analyst extension for ArcGIS Pro to perform this calculation. The 3D Analyst extension is an add-on for ArcGIS Pro that provides increased functionality to perform analysis and calculations with 3D data. To learn more about the 3D Analyst extension go to `https://www.esri.com/en-us/arcgis/products/arcgis-3d-analyst/overview`.

In the **Polygon Volume** tool, you specified the surface that represented the current elevations in the area of the new reservoir. This surface had to be a TIN, Terrain, or LAS format. Other surface types such as DEM were not allowed. Next, you specified the polygon that formed the plane representing the fill elevation of the reservoir, which was used to overlay the surface. You used the `Full Pool Level` layer for this input.

When the **Polygon Volume** tool ran, it created two fields: `Volume` and `SArea`. Because the desired unit for the volume was gallons, you created a new field called `Gallons` and used the **Field Calculator** tool to convert cubic feet into gallons.

Index

Symbols

3D features
  creating 507-511
  creating, in Scene 511-513
  height, adding to 513, 514
  new 3D observation points feature class, creating 514, 515
  new point features, digitizing 516-519

3D multipatch features
  converting to 505

3D scene
  2D layer, converting to 3D 486-489
  configuring 481-484
  creating 477-480
  database connection, adding 480, 481
  layers, adding to 484, 485
  layer symbology, adjusting 485, 486

A

Add Data button
  used, for adding map layers 31, 32

annotation
  labels, converting to 105-107

Append geoprocessing tool 380
  results, verifying 383-385
  running 381-383

Append tool
  used, for adding data to layer 494, 495

ArcGIS Online
  new layer, sharing to 258-263

ArcGIS Online, layer
  adding 41-44
  hosted basemap, changing 42-44
  hosted basemap, verifying 42-44
  reference link 41
  symbology, changing 45, 46

ArcGIS Online Map Viewer
  using, to verify results 264, 265

ArcGIS Pro
  Excel spreadsheet, importing into 393-401
  project, opening 30, 31
  required specifications, verifying by computer 2-5
  stock project templates 24, 25
  tables, joining 95

ArcGIS Pro 3.0
  requisites 2

ArcGIS Pro license level
  determining 5-8

area drain (AD) 367

Index

attribute rules
  creating 310-321
  creating, to calculate mapped acreage 315-317
  new field, adding 312-314
  testing 317-320

attributes 199

Attributes pane
  individual attributes, editing with 199-204
  multiple attributes, editing with 204-210

Autocomplete Polygon tool
  polygon features, creating with 192, 193
  used, for creating new parcel 193-197

B

buffers
  creating 417-419

Buffer tool
  using 420-424

Building Information Models (BIM) 475

C

Calculate Field tool
  using 218-222
  using, to populate multiple features 216-218
  working 222

Calculate Geometry tool
  area in square meters, calculating 225-227
  used, to populate values for multiple features 223-225

cardinality 90
  works, verifying 94

Catalog pane
  dragging, to new layer 33, 34
  dropping, to new layer 33, 34
  used, for adding new layer 33

catch basin (CB) 367

central feature, based on geographic distribution
  identifying 458-464

Central Feature (Spatial Statistics)
  reference link 458

cluster 441

coded values
  adding, to domain 283, 284

coded values domain
  creating 281-283

Code Values 281

composite relationship 112

computer-aided drafting and design (CADD) 499

Computer Assisted Mass Appraisal (CAMA) 66

contingent values
  creating 321-332
  new field and domain, adding 324, 325
  testing 328-331

Convert to Annotation tool 112

coordinate system
  assigning, to data 248-251
  data, projecting to 253-257
  data, projecting to different 252, 253
  defining, for data 245-248
  determining, for map 231-235
  modifying, of map 240-245
  setting, for new map 235-240

D

data
  coordinate system, assigning 248-251
  coordinate system, defining 245-248
  editing, with map topology 353-362
  enabling, to store Z coordinates (elevation) 490-493

Index 533

exporting, to feature class 399-401
merging 376-378
projecting, to coordinate system 253-257
projecting, to different coordinate system 252, 253
querying, in joined table 82-87
validating, with topology rules 345-348
data, with map topology
new layer, adding with layer file 355-357
Digital Elevation Model (DEM) 477, 499
Dissolve Option 421
domain 283
assigning, to attribute field 284-286
drop inlet (DI) 367

E

editing options
ArcGIS Pro options, accessing 124-128
ArcGIS Pro options, configuring 124-128
configuring 122-130
snapping settings, changing 128, 129
editor tracking
enabling 298-303
enabling, verification 301-303
event layer 47
Excel
street centerlines, exporting to 389, 390
Excel spreadsheet
exporting, to standalone table in GIS database 402-404
importing, into ArcGIS Pro 393-401
issues, fixing to import into geodatabase 396
results, verifying 392, 393
street centerlines, exporting to 390-392
tabular data, exporting to 385-387
existing ArcGIS Pro project
opening 8-11

existing feature
reshaping 130-136
selecting 133-136
existing layer
selected features, importing into 406-410
existing layer, in map
spreadsheet data, joining to 397-399
Export Table tool 387
reference link 387
results, verifying 387-389

F

feature class 29
creating 268-273
data, exporting to 399-401
fields, adding to attribute table 274-276
Feature Class to Geodatabase conversion tool 374
feature dataset
creating 277-369
feature class, creating 279, 280
working 280
feature distance
calculating, with Generate Near Table tool 431-440
feature-linked annotation 102
creating 101-104
results, verifying 108-112
features
aligning 153
alignment issues, fixing 155-160
gap, determining 154, 155
joining spatially 96-99
labeling, with joined table 74-82
layers, setting as editable 148, 149
layers, setting as selectable 148, 149
merging 146-151

overlap, determining 154, 155
overlap, fixing 155-160
results, verifying 151, 152
selecting 149-151
selecting, within specific distance 442-450
features, within specific distance
  selecting 411-417

G

Generate Near Table tool
  feature distance, calculating with 431-440
geocoding 51
geocoding addresses 51-62
  address locator, creating 53-57
  address locator settings, adjusting 57
  ArcGIS Pro, launching 53
  Excel spreadsheet, adding 53
  spreadsheet data, geocoding 58-61
geodatabase
  creating 268-270
  Excel spreadsheet issues, fixing to import into 396
geodatabase feature class
  shapefile, exporting to 369-372
  shapefiles, converting to 364-375
geodatabase format
  reference link 29
geodatabase layer
  adding 29, 30
  methods 40
geodatabase topology
  creating 334-339
  feature class and rules, adding to 339-342
  topology and layers, adding to map 343-345
  used, for validating spatial data 342, 343
geographic dispersion of data
  calculating 464-473

Global Navigation Satellite Systems (GNSSs) 304
GPS metadata capture
  adding 304-310
  challenges 309, 310
  fields tool results, verifying 307-309
graphics processing unit (GPU) 478

H

Heavy Industrial (HI) 432
hotspot analysis 441

I

Import menu and Feature Class(es) option
  accessing 373
individual attributes
  editing, in table view 210, 211
  editing, with Attributes pane 199-204
individual attributes, editing in table view 210, 211
  attributes, editing 214, 215
  layers for editing, configuring 213
  manholes, selecting 214, 215
  symbology, setting for manhole layer 211, 212

J

join 64
  adding 404, 405
  creating 71-74
  creating, in ArcGIS Pro 64-71
  removing 404, 405
joined table
  data, querying 82-87
  features, labeling with 74-82

Index   535

L

label placement
adjusting 78-81
labels
converting, to annotation 105-107
Light Detection and Ranging (LiDAR) 475
line feature
creating 141-143
new layer, adding to map 138-140
new layer, adding with layer file 140
splitting 136-146
symbology, changing 138-140
lines of sight
calculating 520-522
obstruction, determining 522-524

M

map
additional layer, adding to 239, 240
coordinate system, determining for existing 231-235
coordinate system, modifying 240-245
coordinate system, setting for 235-240
layers, adding with Add Data button 31, 32
navigating 12-17
new layer, adding to 237-239
opening 12-17
topology and layers, adding 343-345
mapped acreage
attribute rule, creating to calculate 315-317
map topology
data, editing with 353-362
mean center, of geographic distribution
finding 451-458
Merge geoprocessing tool 380
results, verifying 379-385

multipatch 499
multipatch features, creating from 2D
features, extruding 504
layer 3D, making 504
layers, adding to scene 502, 503
results, verifying 506
multipatch features, from 2D
creating 499-502
multiple attributes
editing, with Attributes pane 204-210
multiple attributes, editing
attributes for multiple features, editing 207-209
edit results, verifying 209, 210
selectable and editable layers, selecting 205-207
multiple shapefiles
merging, into single geodatabase feature class 375-380

N

nearest feature
determining, with Near tool 424-431
Near tool
nearest feature, determining with 424-431
new layer
adding, from Catalog pane 33
new line features
based on specific measurements, creating 179-182
creating 168-171
creating, by tracing source document 172-179
creating, with Popular Circle 182, 183
editable layers and snapping settings, setting up 172
working 184

new point features
   creating 162-168
   snapping options and editable layers, setting 165

non-weighted center
   versus weighted center 455

P

Parcel Identification Number (PIN) 66
patches 499
polygon features
   building polygons, creating 187-191
   creating 184-187
   creating, with Autocomplete Polygon tool 192, 193

Polygon Volume
   calculating 525-530
   volume in gallons, calculating 528, 529

polyline 168
project
   creating, without template 25-28
   creating, with template 17-24

R

Random Access Memory (RAM) 127
Range domains 281
relate
   creating 88-91
   using 88-90
   versus join 94
   works, verifying 91-94

relationship class
   creating 112-117
   properties, verifying 117-119

S

scenes 475
selected features
   copying and pasting 409, 410
   importing, into existing layer 406-410

shapefiles
   converting, to geodatabase feature class 364-375
   exporting, to geodatabase feature class 369-372

single geodatabase feature class
   multiple shapefiles, merging into 375-380

spatial 199
spatial data
   validating, with geodatabase topology 342, 343

Spatial Database Engine (SDE) 20
spatial join 96
   verifying 99-101

spreadsheet data
   joining, to existing layer in map 397-399

Standard Distance (Spatial Statistics)
   reference link 464

store Z coordinates (elevation)
   data, adding to layer with Append tool 494, 495
   data, enabling to 490-493
   results, verifying 498, 499
   Z values, calculating 495-498

street centerlines
   exporting, to Excel 389, 390
   exporting, to Excel spreadsheet 390-392

Structured Query Language (SQL) 84

subtypes
  creating 286, 287
  feature class, creating with model 288-291
  setting up 291-293
  values, populating 293-296
symbology settings
  changing 35
  importing, from layer file 37-40
  unique attribute values, using 36, 37

T

tabular 199
tabular data
  exporting, to Excel spreadsheet 385-387
topology tools
  spatial features, correcting with 349
  used, for fixing errors 350, 351
  used, for fixing multiple errors
      at same time 351, 352
Triangulated Irregular Network (TIN) 477

U

Universal Transmercator (UTM) 245
**Unmanned Aerial Vehicles
    (Drones/UAS/UAV)** 475
user-defined fields (UDFs) 366

W

weighted center
  versus non-weighted center 455

X

X data
  plotting, from table 47-50

Y

Y data
  plotting, from table 47-50

⟨packt⟩

packtpub.com

Subscribe to our online digital library for full access to over 7,000 books and videos, as well as industry leading tools to help you plan your personal development and advance your career. For more information, please visit our website.

Why subscribe?

- Spend less time learning and more time coding with practical eBooks and Videos from over 4,000 industry professionals
- Improve your learning with Skill Plans built especially for you
- Get a free eBook or video every month
- Fully searchable for easy access to vital information
- Copy and paste, print, and bookmark content

Did you know that Packt offers eBook versions of every book published, with PDF and ePub files available? You can upgrade to the eBook version at packtpub.com and as a print book customer, you are entitled to a discount on the eBook copy. Get in touch with us at customercare@packtpub.com for more details.

At www.packtpub.com, you can also read a collection of free technical articles, sign up for a range of free newsletters, and receive exclusive discounts and offers on Packt books and eBooks.

Other Books You May Enjoy

If you enjoyed this book, you may be interested in these other books by Packt:

Geospatial Analysis with SQL

Bonny P McClain

ISBN: 978-1-83508-314-7

- Understand geospatial fundamentals as a basis for learning spatial SQL
- Generate point, line, and polygon data with SQL
- Use spatial data types to abstract and encapsulate spatial structures
- Work with open source GIS combined with plug-ins
- Visualize spatial data and expand QGIS functionality with Postgres
- Apply location data to leverage spatial analytics
- Perform single-layer and multiple-layer spatial analyses

Learning Geospatial Analysis with Python

Joel Lawhead

ISBN: 978-1-83763-917-5

- Automate geospatial analysis workflows using Python
- Understand the different formats in which geospatial data is available
- Unleash geospatial tech tools to create stunning visualizations
- Create thematic maps with Python tools such as PyShp, OGR, and the Python Imaging Library
- Build a geospatial Python toolbox for analysis and application development
- Unlock remote sensing secrets, detect changes, and process imagery
- Leverage ChatGPT for solving Python geospatial solutions
- Apply geospatial analysis to real-time data tracking and storm chasing

Packt is searching for authors like you

If you're interested in becoming an author for Packt, please visit `authors.packtpub.com` and apply today. We have worked with thousands of developers and tech professionals, just like you, to help them share their insight with the global tech community. You can make a general application, apply for a specific hot topic that we are recruiting an author for, or submit your own idea.

Share Your Thoughts

Now you've finished *ArcGIS Pro 3.x Cookbook*, we'd love to hear your thoughts! Scan the QR code below to go straight to the Amazon review page for this book and share your feedback or leave a review on the site that you purchased it from.

`https://packt.link/r/1-837-63170-0`

Your review is important to us and the tech community and will help us make sure we're delivering excellent quality content.

Download a free PDF copy of this book

Thanks for purchasing this book!

Do you like to read on the go but are unable to carry your print books everywhere?

Is your eBook purchase not compatible with the device of your choice?

Don't worry, now with every Packt book you get a DRM-free PDF version of that book at no cost.

Read anywhere, any place, on any device. Search, copy, and paste code from your favorite technical books directly into your application.

The perks don't stop there, you can get exclusive access to discounts, newsletters, and great free content in your inbox daily

Follow these simple steps to get the benefits:

1. Scan the QR code or visit the link below

   `https://packt.link/free-ebook/9781837631704`

2. Submit your proof of purchase
3. That's it! We'll send your free PDF and other benefits to your email directly

Printed in Great Britain
by Amazon